GUMS AND STABILISERS FOR THE FOOD INDUSTRY 5

Proceedings of the 5th International Conference held at
Wrexham, Clwyd, Wales, July 1989

GUMS AND STABILISERS FOR THE FOOD INDUSTRY 5

Edited by

Glyn O.Phillips, P.A.Williams

The North East Wales Institute of Higher Education, Deeside, Clwyd, Wales

and

David J.Wedlock

Shell Research, Sittingbourne, Kent

Oxford University Press
Walton Street, Oxford OX2 6DP

Oxford is a trade mark of Oxford University Press

Published in the United States
by Oxford University Press, New York

British Library Cataloguing in Publication Data

Data available

Library of Congress Cataloging in Publication Data

Data available

ISBN 0–19–963061–5

Printed by Information Press Ltd, Oxford, England

PREFACE

This series of books on Gums and Stabilisers, of which this is the fifth, has always had a strong relation to industrial needs. Here this aspect has been strengthened, with emphasis given to the practicalities of the subject. This entire volume is product led. Discrete sections are devoted to major groups of current industrial products:

- GUM ARABIC AND OTHER EXUDATES
- STARCH
- GELATIN
- PECTIN
- MICROBIAL POLYSACCHARIDES
- CELLULOSICS AND SEED GUMS
- MARINE POLYSACCHARIDES

The objective is to marry the technical and industrial considerations with more basic interpretation of functionality. The producer and the user have been given prominence, but the material is presented with due consideration to the underlying scientific principles which control the usefulness of the products. Structure-function relationships, for example, are repeatedly emphasised. This concentration on product functionality is based on the strong belief of the organisers that the best results with individual hydrocolloids can only be obtained if the user truly understands what he is dealing with, and why they work in practice. This may well be stating the obvious, but we have repeatedly noticed undue emphasis on the commodity characteristics of the raw materials, as distinct from their technical functionalisation.

Academic scientists will find this book particularly helpful, since it educates them about the capabilities of the products and the problems still requiring solution. They will see that many of these still remain. The integration of industrial need with academic provision is a constant objective of this series of now well-established Conferences. This book is thus a new standard text-book which all working in this field will need.

PROFESSOR GLYN O.PHILLIPS
CHAIRMAN

ACKNOWLEDGEMENTS

The Fifth Meeting owed its success to the invaluable assistance of the Organising Committee.

Members of the Organising Committee

Dr J.C.Allen	The North East Wales Institute
Dr R.Ashton	Star Blends Inc.
Mr G.A.Barber (*Honorary Treasurer*)	Technical Consultant
Mr P.Cowburn	National Starch Ltd
Mr D.Gregory	Grindsted Products Ltd
Dr P.Harris	Unilever Research
Dr R.Harrop	The North East Wales Institute
Dr I.Hodgson (*Vice-Chairman*)	Kelco International Ltd
Mr R.M.W.Hopkins	Meyhall Chemical (UK) Ltd
Mr H.Hughes (*Secretariat*)	The North East Wales Institute
Dr J.Mlotkiewicz	Spillers Foods Ltd
Dr R.G.Morley	Delphi Consultant Services Inc.
Professor E.R.Morris	Cranfield Institute of Technology
Dr V.J.Morris	AFRC Food Research Institute
Dr J.C.F.Murray	PFW (UK) Ltd
Dr A.Onions	Honeywill and Stein Ltd
Mr G.Owen	General Foods Ltd
Professor G.O.Phillips (*Chairman*)	The North East Wales Institute
Mr A.Procter	Cerestar UK Ltd
Dr D.J.Wedlock	Shell Research Ltd
Dr P.A.Williams (*Secretariat*)	The North East Wales Institute

The Editors would like to express their appreciation to Mr Haydn Hughes and Miss Linda Sneddon for assisting in manuscript preparation.

CONTENTS

Part 4: Pectin

Part 5: Microbial polysaccharides

Part 6: Cellulosics and seed gums

Part 1

GUM ARABIC AND OTHER GUM EXUDATES

Structure and properties of exudate gums

ALISTAIR M.STEPHEN

Chemistry Department, University of Cape Town, Rondebosch 7700, South Africa

ABSTRACT

Industrially-important plant gum exudates differ in molecular structure, and there are small variations in this respect among different specimens of gum from a particular species. All *Acacia* gums contain similarly-bound sugar and uronic acid units, but their molecular-weights and protein contents vary with species. Gum arabic may be fractionated by affinity chromatography into a major polysaccharide component, a minor one which contains as much protein as carbohydrate, and an intermediate fraction. Gum tragacanth consists of a neutral arabinogalactan and modified pectin; gum karaya is an acetylated rhamnogalacturonoglycan with neutral and acidic side chains. In gum ghatti a complex, acidic arabinogalactan moiety is attached to a core of alternating glucuronic acid and mannose residues. There are indications, apart from the limited numbers of modes of linkage between the monosaccharide units, that similar, large blocks of sugars become joined in the biosynthesis of plant gums. The analysis of plant gums requires a range of spectroscopic and chromatographic techniques, and the application of methods of chemical breakdown specifically adapted to their complex molecular structures. Availability, and acceptability as edible, limit the use of gums in foodstuffs. Satisfactory rheological, emulsifying, stabilizing and adhesive properties are sought for different commercial purposes, and possessed by plant gums in varying degrees. Molecular features may be correlated with physical.

INTRODUCTION

Although gum exudates from several hundred plant species are known, as acidic polysaccharides, they come from relatively few (fig. 1) of the ninety-two orders of angiosperm or from gymnosperms (1,2). The important industrial gum arabic (from *Acacia senegal*), tragacanth (*Astragalus* spp), karaya (*Sterculia urens*), ghatti (*Anogeissus latifolia*) and larch gum (*Larix* spp) have been examined in detail with respect to their composition, molecular structure, physical properties and applications (3-6). The structures of these polysaccharides have been classified A-D(2). Type A is based on a ramified 3,6-linked D-galactopyranose core to which are bound L-arabinofuranosyl units singly or in chains; exterior to the core L-rhamnopyranosyl groups are attached at O-4 to D-glucopyranuronic acid, which is linked to O-6 or O-4 of

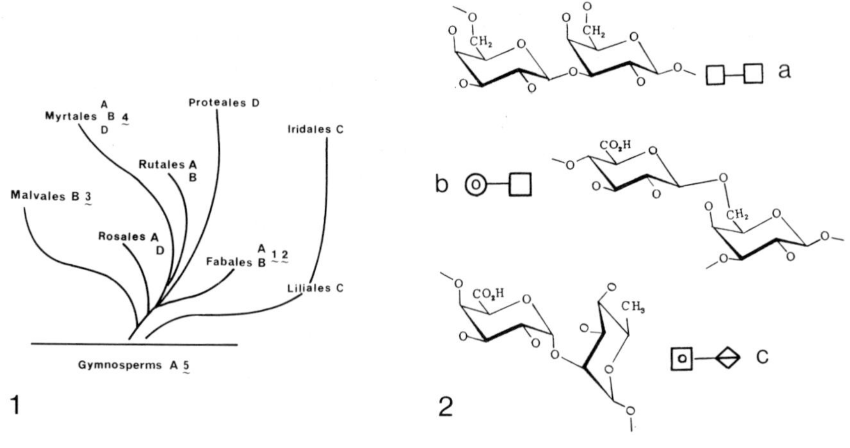

Figure 1. Major orders yielding gum exudates. Capital letters show structural types; gums arabic 1, tragacanth 2, karaya 3 and ghatti 4 and larch-gum 5 from the taxa indicated. Figure 2. Linkage modes between sugar units in gums, a →6)-β-D-Galp-(1→3)-β-D-Galp-(1, b →4)-β-D-GlcA-(1→6)-β-D-Galp-(1, and c →4)-α-D-GalA-(1→2)-α-L-Rhap-(1.

D-Gal by β and α linkages respectively. Fig. 2 illustrates some of the detail, and fig. 3 a putative display of the sugars and uronic acids in gum arabic (7,8); the symbols used (□ for D-Galp, etc.) furnish a general impression of the complexity of the molecular structure. Inspection reveals the conservatism of inter-sugar linkage that is maintained in this, as in all other known examples of plant gum structures, though the disposition of side chains is arbitrary, and the sugar residues displayed are only a microcosm of the molecule as a whole. Larch gum, and those

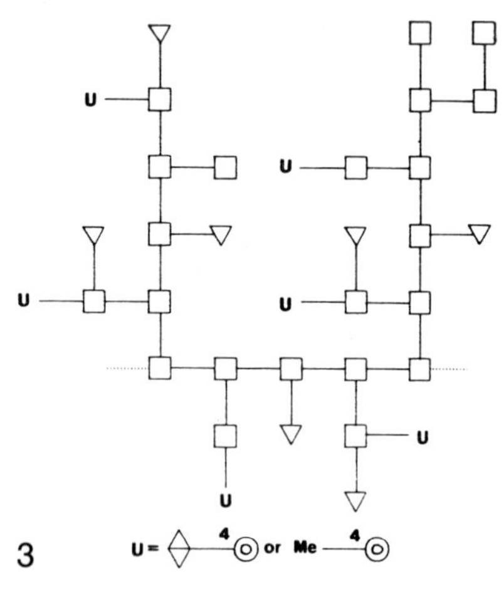

Figure 3. Representative segment, the major polysaccharide component of gum arabic; sugar symbols as in fig. 2, ▽ L-Araf, or L-Arap-terminated short chains, (1→3)-linked, or α-D-Galp-(1→3) -L-Araf.

from certain <u>Acacia</u> spp, are less complicated.

Tragacanthic acid (fig. 4) is the major component of gum tragacanth (9,10), the other being a nearly neutral arabinogalactan (11). Whereas the carbohydrate core of pectins consists principally of (1→4)-linked α-D-GalpA units, interrupted only occasionally by L-Rhap, type B gums include higher proportions of Rhap, with concomitant modification of conformation, and there are side chains, usually short, which include D-Gal, L-Araf, D-Xylp, L-Fucp, and D-GlcpA residues that further complicate the molecular structure. In tragacanthic acid the Rha content is minimal, but in gum karaya (fig. 5) such residues alternate with the acid.

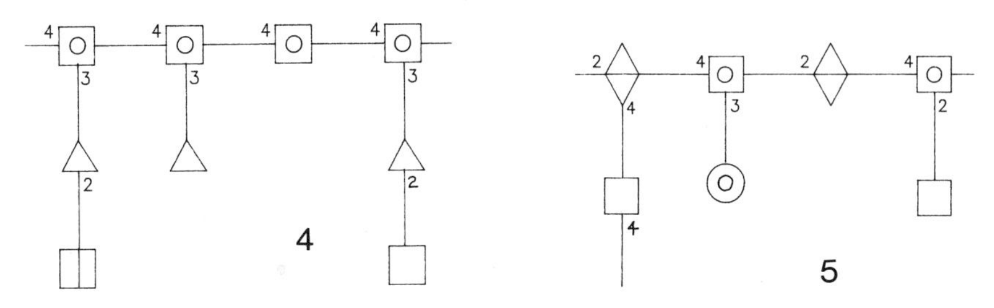

Figure 4. Gum tragacanth, pectic component; Δβ-D-Xylp, ◫α-L-Fucp-.
Figure 5. Structural elements in <u>Sterculia</u> gum. Side chains β-, main chain α-linked.

There are few recorded examples of gums of type C, which are based on a (1→4)-linked β-D-xylan core but with extensive modification, resulting in enhanced water-solubility, in the form of numerous single units of L-Araf and D-GlcpA, and some D-Gal, attached at both O-2 and O-3. However, increasing numbers of gums are being found of structural type D, e.g. the exudates of <u>Anogeissus latifolia</u> (ghatti) and <u>A. leiocarpus</u> (11). Complex arrays of sugar units are attached to a chain of alternating (1→4)-β-D-GlcA and (1→2)-α-D-Manp units which comprises the mannoglucuronoglycan core; many type A features appear in the side chains, which are heavily branched and contain additional GlcA residues (fig. 6).

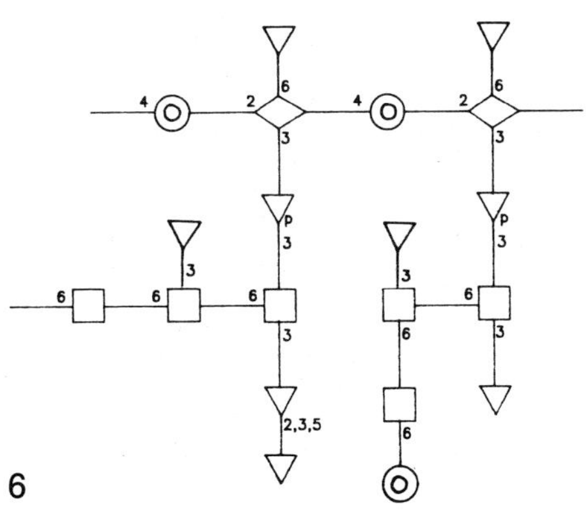

Figure 6. Some structural features of <u>Anogeissus latifolia</u> gum (gum ghatti); ◇ α-D-Manp.

Before discussing the physical properties of gums, and the applications to food products which are dependent thereon, analytical procedures which have led to the formulation of their molecular structures will be considered. A balanced view of the degrees of certainty with which various aspects of structure are known may be obtained thereby. Whereas the proportions and modes of linkage of the individual monosaccharide units are well-defined, and there is a growing awareness that in all Acacia gums there are quite complex but relatively uniform sub-units of molecular structure within the macromolecules, the problems associated with variability in composition of sample, the homogeneity question and the special difficulties of analysis which arise from units which vary in their sensitivity to acid and in the degree of firmness of their mutual bonding, should not be underestimated.

MOLECULAR STRUCTURES OF ACACIA GUMS

This topic is emphasized on account of the predominance of gum arabic as an industrial commodity, the fact that Acacia gums generally possess sets of similar molecular features yet enjoy individual structural characteristics dependent on species of origin, and the relatively large numbers of examples that have been studied. The genus is divided according to Bentham (12). Series 1 and 2 yield gums of low molecular weight which are essentially arabinogalactans, with <10% of GlcA and Rha combined (13). Series 4 includes A. arabica, A. seyal and other spp, the exudates of which have been studied extensively by D.M.W. Anderson et al. They have high molecular weights, are relatively high in Ara*f*, contain more uronic acid, and are sometimes associated with significant amounts of protein (>50%); high positive $[\alpha]_D$ in water is a useful diagnostic parameter. Gum arabic is among those from trees classified in Series 5, average values for which gum approximate to: average mol. mass 5×10^5, $[\alpha]_D$ -30°, Gal 36, Ara 31, Rha 13, GlcA 18 (includes some 4-Me ether), and protein 2 (it is 50% in a minor component). Special features of some Acacia gums follow.

(i) Structural sub-units. The variability of plant gums generally, which seems to be dependent on such factors as growth conditions and age of sample, does not disqualify the use of parameters as listed above in typifying the gum from any particular species. Although molecular weights, for example, may vary over wide ranges, this is not entirely due to physical aggregation or random biosynthetic growth. Gel chromatography shows that the molecular weight distribution of A. podalyriaefolia (Series 1) gum (average 16000) varies with samples taken from different localities at different seasons; peaks occur as multiples of ~4000, however, suggesting that sub-units of this mol. mass are joined, and that samples contain differing proportions of these components.

That aggregation is not the explanation for this and other Acacia gums of low molecular weight has been demonstrated by Smith degradation. Despite their polydispersity, the polysaccharides are cleft by this process into what are essentially (1→3)-linked D-galactans of relatively uniform, small size (mol. wt. as low as 2000). Fig. 7 outlines the chemical degradation process, and in fig. 8 two Acacia gums from Series 1 of higher mol. wt. (~5×10^5) are seen to require two or three successive degradations. The Smith degradation is in three steps, $NaIO_4$ oxidation of the gum, $NaBH_4$ reduction of sensitive aldehyde groups generated, and specific cleavage by cold acid at what was the glycosidic bond of the sugar unit oxidized. Use of iodate in

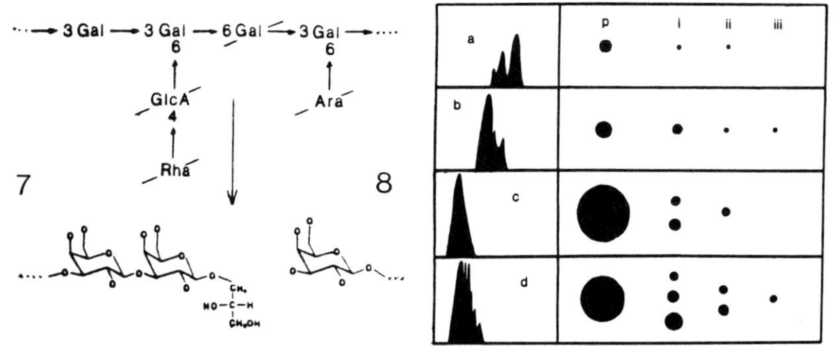

Figure 7. Smith degradation of part of a hypothetical, substituted arabinogalactan of Type II (14). Fragments from Rha, GlcA, Ara become detached, but unoxidised (e.g. →3-linked) Gal units remain attached to each other (ref. 15, published by permission).
Figure 8. S.e.c. of gums and derivatives. a Acacia cyclops, b A. implexa, c A. longifolia, d A. saligna (formerly cyanophylla). Relative mean sizes of polysaccharides (p), and of their Smith degradation products (i, ii, iii); scale, a p $1{,}2 \times 10^4$.

this sequence of reactions resulted in the molecular size distribution of A. mabellae gum being unaffected (16) so that the chemical operation leading to sensitising, in particular, the periodate-vulnerable (1→6)-linked Gal units in main chains is the root cause of breakdown.

For A. pycnantha (Series 1) and A. mabellae (Series 2) gums the lowest molecular weight component was separated from the rest, and gave on Smith degradation a compound similar to that from the remaining, combined components of the gums (2). From observations on these and many other Acacia gums, it appears they may be formed biosynthetically from blocks containing (1→3)-linked D-Galp residues, which undergo polymerisation through O-6 of non-reducing terminals. Peripheral sugars may be attached at a later stage; it should furthermore be noted that first Smith degradation products are usually ~one-half the size of the natural "repeating" units.

(ii) Peripheral units. The nature of the sugar residues in peripheral regions around the galactan core (which itself may be branched) is relatively constant among Acacia gums and others of type A (and D). A predominant mode of linkage of the (4-Me) glucuronic acid units is as shown in fig. 1, though (α1→4)-linkages to D-Gal occur in gums from Series 4. Of equal importance (see section on properties) is the attachment of L-Rhap, mainly as end-group, at O-4 of the GlcA, bringing ionic and relatively hydrophobic centres in juxtaposition. Acid-labile Araf units are linked to the core, either singly or as (1→3)-linked chains.

(iii) Protein. Whereas in some *Acacia* gums the protein content may be up to 20%, and covalent attachment to the polysaccharide may be inferred from co-chromatography on s.e.c., even after enzymatic treatment, short carbohydrate chains, resembling those found externally to the galactan core in the polysaccharides, have been shown to be attached as O-glycosides to hydroxyproline residues (17-19); serine is another site for glycosidation, though aspartic and glutamic acids could also be implicated.
(iv) Gum arabic. The term heteropolymolecular has been applied (20) to gum arabic, there being variations in the chemical composition of fractionated material as well as in molecular size. Although major structural aspects are depicted in fig. 3, this does not take into account the extended ramification and coiling of the polysaccharide which results in its overall, spheroidal shape and concomitant low viscosity in solution, nor the evidence that sub-units of structure exist at levels comparable with those (fig. 8) revealed by Smith degradation of *A. longifolia* gum, and the recently-confirmed (21) finding that protein rich components accompany the polysaccharide. Smith degradation of whole gum (2) yields two major products (mol. wts ~6 and 3×10^4) after the first and two (3 and $1,6 \times 10^4$) after the second Smith degradation; at this stage periodate-vulnerable units are exposed at such regular intervals that the next degradation yields blocks of ~12 sugar units. Prior, mild acid hydrolysis of gum arabic removes mainly Ara*f* and Rha*p* groups, and a single Smith degradation then gives a similar, low mol. wt. (2000) product. Regularity of structure exists therefore at both low and high levels of molecular weight. The picture is elaborated by the separation of gum arabic using hydrophobic affinity chromatography (21) into fractions of which the bulk is substituted acidic arabinogalactan, ~10% contains identical inter-sugar linkages in much the same proportions but with 10% protein (an "AG-P", 22), and ~1% which has qualitatively similar sugar residues and 50% protein; in the last-named there may be a preponderance of terminal sugar groups indicating shorter carbohydrate chains attached to amino acids. The AG-P (second fraction) is envisaged as comprising ~five carbohydrate moieties averaging $2,8 \times 10^5$ in mol. wt. bound to the protein, in which Hyp and Ser greatly predominate.

GUMS FROM *STERCULIA* SPP. GUM KARAYA

Gum karaya, an exudate from the bark of *Sterculia urens*, is a heavily acetylated acidic polysaccharide containing α-D-GalA and α-L-Rha*p* residues in chains, the linkages being to O-4 of the acid and O-2 of the Rha. O-2 of the acid is glycosylated by β-D-Gal*p* or O-3 by β-D-GlcA, whereas one-half of the Rha units carry at O-4 (1→4)-linked β-D-Gal units as side chains (fig. 5). This structural formula is based on the identification of fragments produced by acetolysis of acetylated polysaccharide that had first been carboxyl-reduced or submitted also to Smith degradation (core units being largely protected against oxidation by their modes of linkage), or alternatively by partial acid hydrolysis which favours the fission of Rha*p* linkages. Additionally, specific methods of degradation developed by G.O. Aspinall et al. have determined the nature of the sugar residues exterior (towards the non-reducing end) to the GalA units; in particular, the hex-5-enose degradation (in which fully methylated gum is so modified as to replace the carboxyl function by methylene) leads to the detachment of the internal residue, and the subsequent identification of derived, substituted oligosaccharides. Earlier formulations involving (1→4)-linked α-D-Gal in the main chain are thus superseded, but the structure is undoubtedly complex (23).

GUMS OF MANNOGLUCURONOGLYCAN TYPE (D)

It has long been known that minor proportions of (1→2)-linked Man*p* are present in certain plant gums, notably from *Prunus* spp (Rosales), and the first convincing demonstration that Man residues are combined with GlcA to form the core structure of a plant gum came from Aspinall's group (11). Gums from two *Anogeissus* spp, of which one (*A. latifolia*) yields gum ghatti, were consequently formulated as indicated (fig. 6). The side-chains of gum ghatti comprise several (1→6)-linked D-Gal units, some terminated by GlcA, joined to the Man*p* units at O-3 through (1→3)-linked L-Ara*p* residues. There are also terminal Ara groups, some at O-6 of the Man units (double branch points), and other short sequences of L-Ara*f*. The gum from *A. leiocarpus*, fractionated by differential Cetavlon precipitation, gave a major polysaccharide component in which the basal chains were established by the characterisation of a series of oligosaccharides containing 4Glc→2 Man isolated after acetolysis of carboxyl-reduced material. Smith degradation afforded additional information (the hydrolysis step in such cases slowly reaching a second limit only after fission of resistant erythronic acid-glycollic acetal linkages), the conclusion being reached that apart from there being appreciable chains of (1→3)-linked Gal*p* in the *A. leiocarpus* gum, this polysaccharide is substantially similar to gum ghatti; β-D-Xyl*p* has, however, been placed at O-6 of D-Man and β-L-Ara*f* at O-3 by use of the $Pb(OAc)_4$ procedure applied to a methylated fraction from *A. leiocarpus* gum (24).

ANALYTICAL CONSIDERATIONS

Plant gum exudates are comparatively easily converted by dissolving in water, centrifuging and freeze drying into material that is satisfactory for inspection by colour and immunological tests, ^{1}H- and ^{13}C-n.m.r., $[\alpha]_D$ measurement, combustion analysis, and acid hydrolysis to give the proportions of sugar and uronic acid residues present. These techniques have been extensively described (25,26). However, the presence of more than one component is not normally revealed (although occasionally fractional precipitation from water by ethanol, salts or Cetavlon suffices) except by dialysis, electrophoresis or chromatographic analysis. Steric exclusion (gel) chromatography is useful, and indeed indispensable in following the course of breakdown of gum structures by hydrolytic methods, before and after chemical modification.

Given a gum sample of satisfactory homogeneity (i.e., at the practical limit of fractionation), structural analysis depends primarily on either total or controlled, partial hydrolysis, usually in acid media, and the quantitative analysis of the components. As pentoses are liberated first and are comparatively acid-labile, and uronic acid units require drastic treatment to effect their detachment from interior sugars, the production of hydrolysates may have to be carried out in stages. G.l.c. methods of analysing hydrolysates continue to cause difficulty, as the necessary processes of derivatisation and assay depend on a standardised approach and reliable response factors.

Modes of inter-sugar linkage are determined after successful per-*O*-methylation of the gum, something which cannot be achieved by following a standard recipe but requires careful assessment of the methods available; insolubility in water or DMSO, and the presence of much (>10%) uronic acid, cause problems. As industrial gums are plentiful the macro-method of Haworth is satisfactory, though subsequent treatment of the product (now soluble in $CHCl_3$) by alternative methods is essential to ensure conversion of all OH groups to OMe. Hakomori's method using anhydrous DMSO and the powerful base $^-CH_2SOMe$ in conjunction with MeI may well cause loss of sugars exterior to uronate by β-elimination and modification of interior units, though adaptation of this process establishes the bonding of such sugars through D labelling. The Purdie method (MeI, on Ag_2O surface) is therefore invaluable in the last stage of per-*O*-methylation, though the Kuhn variant, and other, modern methods should be considered. Thereafter the hydrolysis step and subsequent g.l.c. analysis of derivatives of the sugar methyl ethers is subject to the same difficulties as the determination of the sugars themselves. A useful test lies in the balancing of terminal groups and branch points. G.l.c.-m.s. (including selected-ion monitoring) is indispensable for identification of small amounts; and there are obvious advantages in working on such a scale as to permit semi-quantitative separation of the methyl ethers, which can be identified and employed as standards.

Partial hydrolysis (using acids or enzymes) or acetolysis is invaluable in sequencing pairs of sugars, especially between uronic acid and interior units, and where possible tri- or higher saccharides, neutral and acidic, are isolated and characterised by methylation and 2D n.m.r. techniques. The differences in rate of glycosyl cleavage within polysaccharides are exploited in this approach, these rates being modified by the Smith procedure, or by prior reduction of carboxyl groups (to Glc or Gal) which obviates the firm binding between the respective acids to interior sugars. Acetolysis rates differ from those of aqueous acid (TFA and H_2SO_4) hydrolysis, and this has been exploited in the case of gum arabic; carboxyl-reduced gum was acetylated and acetolysed, the relatively firm Rha to Glc bond surviving and enabling the presence of a (1→4)-linkage to GlcA to be established.

STRUCTURAL AND FUNCTIONAL PROPERTIES OF PLANT GUMS, AND THEIR USES

Aspects of the structure and physical properties of gums, and mucilages, may be related to their practical use and industrial application. The very terms imply physical characteristics which involve water retention, high viscosity, adhesiveness, and a slimy or tacky feel; the applications of gums and mucilages in the food, pharmaceutical and mining industries depend on these and other related properties (4,5). Gums have been termed hydrocolloids and polyelectrolytes, as befits their acidic, polysaccharide character, the relevant molecular properties being size distribution, shape, degree of aggregation, and changes in conformation with concentration, temperature, ionic strength of the medium, and pH. The distribution of charged sites, hydrophobic regions and points of attachment to protein, as well as the degree of regularity of the macromolecules as a whole have all to be taken into account in any attempt to match molecular structural features with physical properties and uses.

Ready availability of raw material in uncontaminated form is an essential prerequisite for the use of gums in the food industry; there should be no hint of unacceptability on the grounds of potential physiological hazard, and to be marketable the material should be of satisfactory appearance and odour. Uniformity of composition and texture of compounded products should be maintained for as long as deemed necessary, and in particular there should be no alteration in flow properties. These and other requirements are exacting.

In summary, Table I lists a number of food products in which gums are incorporated and the function required (27). The versatility of gum arabic as a hydrocolloid is evident. Gum ghatti has found a limited use in syrups, but high concentrations are not permitted in foodstuffs. Tragacanth is relatively expensive. The low calorific values of gum arabic and larch gum are important from the dietetic point of view, and to add to its many other uses gum arabic is used in vitamin supplements.

Table I. GUMS IN THE FOOD INDUSTRY, AND THEIR FUNCTION

Food product	Plant gum		
	Arabic	Tragacanth	Karaya
Bakery (mixes)	a-c		d e
Icings	e-f		
Meringues	c d	c d	c d
Flavourings	b g	b	
Beverages	b c g h		
Confections	a-g	b-f	
Dairy		a-d f	a-d f
Dressings, sauces		a-c	a c
Processed meats	h		d

a Thickener. b Emulsifier. c Stabilizer. d Water binding agent. e Adhesive. f Crystallisation inhibitor. g Encapsulating agent. h Competitive interaction. See refs 28,29.

RHEOLOGY

The behaviour of gum arabic in water, as of other hydrocolloids, is influenced by molecular interactions with solvent, itself, and other solutes (electrolytes, proteins). Most of the useful functions of gums in foods depend on their viscosity, which causes stabilisation and affects texture (30). Viscosity in turn is governed by shape, molecular size and concentration, and is affected by temperature, pH, and accompanying solutes. A 5% solution of gum arabic at 25°C is more mobile than ghatti, tragacanth or karaya, but less so than larch. Most gums other than arabic and larch are non-Newtonian (viscosity decreasing inversely with shear rate), there being some correlation with the type of "mouthfeel" - Newtonian behaviour is associated with "sliminess". High viscosity affects in addition the perception of aroma, the vapour pressure of volatile constituents increasing as water is immobilized (31). Although gel formation is important, there are as yet few dependable, measurable parameters that correlate molecular and gel structures with texture (27).

Gum arabic is extremely soluble, aggregation occurring at 40% concentration and above. Inverse proportionality between viscosity and temperature holds between 10° and 40°C, a levelling off at higher temperatures being due to the destruction of aggregates. The viscosity of gum arabic is highest between pH 4.7 and 7, there being a high degree of ionisation of carboxyl groups in this range; shielding by electrolytes decreases viscosity and aggregation (4). Added citrate has the contrary effect on account of its complexing power, but viscosity is decreased by ageing, application of ultrasound, and u.v. irradiation.

There is a strong positive correlation between intrinsic viscosity [η] and molecular weight, as is shown by fractions of gum arabic (32); viscosity decreases with increasing 4-OMe content of uronic acid groups (33), and increases, disproportionately, with %N due to associated protein (34).

In relation to gum arabic, the most-used, the others show special features.

(i) Gum ghatti. The somewhat higher viscosity of ghatti (35), which varies with source, may relate to its efficacy in stabilizing concentrated emulsions and suspensions. Nodules are not entirely soluble in cold water, but maceration and heating at 90°C enhance solubility; divalent cations cause the opposite, and their removal has a permanent, beneficial effect. No chemical difference is detectable between soluble and gel-forming components. The structure has been described as rod-like (4,5), Newtonian behaviour of solutions (and consequent "sliminess") being observed at concentrations allowable (GRAS).

(ii) Gum tragacanth may also be fractionated on grounds of solubility, but the chemical distinction between arabinogalactan and less-soluble (tragacanthic acid) components has been noted. Accompanying Ca^{2+} and Mg^{2+} may contribute to high viscocity and gelling properties; the gum is rated as relatively "non-slimy" (30). Viscosity increases with OMe content, here due to ester groups; this property is little changed with pH even in the wide range 2-10, hence the value of the gum in salad dressings and sauces. Apart from a potential loss of peripheral groups, the molecular framework is acid-stable and unlikely to undergo undue lowering of molecular weight (36).

(iii) Gum karaya is the least soluble, yielding highly viscous colloids (pH 5-7) at 5% concentration; swelling in water (most rapid if finely ground) takes place, and if autoclaved, solutions of up to 20% may be obtained though with permanent loss of viscosity due to hydrolysis. A major feature, its OAc content, is affected by alkaline treatment, the dispersion being converted into a ropy mucilage (4). Salts cause viscosity loss, though Ca^{2+} ions

may produce a slight increase if present in very high concentration. Gum karaya solutions deviate from Newtonian behaviour, hence a low rating on the "sliminess" scale (30), and show viscoelasticity. Resistance to flow of dispersions of karaya is indicative of the complex molecular network which absorbs water strongly, so immobilising it.
(iv) Larch gum. In larch arabinogalactan the molecular weight and proportion of uronic acid groups are low, measurable viscosity being attained only at concentrations as high as 10% (4); however, up to 40% solutions can be prepared, and the gum is used in beverages. Slimy mouthfeel is a predominant characteristic. No irreversible viscosity loss occurs on heating, and this property is relatively unaffected by pH change and added electrolytes.

GUMS AS EMULSIFIERS AND STABILIZERS

Thickening and texture properties are a consequence of the rheological behaviour of gums; additionally, the important function of stabilizing (prevention of flocculation or coalescence) two-phase dispersions (emulsions, suspensions, and foams) is possessed by all the commercial gums in some degree. The property of emulsification, the active promotion of liquid-liquid dispersions, is less common; this depends on reduction of surface tension.
(i) Emulsifiers. The role of gums as emulsifiers has not been considered prominent, but some, gum arabic among them, appear to align at the oil-water interface; the spheroidal shape facilitates such close-packing, though the fact that the less abundant proteoglycan portion of the exudate is predominant in surfactant ability (37) points to the need for considering the emulsifying power of this type of molecule when present in other gums, e.g. tragacanth. It would appear that spatial separation of relatively hydrophobic portions of the gum molecule, such as the Me groups of Rha*p* in terminal positions, from the hydrophilic hydroxyl and ionic regions may be implicated (38). Perhaps the minor polyphenolic component of larch gum is efficient in the same sense (39), though there are indications that other hydrophobic associations may also be responsible. Gum tragacanth is an effective emulsifier at as low a concentration as 0,25%, hydrophobicity being attributed to the terminal deoxyhexosyl (in this case Fuc) groups of tragacanthic acid, while the spheroidal arabinogalactan component acts like the polysaccharide component of gum arabic.
(ii) Stabilisers. By increasing the viscosity of the aqueous phase, plant gums impede the migration of dispersed globules or particles, and so inhibit coalescence, flocculation and sedimentation. The superior properties of gum ghatti and gum tragacanth relative to arabic follow from high viscosity at low concentration (see Table). However, gum karaya is a hydrocolloid having a yield stress and is useful in stabilising suspensions (40). Another mode of action involves adsorption of the dispersed globule or particle (31), where gum arabic is effective despite the low viscosity of its solutions; both the time taken for adsorption to take place and the permanence of encapsulation are significant, and once again the molecular weight and spatial distributions of charged groups are implicated. Addition of surfactants facilitates the displacement of adsorbed air prior to encapsulation, so that gums

arabic and tragacanth (both having surfactant properties) are particularly effective (40).

OTHER ADSORPTIVE INTERACTIONS

The unique water-holding capacity, and consequent swelling, of gum karaya leads to its use in ground meat and cheese products, and in the retardation of staling in baked goods (5). Inhibition of ice crystal formation in ice-cream is a valuable property of gum tragacanth (36,41), which has emulsifying and aerating powers in addition. The high affinity of gums for water gives adhesive properties, shown best by gum arabic, which is widely applied in icings. The major importance of gum arabic in confectionary, however, stems from its interaction with other carbohydrate or polyol constituents, through hydrogen bonding, particularly at the high concentrations used in pastilles and lozenges, where crystallisation is inhibited. Similarly, fats remain evenly distributed in caramels as a consequence of the emulsifying property of gum arabic, and the interaction with flavouring ("aromatic") compounds enables efficient encapsulation to be maintained even in dry mixes, spray dried and stored, and in the production of clouding agents in beverages.

Finally, the interaction of plant gums with proteins is of importance in view of the multiplicity of protein-containing foodstuffs. Controlled reaction (at pH 8, in presence of salts) between oppositely charged gum arabic and gelatin yields a coacervate (a complex dispersed in the liquid phase that can encapsulate oil droplets), cooling producing a gel used in such diverse applications as cake mixes, confections, and soluble coffee (40).

CONCLUSION

All plant gums have high molecular weights and contain uronic acid. In addition to the multiplicity of hydroxyl groups, there are regions of relative hydrophobicity in the periphery,and core. Shapes range from globular to rigid and rod-like, and there is a growing awareness of the significance of bound protein. Dispersions of gums engage water molecules through extensive hydrogen bonding, thus profoundly influencing their rheology. Viscosity varies greatly from one gum to another, aggregation, gel formation and flocculation resulting from this intrinsic property and the influence of external agents such as acid, alkali, and other electrolytes. Associations with sugars, lipids and proteins depend on ionic as well as hydrophobic regions in the gums, emulsifying and stabilising power perhaps arising from such amphilicity. Whereas there is a growing understanding of the control exercised by molecular features upon the physical properties of gums and consequently their application in foodstuffs, much has yet to be learned about disaggregation, depolymerisation and dehydration which are among the changes that accompany heat treatment (cooking), not all of them desirable.

ACKNOWLEDGEMENTS

I am most grateful to the organisers of the conference for enabling me to participate, to the FRD/CSIR and the University of Cape Town for financial assistance, and to my colleagues, past and present, who have co-operated in exploring many fascinating aspects of plant gum chemistry and in the production of manuscripts.

REFERENCES

1. Stephen, A.M. (1980) Encyl. Plant Physiol., New Ser. 8, 555-584.
2. Stephen, A.M. (1983) in The Polysaccharides, (ed. Aspinall, G.O.) Vol. 2, pp 97-193. Academic Press, New York.
3. Smith, F. and Montgomery, R. (1959) The Chemistry of Plant Gums and mucilages, Van Nostrand-Reinhold, Princeton, New Jersey.
4. Whistler, R.L. and BeMiller, J.N., eds. (1973) Industrial Gums, 2nd edn., Academic Press, New York.
5. Glicksman, M. (1983) in Food Hydrocolloids, (ed. Glicksman, M.) Vol. II. CRC Press, Boca Raton, Florida.
6. Sandford, P.A. and Baird, J. (1983) in The Polysaccharides, (ed. Aspinall, G.O.) Vol. 2, pp 411-490. Academic Press, New York.
7. Anderson, D.M.W., Hirst, E. and Stoddart, J.F. (1966) J. Chem. Soc. C pp 1959-1966.
8. Street, C.A. and Anderson, D.M.W. (1983) Talanta, 30, 887-893.
9. Aspinall, G.O. and Puvanesarajah, V. (1984) Can. J. Chem., 62, 2736-2739.
10. Aspinall, G.O. (1987) Accounts Chem. Res., 20, 114-120.
11. Aspinall, G.O. (1969) Advan. Carbohydr. Chem. Biochem., 24, 333-379.
12. Bentham, G. (1875) Trans. Linn. Soc. (London), 30, 335-664.
13. Anderson, D.M.W. and Dea, I.C.M. (1969) Phytochemistry, 8, 167-176.
14. Aspinall, G.O. (1973) in Biogenesis of Plant Cell Wall Polysaccharides, (ed. Loewus, F.) pp 95-115. Academic Press, New York.
15. Stephen, A.M. (1987) S.Afr.J.Chem., 40, 89-99.
16. Churms, S.C., Merrifield, E.H. and Stephen, A.M. (1978) Carbohydr. Res., 63, 337-341
17. Connolly, S., Fenyo, J-C. and Vandevelde, M-C. (1988) Carbohydr. Polymers, 8, 23-32.
18. Akiyama, Y., Eda, S. and Kato, K. (1984) Agric. Biol. Chem., 48, 235-237.
19. Gammon, D.W., Stephen, A.M. and Churms, S.C. (1986) Carbohydr. Res., 158, 157-171.
20. Anderson, D.M.W. and Stoddart, J.F. (1966) Carbohydr. Res., 2, 104-114.
21. Williams, P.A., personal communication.
22. Clarke, A.E., Anderson, R.L. and Stone, B.A. (1979) Phytochemistry, 18, 521-540.
23. Aspinall, G.O., Khondo, L. and Williams, B.A. (1987) Can. J. Chem., 65, 2069-2076; c.f. Raymond, W.R. and Nagel, C.W. (1973) Carbohydr. Res., 30, 293-312.
24. Aspinall, G.O. and Puvanesarajah, V. (1983) Can. J. Chem., 61, 1864-1868.
25. Aspinall, G.O. (1982) in The Polysaccharides, (ed. Aspinall, G.O.) Vol. 1, pp 35-131. Academic Press, New York.
26. Stephen, A.M., Churms, S.C. and Stephens, D.C. (1989) in Methods in Plant Biochemistry, (ed. Dey, P.M.) Vol. 1. Academic Press, London. (In publication).

27. Szczesniak, A.S. (1986) in Gums and Stabilizers for the Food Industry, (eds. Phillips, G.O., Wedlock D.J. and Williams, P.A.) Vol. 3, pp 311-323. Elsevier, Barking, U.K.
28. Glicksman, M. (1969) Gum Technology in the Food Industry, Academic Press, New York.
29. Stephen, A.M. and Churms, S.C. (1989) in Food Polysaccharides, (ed. Dea, I.C.M.) Chap. 11. Marcel Dekker, New York. (In publication).
30. Mitchell, J.R. (1979) in Polysaccharides in Food, (eds. Blanshard, J.M.V. and Mitchell, J.R.) pp 51-72. Butterworth, London.
31. Walker, B. (1984) in Gums and Stabilizers for the Food Industry, (eds. Phillips, G.O., Wedlock, D.J. and Williams, P.A.) Vol. 2, pp 137-161. Pergamon Press, Oxford, U.K.
32. Anderson, D.M.W. and Rahman, S. (1967) Carbohydr. Res., 4, 298-304.
33. Anderson, D.M.W., Cree, G.M., Herbich, M.A., Karamalla, K.A. and Stoddart, J.F. (1964) Talanta, 11, 1559-1560.
34. Vandevelde, M-C. and Fenyo, J-C. (1985) Carbohydr. Polymers, 5, 251-273.
35. Jefferies, M., Pass, G. and Phillips, G.O. (1977) J. Sci. Food Agric., 28, 173-179.
36. Stauffer, K.R. (1980) in Handbook of Water-Soluble Gums and Resins, (ed. Davidson, R.L.). McGraw-Hill, New York.
37. Randall, R.C., Phillips G.O. and Williams, P.A. (1988) Food Hydrocolloids, 2, 131-140.
38. Anderson, D.M.W. (1978) Kew Bull., 32, 529-536.
39. Ettling, B.V. and Adams, M.F. (1968) Tappi, 51, 116-118.
40. Glicksman, M. (1982) in Food Hydrocolloids, (ed. Glicksman, M.) Vol. I. CRC Press, Boca Raton, Florida.
41. Cottrell, J.I.L., Pass, G. and Phillips, G.O. (1979) J. Sci. Food. Agric., 30, 1085-1088.

Physico-chemical properties of gum arabic in relation to structure

J.C.FENYO and M.C.VANDEVELDE

Laboratoire de Physiologie Cellulaire, URA 203 CNRS, SCUEOR, UFR des Sciences et Techniques de l'Université de Rouen, BP 118, 76134 Mont Saint-Aignan Cedex, France

ABSTRACT

Even today, the origin of variability of measured molecular weights of gum arabic has not been clearly elucidated. Recent developments have evidenced that the bulk gum is a mixture of at least four compounds: two minor species (a few percent of an insoluble mucilage and a glycoprotein) and two major ones which consist of both protein-linked and free well defined polysaccharide chains of molecular weight around 200,000, the actual M_w depending on their relative ratio. This can be well represented by a "wattle-blossom" model partly consistent with previous proposed block or repetitive structures. It is thus demonstrated that the current meaning of molecular weight is not suitable for gum arabic, contrary to its other well-established analytical characteristics. Desirable properties could be ensured only if M_w and its distribution are adequate. Incidently, spectroscopic measurements have been proved to be helpful to study the heterocompounds which are involved in the molecular structure and the biosynthesis process.

INTRODUCTION

Analytical data for many gums exuded from species of the genus *Acacia* have been extensively published over more than twenty years, chiefly by Professors D.M.W. Anderson, A.M. Stephen and coworkers. These fingerprints generally include parameters for which it was definitively demonstrated by large scale investigations that their variations are but little affected by edaphic factors. **This is true** for the gum from *Acacia senegal* (1-4) **It was shown that samples originating from Sudan and from West Africa had the same characteristics when the origin of seeds, nature of soil, exposition, dates of plantation, wounding and harvesting etc. were safely controlled (5).**

However, some fascinating challenges still remain: (i) a convincing explanation of the large variations in viscosities and molecular weights is **still** lacking after the earlier determination by Oakley (6); (ii) in spite of important progress (7-10), the exact molecular structure and the biosynthesis process are not known.

In the five last years, a renewal of academic research was observed in these fields due to techniques such as size exclusion chromatography, light scattering, NMR and the new interest **in** amino acids composition of several gums (11). In particular, **better** understanding of the connection between macromolecular weights and **has been achieved (12-16).**

The aim of this paper is to review these last considerations and to try to answer to the question: what is the meaning of actual molecular weight of gum arabic ?. This may have consequences on technical uses as it **has been** demonstrated (17-19). Moreover, some spectral data will be discussed **in** the light of these considerations and new paths of research for a better understanding of the biosynthesis will be presented.

MACROMOLECULAR WEIGHTS AND HETEROGENEITY OF GUM ARABIC

Due to the lack of solubility in usual organic solvents and its polyelectrolyte character, macromolecular weight measurements of gum arabic have to be performed in salt aqueous solutions.

Osmometry is rather technically difficult as the molecular weights are high (20) and light scattering studies delicate as dust-free aqueous solutions are very difficult to obtain (21). New low angle laser light scattering photometers avoid this problem (12).

The main published data are recorded in Table I. Number-average molecular weights, M_n, are reported around $2x10^5$ whatever the medium. Larger values are observed for weight-average molecular weights, M_w, by light scattering, in a broad range from *ca* $4x10^5$ to largely more than 10^6. The intrinsic viscosities (not reported in the Table) are well correlated with the molecular weights by the Mark-Houwink relationship (22) which was confirmed later (12).

Table I. MACROMOLECULAR WEIGHTS OF GUM ARABIC

Authors	M_n	M_w
Oakley (1935) (6)	$2x10^5$	
Veis and Eggenberger (1954) (23) equivalent radius 555 Å when uncharged		$1.0x10^6$
Mukherjee and Deb (1962) (24) root-mean-square end to end distance 1100 Å in salt		$5.8x10^5$
Anderson et *al.* (1967) (21) mean radius of gyration 339 Å on unfractionated sample; from 387 Å to 186 Å on fractions		$5.8x10^5$
Swenson et *al.* (1968) (20) radius of gyration 101 Å in salt-EDTA solutions	$2.5x10^5$	$3.6x10^5$
Acharya and Chattoraj (1975) (25)		$6.5x10^5$
Churms et *al.* (1983) (7) by size-exclusion chromatography on Sepharose-4B		$5.6x10^5$
Vandevelde and Fenyo (1985) (12) fractions obtained on Sephacryl S-400 and S-500		$2\text{-}40x10^5$
Nakamura (1986) (26) on 15 industrial samples	$1.7x10^5$	$2.5x10^5$
Snowden et *al.* (1987) (18)		$1.0x10^6$
Connolly et *al.* (1988) (15) reduced to $1.8\text{-}2.0x10^5$ after pronase treatment		$5\text{-}7x10^5$
Randall et *al.* (1989) (16) fractions from $1.4x10^6$ to $2.8x10^5$; R_h from 141 Å to 92 Å		$4.6x10^5$
Duvallet et *al.* (1989) (27) M_n and M_w $1.8x10^5$ after pronase treatment	$1.9x10^5$	$7.2x10^5$

Two points have to been underlined: authors generally agree for M_n around $2x10^5$ whatever the samples, and naturally M_n are lower that the M_w. However the origin of the discrepancies for M_w when compared with the first published data (23) was attributed to various factors: polydispersity (24), pH and media effects (20,25). It was concluded that a value of M_w = 580,000 is generally acceptable (29).

Evidence of heterogeneity of gum arabic was observed early (30,31). Fractional precipitation by sodium sulphate and subsequent chemical analysis and molecular weight determinations (21,28) have evidenced both dispersity in the molecular weights and a net increase in nitrogen content of the upper molecular weight fractions. The authors have also quoted that differences in the sugar compositions between fractions were a proof of heterogeneity but to our knowledge it has never been demonstrated that these variations were significant.

It is rather surprising that in the following years so little attention has been paid to these observations as well as to the amino acid compositions which for *Acacia senegal* were first published in 1972 (3). Interpretation of polydispersity by fractional precipitation with organic solvent is also incomplete (20, 32).

A renewal of research appeared some years ago in two directions. The amino acid composition of gum arabic has been throughly investigated (11,34-37) and efficient fractionation on Sephacryl S-400 and S-500 gels (Pharmacia) was achieved (12,38). Fractions of M_w from $1.8x10^5$ up to $4x10^6$ were evidenced on a sample of mean M_w $9.1x10^5$ and from $1.8x10^5$ to $3.4x10^6$ for an other sample of mean M_w $4x10^5$. Physicochemical data and sugar analysis of fractions have shown that around 70% of the material is composed of homogeneous polysaccharide chains with a very low nitrogen content, the remaining material being an arabinogalactan-protein. Further investigations (13,15,39) have confirmed these results and by using the proteolytic enzyme pronase, gum arabic was described by the "wattle-blossom" model (13) for which a varying number of polysaccharide units of molecular weight *ca* $2x10^5$ are linked to a protein core, the amino acid composition of which being remarkly constant (14,16,34,35). Typical HPLC for a specimen originated from M'Biddi (Senegal) (specific rotation at 589 nm -31°; %N: 0.32%; intrinsic viscosity 23 $ml.g^{-1}$ in 1M NaCl) are represented in Figure 1. The model was also recently confirmed by comparison between M_n and M_w before and after pronase treatment (27). In the last situation, M_n and M_w were both close to $1.8x10^5$ which confirms the symmetry of chromatographic peaks and the macromolecular homogeneity of this moiety. It is striking that this value is near to the earlier published M_n values which were observed around this value. These results have been confirmed in the North East Wales Institute by fractionation using hydrophobic affinity gels (16). Moreover a glycoprotein which is eluted near V_t on Sephacryl has been found by the authors. Also the role of the proteinaceous component on emulsifying properties has been demonstrated (17-19).

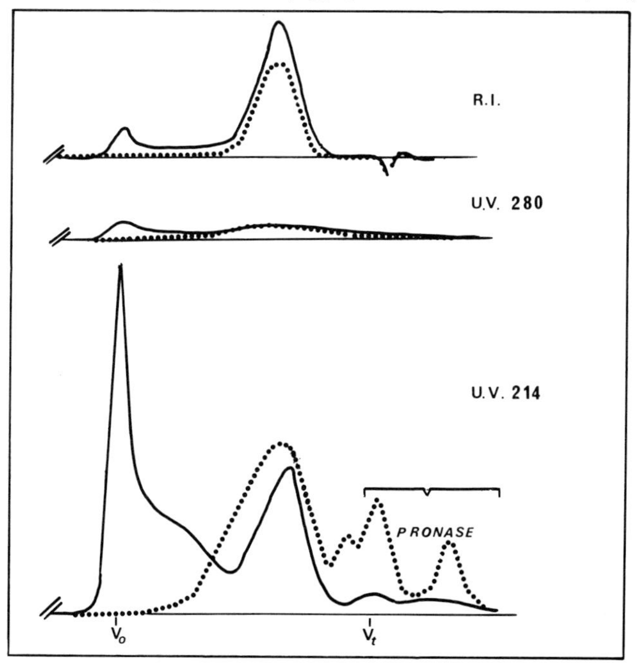

Figure 1. Elution profiles on LKB ultropac TSK G-5000 PW of a specimen of gum arabic from M'Biddi (Senegal). Full lines before and point lines after pronase treatment.

An attempt to recalculate the molecular weight using the last structure presented by Street and Anderson (8) has been done using their number of sugar residues after several Smith degradations. We found a molecular weight of 1.44×10^5. This value is not so far from the measured one taking into account the hypothesis.

Dimensions of the chains can be deduced from macromolecular weight measurements. It can be seen in Table I that radius of gyration vary from 100 Å to more than 500 Å. Reliable discussion must take into account the polydispersity and it is interesting that the relatively low molecular weights around 3×10^5 lead to radii from 100 Å to 200 Å. This order of magnitude is the same as recent data obtained by photon correlation for the hydrodynamic radius (16).

A relationship has been derived between the diffusion coefficient and the weight-average molecular weight of fractions of the form $D_t = 1.48 \times 10^{-4} M^{-0.55}$ (40). The hydrodynamic radius of the equivalent sphere R_h is related to D_t by $R_h = 2.1 \times 10^{-13} D_t^{-1}$ (R_h in cm). With $M_w = 2 \times 10^5$, $R_h = 120$ Å. All these data are consistent and it can be concluded that they represent well the dimensions of individual polysaccharide chains and agree with the very highly branched structure for such high molecular weights. Also viscosity and dimensions of samples after several Smith degradations are not drastically reduced as they depend principally on the main chain (38).

EQUILIBRIA AND SPECTROSCOPIC PROPERTIES

New interesting data have been recently published. Equilibrium properties such as binding and counterion activities (41,42) and potentiometric data (43) confirm that interactions with multivalent counterions are only of electrostatic origin and that the equivalent spatial intercharge distance is around 7 Å to 11 Å.

Studies by ^{13}C NMR (10,44) on undegraded and degraded gums by successive Smith reactions gave new information on the internal structure suggesting repetitive sequences but the influence of sample origin and macromolecular weights is not sensitive to this method.

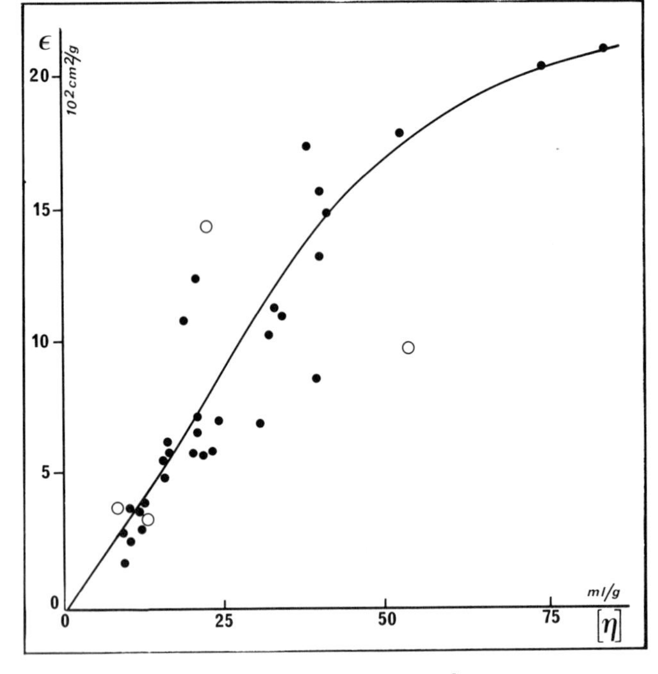

Figure 2. Dependence of the extinction coefficient at 214 nm (10^2 $cm^2.g^{-1}$) • on intrinsic viscosity ($ml.g^{-1}$ in 1M NaCl) for unfractionated and fractionated gum arabic. ○ from reference 19 at 218 nm.

We have **shown** that Ultraviolet, Optical **Rotary** Dispersion and Circular Dichroism measurements could be helpful to demonstrate the presence of the heterocompounds (12). At the investigated scale, **UV** and **ORD** do not exhibit extremum until the limit of investigation (200 nm) but **CD** of high molecular weight fractions **highlighted** the high proteinaceous content. A correlation between absorption at 214 nm and viscosity can be observed (Figure 2) and it has been demonstrated that **UV** can only be considered as a qualitative tool in detection of gum arabic fractionation (13).

DISCUSSION

These results respond to a question raised at the last meeting by E. Dickinson (45) about the correlation between nitrogen content and molecular weight of gum arabic.

It seems now well established that gum is at least a mixture of four chemically different compounds: (i) an insoluble mucilage which represents at **most 1%** of the bulk sample (3) always observed at the first filtration on Millipore membranes; this phase seems **never to have been analyzed**; (ii) a mixture of well defined polysaccharide chains of molecular weight *ca* $2x10^5$ and (iii) linked chains to a protein; and (iv) moreover a glycoprotein recently evidenced (16) which may be eluted near the null exclusion volume on Sephacryl S-400 columns. Structural studies must therefore take into account these various compounds and it should be preferable to fractionate the gum before these studies.

Gum exudates contain more or less high proportions of proteoglycans and glycoproteins (46,47) and the "wattle-blossom" model for gum arabic is difficult to generalize to other exudates. For example, we have observed the same behaviour towards pronase for the gum from *Acacia mearnsii* but not for the gum from *Combretum nigricans* (39).

Also the nature of linkage between sugars and amino acids has to be investigated. We have shown that moderate heating can lower the viscosity of samples towards usual values for industrial uses but this effect is **not as complete as** that of pronase and cannot be compared to degradation by autohydrolysis or mild acid hydrolysis (37) (Figure 3 and Table II). The linkages are more or less resistant to attack.

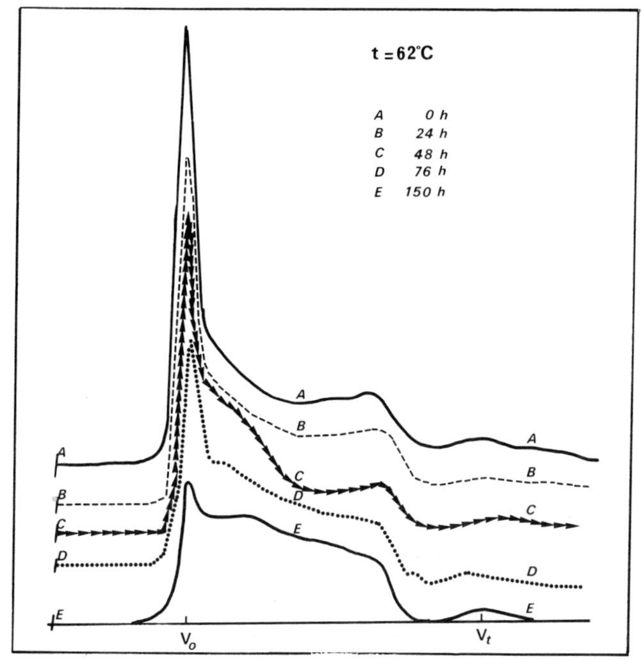

Figure 3. Elution profiles on Sephacryl S-400 gel of a sample of gum arabic heated at 62°C for several hours.

The role and function of arabinogalactan-proteins and glycoproteins have been yet largely discussed (48,49) and it would be of great interest forbetter knowledge of the biosynthesis process to analyze if possible the precursors of gum prior to formation of the nodules.

Also it is now demonstrated that amino acids play a major role in functionality and if only *ca* 20% of gum remains still linked to the protein, one can ask the usefulness of the remaining gum in technical processes.

Table II. INFLUENCE OF HEATING ON THE VISCOSITY OF GUM ARABIC

upper line: viscosity (ml.g^{-1} in 1M NaCl)
lower line: molecular weight by the Mark-Houwink relation (12)

Time / Temperature	24 h	48 h	76 h	150 h
20°C		22.1 7.41×10^5		
48°C	21.4 7.00×10^5	19.7 5.97×10^5	19.8 6.05×10^5	18.3 5.22×10^5
62°C	20.0 6.18×10^5	17.8 4.62×10^5	16.8 4.46×10^5	15.9 4.00×10^5

REFERENCES

1. Anderson, D.M.W., Dea, I.C.M., Karamalla, K.A. and Smith, J.F. (1968) Carbohydr. Res., 6, 97-103.
2. Anderson, D.M.W. and Dea, I.C.M. (1968) Carbohydr. Res., 6, 104-110.
3. Anderson, D.M.W. and Dea, I.C.M. (1971) J. Soc. Cosmet. Chem., 22, 61-76.
4. Anderson, D.M.W. (1978) in Gommes et Colloïdes Végétaux Naturels Hydrosolubles, 4, pp 115-130. Iranex S.A., Neuilly-sur-Seine, France.
5. Sene, A. (1988) Thèse de 3ème Cycle, Université Paul Sabatier, Toulouse, France.
6. Oakley, H.B. (1935) Trans. Faraday Soc., 31, 136-148.
7. Churms, S.C., Merrifield, E.H. and Stephen, A.M. (1983) Carbohydr. Res., 123, 267-279.
8. Street, C.A. and Anderson, D.M.W. (1983) Talanta, 30, 887-893.
9. Ullmann, G. (1983) Thèse de 3ème Cycle, Université Scientifique et Médicale de Grenoble, France.
10. Defaye, J. and Wong, E. (1986) Carbohydr. Res., 150, 221-231.
11. Anderson, D.M.W. (1988) in Gums and Stabilisers for the Food Industry 4, (eds. Phillips, G.O., Wedlock, D.J. and Williams, P.A.) pp 31-37. IRL Press, Oxford, UK.
12. Vandevelde, M.C. and Fenyo, J.C. (1985) Carbohydr. Polym., 5, 251-273.
13. Connolly, S., Fenyo, J.C. and Vandevelde, M.C. (1987) Food Hydrocolloids, 1, 477-480.
14. Connolly, S., Fenyo, J.C. and Vandevelde, M.C. (1987) C. R. Soc. Biol. Fr., 181, 683-687.
15. Connolly, S., Fenyo, J.C. and Vandevelde, M.C. (1988) Carbohydr. Polym., 8, 23-32.
16. Randall, R.C., Phillips, G.O. and Williams, P.A. (1989) Food Hydrocolloids, 3, 65
17. Snowden, M.J., Phillips, G.O. and Williams, P.A. (1987) Food Hydrocolloids, 1, 291-300.
18. Snowden, M.J., Phillips, G.O. and Williams, P.A. (1988) in Gums and Stabilisers for the Food Industry 4, (eds. Phillips, G.O., Wedlock, D.J. and Williams, P.A.) pp 489-496. IRL Press, Oxford, UK.
19. Randall, R.C., Phillips, G.O. and Williams, P.A. (1988) Food Hydrocolloids, 2, 131-140.
20. Swenson, H.A., Kaustinen, H.M., Kaustinen, O.A. and Thompson, N.S. (1968) J. Polym. Sci., A-2(6), 1593-1606.
21. Anderson, D.M.W., Hirst, E., Rahman, S. and Stainsby, G. (1967) Carbohydr. Res. 3, 308-317.
22. Anderson, D.M.W. and Rahman, S. (1967) Carbohydr. Res., 4, 298-304.
23. Veis, A. and Eggenberger, D.N. (1954) J. Am. Chem. Soc., 76, 1560-1563.
24. Mukherjee, S.N. and Deb, S.K. (1962) J. Indian Chem. Soc., 39, 823-826.
25. Acharya, L. and Chattoraj, D.K. (1975) Indian J. Chem., 13, 569-573.
26. Nakamura, M. (1986) Yukagaku, 35, 554-560.
27. Duvallet, S., Fenyo, J.C. and Vandevelde, M.C. (1989) Polymer Bulletin, in press.

28. Anderson, D.M.W. and Stoddart, J.F. (1966) Carbohyd. Res., 2, 104-114
29. Glicksman M. and Sand R. E. (1973) in Industrial Gums, (ed. Whistler, R. L.) p 207. Academic Press, New-York, USA.
30. Lewis, B. A. and Smith F. (1957) J. Am. Chem. Soc., 79, 3929-3931.
31. Jermyn, M.A. (1962) Aust. J. Biol. Sci., 15, 787-791.
32. Alain M. and McMullen, J.N., (1985) Int. J. Pharm., 23, 265-275.
33. Anderson, D.M.W., Hendrie, A. and Munro, A.C. (1972) Phytochemistry, 11, 733-736.
34. Akiyama, Y., Eda, S. and Kato, K. (1984) Agric. Biol. Chem., 48, 235-237.
35. Anderson, D.M.W., Howlett, J.F. and McNab, C.G.A. (1985) Food Additives and Contaminants, 2, 159-164.
36. Anderson, D.M.W. and McDougall, F.J. (1987) Food Additives and Contaminants, 4, 125-132.
37. Anderson, D.M.W. and McDougall, F.J. (1987) Food Additives and Contaminants, 4, 247-255.
38. Vandevelde, M.C. (1986) Thèse de Doctorat ès Sciences, Université de Rouen, France.
39. Connolly, S. (1988) Thèse de Doctorat, Université de Rouen, France.
40. Muller, G. (1989) Université de Rouen, unpublished results.
41. Yomota, C., Okada, S., Mochida, K. and Nakagaki, M. (1984) Chem. Pharm. Bull. Japan, 32, 3793-3802.
42. Yomota, C. and Nakagaki, M. (1987) Chem. Pharm. Bull. Japan, 35, 933-940.
43. Vandevelde, M.C. and Fenyo, J.C. (1987) Polymer Bulletin, 18, 47-51.
44. Artaud, J., Zahra, J.P., Iatrides, M.C. and Estienne, J. (1982) Analusis, 10, 124-131.
45. Dickinson, E. (1988) in Gums and Stabilisers for the Food Industry 4, (eds. Phillips, G.O., Wedlock, D.J. and Williams, P.A.) pp 249-263. IRL Press, Oxford, UK.
46. Churms, S.C. and Stephen, A.M. (1984) Carbohydr. Res., 133, 105-123.
47. Gammon, D.W., Stephen, A.M. and Churms, S.C. (1986) Carbohydr. Res., 158, 157-171.
48. Clarke, A.E., Anderson, R.L. and Stone, B.A. (1979) Phytochemistry, 18, 521-540.
49. Fincher, G.B., Stone, B.A. and Clarke, A.E. (1983) Ann. Rev. Plant Physiol., 34, 47-70.

Structure-function relationships of gum arabic

P.A.WILLIAMS, G.O.PHILLIPS and R.C.RANDALL
Faculty of Research and Innovation, The North East Wales Institute, Connah's Quay, Deeside, Clwyd CH5 4BR, UK

Abstract

Recent developments on the structure of gum arabic are reviewed and are considered in relation to its rheological and emulsification properties. The effect of heat on the functional characteristics is also discussed.

1. Introduction

Gum arabic (Acacia senegal) is a unique polysaccharide in that it has excellent emulsifying properties and despite its relatively high molecular mass (~ 400,000), yields solutions of surprisingly low viscosity. Such behaviour is uncharacteristic of polysaccharides generally and is a consequence of the gum's molecular characteristics.

Studies on the structure of the gum using Smith degradation procedures (1,2) indicated that the molecules consist of a $\beta 1 \to 3$ linked galactopyranose backbone chain with numerous branches linked through $\beta 1 \to 6$ galactopyranose residues and containing arabinofuranose, arabinopyranose, rhamnopyranose, glucuronic acid and 4-O-methyl glucuronic acid. Although Anderson and Stoddart(3) reported that the gum contained a small amount of proteinaceous material as an integral part of the structure it was Fenyo and coworkers(4-6) who recently recognised its significance to the overall molecular characteristics. They showed that the gum consisted of two distinct fractions, namely a high molecular mass arabinogalactan-protein complex (AGP) representing about 30% of the total and a lower molecular mass fraction which was protein deficient. They described the structure of the

AGP in terms of the "wattle blossom" model where carbohydrate blocks of molecular mass ~200,000 are linked together by a main polypeptide chain.

This paper now reviews some further developments concerning the structural characterisation of the gum and gives an insight into the relationship with its functional properties.

2. Molecular characterisation

We have recently shown(7) using hydrophobic affinity chromatography that gum arabic can be separated into three principal fractions, namely an arabinogalactan (AG), an arabinogalactan-protein complex (AGP) and a glycoprotein (Gl). The physico-chemical characteristics of the various fractions are summarised in Table 1.

Table 1 - Physico-chemical characteristics of gum arabic and its fractions

	WHOLE GUM	AG	AGP	Gl
% wt of total recovered		88.4	10.4	1.2
% galactose	36.2±2.3	34.5±2.2	29.3±0.7	12.3±0.5
% arabinose	30.5±3.5	27.6±1.9	31.4±1.0	15.0±1.3
% rhamnose	13.0±1.1	11.8±2.2	12.9±1.0	6.7±1.1
% glucuronic acid	19.5±0.2	23.1±0.4	17.6±0.1	11.2±0.3
% protein	2.24±0.15	0.35±0.10	11.8±0.5	48
protein expressed as a % of the total protein in the whole gum	100	20.1	49.5	30.4
relative molecular mass	460,000[a]	279,000[a]	1,450,000[a]	250,000[b]
hydrodynamic radius/nm[c]	14.1	9.2	22.8	-

All wt. expressed on a dry wt basis.

a - obtained from light scattering data; solvent 0.5 mol dm^{-3} NaCl, 25°C.

b - obtained by gel permeation chromatography; solvent 0.5 mol dm^{-3} NaCl, 25°C

c - obtained using photon correlation spectroscopy; solvent 0.5 mol dm^{-3} NaCl, 25°C.

The findings are in accordance with those of Fenyo et al(4-6) in that the bulk of the gum consists of a lower molecular mass arabinogalactan fraction (AG) which contains very little protein (only 20% of the total) and a higher molecular mass arabinogalactan-protein complex (AGP) which contains about half of the total protein. In addition, these studies have also identified the presence of a glycoprotein fraction (Gl).

The hydrodynamic radii of the AG and AGP fractions reported in Table 1 are very small in relation to the molecular masses as illustrated in Table 2, which gives the molecular dimensions of a number of other polymers for comparison. The results, therefore, support the concept of a highly branched compact structure as suggested by previous investigations(1,2).

Table 2 - Radius of gyration and average molecular mass of some hydrocolloids

Polymer	Solvent	Radius of gyration/nm	Molecular Mass	Reference
xanthan gum	0.1 mol dm^{-3} NaCl, $20^{o}C$	289.5	2,940,000	8
konjac mannan	water, ambient temp.	105	680,000	9
bovine serum albumen	tris buffer 0.04 mol dm^{-3}	3.7*	69,000	10
sodium carboxy-methyl cellulose	0.5 mol dm^{-3} NaCl	94	440,000	11

* hydrodynamic radius

The AGP fration has been shown to be degraded by proteolytic enzyme(7) with the molecular mass decreasing to a similar value as the AG fraction and this is illustrated by the gel permeation chromatograms given in Fig.1. The results, therefore, support the model put forward by Fenyo and coworkers(4-6) and suggest that the AGP fraction consists of about five carbohydrate blocks of molecular mass 2.8×10^5 linked together by a polypeptide chain which may contain as many as 1600 amino acid residues. It is likely that the polypeptide chain is located at the periphery of the molecule, thus facilitating its adsorption onto hydrophobic substrates. The AG fraction is not degraded by proteolytic enzyme(7) and may simply be a fragment of the AGP fraction.

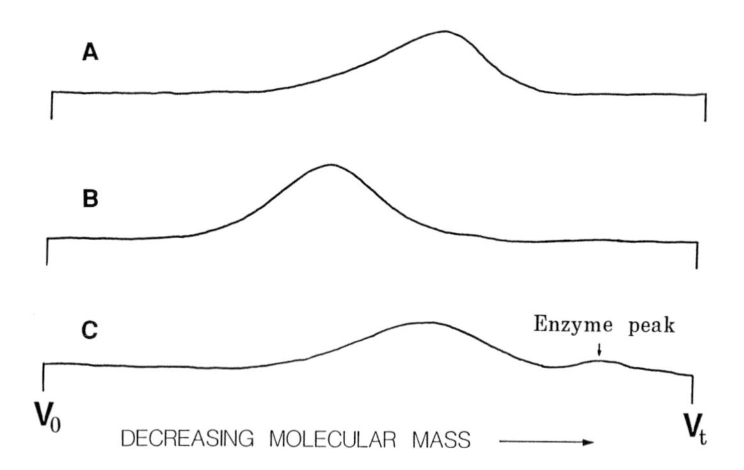

Fig.1. Gel permeation chromatograms (gpc) of : A, arabinogalactan(AG); B, arabinogalactan-protein complex(AGP) and C , AGP after incubation with Pronase . Chromatograms obtained using a 2.6x90 cm column containing Sephacryl 500 ; solvent ,0.5 mol/L NaCl ;eluent monitored by UV at 218nm

3. Functional Characteristics

3.1 Rheological Properties

At relatively low concentrations, gum arabic yields solutions which are essentially Newtonian in behaviour and have very low viscosities compared to other polysaccharides of similar molecular mass. This is shown in Fig.2 which compares the viscosity of 1% solutions of guar gum, locust bean gum, sodium carboxymethyl cellulose and gum arabic as a function of shear rate. Indeed, as with globular proteins, viscous solutions are only obtained at much higher concentrations (>30%) above the so-called "interactive volume"(12) where effective molecular overlap occurs. This is illustrated in Fig.3 which compares the viscosities of gum arabic and β-lactoglobulin(13), as a function of concentration and lends further support to the concept of the gum molecules having a highly branched compact structure. Gum arabic is a polyelectrolyte and hence the viscosity of solutions will vary as a function of ionic strength and pH(14), and this is illustrated in Figs. 4 and 5. The viscosity is seen to decrease on addition of electrolyte and this is explained by a reduction in the effective volume due to suppression of the electrostatic charge. The viscosity of gum arabic solutions achieves a maximum at pH 5-5.5. Below this range the caboxylate groups become increasingly more undissociated reducing the overall charge on the molecules and hence decreasing the viscosity. At higher pH values the viscosity gradually decreases due to the increase in the solution ionic strength, brought about by the addition of sodium hydroxide necessary to adjust the pH.

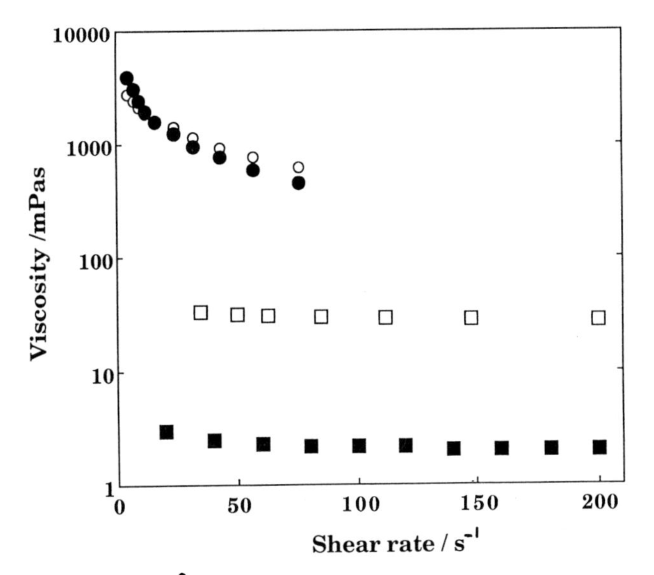

Fig.2. Viscosity at 25°C of 1% w/w aqueous gum solutions: ○, locust bean gum; ●, guar gum; □, 7M CMC; ■, gum arabic, as a function of shear rate determined using a Contraves Rheomat 15 FC rheometer.

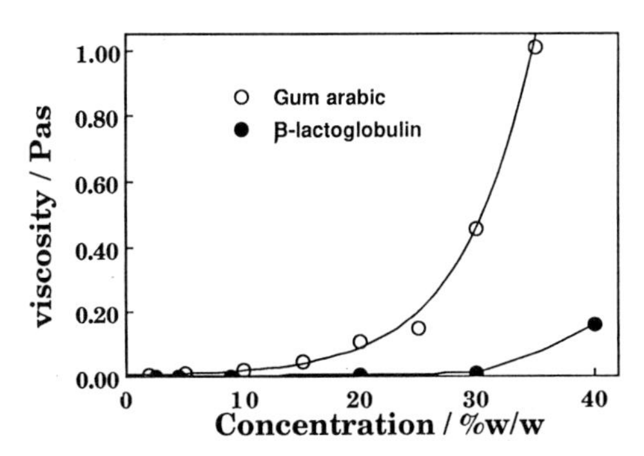

Fig.3. A comparison of the viscosities of gum arabic (shear rate 100s) and β-lactoglobulin (Ref 13) as a function of polymer concentration.Gum arabic viscosities measured using a Carrimed Controlled Stress Rheometer

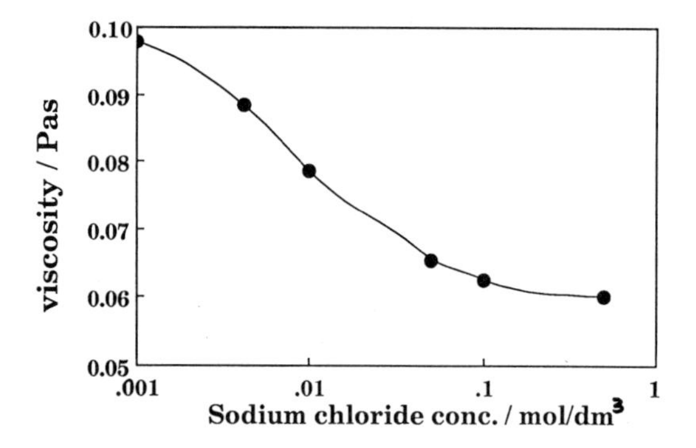

Fig.4. The viscosity (shear rate $100s^{-1}$) at 20^{0}C of 20% w/w gum arabic solution as a function of sodium chloride. Measurements made using the Carrimed Controlled Stress Rheometer.

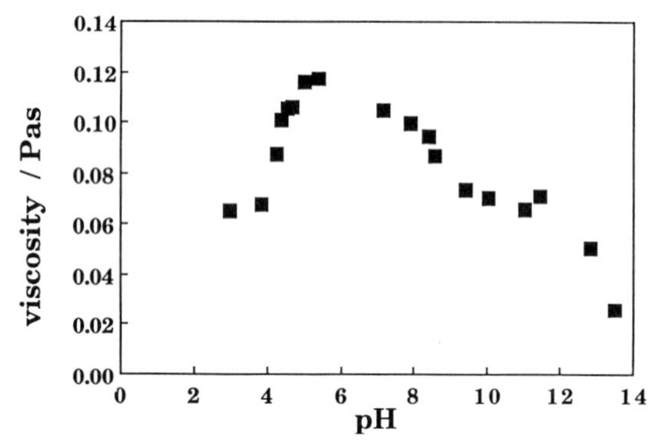

Fig.5. The viscosity (shear rate $100s^{-1}$) at 20^{0}C of 20% w/w gum arabic solution as a function of pH. Measurments made using the Carrimed Controlled Stress Rheometer.

3.2 Emulsification properties

Although gum arabic has been widely used for many years to stabilise oil in water emulsions(15,16), it has only recently been demonstrated that it is the high molecular mass AGP fraction that is responsible for the gum's emulsifying ability(17). This is evident from figure 6 which shows the gel permeation chromatogram of a solution of gum arabic monitored before preparing an orange oil emulsion and also the chromatogram of a solution of the continuous phase recovered following emulsification. The difference between the two profiles represents the portion of gum arabic that is adsorbed onto the oil droplets. Only a small amount of the gum does actually adsorb, as has been discussed in detail previously(17), and it can be seen that the adsorbing species is the high molecular mass AGP fraction. The AG fraction has very little, if any, affinity for the oil surface, and hence has little role to play in the emulsification process apart from enhancing the viscosity of the continuous phase. Consequently, relatively high concentrations of gum arabic are required to produce stable emulsions of relatively small droplet size. This is illustrated in Fig.7 which shows the average droplet size of orange oil emulsions produced using various concentrations of gum arabic. In this particular example using 20% w/w orange oil and shearing with a Silverson mixer, over 10% w/w gum arabic solution is required to produce emulsions of minimum droplet size. This indicates that at lower gum arabic concentrations there is insufficient surface active material present to fully coat all the droplets produced on shearing the system and hence droplet flocculation and coalescence occurs yielding emulsions of higher average droplet size. Obviously the optimum concentration will vary with the nature of the oil, the type of mixer and the conditions used since increased shearing of the emulsion would generate even smaller droplets thus requiring increased amounts of gum arabic.

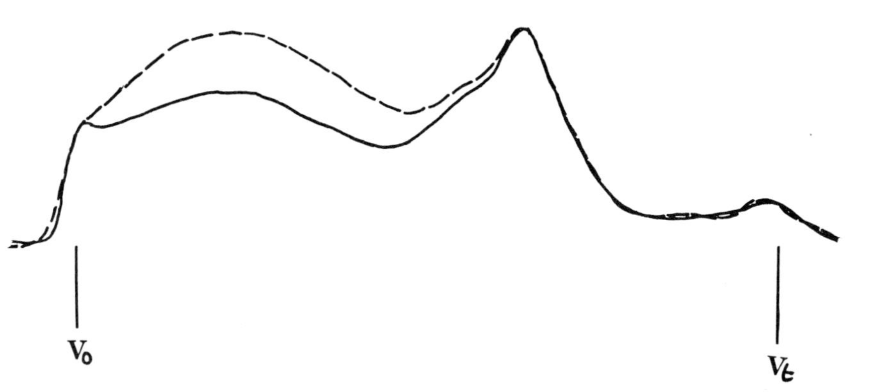

Fig.6. Gel permeation chromatograms of gum arabic in solution before — — ,and after emulsification ———— ;(19% w/w gum solution in distilled water,20% w/w orange oil).Solutions prepared in 0.5 mol/L NaCl and conditions used as in fig 1.

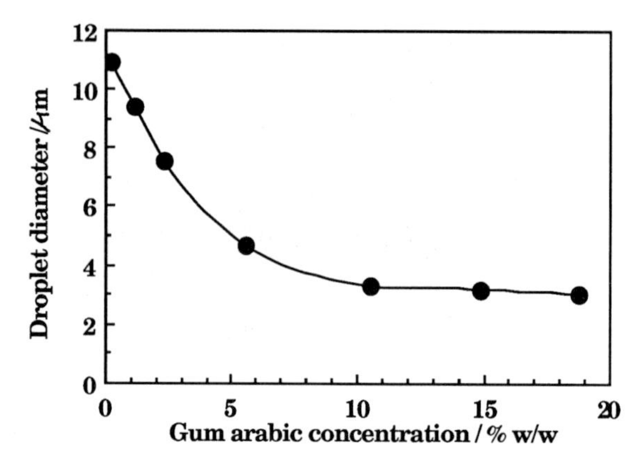

Fig.7. The average droplet diameter of emulsions (20%w/w orange oil) as a function of gum arabic composition. Measurements were performed on diluted dispersions(1/10000) with distilled water using a ZM model Coulter Counter equipped with a 50 μm orifice electrode.

The ability of the AGP fraction to stabilise emulsions arises from its unusual molecular characteristics. It is believed that the molecules behave much like graft co-polymers with the hydrophobic amino acid residues anchoring them to the interface and the hydrophilic carbohydrate blocks extending out into solution preventing droplet flocculation and coalescence through steric repulsive forces. This is schematically illustrated in Fig.8. Evidence for the ability of gum arabic to act as a steric stabiliser has been gained from model studies using polystyrene latices(18,19). Although the bare latex particles were found to flocculate in the presence of 0.5 mol dm^{-3} NaCl, aggregation was prevented on the addition and adsorption of gum arabic at a concentration corresponding closely to the adsorption capacity, i.e., ~5-7mg m^{-2}(18,19). Since the electrical double layers would be compressed at this ionic strength, stabilisation could only occur as a consequence of the repulsive interactions between adsorbed polymer layers.

3.3 Effect of Heat

Gum arabic has been shown to be heat sensitive and some precipitation will occur on prolonged heating,(20,21). The precipitate has been shown to consist essentially of proteinaceous material and the gel permeation chromatograms obtained before and after heating a gum arabic solution which are given in figure 9 , indicate that it is both the AGP and Gl fractions that are removed from solution. As a consequence of the loss of the high molecular mass AGP fraction the solution viscosity decreases considerably as shown in Fig.10. Even more importantly perhaps, the gum also loses its emulsifying ability. This is demonstrated in Fig.11, which shows that the average droplet size of an orange oil emulsion (20% w/w) prepared using a 19% gum arabic solution increases, as the time of heat treatment of the gum arabic solution increases, thus reflecting a decrease in emulsification efficiency.

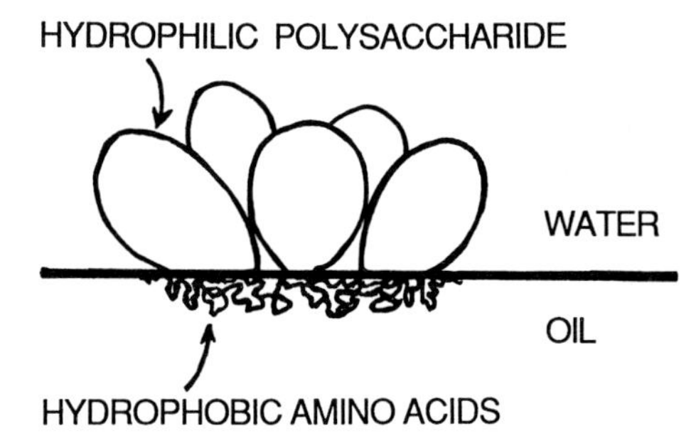

Fig.8. Schematic illustration of the structure of the arabinogalactan-protein complex at the oil-water interface.

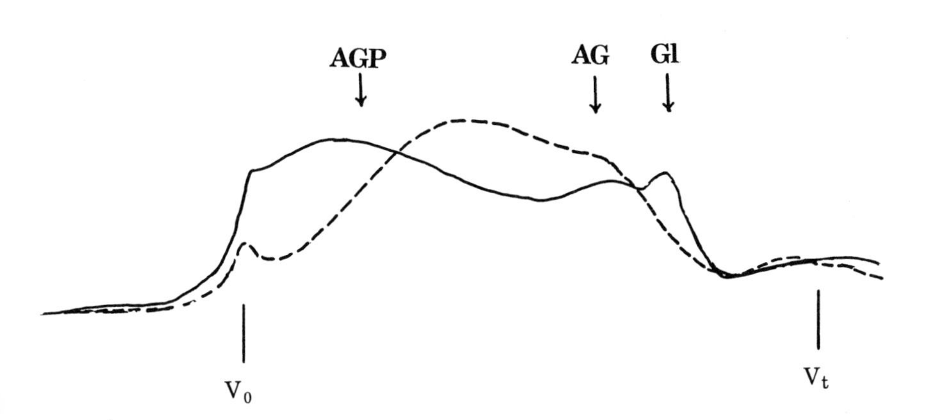

Fig.9. Gel permeation chromatograms of 1.5%w/w spray dried gum arabic solutions before ——— and after heat treatment (102° C for 4 hrs) — — — — (see fig. 1 for conditions).

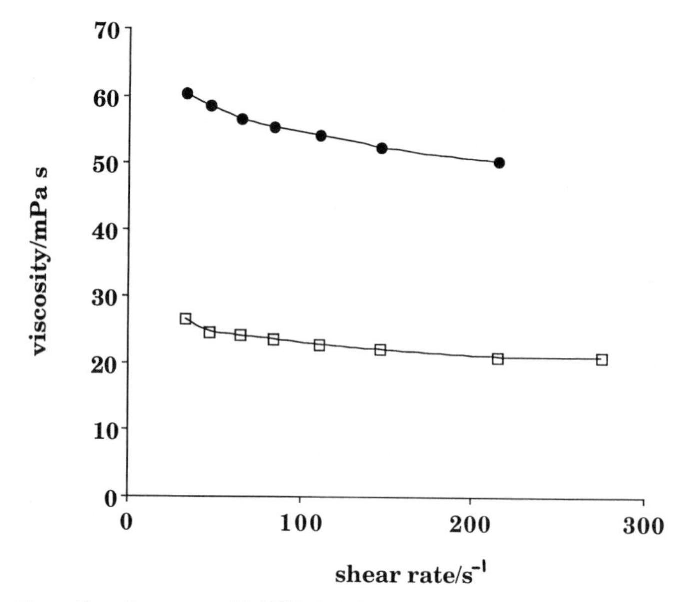

Fig.10. Viscosity of gum arabic(19%w/w) before ● and after □ heat treatment(102°C for 16 hrs) using the Contraves Rheomat FT 15RC rheometer at 25°C.

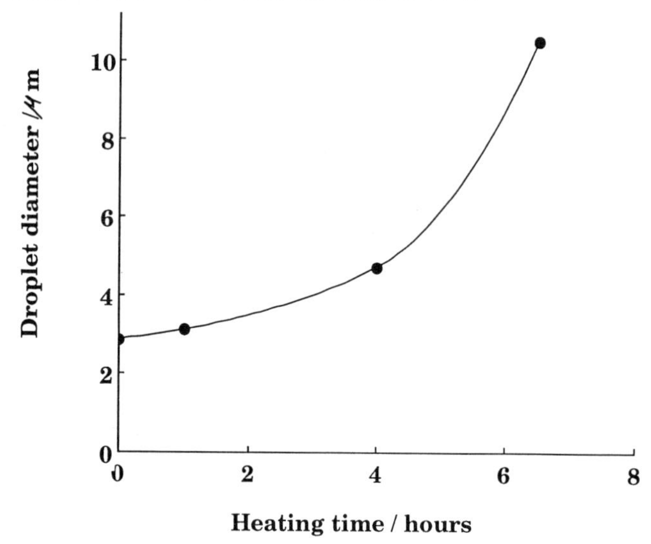

Fig.11. The average droplet diameter in orange oil (20%w/w) emulsions using gum arabic (19%w/w) heated for various times at 100°C (Conditions used as in fig.7.)

4. Summary

Gum arabic consists of three principal fractions, i.e., an arabinogalactan, an arabinogalactan-protein complex, and a glycoprotein. The highly-branched compact structure of the AG and AGP results in solutions of particularly low viscosity. The complex structure of the AGP, where hydrophilic carbohydrate blocks are linked to a main polypeptide chain, enables strong adsorption at the oil-water interface and promotes emulsion stability. The AGP and Gl components are denatured on heating resulting in a loss of emulsification efficiency and a reduction in solution viscosity.

Acknowledgments

The authors are indebted to Agrisales Ltd. for their support during the course of this work and also to Michael Langdon who assisted with some of the experimental work.

References

1. Street, C.A. and Anderson, D.M.W., (1983) Talanta 30, 887.

2. Churms, S.C., Merrifield, E.M. and Stephen, A.M., (1983) Carbohydr. Res. 123 267.

3. Anderson, D.M.W., and Stoddart, J.F., (1966) Carbohydr. Res. 2 104.

4. Vandevelde, M.C., and Fenyo, J.C., (1985) Carbohydr. Polymers 5 251.

5. Connolly, S., Fenyo, J.C., and Vandevelde, M.C., (1987) Food Hydrocolloids 1 477.

6. Connolly, S., Fenyo, J.C., and Vandevelde, M.C., (1988) Carbohydr. Polymers 8 23.

7. Randall, R.C., Phillips, G.O., and Williams, P.A., (1989) Food Hydrocolloids 1, 65

8. Coviello, T., Kajiwara, K., Burchard, W., Dentini, M., and Crescenzi, V., (1986) Macromolecules 19 2826.

9. Sugiyama, N., Shimahara, H., Andoh, T., Takemoto, M., and Kamata, T., (1972) Agri. Biol. Chem. 36 1381.

10. Nynstrom, B., Roots, J., (1985) In "Physical Optics of Dynamic Phenomena and Processes in Macromolecular systems", Ed. B.Sedlack, W.de Grugter N.Y. P305.

11. Sneider, N.S., and Doty, P., (1954) J. Phys. Chem. 58 762.

12. Menjivar, J.A., and Rha, C.K., (1980) In "Rheology : fluids" Vol.2 eds Astarita, G., Marrucci, G., and Nicholas, L., Plenum Press, New York, p293.

13. Pradipasena, P., and Rha, C.K., (1977) Polymer Eng. Sci. 17 861.

14. Thomas, A.W., and Murray, (Jr) H.A., (1928) J. Phys. Chem. 32 676.

15. Glicksman, M., "Gum Technology in the Food Industry" (1969), Academic Press, New York, p.104.

16. Glicksman, M., and Sand, R.E., (1973) In Whistler, R.L., (ed), "Industrial Gums", Academic Press, New York, p208.

17. Randall, R.C., Phillips, G.O., and Williams, P.A., (1988) Food Hydrocolloids 2, 131.

18. Snowden, M.J., Phillips, G.O., and Williams, P.A., (1987) Food Hydrocolloids 1, 291.

19. Snowden, M.J., Phillips, G O and Williams, P.A., (1988) In "Gums and Stabilisers for the Food Industry 4" eds Phillips, G.O., Wedlock, D.J., and Williams, P.A., IRL Press, Oxford p489.

20. Anderson, D.M.W., and McDougall, F.J., (1987) Food Additives and Contaminants 4 247.

21. Randall, R.C., Phillips, G.O., and Williams, P.A., (1989) in "Food Colloids" R.S.C. Publ. p.386.

The processing of gum arabic to give improved functional properties

G.R.WILLIAMS
Masterdry Ltd, Bason Bridge, Highbridge, Somerset TA9 4RP, UK

ABSTRACT

Gum Arabic has advanced dramatically during the past 30 years with the advent of processed gums. Gums are cleaned prior to processing to produce a pure product, free of impurity. It has also been shown that excessive temperatures during processing damage the product. Care must be taken to ensure that no deterioration of gum properties occurs during processing. The future suggests that Gum Arabic will become a specialised process product in powder form of varying particle size and bulk density, with specific enhanced solution attributes to meet end user requirements.

INTRODUCTION

The processing of Gum Arabic has advanced a great deal over the past 20 to 30 years. Historically, gum was sold in its natural form with little or no processing. If any processing was carried out, this was either a removal of sand and fine gum, a granulation without any pre-cleaning, or a grinding of gum into powder, again without pre-cleaning.

Quality control was minimal, as many importers did not maintain any laboratory facilities. However, a visual comparison of gum was often undertaken, and comparisons made with retained samples. Gum solutions were also examined for colour, cleanliness and ropiness. Occasionally, simple viscosity measurements were carried out and if a granulation had been effected, particle size was checked.

In the mid 70's, there developed a need for a more pure, cleaned gum. The end user did not want material with excessive dissolving periods of 24 hours or more as with lump gum. Thus were introduced the processed gums, produced to high standards of purity and microbiological cleanliness together with a greatly reduced dissolution time. The advent of these processes meant that the competitive product (i.e. starches, dextrins etc.) was being directly challenged by a new form of gum that had the properties of the traditional material with the advantages of competition.

GUM ARABIC PROCESSING

Raw gum is still sold in substantial quantities, but pre-cleaning of the gum for removal of sand, fines and bark is required. This mechanical cleaning of gum is summarised in Figure 1. If gum is required in kibbled, granular or powder form, the material is cleaned as a pre-requisite. The advantage in kibbling material as seen by a reduction of dissolving time is demonstrated in Figure 2.

The production of both spray and roller dried gums is now commonplace and both processes take kibbled gum as its start point. Due to recent advances, it is now known that excessive temperatures and long holding times should be avoided during processing if the gum is to retain its original properties. Once dissolved, the gum solution is exposed to several sieving, decanting and centrifugation systems prior to spray drying (see Figure 3) with the aim of removing fine insoluble material, and the gum is then pasteurised immediately prior to drying to remove microbial contamination.

In roller dried gum, the product is distributed onto rotating, steam heated rollers. On coming into contact with the hot roll surface, water will be evaporated and removed by a flow of air. However, high temperatures used on heating surfaces can denature proteins and destroy the gum. A knife is used to scrape the gum film continuously off the roll. The thickness of the film can be varied by adjusting the gaps between the rolls. Roller drying produces large particles (flakes) which are easily dissolved in water.

In spray dried gum, the gum solution is sprayed into fine droplets (atomisation) into a stream of hot air which quickly evaporates the water. Cyclones are used to separate dry powder from drying air. Powder particles do not exceed the temperature of the exhaust air, usually 80°C to 90°C. By altering atomising conditions, powder can be produced with varying particle sizes and bulk densities.

Important parameters for spray dried powders are wettability (the ability of powder to be wetted, either by sinking under the surface of the water, or by water penetrating into the powder layer), dispersibility (the amount of powder dispersed into water and forming particles under given conditions), solubility (the amount of insolubles) and flowability (the ability of the powder to flow). Spray dried powder can now be specifically tailored for particular customer requirements.

Packaging requirements have also altered and there is now available a variety of packaging materials, such as polymer lined multi-walled paper sacks, cardboard boxes with polymer liners, fibre drums and bulk bags. These are available in a variety of nett weights to accommodate customer recipes.

Quality Control demands have also increased with the advent of processed gums. Q.C. starts with the raw gum, and a number of samples are taken to establish the characteristics of the parcel prior to processing. Optical rotation is undertaken to ensure gum purity and can also be backed by TLC, spectroscopy and chemical identifications.

Monitoring of the mechanical processing of cleaning and granulation are carried out for product cleanliness at all levels.

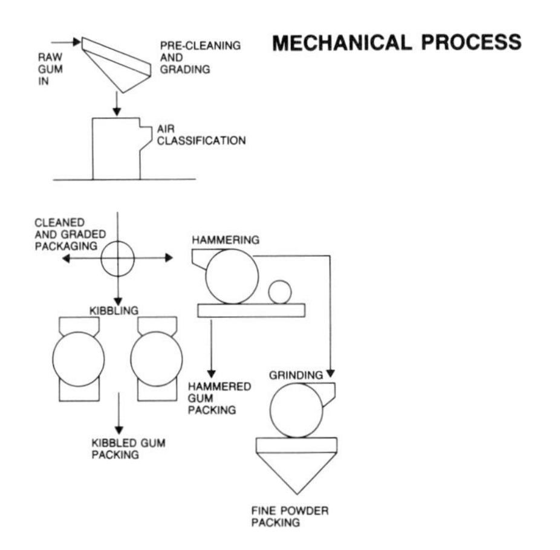

FIGURE 1:

THE MECHANICAL CLEANING OF GUM.

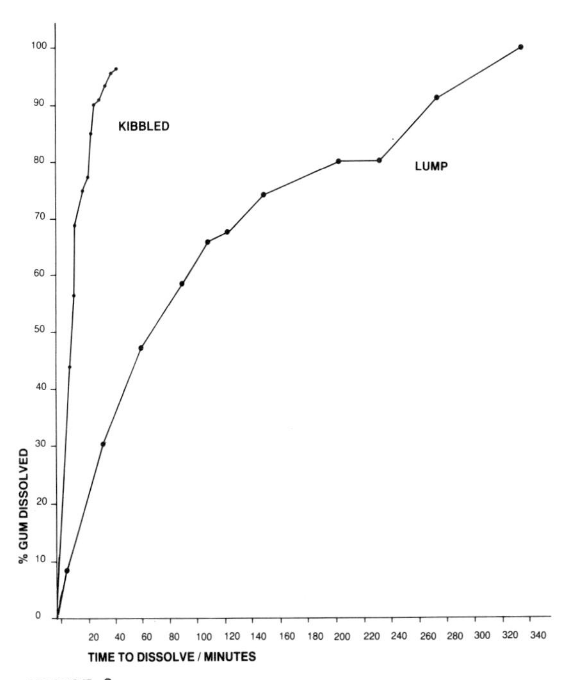

FIGURE 2:

COMPARATIVE SOLUBILITIES OF RAW AND KIBBLED GUM.

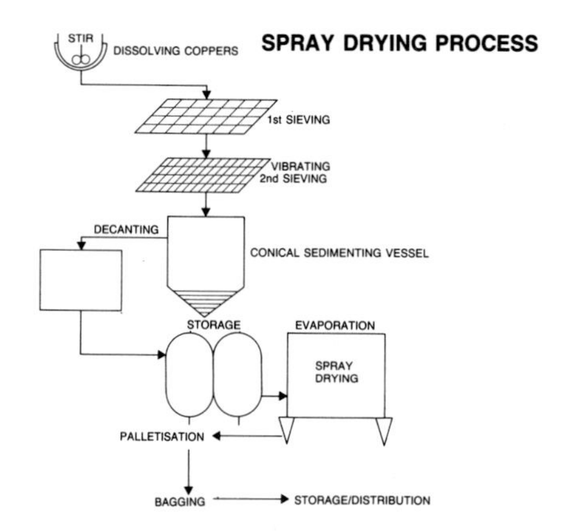

FIGURE 3:

THE PRODUCTION OF PROCESSED GUM.

Solution characteristics, insoluble content and ash content are now commonplace analyses.

In processed gums (spray and roller dried) Q.C. begins with dissolving times and temperatures by way of microprocessor controls and chart recording. Solution examination is carried out at all stages of liquor cleaning to ensure efficiency of the process, and microbiological assays are commonplace to determine loading both before and after heat treatment. On finished powders, - colour, moisture, particle size, sediment and solubility are now combined with identification testing, viscosities, functionality, emulsification, formulation and microbiological assays. Pharmacopeal analysis certification is also essential for each batch of gum produced. Studies on the molecular profiles of the gum can also be carried out to ensure that the essential molecular characteristics are retained and unaffected by any gum processing.

THE FUTURE PROCESSING OF GUM ARABIC

It can be foreseen that gum arabic will evolve into a series of products to meet a particular usage sector. There will be a continuation of the present trend towards a highly refined product. Current research into gum structure has suggested which parts of the molecule are important to the various end usages to which the gum is employed. This research, for instance, indicates that the protein area is the area of main interest for emulsification. It would therefore be possible for the production of protein isolates from gum for emulsification usage. Other work may indicate molecular areas of importance for manufacturing of confectionery with subsequent specialist isolations.

It is now possible to produce materials with enhanced characteristics. If this trend continues the convenience and enhanced properties of processed gums indicate that the use of raw gums could be a dying art.

Emulsifying behaviour of acacia gums

ERIC DICKINSON, VANDA B.GALAZKA and
DOUGLAS M.W.ANDERSON[+]

Procter Department of Food Science, University of Leeds, Leeds LS2 9JT, UK
[+]Chemistry Department, University of Edinburgh, Edinburgh EH9 3JJ, UK

ABSTRACT

Time-dependent droplet-size distributions are reported for D-limonene-in-water emulsions (1 wt% gum, 10 vol% oil, pH 7) made under standard conditions with six Acacia gum samples with nitrogen contents in the range 0.1—7.5%. The best emulsion stability was found with the two Acacia gum samples with the highest nitrogen content. The relative emulsifying abilities of the different samples are broadly in conformity with those reported previously for hydrocarbon-in-oil emulsions, though the mechanism of droplet growth is distinctly different for the two cases.

INTRODUCTION

Gum arabic (Acacia senegal) consists of a mixture of highly branched arabinogalactan heteropolymers with a small amount of proteinaceous material (typically ca. 2 wt%) covalently attached to the polysaccharide component. A particular feature of gum arabic is its ability to adsorb at the oil—water interface to form a stabilizing film whose surface viscoelasticity is quite insensitive to dilution of the aqueous phase (1). It appears that almost all the nitrogen content of an Acacia gum (ca. 0.3% N equivalent in Acacia senegal) is associated with a high-molecular-weight fraction representing only 20—30% of the total gum. It is this protein-rich fraction which adsorbs strongly at the oil—water interface, and which is mainly the source of its emulsifying and stabilizing action in the formulation of soft drinks.

We recently reported (2-4) measurements of the surface activity, surface viscosity and emulsifying properties of samples of various Acacia gum species with nitrogen contents up to 7.5%. With n-hexadecane as the oil phase at neutral pH, it was shown (3) that there is a good correlation between the amount of proteinaceous material in the gum and its limiting long-time surface activity. The relationship between nitrogen content and emulsifying behaviour is more complicated, however, because of the need to distinguish between initial droplet-size distribution (emulsifying capacity) and change of droplet-size distribution with time (emulsion stability).

Essential oils are chemically very different from hydrocarbon oils. So there is a need to establish whether fundamental studies on n-hexadecane-in-water emulsions are at all relevant, say, to orange oil-in-water emulsions. In what follows, we use D-limonene to compare the emulsifying behaviour of various Acacia gum samples at neutral pH.

MATERIALS AND METHODS

Materials

Six special Acacia gum samples having a wide range of nitrogen contents were investigated: A. seyal (0.1% N), A. senegal (0.33% N), A. resinomarginea (0.83% N), A. irrorata (1.57% N), A. gerrardii (1.86% N), and A. difficilis (7.5% N). Full analytical data have been reported elsewhere (3) for the last four of these samples. The D-limonene was AnalaR grade from Sigma Chemicals. Gums were dissolved in phosphate buffer solution (pH 7, ionic strength 0.05 M) containing 0.1 wt% sodium azide as bactericide. Buffer salts and azide were AnalaR grade. Water was double-distilled.

Emulsion Formation and Stability

Emulsion samples of small volume (ca. 10 ml) were produced using the Leeds one-stage valve minihomogenizer (5). To a 1 wt% solution of the gum in phosphate buffer at 25 °C was added an appropriate volume of D-limonene to give a premix containing 10 vol% oil. After blending, the coarse premix emulsion was homogenized at a pressure of 300 bar. The resulting emulsion was separated into two samples which were stored in a waterbath at 25 °C: one half was inspected visually at regular intervals to record creaming and serum separation, and the other half was sampled, after thorough mixing, for droplet-size determination using a Coulter counter model TAII with a 30-μm orifice tube and 0.1 M NaCl as suspending electrolyte.

RESULTS AND DISCUSSION

Figure 1(a) shows the droplet-size distributions of the freshly made emulsions. In terms of the position of the maximum in the size distribution

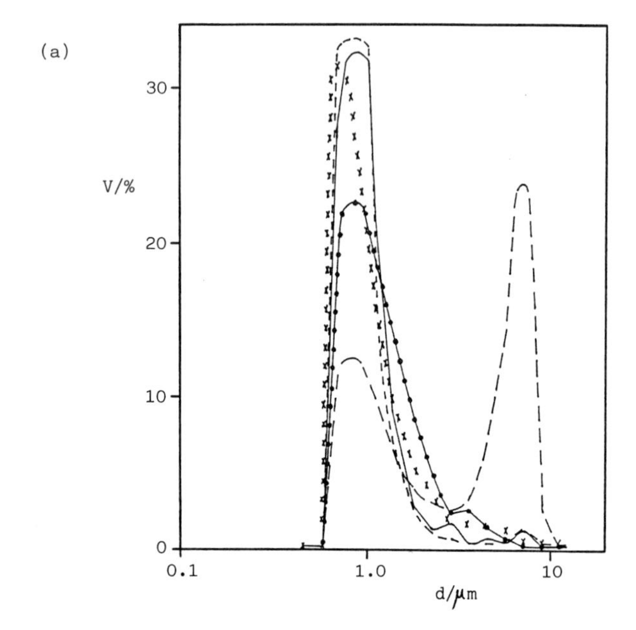

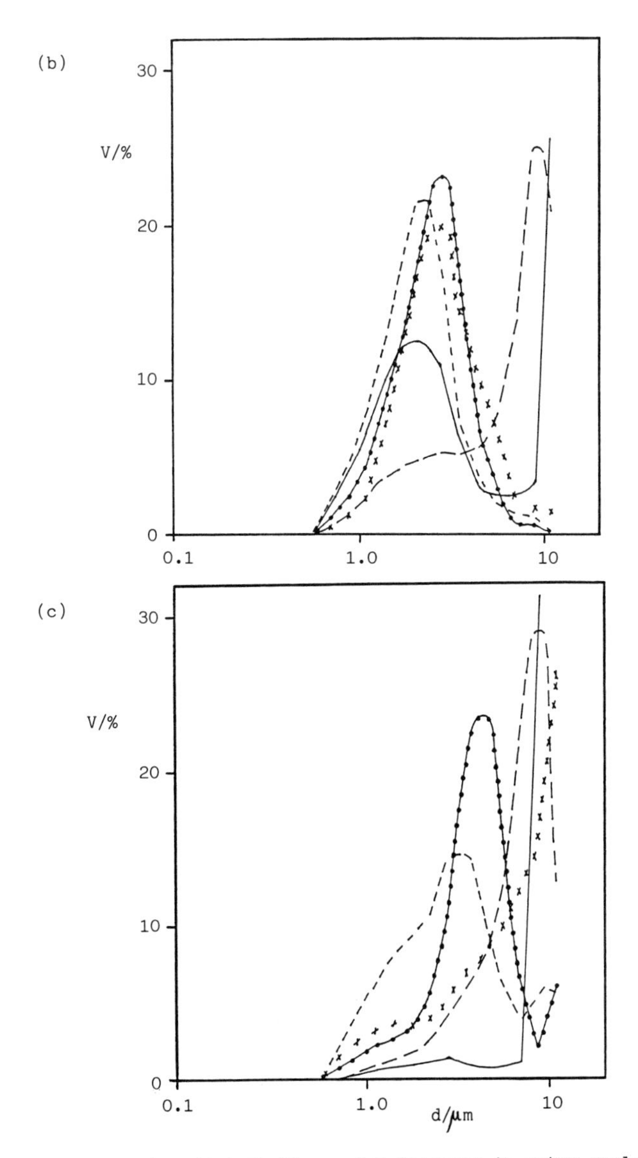

Figure 1. Droplet-size distributions of D-limonene-in-water emulsions measured by Coulter counter (a) immediately after preparation, (b) after 24 hours, and (c) after 7 days. The smoothed percentage differential volume V is plotted against particle diameter d: ———, A. seyal; — — —, A. senegal; x x x x x, A. irrorata; —●—●—●—●—, A. gerrardii; - - - - - -, A. difficilis.

function V(d), the emulsifying capacities of the gum samples lie in the order: *A. irrorata* > *A. difficilis* > *A. seyal* > *A. gerrardii* > *A. senegal* > *A. resinomarginea*. In fact, the droplet-size distribution for the last mentioned gum is not shown in figure 1 since a stable emulsion could not be formed at all with 1 wt% gum and 10 vol% oil. The reason for the particularly poor emulsifying capacity of *A. resinomarginea* may be related to its low intrinsic viscosity, and therefore low average molecular weight, as compared with the other *Acacia* gum samples investigated (3).

Figures 1(b) and 1(c) show the droplet-size distributions of the emulsions after 24 hours and 7 days respectively. We see that the position of the maximum in V(d) gradually moves to larger droplet diameters with increasing storage time. Comparison of the results in figure 1 with similar data for n-hexadecane-in-water emulsions shows the same general trend of relative emulsifying behaviour for the different gum samples, although the nature and rate of growth of D-limonene droplets is not the same as for hydrocarbon oil droplets. This is mainly due to the process of Ostwald ripening (6), which is significant with the former, because of its appreciable solubility in water, but not with the latter, which is essentially insoluble.

Based on the data in figure 1(c), the stabilizing capacities of the gum samples lie in the order: *A. difficilis* > *A. gerrardii* > *A. senegal* > *A. irrorata* > *A. seyal* > *A. resinomarginea*. We note that the two gum samples with the highest nitrogen content are the ones with the smallest emulsion droplets after 7 days storage. While *A. irrorata* gives the smallest droplets initially, the emulsion is rather unstable, giving larger droplets than *A. senegal* after 7 days. This particular sample of *A. senegal* had poor emulsifying properties, as was found also with n-hexadecane previously (3). Good commercial samples of gum arabic (*A. senegal*) give droplet sizes for D-limonene-in-water emulsions similar to those found for the best of the gum samples investigated here (i.e. *A. difficilis*).

Finally, as expected, we have observed a good inverse correlation between the relative rates of emulsion creaming and the corresponding droplet-size distributions. In terms of the extent of serum layer separation observed visually after 4—5 days, the relative emulsion stabilities lie in the order: *A. difficilis* > *A. gerrardii* > *A. irrorata* > *A. seyal* > *A. senegal* > *A. resinomarginea*. Again, the consistency with the earlier hydrocarbon oil emulsion stabilities is noteworthy.

REFERENCES

1 Dickinson, E., Elverson, D. J. and Murray, B. S. (1989) Food Hydrocolloids, 3, 101-114.

2 Dickinson, E. (1988) In Phillips, G. O., Wedlock, D. J. and Williams, P. A. (eds), Gums and Stabilisers for the Food Industry. IRL Press, Oxford, vol. 4, pp. 249-263.

3 Dickinson, E., Murray, B. S., Stainsby, G. and Anderson, D. M. W. (1988) Food Hydrocolloids, 2, 477-490.

4 Dickinson, E., Murray, B. S. and Stainsby, G. (1988) In Dickinson, E. and Stainsby, G. (eds), Advances in Food Emulsions and Foams. Elsevier Applied Science, London, pp. 123-162.

5 Dickinson, E., Murray, A., Murray, B. S. and Stainsby, G. (1987) In Dickinson, E. (ed.), Food Emulsions and Foams. Royal Society of Chemistry, London, pp. 86-99.

6 Dickinson, E. (1986) Annu. Rep. Prog. Chem., Sect. C, Royal Society of Chemistry, London, pp. 31-58.

Indicators for present and future supply of gum arabic

EL HAG MAKKI AWOUDA
The Gum Arabic Company, Khartoum, Sudan

Abstract

Gum Arabic is a unique product - whether in its natural habitat of a tree and part of a dynamic ecosystem, providing income to subsistence farmers at a time when it is most needed or as a commodity used in a wide spectrum of industries, produced in Africa and consumed by the Western World.

Drought in recent years has had a marked adverse effect on production and supply of gum arabic. Such a decline has induced serious implications both locally and internationally.

The Sudan is aware that gum arabic is a national wealth that needs to be conserved and developed, not only to provide a sustained yield of gum, but also for the other important socio-economic benefits.

Ecologically the Sudan is also aware that the gum belt is a natural buffer zone between the desert proper in the North and the agricultural tall grass savanna in the South. Therefore destruction or misuse in the zone will induce desert encroachment and consequently threatens agricultural production.

The region also supports an enormous population of animal wealth totalling up to about 20 million head of camels, cattle, sheep and goats.

All this explains the high priority given by the Sudan Government for the restocking of the gum belt and the improvement of gum developments in all fields of production, marketing and research with the main aim of security of supplies and continuity of benefits.

This paper highlights the continuous efforts made by the Sudan in developing the species quantitatively and qualitatively, thus securing supplies, maintaining confidence and creating a strong baseline for the long term involvement of both producers and consumers.

The Sudan has been known since the time of the Pharaohs (4000 years ago) as the biggest producer and exporter of the best quality of gum arabic, and of course will continue to do so indefinitely. This is not wishful thinking but a mere fact historically proven through centuries. The Sudan managed to exercise such a monopoly for such a long time through many factors - some related to the resource itself, others related to the people who work the resource while many are attributed to sound long term planning. In this study we should try to analyse these factors and show efforts made to keep the Sudan as always the biggest producer and exporter of the best qualities.

RESOURCE FACTORS :

1) Unique Product

Gum Arabic as displayed in the forthcoming chapter, is unique, whether in its natural habitat as a tree and part of a dynamic ecosystem or as a commodity, used in a wide spectrum of industries and providing incomes to subsistence farmers at a time when it is most needed. It has also so many social, economic and ecological benefits that makes it highly desirable under all circumstances:-

(a) From a biological point of view, it can regenerate naturally or artificially, by seeds or seedlings or even by copice. Consequently it can be established at a very low cost.
(b) From a protection point of view and in its natural habitat, it protects the sandy soil from being blown off and hence checks desert encroachment.
(c) The tree is also a natural soil improver because it fixes nitrogen to the soil and also adds phosphorous and ammonium nitrates and organic matter thus contributing to the fertility of the soil.
(d) Consequently it is safe to conclude that it helps to increase agricultural production either through protection of soil and crops or through improving and adding to its fertility.
(e) Like any other tree, it also contributes to the provision of forest products in rural areas e.g. fuelwood, charcoal, and also round building poles.
(f) From a socio-economic point of view; it reduces the rate of migration by :
    (i) providing employment at idle time; thus making the farmer stick to the land.
    (ii) completion of the annual work cycle and providing a sustained income for the farmer well distributed through the dry months of the Summer.
(g) From an ecological point of view it provides a well balanced land use system - well suited to the area (the gum cultivation cycle provides an ideal solution for all the ecological, socio-economic imbalances that are now evident in the area).

After all this, comes gum arabic production to provide extra sustained income to the farmer and the state.

The protective role of the Hashab tree is the most important role of all. The gum areas of the Sudan, are mostly sandy, unstable and highly vulnerable to erosion. If there is no gum tree in such areas it is possible to conclude that in the long run, there would be no agriculture. Consequently, the opportunity cost of foregoing the gum trees in their natural habitat is very high; higher than even the State can afford.

On the other hand the product itself is unique being a natural healthy product displaying so many properties that render it useful and safe to into the making of many end products - very well known to the industry.

2) Flexibility In Production

In the Sudan, gum is being produced in two distinct areas namely Western Sudan and Eastern Sudan. The West comprising North Kordofan, North Darfur and part of the White Nile regions, while the east includes Blue Nile and Kassalla and Southern Kordofan regions. The two areas however, exhibit marked differences in soils, climatic conditions and mode of production which do have an important significance in the security of supply. Some of these differences are given in table (1) below

TABLE 1 COMPARISON OF GUM PRODUCTION AREAS IN SUDAN

East Sudan	West Sudan
Blue Nile, Kassalla & S.Kordofan Regions	N.Kordofan, N.Darfur & White Nile Regions
1) Clay plains	Sandy Soils.
2) Rainfall 600 mm/annum	Rainfall 300 mm/annum
3) Large scale mechanized rainfed agriculture with pockets of gum arabic.	Traditional Agriculture. No mechanization - gum is part of agricultural rotation.
4) Large private holdings and government land.	Small individual ownership
5) Semi-modern economic sector	Traditional subsistance sector.
6) Lack of experienced gum producers.	Well experienced gum producers.
7) Gum trees established by mechanical sowing	Establishment of trees by traditional means.
8) Accessible region with relatively easy transport and better water supply	Remote-bad roads-difficult communication and acute drinking water problems.
9) Produces 20% of total	Produces 80% of total

It is clear that those differences have given the Sudan - the latitude to produce gum arabic even in the most difficult and adverse conditions - also to develope the ability to respond very quickly and recover production at the time of serious droughts. This flexibility is a key factor in the security of supplies because a drop in production in one area is normally counterbalanced by an increased production in the other.

3) Comparative Advantages

Gum occurs throughout the length and breadth of Africa south of the Sahara desert and also in some parts of India. Commercial production of gum arabic is, however, restricted to certain zones and almost all of it is produced in Africa, north of the equator.

Further limitation of commercial production within the African continent depends on the degree of uniformity of the species. In the Sudan, and particularly in Kordofan and Darfur provinces the species is uniform and is found in pure stands giving the Sudan the advantage of being the biggest producer and exporter of the best qualities. In other producing countries, the distribution is not so uniform and is often found mixed with other species. Another important comparative advantage is that in the Sudan it occurs both wild and cultivated in a wide area giving the advantages of economies of scale.

In comparison of the Sudan situation with other producing countries, some marked differences are evident which explain the Sudan superiority in gum production, and shown in Table 2.

TABLE 2 COMPARISON OF GUM PRODUCING AREA SUDAN / OTHER COUNTRIES

SUDAN	OTHER COUNTRIES
1) Uniform stands of monoculture of the right species in large tracks.	Mixed species in patches.
2) Well experienced producers, tappers, collectors, cleaners and graders	Lack of experience in all fields.
3) Land tenure settled - recognized private ownership.	Serious land tenure problems gum is produced in the no-man's land.
4) Producers are mainly settled agriculturists	Producers are mostly nomads not interested in gum development.
5) The stable system of gum cultivation cycle is practiced in many areas.	Over exploitation and cutting of trees through shifting cultivation is practiced.
6) Development of gum projects at all levels.	Lack of serious development projects.
7) Produces 85 - 90% of world production.	Produces 10 - 15% world production.

Four serious problems hinder the production of gum in countries other than the Sudan and these are ;-

(a) The mixed nature of the resource.
(b) Lack of security of land tenure.
(c) Inexperienced producers, traders etc....
(d) Nomadic nature of producers.

Two important conclusions follow from this comparison ;-

(i) Because of the mixed nature of the resource coupled with inexperienced producers / traders etc... Quality is never assured.

(ii) Because of lack security of tenure coupled with nomadic nature of the resource - development of gum production is impossible.

HUMAN FACTORS

This factor is a focal point in gum production, it concerns the people who work manage, and develop the resource to produce gum arabic.

Gum production in the Sudan has developed through the centuries into a traditional art handed from generation to generation from father to son. The long experiences gathered among technicians, producers, traders, sorters etc... is not matched anywhere else in the world. Such an experience has also a very strong back-up by a well established research and extension services.

Moreover in the Sudan, gum producers are settled agriculturists and gum is entrenched in a well defined agricultural rotation (gum cultivation cycle). Under such circumstances and contrary to other producing countries - every A. senegal tree is owned by somebody - thus providing traditional well defined rights.

PLANNING & DEVELOPMENT FACTORS

Plans are constantly being developed in all fields to cope with changing situations to sustain production. These plans include :-

(1) Institutional

The first important measure was the reorganization and creation of new institutions to divide the labour according to specialities, work in harmony to stop degradation enhance development, to increase production and to achieve security and continuity of the gum trade.

Under such a state of affairs three main institutions were recognized and created to help bring back the balance :-

The Gum Research Division was established in 1958 mainly to carry on research on gum, but in the absence of a production division, it had to carry out extension work to increase production. Accordingly, the Research Division was rather handicapped because more time was devoted to extension rather than research.

In 1969 the Gum Arabic Exporting Company was established to carry on the business of marketing the commodity. In this arrangement it is clear that the production side of gum arabic has completely been neglected. After the decline in production and to correct such a defect a new division has been established (1974) known as the "Management and Services Division for Gum Arabic " to carry on the production side. By doing so, the three main spheres research, production and marketing have all been catered for.

The main aim of this new Division is to improve, develop and increase gum production in the Sudan; to keep pace with the world demand from both quality and quantity points of view. This could be achieved through the pooling of all efforts and available facilities in all levels; provincial, central, local and international, and making full use of them to the best advantage.

The responsibilities of this new division can be summarized as follows :-

(a) Draw out all development plans for gum arabic in order to ensure continuity and sustain yields.
(b) Follow up and execute all development plans for gum arabic.
(c) Carry out all extension services needed and give advice to producers, commissioners, Forestry staff and G.A.C.
(d) Management of all Forests Department gum arabic reserves both natural and planted, mainly to act as a buffer stock.
(e) Coordination of Forests Department afforestation programme of gum arabic with other national projects (multipurpose plantation).

(2) DEVELOPMENT STRATEGY

Having reorganized the institutional network then serious development has taken place in all spheres. All development projects in the various fields were aimed to achieve three main objectives :-

(i) To introduce ecologically a stable land use system to combat desertificaiton i.e. by introduction of the gum cultivation cycle - a recapitulation on the old system developed by the producer himself and is well suited to the area.

(ii) To improve the producers standard of living by solving production problems, improving marketing conditions and providing essential services and a local community development programme.

(iii) To increase production according to world demand, achieving security of supplies at realistic prices and assuring the industry of a reliable source.

(3) DEVELOPMENT PROJECTS

The restocking of gum belt projects in the Sudan have had an immense response from most of the International Organization and also many governments. This is attributed to two main factors :-

(a) The commodity affects the traditional small farmers and it is a key factor in raising his standard of living.
(b) The commodity is a natural product that goes into the making of so many industrial products - and since man has to follow a course consistent with nature, the future is then for natural products rather than synthetics.

Following from this and bearing in mind the main objectives, a number of projects sponsored by International donors have come to the surface. Some of them have already started while others are still in the pipeline. These projects are :-

(i) UNSO/FAO

The projects aim at helping the small farmer to reafforest his own land by gum trees in parts of Eastern and Central Kordofan. The aim is to establish nurseries and produce seedlings annually to be planted in private holdings, and also to provide extension service and improvement of marketing institutions. The project has already completed its first phase (1980-86) and now in second phase. The project has also a rural community development component which goes side by side with tree plantations to help rural families.

(ii) SHORT TERM AGRICULTURAL CREDIT FACILITIES PROJECT

The project is financed by the World Bank and administered by the Agricultural Bank of Sudan (ABS). It has been operational for four years and aims at providing short term credit facilities for small farmers in the traditional sector. The project started as a small experiment and covered only agricultural crops. It has now been appraised by the World Bank mission and contrary to all expectations it has been found extremely successful with 100% repayment. Now the (ABS) has the green light not only to expand on a large scale but also to include gum arabic - as part of the traditional agricultural station.

(iii) THE E.E.C. PROJECT FOR DEVELOPMENT OF GUM ARABIC IN SUDAN

The project covers most of Kordofan Province and is mainly directed towards the small producer to reafforest his land with gum arabic trees together with introduction of extension service, cooperatives, credit facilities and improvement of marketing and grading of gum arabic.

(iv) THE COOPERATIVE FOR AMERICAN RELIEF EVERYWHERE (CARE)

A plantation project with a rural development component covering parts of Kordofan and Eastern Region.

(v) THE FINNISH PROJECT

Finland is financing the project under a joint Sudanese management. The project had been launched in June 1981 covering parts of the White Nile Province. The project includes a research component for the introduction of modern nursary techniques, improvement of production methods and thus reducing establishment costs. The results will be extrapolated right across the gum belt of the Sudan.

(vi) WFP PROJECT

This is a specialized project donated by the World Food Programme (WFP) of the United Nations. It aims at providing food for work to the farmers involved in the gum projects as an incentive and a relief until income from the new plantations is realized.

(vi) U.S.AID PROJECT

This project covers five rural councils in Northern & Southern Kordofan. It is a plantation project with a breading and community development programmes. It started in 1988.

(viii) U.N.D.P. DARFUR PROJECT

Sponsored and financed by U.N.D.P. Sudan Government and Gum Arabic Company covering both Northern & Southern Darfur with the aim of rehabilitation degraded areas, and improving of both production and marketing system - ready to start before the end 1989.

(ix) JAPANESE PROJECT

It is a proposed project covering two main areas :-

(a) Rehabilitaiton of sandy areas of Eastern Kordofan and parts of the White Nile.

(b) Management of gum forests in the Blue Nile in the Central Region.

(x) I.T.C. PROJECTS

Projects sponsored by I.T.C. are many and varied and mainly directed towards Gum Trade Promotion Activities with regard to improvement of quality, cleaning & grading operations and enhancement and promotion of gum usage. I.T.C. also helps in promotion of other gums e.g. Guar & Karaya Gum

In addition to all these projects the Gum Arabic Company traditionally helps in participating in solving production problems such as, drinking water, pest control, social welfare community development programmes.

(4) PRICING POLICIES

Pricing policies are based on the important fact that the producers must be assured of a remunerative return to encourage them to promote gum production. Prices as such have been rising steadily and have more than tripled in five years.

Such price increases are mainly achieved and maintained through:-

(a) Reflection of upward readjustments in international prices.
(b) Revision of taxes whenever need arises - and adjusting them to suit conditions.
(c) Creation of stabilization fund especially to cater for price fluctuations and assure the producer a rewarding stable income.

On the other hand the pricing policy was directed to be flexible, realistic, geared to maintain and promote the marketing of the commodity. The principles behind the policy can be summarized as follows :-

(i) To secure the interests of both producers and consumers.
(ii) To curb the encroachment of substitutes.
(iii) To regain lost grounds and to acquire new fields.
(iv) To promote and help in short, medium and long term planning of the industry.

(5) ORGANIZATION OF PRODUCERS

The most important factor in the future development of production and marketing of gum arabic is the organization of producers. The idea was executed five years ago as follows :-

(1) Formation of producer associations in all important production centres.
(ii) Facilitate their participation as shareholders in the gum arabic company.
(iii) Have them represented in the Board of Directors of the gum arabic company.

This will achieve the following :-

a) Extra income to producers would be realized from dividends of their shares.
b) They will be part of the decision making machinery, enabling them to influence price, production and marketing policies.
c) The associations will be one of the main channels for the execution of gum development projects.

Now the producers are full shareholders in the G.A.C. and they take share in the formulation of all policies.

(6) IMPROVEMENT OF INFRASTRUCTURE

This is geared to reduce production and handling costs. The programme is implemented by the Local Government to improve access roads and transport facilities - creation of new auction markets etc... Such programmes are already underway and each year improved facilities are being introduced. The Gum Arabic Company is also now completing an extensive field network - by establishing offices and warehousing facilities in all production regions.

(7) PROMOTION OF RESEARCH

The research policy is directed into three main spheres :-

(i) Field Research to improve methods, introduce new techniques and reduce production costs are already in progress, catered for by the International Aid Community. The results of such programmes will of course be available to all plantation projects to make use of.

(ii) Tissue Culture techniques have been introduced to produce high yielding - drought resistant strains. This is a long term security for sustained gum production.

(iii) Industrial Research to expand future markets, discover new uses and combat substitutes. Such fields are hoped to be covered in collaboraiton with the International consumers - by creating a closer link - _

(8) CREATION OF A TRIPLE SECURITY SYSTEM

The outcome of all these devclopment programmes is to provide a triple security system :-

a) Buffer stock in the form of gum
b) Buffer stock in the form of tree reserves - to be tapped when needed.
c) Buffer stock of trees in the form of improved varieties high yielding - drought resistant.

(9) QUALITY ASSURANCE

The concept of quality in gum follows a much wider concept than what is normally practiced. It contains all the elements of good manufacturing practice (GMP) and introduces the philosophy of quality assurance (Q.A.) rather than just quality control. Under this concept the quality of gum is being considered and determined even before the tree is being planted and each technical operation that follows after planting has a bearing on the final product. Some of these important operations include :-

(a) Seed Selection

Before sowing or planting, seeds must be acquired from an authenticated source - better still from improved seed varieties high yielding strains - a factor that has an important bearing on the economics of production.

(b) Spacing

Too wide spacing gives fewer trees per unit area and allows for the growth of other species - thus affecting production economics and purity of the product - likewise closer spacing has also disadvantages of undeveloped crowns and difficulty in accessability. Therefore the optimum spacing as developed

Current and future prospects for starch

S.A.BARKER
Department of Chemistry, The University of Birmingham, PO Box 363, Birmingham B15 2TT, UK

ABSTRACT

Some current starch derivatives are cited and the way forward to future food acceptable derivatives suggested. Such derivatives will be largely produced by enzymic modification or if by chemical means then by using reagents that are precursors of food acceptable ingredients like citric acid anhydride. Such approaches should more readily meet existing legislation.

FOOD ACCEPTABLE STARCH DERIVATIVES

Current starch derivatives have closely mimicked those developed from cellulose except in three major respects (a) starch according to its source can contain appreciable to a very large proportion of the branched component amylopectin (b) the α–linkages in starch molecules endow them with a different spectrum of water solubility from those of cellulose molecules which are β–linked and finally (c) in humans cellulose is not digested whereas starch is unless it is so substituted as to impede the action of α–amylase and/or phosphorylase. In some respects these differences in starch and cellulose can lead us to define their market outlets in the food industry and in part can define their future potential markets.

Table I. RECENT APPLICATIONS OF STARCH DERIVATIVES IN FOOD.

Starch Derivative	Application
Carboxymethyl ether	Sour cream stabilizer A milk protein stabilizer Stabilizers for ice–cream Disintegrating agent for the sustained release of pharmaceuticals Use in tablets of aspartame.
2–Hydroxyethyl ether	A cryo protectant A blood substitute Incorporation in fat emulsions
Hydroxypropyl ether	Instant dry mixes for puddings Use in salad dressings Use in fish pastes

Acetate ester	Manufacture of cheese substitutes For noodle shelf-life improvement
Hexadecanoate ester	Thermally reversible gelation
Phosphate ester	Manufacture of cheese substitutes Thickeners Instant fruit jelly preparations with amylopectin phosphate Formation of foam with sodium caseinate.
Hydrogenated starch	Manufacture of fruit and vegetable snacks Incorporation in low calorie hard candy composition Marsh mallow and nougat containing it. Preservative gel for mackerel and pollack during frozen storage

THE WAY FORWARD

Thus it is obvious that with restrictions on the introduction of new starch derivatives there are only a limited number of ways forward. First and foremost we can make maximum use of those food quality enzymes available like amyloglucosidase, β-amylase, pullulanase, isoamylase, α-amylase and introduce those two enzymes neglected for forty years namely phosphorylase and the branching enzyme. Second, we can consider using transglycosylase action to explore the introduction of new sugars into the 6-position of the α-1:4 glucosan chains. Third, we can note that nature elsewhere has created comb structures where carbohydrate chains are substituted with fatty acid derivatives. Fourth, we can screen for oxidases of three fundamental types (i) those that could oxidise the 6-hydroxymethyl groups in starch to carboxylic acid groups (ii) those that would oxidise the reducing group or groups present to σ-lactones (iii) those that oxidise secondary alcohol groups to keto groups. Fifth, we can envisage making synthetic glycoproteins although these have already been created by reaction of starch cyclic carbonate with protein enzymes. Sixth, starch or its dextrins can be used as a substrate for transglycosylase action and transferred to receptors like sorbitol or sucrose. Because the reduced starch derivative polysorbitol is food permitted we can consider the unique sorbitol end as a receptor of fatty acids.

Chemically we could get new derivatives of starch created with the precursors of food permitted additives. Examples of this philosophy are (a) citric acid anhydride which will citrylate and only give citric acid as a byproduct and (b) cysteine thiolactone which will introduce thiol groups as it esterifies and yield cysteine as a byproduct. It is obvious that once this strategy is appreciated it can be extended to other molecules.

If we turn to those starch derivatives already permitted particularly those with negative charges, polyion complexes, with proteins or chitosan, or sugars like glucosamine, become distinct possibilities with starch phosphate or carboxymethyl starch.

I was of the generation that believed that amylopectin was soluble in water directly while amylose dissolved from the dried state required a spot of alkali to get it into aqueous solution and that often it came out of solution on neutralisation except at very low concentrations. Professor Gilbert (1) in our laboratories proved that man had created these artefacts. He showed that careful extraction of the starch grain with alkali selectively extracted amylose leaving insoluble amylopectin. This little known fact has never been commercially exploited as far as I am aware.

A sequel of this is to go back to the cereal itself and attempt to cut costs by not isolating the starch grain itself by a capital intensive process but by derivatising it in situ so that starch derivatives can be selectively extracted. Thus it has been claimed that propylene glycol alginate can be prepared by direct derivatisation of the alginate without prior extraction of seaweed. Perhaps the potato would be more amenable to such an approach. We achieved selective solubilisation and hydrolysis of the starch directly from wheat by treatment with $CaCl_2$/HCl or LiCl/HCl (2). This environment contained little water and so would be amenable to methanolysis and alkanolysis in general. For food use sorbitol, xylitol or other polyhydric alcohols would be more suitable.

Another approach is to practice derivatisation during cooking. Already it is well established that 1,6-anhydroglucose is formed when starch is submitted to microwave action. Further a recent paper (3) shows that organic compounds can be synthesised up to 1240 times faster in sealed Teflon vessels with a pressure release cap in a microwave oven than by conventional reflux techniques. Polar molecules such as primary alcohols (starch) and acids absorb microwave energy readily whereas nonpolar molecules do not. The microwave technique is equally effective in enhancing the rates of both homogeneous and heterogeneous reactions.

Many years ago it was demonstrated that provided a small amount of α 1:4 linked oligosaccharide was present, phosphorylase would synthesise linear amylose chains from glucose 1- phosphate (4). The latter was the product obtained from starch by phosphoylase action in the presence of phosphate. Originally the enzyme was heavily impeded by α 1:6 branch points in starch but these can be removed by addition of the debranching enzymes isoamylase and pullulanase. Conversely branched structures can be synthesised by the combined action of phosphorylase and the branching enzyme on α-D-glucose-1-phosphate (5).

Four major factors will often permit a quantum jump in the price of a food constituent. These are texture, taste, shape and colour. All four could be achieved by admixture with sodium alginate and subsequent addition of food permitted additives. Further in some cases it can be applied to the source of the starch like instant potato powder or its equivalent undried cooked version. Much of living matter is held together with calcium ions and their addition to alginate - starch mixtures permits (a) them to be formed into any shape and (b) entrapment of ingredients conferring changes in texture, taste and colour. Additionally, very small amounts of phytic acid present to the extent of 10% in bran, enable many other changes to be wrought since phytic acid chelates with metal ions and proteins. The imminent arrival of phytase from Gist-Brocades (6) for use in animal feed may also provide a human permitted version to change the chelating pattern of the phytate. Exciting developments could also stem from the advent of amylases (α- and gluco-) largely being developed in Japan (7) which permit attack on raw starch organised granules without pregelling of the starch. Aspergillus SP K-27 (8) is a rich source of such enzymes and was located by screening for attack on amylose - lipid complex resistant to many other α-anylases.

The dietetic food market requires new food permitted ingredients to make it exciting just as the advent of aspartame and other high intensity sweetners require novel bulking agents (9). Maltodextrins can be generated rapidly by α-amylase action on starch and then could be used as receptors to a transfer of α 1:6-linkages to the non- reducing terminal ends of the dextrins using dextransucrase and sucrose. Dextransucrase is currently used by Fisons and other companies to manufacture dextran for intravenous transfusion.

Such a bulking agent would become even more attractive if the sweet fructose byproduct was left within the mixture. A reactor separator device has recently been devised (10) that would separate pure polyglucosans of a defined size from pure fructose continuously produced by dextransucrase with a sucrose feed. It offers solutions to a major industrial problem, namely end product inhibition encountered with enzyme catalysed processes.

Important new products could be created from starch by formation of comb polymers (11) where the C6 position of the D-glucopyranosyl units are attached to fatty acid residues via ester links. Eventually this should be feasible both chemically and enzymatically. Not only could such products be valuable nutritionally as low calorie foods but they would have other valuable physical properties based on their unusual surfactant like structure. Currently the lipases launched by Novo, Gist Brocades and Biocatalyst are exciting wide interest in their synthetic mode whereby they can produce esters from polyhydric alcohols like glycerol and fatty acids or catalyse transesterification between an ester and an acid. Already sorbitol and other polyhydric alcohol esters have been synthesised (12) using the Finn sugar lipase. There is hope therefore that the terminal sorbitol in polysorbitol like mixtures could act as an acceptor of a fatty acid residue.

LEGISLATION

All starch products (except starches, maltodextrins or glucose syrup), when they are derivatives of starch, are assessed for their suitability to be incorporated into food before being assigned an EEC code number. This means that the process whereby they are derived is scrutinized closely, including those using enzymes rather than chemical procedures to prepare the derivative. It is implicit that when designing such processes it is desirable wherever possible to utilise food permitted additives and to eliminate any toxic chemical from the process. This means that for food use enzymes should be immobilised with food permitted additives also eg. calcium alginate.

Table 2. SOME POTATO STARCH DERIVATIVES AVAILABLE COMMERCIALLY

Starch Derivative	EEC Code	Commercial Name
Pregelatinised slightly oxidised starch	E1403	Pregel PA5 or 97
Oxidised starch	E1404	Perfectamyl A3108
Distarch phosphate	E1411	Extrusamyl MSP
Pregelatinised distarch phosphate	E1411	Paselli BGD
Acetylated distarch phosphate	E1414	Perfectamyl AC75 or 100
Pregelatinised acetylated distarch phosphate	E1414	Paselli BC or PAC
Starch acetate	E1420	Perfectamyl AC or Gel 45
Pregelatinised starch acetate	E1420	Amylogum CLS, FS or NFS
Hydroxypropyl distarch phosphate	E1422	Farinex VA70, VA70C and VA70WM

Table 3. MAJOR ADVANTAGES OF STARCH MODIFICATIONS.

Modification	Properties
Partial acid hydrolysis	Produces increased degree of retrogradation, increased set back and rigid gel formation in confectionery.
Oxidation	Produces reduced tendency to retrograde, soft bodied gels of high clarity, good for clear canned soups.
Pregelatinsation	Produces cold water dispersibility, useful in instant puddings, desserts, soups etc.
Dry heat degradation (dextrinisation)	Produces white and yellow dextrins used in glazing baked goods.
Ester cross-linking	Stabilizes the granules and makes swelling more resistant to processing conditions particularly waxy maize starch.
Anionic substitution	Produces freeze-thaw stabilised starches important in the frozen food industry.

Cerestar, National Starch and Chemical and other companies supply a comprehensive range of such modified maize, waxy maize, wheat, tapioca and potato starches within the EEC classes A-R.

In addition to E numbers there are Benelux code letters eg. E1403 is D, E1411 is I, E1414 is J, E1420 is G and E1422 is R. In the United States the regulatory authority is the Food and Drug Administration. Thus Avebes Primojel meets the USP XX- NF XV specifications for sodium starch glycolate, the sodium salt of a carboxy methyl ether of starch. It is used as a disintegrant in pharmaceutical tablets. In the code of Federal Regulations: Title 21 Food and Drugs code letters are assigned in the survey according to 172.892 Food starch modified. Thus E1403 is b, E1404 is c but products covered by E numbers 1411, 1414, and 1420 are all d with product EE1442 as f. Products are sold for their functional properties and often may be mixtures in the chemical sense eg. contain a molecular weight spectrum of related derivatives. This is best reflected in the use of the term D.E. (dextrose equivalent). Only in recent years has it been possible to translate that into a realistic assessment of composition using liquid chromatography or HPLC. Again the distribution of substituents in the longer chain molecules will not be uniform and only NMR and detailed studies will unravel that mystery. ^{13}C NMR is much surer in detailing information on the nature of glycosidic linkage and branching, and may eventually be used to fingerprint the molecules.

REFERENCES

1. Gilbert G. Nature (1958) 185, 231-2.
2. Barker S.A., Somers P.J., Beardsmore A.J. and Rodgers B.L.F. (1983) European Patent Application 0096497.
3. Gedye R.N., Smith F.E. and Westaway K.C. (1988) Can. J. Chem. 66, 17-26.
4. Barker S.A., Bourne E.J., Wilkinson I.A. (1950) J. Chem. Soc. 84-92.
5. Barker S.A., Bourne E.J., Peat S. and Wilkinson I.A. (1950) J. Chem. Soc. 3022-27.
6. Bresser G. (1988) The Gist, No.35, 1-3.
7. Bergmann F.W., Abe J. and Hizukuri S. (1988) Appl. Microbiol. Biotechnol. 27, 443-446.
8. Abe J., Nakajima K., Nagano N. and Hizukuri S. (1988) 175, 85-92.
9. Tuley L. (1989) Food manufacture, January, 17-18.
10. Barker P.E., Zafar I. and Alsop R.M. (1987) Int. Conf. Bioreactors and Biotransformations, Gleneagles, Scotland.
11. Cooper D.G. and Zajic J.E. (1980) Adv. Appl. Microbiol. 26, 229-34.
12. Chopineau J., McCafferty F.D., Therisod M. and Klibanov A.M. (1988) Biotech. Bioeng. 31, 208-214.

Pea starch – a new exciting food ingredient

D.GREGORY and H.BILIDT[+]

Grindsted Products, Northern Way, Bury St.Edmunds, UK

[+]DDS Nutrio, Langebrogadel, PO Box 17, DK-1001 Copenhagen, Denmark

Abstract

Pure starch derived from the yellow field pea (Pisum sativum) is a new, commercially available food ingredient. The starch is natural and characterized by a high resistance to heat and mechanical treatment. The starch is also stable at low pH and when used in food products there is no loss of viscosity in the presence of sugar and salt. The production of pea starch, its characteristics and practical application are described.

Introduction

The pea is a well-known vegetable which is traditionally used in a variety of food products. On the food ingredient market many suppliers are offering split peas, ground peas (pea flour) and pea hulls. However, by separating the main compounds of the pea,a family of powerful food ingredients can be obtained. A typical analysis of yellow peas will show a content of 50% starch, 22% protein, 15% fibre, 8% water and approximately 5% ash, lipids etc. and by by careful selection from the vast range of pea varieties available, it is possible to identify pea types which offer optimum raw materials for the production of functional food ingredients. A production process has now been developed by which a pea protein isolate, a native pea starch and a fibre product can be produced.

Production of pea starch

The production of food ingredients from peas follows the principles shown below. A process based on wet milling, followed by a number of separation units gives the following products:

Pea Protein Isolate	88% protein on dry matter and no content of low molecular weight carbohydrates
Pea Fibre	A fibre product derived from the inner part of the pea
Pea Starch	A native starch containing 98% carbohydrates on dry matter
Pea Hulls	External shell of the pea which is washed and ground
"Molasses"	A fluid by-product which contains all the dry matter substances from the raw materials which are not contained in the above fractions

The pea starch is processed through a range of separation units, washed and dried. The final starch has a moisture content of approximately 8% and is available as a white, free-flowing powder.

Characteristics of pea starch

As mentioned above pea starch is a native starch, characterised by a range of functional properties normally associated with modified starches.

An analysis of the commercially available product is shown in Figure 1.

The particle size distribution is shown in Figure 2 and as can be seen the greater percentage of starch granules lie between 30-40 µ.

A Brabender Viscometry Test of the pea starch - with corn and wheat starches as references - is shown is Figure 3. Pea starch gelatinizes earlier than corn and wheat starches and pea starch is also stable in the higher temperature zone.

In Figure 4 it can be seen that pea starch is very stable even at pH 4.

In Figure 5 the effect of salt is shown - as can be seen,sodium chloride does not have any significant effect in the concentrations that are normally used in food products.

Pea starch gels in concentrations from 4.5%. The gels are very strong, but are not stable in freeze/thaw situations.

Typical Analysis

Carbohydrates in dry matter		98.5%
Carbohydrates	>	90.0%
Protein	<	0.5%
Fat	<	0.5%
Ash	<	0.5%
Moisture		8.5%

Particle Size Distribution

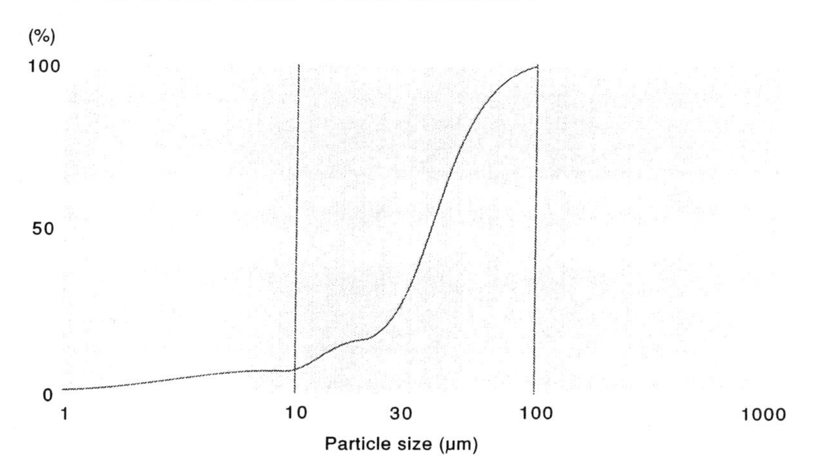

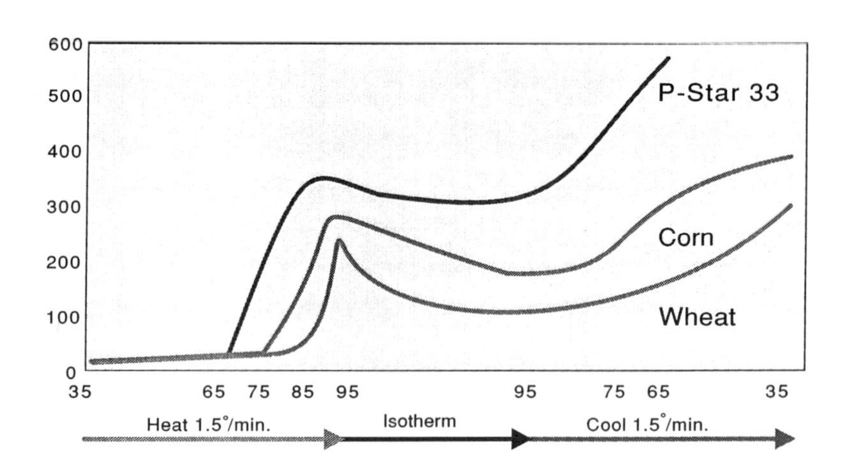
Viscometry Test
600
500
400
300
200
100
0
35
65
75
85
95
95
75
65
35
P-Star 33
Corn
Wheat
Heat 1.5°/min.
Isotherm
Cool 1.5°/min.

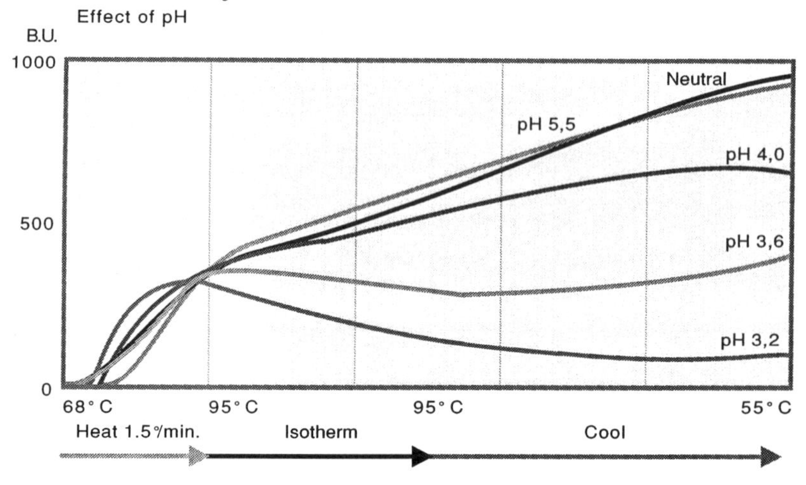
Viscosity
Effect of pH
B.U.
1000
500
0
Neutral
pH 5,5
pH 4,0
pH 3,6
pH 3,2
68° C
95° C
95° C
55° C
Heat 1.5°/min.
Isotherm
Cool

Viscosity

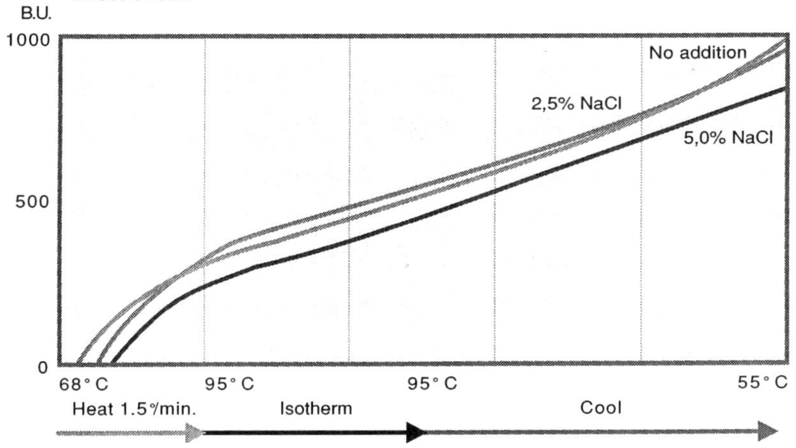

Application of pea starch

Since its commercial introduction, pea starch has mainly been used in the processed meat industry, where heat and mechanical stability is of great interest. In canned products, cooked sausages, pâtés etc. pea starch can substitute traditional starches with very good results.

As a thickening agent in soups and sauces, pea starch gives a pleasant mouth-feel - a property which might also be of interest in a number of other products.

A new area which is presently being tested is the use of pea starch in extruded products. The heat stability and the good expansion property of pea starch seem very promising.

Starches in heat-processed particulate foods

V.DE CONINCK and J.VANHEMELRIJCK
Cerestar Euro Centre Food, Havenstraat 84, B-1800 Vilvoorde, Belgium

ABSTRACT

Continuously processed foods containing particulate components e.g. vegetables have gained increased popularity over the last few years. "In-Flow" processing of liquid foods with solid particulates can be performed at higher temperatures and in shorter time, than processing inside a container. This treatment results in considerably less thermal degradation compared to canning. Scraped-surface heat exchangers are therefore currently receiving the most attention, because of the characteristic high viscosity of the carrier liquids and the need to agitate the liquid/particle mixture during heating and cooling. In both types of aseptic thermal processes however the holding time for the carrying liquids needs to be extended to obtain sufficient microbiological inactivation in the centre of particulates. To eliminate this drawback, new systems for the continuous sterilisation of particulate food products have been developed, such as the APV Ohmic heating and the STORK Steripart System. Using a neutral white sauce with peas as a model system, different modified starches (in terms of degree of modification and starch type) were tested in order to assess the influence of the heating system on the rheological behaviour of the liquid. The results, based on pilot-scale experiments, show the clear advantages of both the scraped-surface heat exchanger and the tubular heat exchanger over traditional canning. Moreover the newly-developed tubular heating systems, such as the STORK Steripart and the APV Ohmic heating add a new dimension to the quality of sauces containing particulates on condition that the proper modified starch is selected.

INTRODUCTION

The continuous processing of foods containing particulates is attracting increasing interest in the food industry. Examples are ready meals, beef chunks in gravy, fruit and vegetable pieces in syrup and sauces, custards with fruits and soups. In most of these products starches are required to provide the essential texture. In our investigation, a neutral white sauce with peas as particulates was used as a model system in order to study and to compare different modified starches. The processing techniques which we selected were canning, a scraped-surface heat exchanger and a tubular heating system.

MATERIALS

1. Starches

Six different modified starches manufactured by Cerestar were selected based on crossbonding level and starch origin (table 1).(1)

Table 1
Modified starches used

Trade Name	Description
Cerestar SF 06309	Acetylated waxy maize adipate
Cerestar SF 06205	Acetylated waxy maize adipate
Cerestar SF 06304	Acetylated waxy maize adipate
Cerestar SF 06203	Acetylated waxy maize adipate
Cerestar SF 06316	Hydroxypropylated waxy maize phosphate
Cerestar SF 30216	Hydroxypropylated tapioca phosphate

The Brabender viscosity profiles clearly demonstrate the broad viscosity range chosen for the trials (Fig. 1)

Fig.1:
Brabender viscosity profiles of modified starches.
Conc.:35 g starch/450 g distilled water (pH: 5.5)
Cartridge: 350 cmg. Cup rotation: 75 min^{-1}

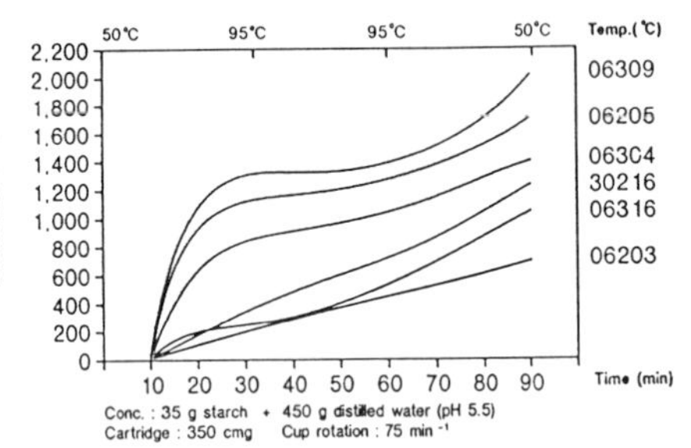

2. Full cream milk powder

Supplied by DOMO, The Netherlands. Fat content: 26%

3. Particulates

Deep frozen, very fine quality peas.

METHODS

Preparation of the neutral white sauce containing peas as particulates

The composition is given in table 2.

Table 2

Ingredients	Weight %
Water	76.5
Modified starch	3.5
Full cream milkpowder (26% fat)	15.0
Peas (only for SSHE trials)	5.0

Canning (rotary process)

The canning trials were carried out at STORK R&D, The Netherlands, using the HYDROMATIC-S Simulator. By means of the STORK simulator it is possible to carry out heat penetration tests, the process temperatures being registered through thermocouples and temperature recorders. The readings can be plotted in a temperature - time diagram giving an insight into the lethal effect (F_0) achieved by the process involved. The full cream milkpowder is mixed in water at 40°C using a high speed mixer. Subsequently the starch is added and mixing continued for 30 min. until a homogeneous liquid is obtained. 450 g of the mixture is then filled into 0.5 l cans. After closing the cans, thermocouples are installed in two cans in order to measure the heat penetration during sterilisation. The cans are placed in the STORK Hydromatic-S Simulator retort and steam is supplied while rotating the cans in axial position at 10 r.p.m. Heating, holding and cooling cycle parameters are adjusted to obtain an F_0 value of 7-9. Cooling of the cans is carried out with water.

Scraped-surface heat exchanger (SSHE)

Trials with the APV CREPACO Rota-Pro scraped-surface heat exchanger (SSHE) took place at C.E.R.I.A., Belgium. The Rota-Pro consists of a jacketed outer cylinder and an inner contact cylinder, between which a heating or cooling medium flows. Adhesion of the product to the inner heat transfer surface is prevented by continuous scraping, thus maintaining uniform heat exchange with a high blending efficiency (fig. 2). The water is preheated at 60°C. The full cream milkpowder is added while stirring at 240 r.p.m. Subsequently the starch is added, then after 15 min the peas. Fig. 3 shows the process scheme for the APV CREPACO Rota-Pro. A constant flow rate of 160-170 l/h is maintained. The scraper blades rotate at a speed of 150 r.p.m. for both the heating and cooling cylinders. The total residence time in both cylinders is 3.2 to 3.7 min.

Fig. 2 Tubular SSHE   Fig. 3 Process scheme APV Crepaco Rota-Pro

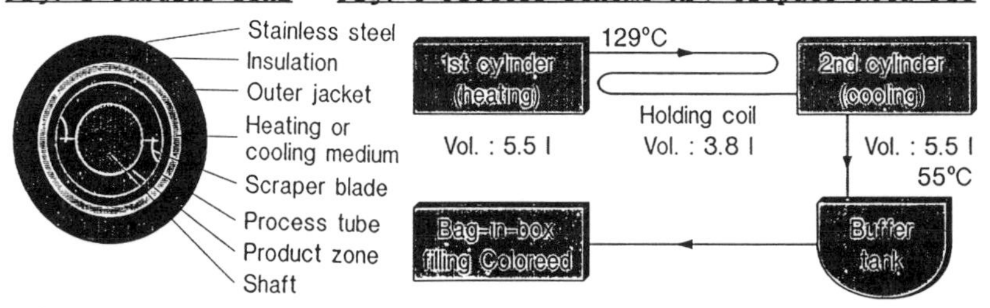

Tubular heat exchanger system

The STORK STERIDEAL tubular U.H.T. indirect heat exchanger was evaluated at the STORK R&D pilot plant in The Netherlands. Trials were carried out to simulate the newly developed STORK STERIPART System specially designed for the thermal processing of foods containing particulates of a recognizable size. The STORK Steripart pilot plant installation with a capacity between

500 and 1500 l/hour is still under construction and will be ready within a few months. This so called Single-Flow Fraction Specific Thermal Processing ("Single-Flow FSTP") is developed in such a way that the processing conditions for the liquid fraction as well as for each of the different size-fraction particulates can be independently designed and controlled while the entire product still remains transported in one single flow. The "Single-Flow FSTP" requires one or more "Selective Holding Sections" ("SHS") to separate either a particulate fraction from the liquid fraction or to separate bigger particulates from smaller ones including the liquid fraction. One type of "Selective Holding Section" is the STORK Rota-Hold (fig. 4).

Fig. 4: Stork Rota-Hold

It consists of a cylindrical vessel provided with a supply and discharge port. Inside this vessel a number of fork blades are mounted on the central shaft which can be rotated slowly within a chosen speed range. The forkblades form cages in which particulates are captured to be conveyed around the cylinder until they are released at the discharge port. The liquid can flow freely between the forkpins and around the encaged particulates finding its way to the discharge port in a much shorter time. Fig. 5 shows the t-T profiles for "Single Flow FSTP" using two STORK Rota-Holds at 135°C.

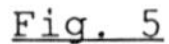

Fig. 5

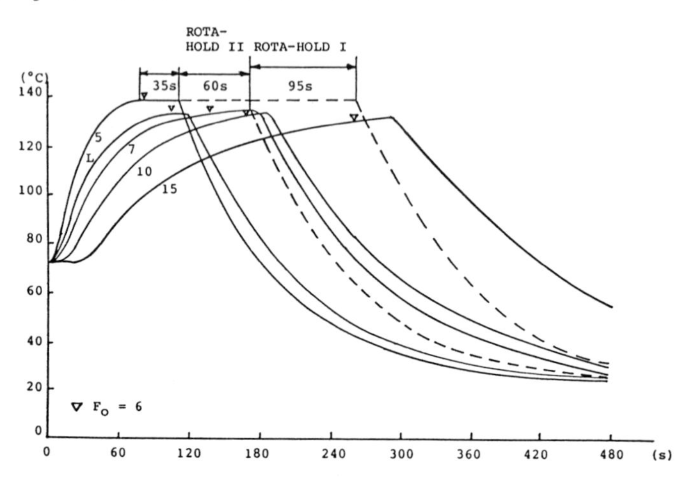

t-T profiles in a Stork Steripart with 2 Rota-Holds in series:
Processing temperature 135°C
Holding times: 35 s for particulates ≤5 mm
95 s for particulates between 5 and 10 mm
190 s for particulates between 10 and 15 mm

The first "SHS" separates the particulates bigger than 10 mm from the smaller ones, the second "SHS" separates the particulates bigger than 5 mm. In this way it is possible to apply specific residence times for the different size fraction particulates.

Fig. 6 shows the calculated F_0 processing values for "Single-Flow FSTP" including two STORK Rota-Holds.

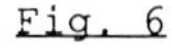

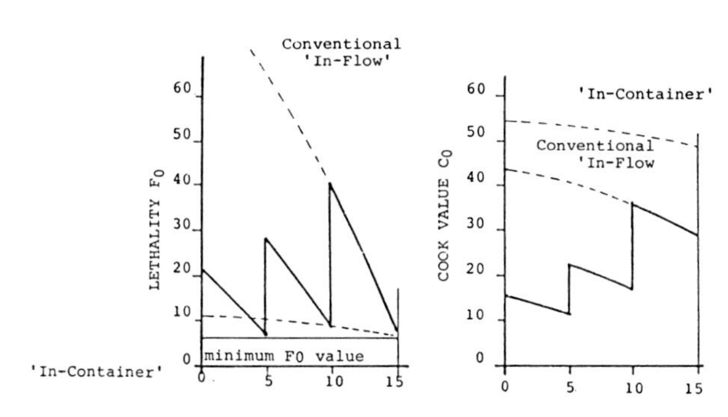

Fig. 6: Calculated F_0 values

In this way it is possible to reduce the considerable difference in process values between liquid, small and big particulates resulting in quality improvement of the treated product. (2)

Fig. 7

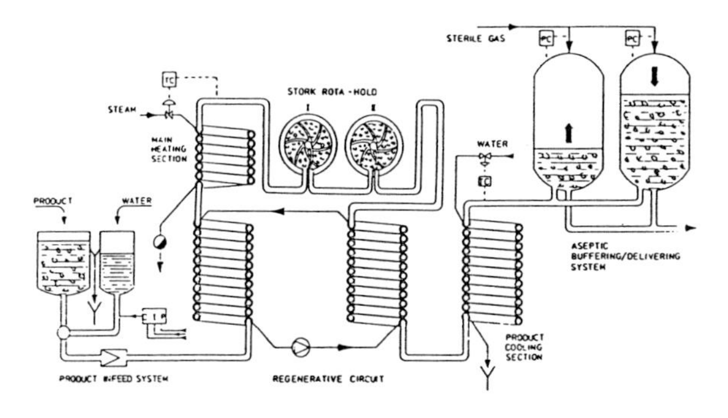

Fig. 7: The STORK Steripart System including an aseptic buffering and delivering system.

The whole milkpowder is mixed in water at 40°C using a high speed mixer (Motovario SPA (IT) type DSM 11002 var/V1-2) at 1450 r.p.m. After a mixing time of 10 min. the starch is stirred in. Subsequently the mix is pumped into a holding tank and stirred at 60 r.p.m. Using a monopump via a homogeniser unit (no pressure) the mix is then pumped to the tubular STORK Sterideal heating system. Heaters 1, 2 and 3 heat the mixture to 60, 105 and 130°C respectively. The residence time for each heater is 15 s. To reach F_0 = 7 the holding time at 130°C is 54 sec. at a flow rate of 200 l/hour. The mixture is then cooled to 90, 70 and 50°C by cooler 1,2 and 3 respectively. After exactly 10 min. from the start of the process, non-aseptic samples are taken.

OHMIC heating

Another new development gaining increasing interest is OHMIC HEATING developed by APV, United Kingdom. We had planned to run pilot plant trials at the APV Food Process Centre at Crawley (U.K.) in parallel with the three previously mentioned techniques. However, relocation of the Ohmic heating pilot plant prevented this. Nevertheless it is worthwhile to give a short description of this revolutionary breakthrough in food manufacture.

Principle of OHMIC heating

Heating occurs when electric current is passed through an electrically conducting food product. The heating effect is similar to that obtained with microwaves in the way that electrical energy is transformed into thermal energy volumetrically throughout the product. However, unlike microwave heating, the depth of penetration is virtually unlimited and for most practical purposes the product does not experience a large temperature gradient within itself as it heats. This simple concept has been developed into a continuous in-line heater for cooking and sterilisation of viscous and particulate-containing liquid food products.

Design of the OHMIC heater

The tubular system consists of a minimum of four electrode housings made from solid blocks of PTFE each containing a single electrode (fig. 8). The food product is pumped upwards through the column where it is progressively heated to full sterilisation temperatures. In practice, heating rates are between 1 and 10°C sec^{-1} depending upon electrical conductivity. Hence the residence time in the heater column is less than 10 s for a temperature rise of 100°C. With this gentle treatment very delicate products can be processed. (3)

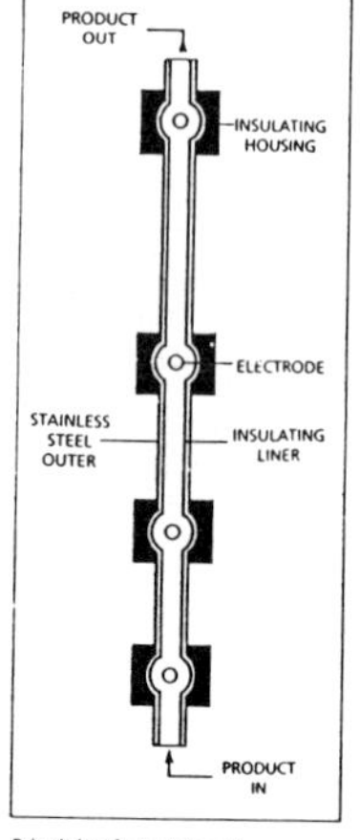

Principle of ohmic heating

RESULTS AND DISCUSSIONS

Canning (rotary process)

Heat penetration. As shown in fig. 9, the different starches clearly influence the heat penetration rate. The starch with the lowest crossbonding level SF 06309 gave a very slow heat penetration until viscosity breakdown during retorting occurred. The somewhat higher crossbonded starch SF 06205 required too long a sterilisation time to obtain the required F_0 value of 7-9. Under the sterilisation conditions for SF 06309 an F_0 value of only 4.5 was reached. The acetylated waxy maize starch adipate SF 06304 (medium crossbonding) gave a fast heat penetration and the strongly crossbonded SF 06203 a very fast heat penetration. Fast heat penetration was also obtained with both hydroxypropylated crossbonded starches from waxy maize and tapioca, showing a pronounced thin-thick effect.

Fig. 9: Heat penetration profiles for different modified starches

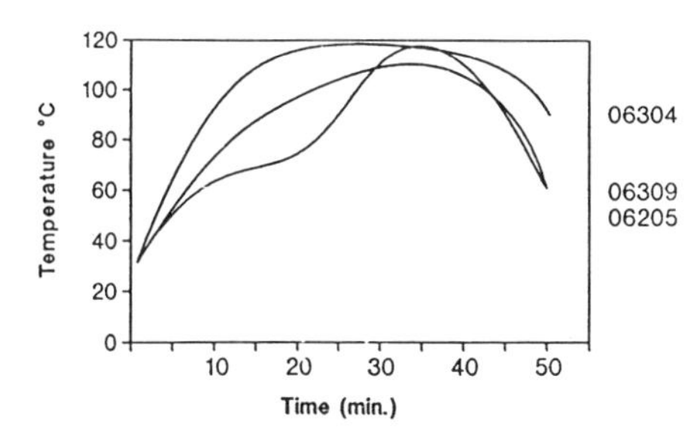

Viscosity determination. Brookfield viscosity values are shown in fig. 10. The most lightly crossbonded starches SF 06309 and SF 06205 gave slightly gelled sauces, whereas the strongly crossbonded waxy maize starch adipate SF 06203 gave too thin a sauce. The three other starches were suitable and gave the desired sauce texture.

Fig. 10: Brookfield viscosity

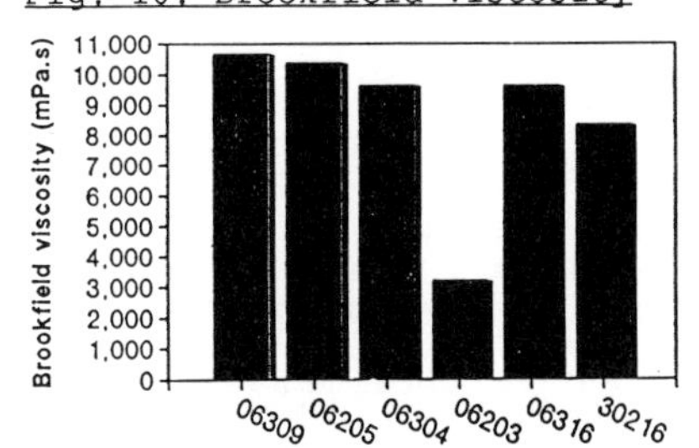

Fig. 11: Colour formation

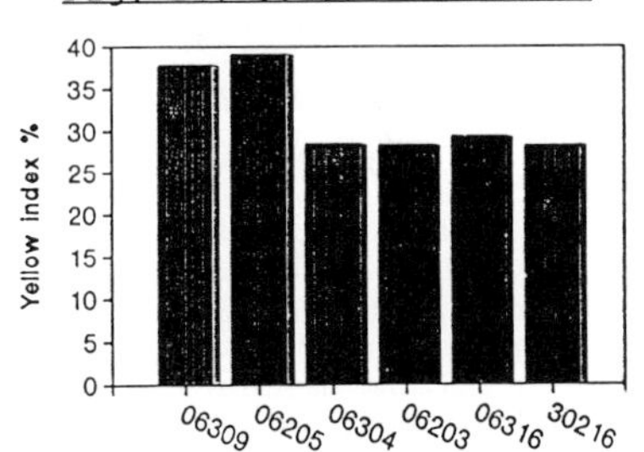

Microscopic examination. All starches were fully gelatinised. The extent of swelling is higher for SF 06309 and SF 06205 compared to the other modified starches. Only SF 06309 showed a slight disintegration of swollen granules.

Colour formation. Colour formation in the white sauce was checked using the Hunter colorimeter (fig. 11). It is obvious that the starches allowing a fast heat penetration give rise to less colour development.

Using the rotary canning process, the modified starches SF 06304, SF 30216 and SF 06316 are preferred; they give white sauces with the desired viscosity texture and colour, ensuring sufficiently fast heat penetration.

Scraped-surface heat exchanger (SSHE)

Viscosity determination. Brookfield viscosity values are shown in fig. 12. Compared to the canning system, more starch degradation occurs as is shown by the lower Brookfield viscosity values for the slightly crossbonded acetylated waxy maize adipates.

Microscopic examination. Again all starches were fully gelatinised. Less swelling is noticed for SF 06203. As for canning SF 06309 shows a slight disintegration of the starch granules.

Colour formation. From fig. 13 can clearly be seen that colour formation is not so pronounced as for the rotary canning system.

Fig. 12: Brookfield viscosity

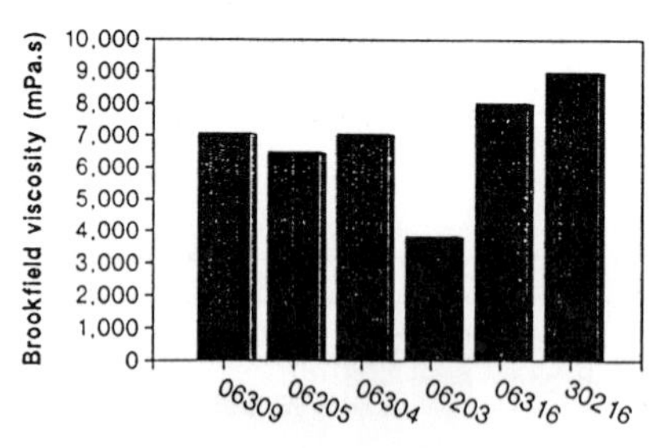

Fig. 13: Colour formation

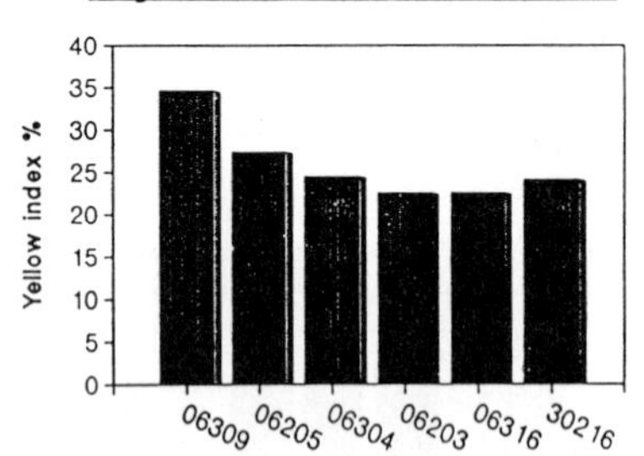

Mechanical degradation of the particulates. The mechanical degradation of the particulates (peas) as a result of the heat treatment in the SSHE is expressed as the percentage of intact peas remaining on a 4.7 mm sieve (table 4).

A minimum mechanical degradation is obtained with both hydroxypropylated crossbonded starches SF 06316 and SF 30216 as well as for SF 06304.

Texture of the particulates. The texture of the peas is measured with the JJ Lloyd Dynamometer expressed in Newtons (table 5). Hydroxypropylated crossbonded starches as well as the medium crossbonded waxy maize starch adipate are recommended as evidenced by the higher texture values.

Table 4: Mechanical degradation of the particulates

SF 06309	75%
SF 06205	79%
SF 06304	84%
SF 06203	72%
SF 06316	91%
SF 30216	93%

Table 5: Texture of the particulates

SF 06309	103 N
SF 06205	115 N
SF 06304	143 N
SF 06203	78 N
SF 06316	135 N
SF 30216	133 N

Ideally, for a SSHE system, the viscosity during heating in the first cylinder should not be too high in order to allow a fast heat penetration and to have less risk of burning, fouling and consequent colour formation. Viscosity build-up should occur only at the cooling step. With respect to the Brabender profiles of the various starches, only SF 30216 and SF 06316 show low initial viscosity and sufficiently high set-back viscosity to obtain a sauce of good consistency. In addition these starches are stable to severe heat and shear conditions. Low crossbonded starches such as SF 06309 and to a lesser degree SF 06205 will develop unwanted high viscosity values directly in the heating cylinder. In this case some of the swollen starch granules will disintegrate during further shearing and holding at high temperatures, but enough viscosity will be retained due to the inherent stability of these starches as a result of their chemical modification.

Tubular heat exchanger

Process parameters for all trials were maintained constant i.e. heating, holding and cooling parameters were always identical. Application of 54 s at 130°C has a different effect on the gelatinisation pattern and subsequently on the viscosity development of the starches. Table 6 shows the calculated F_0 values.

Table 6

Calculated F_0 values

SF 06309	8.1
SF 06205	9.6
SF 06304	7.6
SF 06203	8.1
SF 06316	8.0
SF 30216	8.0

Viscosity determination. Fig. 14 demonstrates the obtained Brookfield viscosity values. Higher viscosity values compared to the scraped-surface heat exchanger are obtained when using the slightly crossbonded waxy maize starch adipates SF 06309 and SF 06205. Again the strongly crossbonded waxy maize starch SF 06203 cannot fully develop viscosity. The hydroxypropylated crossbonded tapioca starch SF 30216 also gives too low a viscosity because the gentle heat and shear treatment prohibits full gelatinisation.

Fig. 14: Brookfield viscosity

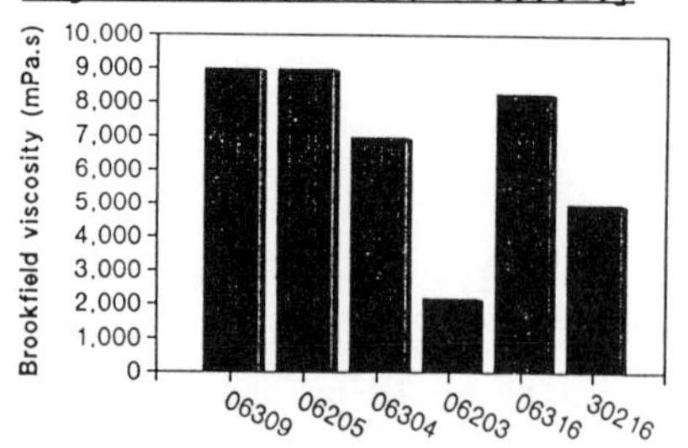

Fig. 15: Colour formation

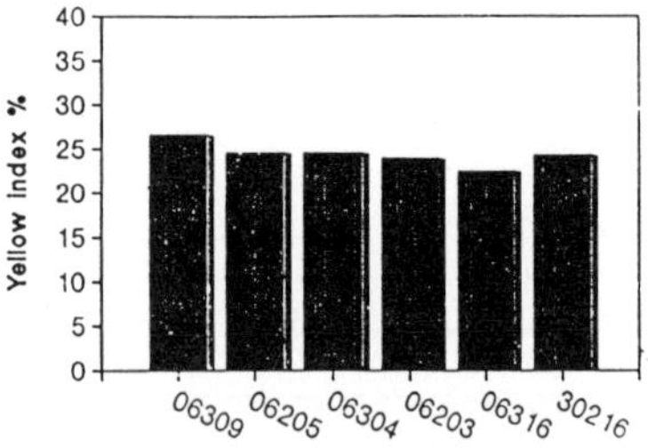

Microscopic examination. With the exception of the trial with SF 06203, all trials show a comparable gelatinisation pattern of the starches i.e. intact swollen starch granules.

Colour formation. Fig. 15 shows the colour formation as measured with the Hunter colorimeter. Generally, much whiter sauces are obtained compared to canning and to some extent to scraped-surface heat treatment.

Compared to the conventional canning system, the tubular heating process is less detrimental to starch viscosity and colour development. This is mainly due to the high heat exchange surface of the tubular heating system combined with the relatively small product content resulting in very short heating and cooling times for the product. With exception of SF 06203, all starches are suitable.

In conclusion, we can state that with respect to quality of a white sauce containing particulates both the scraped-surface and the tubular heat exchanger have advantages compared to the canning system. It is also shown that the tubular heating system performs better than the scraped-surface heat exchanger in terms of quality of the end product. Indeed, a starch with lower crossbonding level can be used with a tubular heating system and at the same time the colour formation in the white sauce is reduced. Finally, it is clear that hydroxypropylated crossbonded starches as well as medium crossbonded waxy maize starch adipate show overall good performance which can be attributed to the thin-thick profile of these starches during heating and cooling.

REFERENCES

1. Cerestar Technical Information on Modified Food Starches.
2. W.F. Hermans, (1989) Voedingsmiddelentechnologie, **7**, 27-31.
3. P.J. Skudder, (1988) C.E.R.I.A. Symposium on Progress in Food Preservation Processes held on April 12-14, 1988 at Brussels, Belgium.

The role and function of starches in microwaveable food formulation

P.COWBURN
National Starch & Chemical, Ashburton Road East, Trafford Park, Manchester M17 1BJ, UK

ABSTRACT

As microwave heating and cooking continues to gain popularity, the microwave food technologist must understand how this unique means of rapidly generating thermal energy can effect starch performance. Likewise, it is essential to become familiar with the many different types of starches that are available, their unique properties and how they can be used to develop food systems ideally suited for the microwave oven. This paper will briefly cover starch structure, solution properties and the importance of modifying starches, followed by the effects of microwave cooking on starch performance, compared to conventional cooking. The paper also offers an insight into the role starches can play in microwave food formulation, particularly frozen foods, dry mix foods and baked goods. The application of new cold water swelling starches will be highlighted.

INTRODUCTION

Starch plays a major role in influencing the overall quality of many of today's foods. Depending upon the type of food, starch can be used to provide different textures, thicken, gel, opacify, retain moisture, inhibit moisture, extend shelf life, form soft coatings or crisp coatings, control expansion, stabilise emulsions or to provide oil resistant films. This demonstrates the truly multifunctional nature of starch as an ingredient in the food industry and the possibilities that abound for the creative food technologist.

As microwave heating and cooking continues to gain popularity and establishes itself as a necessity in today's convenience-orientated culture, the microwave food technologist needs to understand how this unique means of rapidly generating thermal energy can effect starch performance. Likewise, it is essential to become familiar with the many different types of starches that are available, their unique properties and how they can be used to develop food systems ideally suited for the microwave oven.

Recent developments in starch technology have expanded the capabilities of this flexible food ingredient, offering higher levels of performance, while at the same time offering increased ease of use, particularly in the formulation of frozen, shelf stable and dry mix foods for microwave reconstitution.

To understand the benefits of these new ingredients, it is necessary to review some of the general properties of starch and its behaviour.

STARCH PROPERTIES

Starch, a complex carbohydrate, is one of the most prevalent functional ingredients found in foods. The most common sources of food starch are corn, potato, wheat, tapioca and rice. In food systems, most of the desirable properties that can be obtained from starch result from the changes which starch undergoes during and after cooking. Thermal energy is required to weaken hydrogen bonds, swell starch granules and solubilise the functionally different molecular components - amylose and amylopectin.

The building block of all starch is D - glucose. Through enzymatic condensation glucose is polymerised to form two different homopolymers, amylose and amylopectin. Amylose is a linear chained molecule linked together by α, 1-4 glucosidic bonding.

The length of amylose varies with plant source, but in general, the degree of polymerisation will run between 500 and 2,000 glucose units. Amylopectin is also a D - glucose homopolymer, however its structural linkages include 1,6 linkages or branch points in addition to 1,4 bonding. This branching effect results in a molecule which is much more massive in size than amylose and, accordingly, has a much greater water holding capacity. (Figure 1)

FIGURE 1

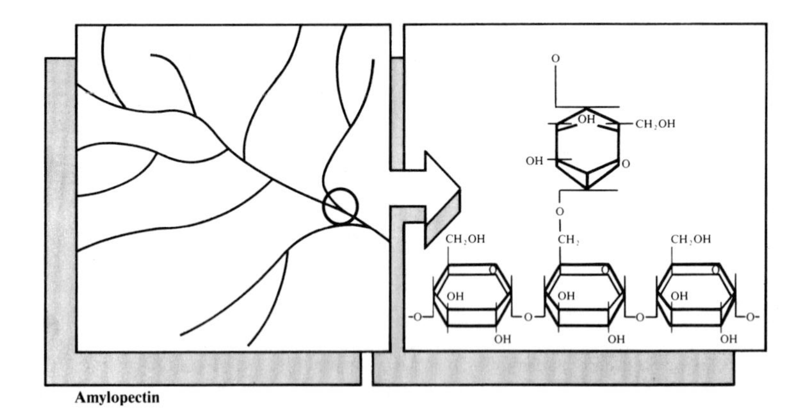

AMYLOPECTIN

All starches are made up of one or both of these molecules but the ratio of one to the other will vary with the starch source. As an example, corn starch has approximately 28% amylose, genetically manipulated high amylose corn starch can contain as much as70% amylose, whereas genetically modified waxy corn contains no amylose.

As plants produce starch molecules, they deposit them in the form of a tightly packed granule. Wherever possible, adjacent amylose molecules and outer branches of amylopectin associate through hydrogen bonding in a parallel fashion to give radially oriented, crystalline bundles know as micelles. These micelles hold the granule together to permit swelling only in heated water without the complete disruption and solubilisation of the individual starch molecules. The gelatinisation temperature of starch is greatly influenced by the binding forces within the granule which vary with granule size, ratio of amylose to amylopectin and species.

Typically, the higher the percentage of amylose and the smaller the granule, the higher the gelatinisation temperature.

When the starch granule is heated in water, the hydrogen bonds are ruptured and the granule swells to many times its original size with progressive hydration. As the granules expand more water is imbibed, clarity is improved, more space is occupied, movement is restricted and viscosity increased.

With the swelling of amylose - containing granules such as maize, the linear amylose molecules are solubilised and leach out into the solution. The molecules, due to their linearity, will then reassociate into aggregates and precipitate at low concentrations or set to a gel at higher starch concentrations and viscosities. (Figure 2)

FIGURE 2

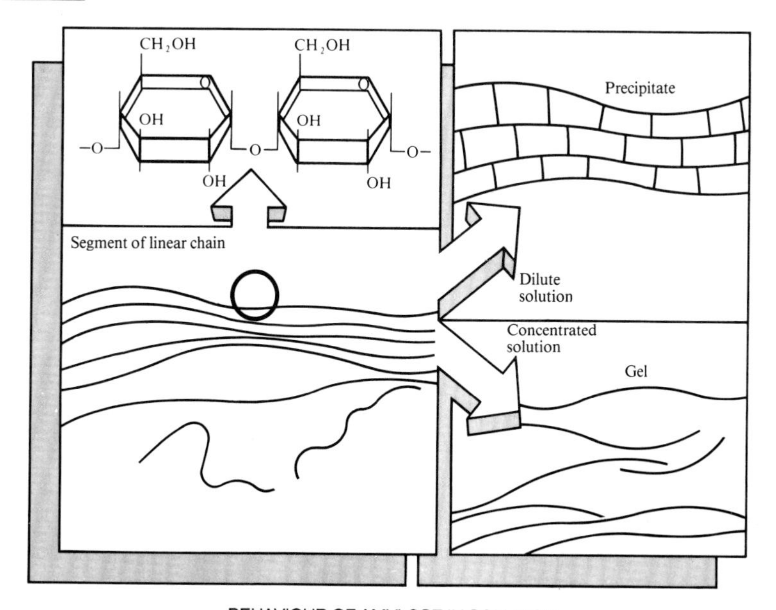

BEHAVIOUR OF AMYLOSE IN SOLUTION

This is referred to as "set back" or retrogradation. The tightly reassociated amylose will tend to expel water resulting in opacity and syneresis. Amylopectin, on the other hand, will tend to bond a larger number of water molecules without setting back, yielding a paste that remains flowable and clear.

THE IMPORTANCE OF MODIFYING STARCHES

In unmodified form, starches have a limited use in the food industry. Waxy maize, a widely used starch, serves as a good example. The unmodified granules hydrate easily, swell rapidly, rupture, lose viscosity and produce weak bodied very stringy and cohesive pastes.

This breakdown in quality is accelerated by high heat, acid, and physical attrition or shear. Since most food systems are thermally processed at high temperatures, and organic acids are common to many food products, today's starches must be tolerant and versatile.

Many food starches therefore are modified by a cross-linking treatment that strengthens the rather labile granules so that their cooked pastes are more viscous and heavy-bodied and are less prone to breakdown with extended or intense cooking, increased acid and severe agitation.

The cross-linking treatment can be varied to develop starches ideally suited for different food systems and different means of production. (Figure 3)

FIGURE 3

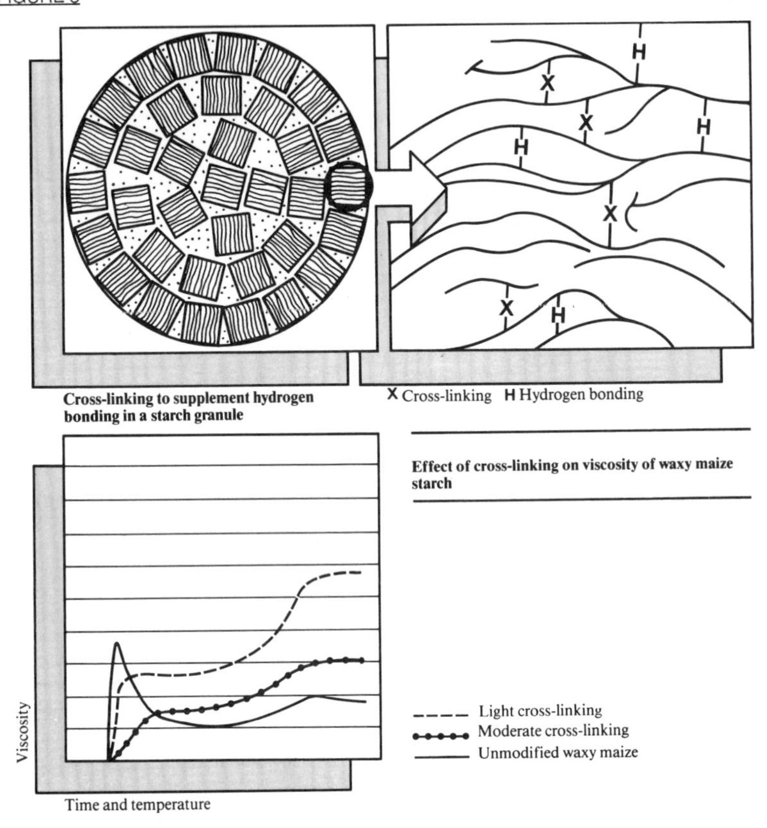

Cross-linking to supplement hydrogen bonding in a starch granule

Effect of cross-linking on viscosity of waxy maize starch

Another limitation inherent with native or unmodified starches or flours is their lack of freeze-thaw stability. Under low temperature or freezing conditions, unmodified starch pastes will become cloudy, chunky and weep. This is caused by reassociation through hydrogen bonding of amylose and the outer branches of amylopectin.

To prevent the occurrence of this condition, monofunctional blocking groups are attached throughout the starch molecules through chemical derivatisation. (Figure 4) The result is a stabilised starch which will produce pastes that will withstand numerous freeze-thaw cycles before the onset of syneresis occurs.

FIGURE 4

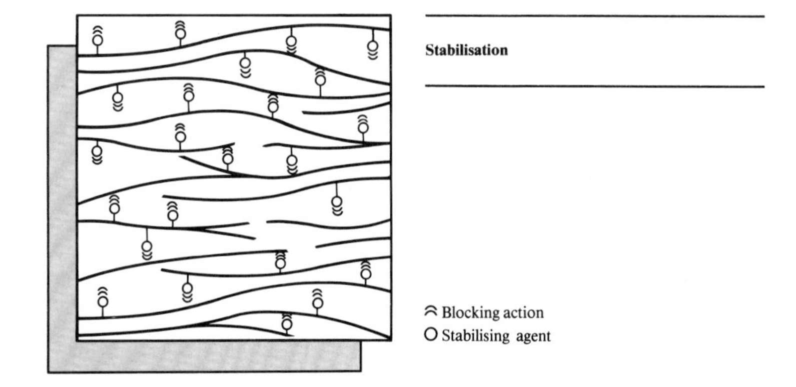

Although cross-linking and stabilisation are two of the most important types of chemical modification for food starches, other types of modified starches are used in certain types of foods. These include converted starches such as thin-boiling starches, dextrins, and oxidised starches.

INSTANT STARCHES

The increased amount of convenience foods that are being manufactured today require ease of preparation not only for the consumer but also for the prepared food manufacturer. As convenience foods continue to grow and parallel the popularity of microwave cooking, the utilisation of instant or precooked starches will also expand.

Instant starches are known for their cold water dispersibility and instant viscosity development. Instant starches are available to yield various textures and viscosities. Their dispersion properties and rate of hydration are normally controlled by altering granulation. Traditional finely ground instant starches typically will hydrate very quickly but also exhibit a strong tendency to lump. These finely ground products although imparting a relatively smooth texture, will normally not be very heat tolerant and will breakdown in texture upon rapid heating.

On the other hand, traditional coarsely ground pregelatinised starches, will disperse better, hydrate more slowly, but also impart an undesirable surface graininess to the food product.

Clearly, the inherent problems with the traditional pregelatinised starches that have been available to the food technologist for years have resulted in prepared food manufacturers having the consumer sacrifice quality for convenience.

However, due to the most recent innovations in instant starch technology, this is no longer required. New developments have led to ease of use of high performance cold water swelling starches which combine the feature of convenience oriented instant products with all of the quality of cook-up preparations. This significant breakthrough in starch technology has led to new unmodified and modified high performance cold water swelling starches which are ideally suited to convenience foods specifically prepared for microwave reconstitution such as instantly prepared and frozen entrees, dinners and pot pies as well as dry mixes. More details on how these new, never before available starches function in different microwave food applications will follow.

MICROWAVE COOKING AND ITS EFFECTS ON STARCH FUNCTIONALITY

Microwave cooking is based on the ability of microwaves to interact with the components of a food product and generate heat energy. The amount of interaction and subsequent heat that is generated is related to the composition of the food and the specific heat of each ingredient. Food molecules which carry a dipolar electrical charge will vibrate as they align themselves with the rapidly fluctuating electric field. This causes heat of friction within the molecules and corresponding heat radiation.

When irregularly shaped foods are heated with microwaves, non-uniform temperature distributions of hot and cold spots are sometimes formed. Also, shape can result in a heating effect common in microwaving, and impossible with conventional cooking - "centre heating". Concentrated centre heating complicates cooking. It makes cooking uniformly difficult, as it is difficult to judge correct serving temperatures as food surfaces may be far cooler than heated centres.

Since water carries a dipolar charge, foods containing high contents of water will generate a great deal of heat due to the reactiveness of water in a microwave.

Fats on the other hand, exhibit little electrical polarity and therefore, will not exhibit much dielectric heat radiation. However fats have a much lower specific heat when compared to water, therefore, they will heat considerably faster than water.

Microwave cooking is also affected by a food product's physical state, density, size, shape and thickness.

Ice is highly transparent to microwaves. Therefore, there is a great difference in the reactivity between ice and water in the microwave. This can result in uneven "runaway" heating once melting has begun.

Since conventional convection cooking heats the exterior of a food product first, starch gelatinisation will proceed from the periphery to the centre. Once gelatinisation begins on the surface, conductive and convective heat transfer is responsible for the cooking that takes place through to the centre.

For most convection heated foods containing starch (ranging from low to high moisture contents), the quality of starch gelatinisation is very uniform with a narrow range of degree of granular swell existing.

Microwave cooking on the other hand, is known for its non-uniform heating which normally results in a broad range of granular swelling which can produce undesirable phase differences and separations within a food product. This broad range of microstructural characteristics decreases at high water contents. However, in intermediate to low moisture foods such as some baked goods, dramatically different patterns of starch transformation can exist, which normally are responsible for poor textures such as toughening and cracking.

It is difficult for crispness to result from microwave cooking, even though a starch film could form on the surface. The main problem is that dehydration and subsequent retrogradation cannot occur at the surface since the water within the product is continually being converted to steam and migrating out, causing evaporative cooling and condensation at the surface.

The ability of ingredients to distribute heat in a microwaveable food system will depend on both their specific heat and their viscosity. Increasing the viscosity of the food system with starch, for example, will decrease the mobility of water, resulting in conductive heating taking place in addition to internal microwave heat generation. Therefore starch thickened pastes will tend to increase in temperature more quickly than an equivalent amount of water in a microwave. In addition, the viscosity that the starch contributes through efficient hydrogen bonding with water enables moisture to be held in the food system, moderating steam migration and rapid boiling.

When starch imbibes water during gelatinisation, increased heating rates will occur in that location, due to the interaction of microwave radiation with water. Therefore phase transitions can continue to advance. Because more intense heat results from microwave energy, a starch's tolerance to textural breakdown is narrowed. Therefore it is imperative that the proper starch selection be made when formulating a good product for the microwave oven. Fortunately, a variety of differently cross-linked products exist, which can tolerate the localised heating that can last during a microwave food's reconstitution.

THE ROLE STARCHES CAN PLAY IN MICROWAVE FOOD FORMULATION

Starch is already an important ingredient in many of today's microwaveable foods. It can be used as a thickener in the gravies, sauces and fillings that can be found in either frozen, refrigerated or shelf-stable microwaveable entrees, dinners or pot pies. It is often the primary thickening agent in microwave reconstituted dry mixes, contributing mouthfeel and providing uniform suspension of ingredients and uniform heating. In addition to providing viscosity in these applications, the proper starch can also be used to impart the desired texture, e.g. smooth, short, pulpy, long.

FROZEN FOODS

The frozen foods segment of the food industry has enjoyed tremendous growth as a result of the convenience which microwave cooking offers. More and more frozen foods are being formulated and packaged for both the microwave and conventional oven, or entirely for microwave preparation.

Firstly, whether the food item is being adapted from a present formulation, or is newly developed, it must be able to withstand the rigours of freezer storage.

Ingredients including conventional thickening agents such as flour or unmodified corn starch are not freeze-thaw stable. Also these native thickeners generally will not be heat, acid or shear stable enough to withstand processing prior to freezing. Formulation with correct cross-linked and stabilised starch is therefore essential.

Many prepared food manufacturers have opted to produce their gravies, sauces, soups and fillings with starches that require cooking prior to freezing. Some manufacturers today want to simplify this procedure and eliminate the need for cooking altogether. This has been possible with traditional pregelatinised starches, however a sacrifice in quality (texture, appearance, viscosity, stability) was always made.

New High Performance Cold Water Swelling Starches have now been developed which for the first time combine cook-up quality with instant convenience. Some of these starches are all-natural, while others would be labelled as "modified starch". It is the modified products which would be recommended for most instantly prepared frozen microwave reconstituted foods.

NEW COLD WATER SWELLING STARCHES

1. Unique Viscosity Profiles.
2. Viscosity Stability.
3. Microwave Tolerance.
4. Excellent Freeze-Thaw Stability.
5. Outstanding Clarity, Sheen and Smoothness.
6. Flavour Advantages.

These new starches contribute all of the qualities necessary for a frozen microwaveable food by offering some distinct advantages over conventional or traditional pregelatinised starches. They will instantly thicken, impart a smooth desirable texture, have excellent freeze-thaw stability and withstand the intense heating that can result during microwave reconstitution. The new high performance cold water swelling starches impart a rapid high cold paste viscosity which is necessary to bind ingredients homogeneously and dramatically reduce the preparation time that would normally be lengthy (due to delayed hydration) with traditional pregels. These products also offer better freeze-thaw stability than conventional pregels together with the microwave heating tolerance of some cook-up starches.

Most conventional pregelatinised starches have little, if any, granular integrity. This broken down structure normally results in grainy textures and poor heat and textural stability. The opened granular integrity of the high performance cold water swelling starches enables them to function like both instant and cook-up starches. (Figure 5)

FIGURE 5

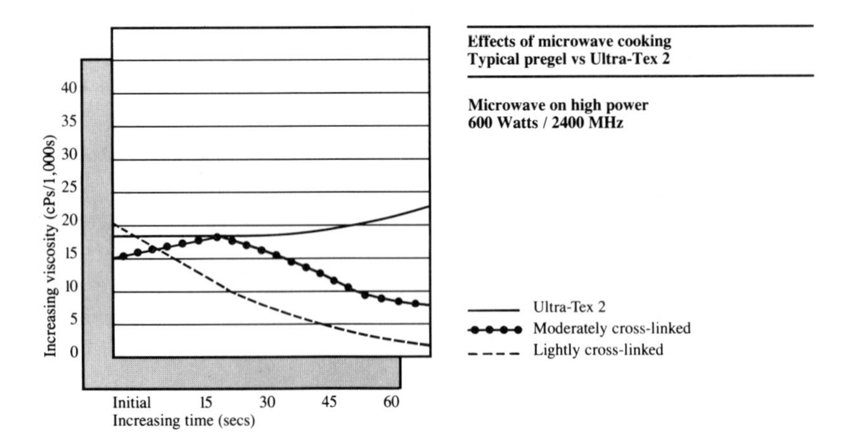

The viscosity that starch imparts to frozen food products allows for a more uniform thawing to take place, also, the mobility of water is reduced preventing steam from escaping. Therefore, by formulating with the correct starch the harsh effects of run-away heating can be reduced.

Although traditional pregelatinised starches have been shown to have their limitations in frozen microwaveable products, some unique instant starches can be used for their distinct texturising properties. The granulation of some specialised modified pregelatinised starches can be altered to impart desirable pulpiness or graininess as in the case of a tomato based or fruit based products.

DRY MIX FOODS

Dry mix foods are certainly orientated items that can be quickly prepared in a single serving size so they are thus logical candidates for microwave oven preparation.

Cook-up mixes depend on cook-up starches and/or flour to build viscosity and thicken the mix. However, viscosity is not obtained until the food reaches approximately 75°C. This can normally take up to two minutes depending on the size of the mix and the microwave power setting. This lag time in viscosity development results in ingredient settling and stratification. Particulate matter, including uncooked starch or flour will settle to the bottom of the container resulting in a heavy gelled precipitate with a thin liquid forming on the top. The consumer could improve upon this by periodically removing the food item and stirring it to ensure homogeneity. This of course, reduces the convenience aspect of the mix.

A pregelatinised starch can provide instant viscosity and help to keep the ingredients in suspension. However, traditional fine grind pregels can be very difficult to disperse and may not exhibit the viscosity and textural stability required during microwaving. Use of this type of starch normally results in scattered lumps (fish eyes) of undispersed starch and a broken down low viscosity.

More coarsely ground pregels will disperse better than fine grinds, however, most of these will impart a grainy appearance to the food product even after microwave heating.

Use of the proper combination of a pregelatinised and cook-up starch can be the solution to the problems with microwave prepared dry mixes. A small amount of the proper pregelatinised starch can be combined with the proper cook-up starch to provide enough instant viscosity to keep the other ingredients in suspension until a high enough temperature is reached to swell the cook-up starch.

The cook-up starch will provide the short, smooth texture desired. While this is certainly a viable means to formulate a high quality microwaveable dry mix, it can take time formulating the correct level and ratio of pregel to cook-up starches.

Another approach to overcome the problems is to use the proper high performance cold water swelling starch. These new products are ideally suited to microwave prepared dry mix foods. (Figure 6)

FIGURE 6

Dry mix cheese sauce for microwave reconstitution using a new high performance CWS starch

	%
Ultra-Tex 2®	**24.75**
N-Zorbit®	**24.75**
Cheese powder	**30.20**
Vegetable oil	**16.50**
Titanium dioxide	**0.83**
Salt	**2.77**
Colour	**0.20**

	100%

Dry mix preparation

1. **Blend Ultra-Tex 2 with the oil.**
2. **Blend in the N-Zorbit.**
3. **Blend in the remaining ingredients.**

Preparation for reconstitution

1. **Add 59.2 grams dry mix to 1 cup of cold water.**
2. **Stir until completely wet and uniform.**
3. **Microwave for 2 minutes.**
4. **Stir and microwave for an additional 2 minutes.**
5. **Stir and use.**

Cold water swelling starches are available that will disperse remarkably into solution without lumping, with a minimum of other dry ingredients required to facilitate dispersion. These starches will develop instant viscosity, uniformly suspending all of the ingredients throughout the mix.

Due to their unique swollen granular structure they will impart an excellent smooth, short texture and provide outstanding microwave heating tolerance.

BAKED GOODS

By far the most difficult challenge that lies ahead for the microwave food technologist is to learn how to produce high quality bakery products in a microwave. Since starch (wheat starch) is the primary constituent of most baked goods, a thorough understanding of the interactions that take place between microwaves and starch in a baked product is necessary.

Advances are being made using special starches to overcome rapid staling and toughening in bread products. By selecting starches with high amylopectin content and low gelatinisation temperature (~52°C) staling can be controlled and water retention improved after rapid heating in the microwave oven.

Microwaveable cakes are continuing to grow in popularity. Microwave cake formulations generally require a reduction in water content to improve cake volume. Also increased batter viscosity will impart a tender, more uniform cell structure, yielding a moist high quality crumb by reducing batter flow which is known to be greater during microwave heating. This can lead to non-uniformity in cake structure.

Selection of the correct finely-ground cross-linked and stabilised waxy maize pregelatinised starch will provide the viscosity and hydration performance required to fulfil the above requirements.

In microwaveable filled pastry products, particularly low pH and frozen products such as fruit pies, the new series of cross-linked and stabilised cold water swelling starches offer several advantages. Water retention in the filling is essential to prevent the crust softening after microwave heating. Freeze-thaw stability, pH resistance and heat stability are all essential requirements for the starch used. Also "boil-out" can be avoided by correct starch selection.

LOOKING AHEAD

Because of the multitude of different starch products that exist and their tremendously diverse functional properties, there is no doubt that microwave product development technologies will capitalise on the virtues of starch in the near future, resulting in exciting new microwaveable foods.

It has been shown that by selecting the proper starch, you can provide both the desired texture and viscosity needed in microwaveable frozen, refrigerated, shelf stable and dry mix foods.

New high performance cold water swelling starches are now available which are ideally suited for many microwaveable foods. Microwave developments and breakthroughs in the years to come will, in all likelihood, involve innovative use of starch technology. These will include:

Dramatic improvements in microwaveable baked goods quality.
Development of microwaveable snacks.
Improvements in microwaveable batters and breadings.

Molecular structures and chain length effects in amylose gelation

MICHAEL J.GIDLEY

Unilever Research Laboratory, Colworth House, Sharnbrook, Bedford MK44 1LQ, UK

ABSTRACT

Notable advances have been made recently in the molecular characterisation of the starch polysaccharides, amylose and amylopectin. In order to relate observed molecular differences in amylose and amylopectin from different botanical sources with functional behaviour of importance to the food industry, it is necessary to develop an understanding of structure/function relationships in molecularly well-defined model systems. This report describes the effect of chain lengths up to 2800 glucose units on amylose aggregation and gelation behaviour. Very short (<80 units) chain lengths function as models for amylopectin branches whereas longer chains model amylose behaviour. All observations of aggregation behaviour (physical state, aggregation rate, rheology) can be explained on the basis of inter-chain double helix formation over chain segments of less than 100 units. Precipitates of amylose aggregates are essentially fully double-helical, and amylose gels are found to contain rigid double-helical junction zones interconnected by more mobile amorphous single chain segments.

INTRODUCTION

Considerable progress has been made recently in the molecular characterisation of the starch polysaccharides amylose and amylopectin from various botanical sources, most notably by Hizukuri and co-workers (1-3). One of the challenges posed by such detailed characterisation is to relate features of molecular structure to functional properties (both beneficial and adverse) of practical importance in the food industry and elsewhere.

Amylopectin (3) is thought to have a periodic clustered arrangement of branch points within its highly branched structure, with the major botanical variable being the distribution of unit branch lengths. This variability can reflect both the periodic distance between clusters of branch points, and (probably more importantly) the number of clusters through which a single unit branch passes (3). Some aspects of the relationships between molecular structure and functional properties in amylopectin can be assessed through studies of pure (1→4)-α-D-glucan oligomers having chain lengths related to amylopectin branch lengths (typically 10-100 glucose units). This is possible at the lower end of this range using oligomers purified by gel permeation chromatography (4,5): independent studies by two groups have shown that the minimum chain length required for crystallisation (retrogradation) is 10 units (4,5) and that glucan chains of 10, 11 and 12 units crystallise into the A-type (6) polymorph whereas chains of 13 units or longer crystallise into the B-type (7) polymorph (4,5). These results provide a rationalisation for the crystallinity of amylopectin in the starch granule, its slow retrogradation once gelatinised, and the previously observed (8) correlation between average amylopectin branch length and the crystalline polymorph adopted in native starch granules. This latter correlation can be explained in terms of known entropic effects in the crystallisation of chain-like molecules (9).

Amylose molecular structures are now thought to involve a mixture of linear and lightly branched (1→4)-α-D-glucans (1,2) with major botanical variables being molecular weight, polydispersity, branching frequency and branch lengths. In order to assess the relationships between molecular structure and functional properties in amylose, the branching heterogeneity and molecular size polydispersity of natural amyloses make them unsuitable subjects for study. However, strictly linear and nearly monodisperse amyloses can be synthesized *in vitro* (10) using phosphorylase enzyme (usually from *potato*). This methodology can be used to prepare amyloses of chain lengths up to several thousand units, and has been used to study the effect of chain length on iodine/iodide complexation (11) and on initial aggregation rates in dilute aqueous solution (11). We have extended these studies by examining the effect of chain length on the aggregation and gelation behaviour of phosphorylase-synthesized amyloses at concentrations (0.5-5.0% w/v) which are related to typical amylose concentrations in starch pastes and gels (12-14).

MATERIALS AND METHODS

Details of amylose synthesis, molecular characterisation, rheological and spectroscopic measurements are described elsewhere (12-14).

RESULTS AND DISCUSSION

1. Physical state of amylose aggregates

Following dissolution in deionised water (12), all amyloses show phase changes on subsequent storage at ambient temperature. For chain lengths up to at least 110 units, precipitation occurs at all practically accessible concentrations. For chain lengths of 250 units and higher, gelation occurs for amylose concentrations greater than ca. 1% w/v. Below this concentration, a precipitate/microgel structure is found for chain lengths of ca. 250, 300, 400 and 660 units whereas a two phase gel + solvent regime is found for 1100 unit chain lengths and higher. These phase properties are represented in Figure 1.

Taken as a whole, these observations suggest two types of amylose aggregation: 1. alignment of short chains into sufficiently large aggregates that precipitation occurs, and 2. cross-linking of longer chains through alignment of chain segments with each other leading to gelation. For many chain lengths (ca. 100 to 1000 units) it is apparent that both processes occur (Figure 1). Indeed, results discussed below suggest that chain alignment without cross-linking probably occurs for only the shortest chain studied (40 units) and that cross-linking without extensive chain alignment only occurs for the longest chains studied (>2500 units).

The molecular form of precipitated amyloses is found to be the B-type crystalline polymorph(7), although the degree of crystallinity decreases markedly with increasing chain length (14). Solid state NMR measurements (Figure 2) show that precipitated amyloses have the same spectral features as B-type amylose with no significant amorphous character, although significant reductions are observed in spectral linewidths with increasing chain length. This shows that the same molecular structures (left handed double helices (7)) are formed in all cases but

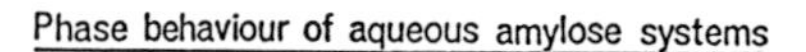

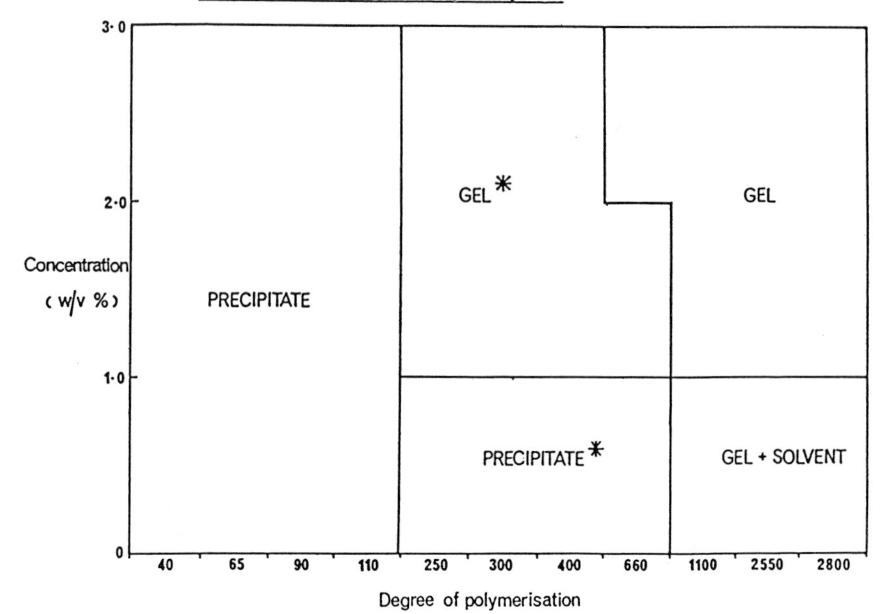

Figure 1. Observed phases on storage of amylose solutions as a function of chain length. GEL* denotes a gelled system with precipitated inclusions and PRECIPITATE* denotes a precipitated system with some microgel inclusions.

that helix packing is more perfect for shorter chain lengths (14,15). This may be interpreted as being due to increasing (but low) levels of cross-linking for chains of 90 and 250 units compared with 40 unit chains. As discussed later and elsewhere (14), amylose gels also contain B-type double helical structures together with interconnecting amorphous chain segments. The same molecular structures are therefore involved in the chain alignment and cross-linking modes of aggregation with the balance between the two determined presumably by thermodynamic factors.

2. Chain length effects on aggregation kinetics

In a pioneering study, Pfannemüller and co-workers (11) showed that the initial rate of aggregation of phosphorylase-synthesized amyloses in dilute (0.1% w/v) solution was maximal for chains of ca. 80 units. In order to extend this work, we have studied

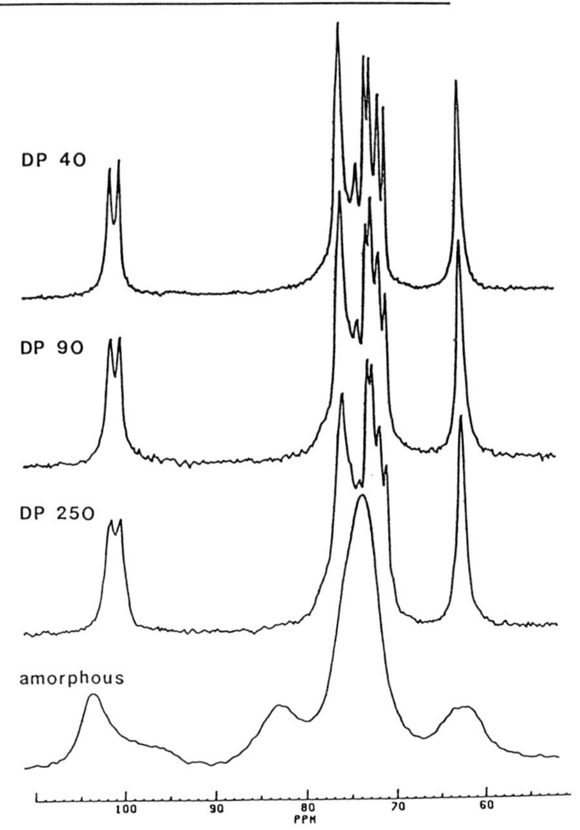

Figure 2. High resolution solid state ^{13}C NMR spectra of amylose precipitates with unit chain lengths (DP) as shown compared with the spectrum of amorphous α-(1→4) glucan. The spectrum of DP40 is essentially identical to that of the B-type polymorph.

aggregation kinetics at higher concentrations using both turbidimetric and rheological techniques. These techniques were chosen as probes of chain alignment/ lateral aggregation and chain cross-linking respectively.

The development of turbidity with storage time for 1.5% w/v amylose solutions of various chain lengths is

considerable practical and academic interest. Usually, it is not possible to utilise various molecular weight fractions in such studies due to the amounts of material required. The availability of well-defined phosphorylase-synthesized amyloses therefore makes them attractive subjects for such a study. For chain lengths of 1100 units or less, pseudo-equilibrium plateau values of G′, the overwhelming component of the shear modulus, are attained by 200 minutes. For a range of amylose chain lengths, values for G′ after 200 minutes storage at 25°C were obtained from data similar to that shown in Figure 5.

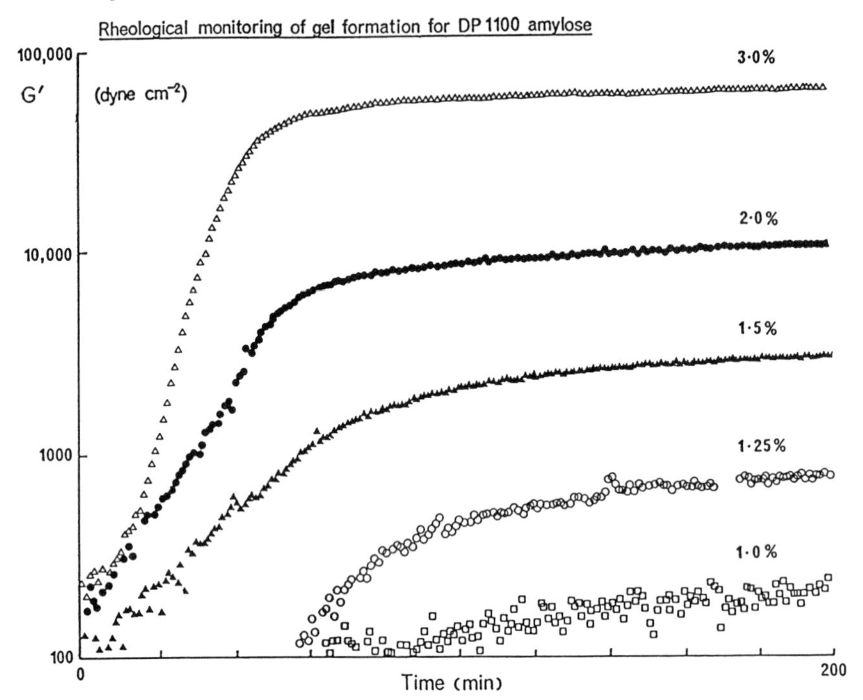

Figure 5. Storage modulus (G′) vs. time for various concentrations of amylose (1100 unit chain length).

Ellis and Ring (16) have recently reported that G′ for pea amylose varies with C^7. For three phosphorylase-synthesized amyloses, G′ apparently varies with $C^{3.9}$ (1100 unit chain), $C^{4.7}$ (660 unit chain) and $C^{4.4}$ (300 unit chain). These different values suggest that a simple power law expression is not sufficient to describe amylose modulus/concentration behaviour. The data obtained (Figure 6) show qualitatively similar behaviour to other gelling biopolymers (17) and can be analysed using the method of Clark and Ross-Murphy (17). Three parameters are involved in this analysis, one of which is the number of cross-linking sites per chain. Making the assumption that this is equal to the unit chain length divided by 100, the data can be fitted (Figure 6) with nearly constant values for the other two parameters (as would be predicted for a constant mechanism). Furthermore, the critical gelling concentrations for all three materials are predicted accurately. The various apparent power law expressions mentioned above are also explicable using this analysis (13).

The successful modelling of amylose gel modulus behaviour on the assumption of a cross-linking functionality of chain length/100 is in good accord with the deductions concerning the competition between chain alignment and cross-linking as a function of chain length described above.

4. Molecular mechanisms involved in amylose gelation

X-ray diffraction studies (eg. ref 18) have shown that amylose gels have a degree of B-type (7) crystallinity following storage. Little information is available however, concerning the non-crystalline component within amylose gels. Furthermore, the driving force behind the sol → gel conversion process is a subject of current discussion. Two limiting models have been proposed, which invoke either gelation as a result of phase separation in a cooled molecularly entangled solution (18), or double helix formation (19) as a driving foce. If gelation occurs as a result of cooling a molecularly-entangled solution (with precipitation occuring at lower concentrations), then there should be a relationship between the critical concentrations for gelation and for molecular entanglement. From measurements on pea amylose, Miles et al (18) suggested that this was indeed the case. Molecular entanglement can be assessed experimentally from viscosity measurements if specific viscosity is plotted on a log-log scale against concentration X intrinsic viscosity (20). Such a plot for various phosphorylase-synthesized amyloses in both water and dimethyl sulphoxide is shown in Figure 7. Two limiting slopes are observed, as for other coil-like polysaccharides (21) at low and high concentration corresponding to 'dilute' and 'semi-dilute' solution conditions (20). The cross--over from one type of behaviour to the other is considered to represent the onset of molecular entanglement. However, there is no sudden change but a smooth transition between the two states for amylose solutions (Figure 7), with a corresponding uncertainty

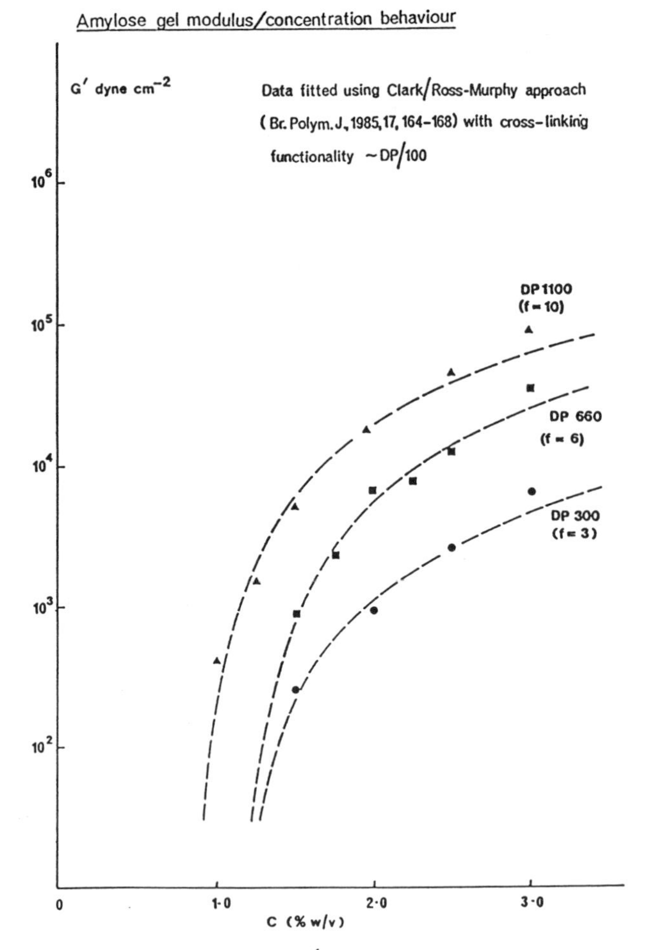

Figure 6. Gel modulus/concentration data fitted by the method of Clark and Ross-Murphy (17) for three amylose chain lengths.

in assigning a molecular entanglement minimum concentration. For practical purposes, two concentrations have been determined, which have been termed (Figure 7) the 'initial overlap' and 'coil overlap' concentrations respectively.

From a knowledge of intrinsic viscosity values (14), these critical concentrations can be calculated for a range of amylose chain lengths and compared with experimentally determined critical gel concentrations (Table 1). These results show a lack of any relationship between parameters of molecular entanglement and critical gelling behaviour. As found for others gelling biopolymers (20), gelation occurs at concentrations significantly below those required for molecular entanglement. The most likely driving force in amylose gelation is therefore the formation of left-handed double-helical structures of the B-type (7).

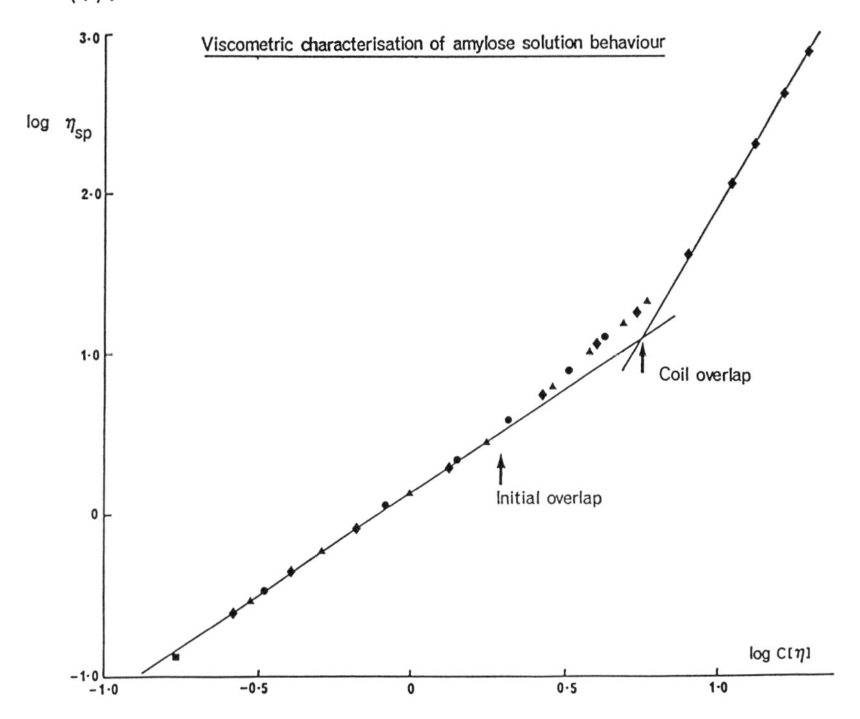

Figure 7. Log of 'zero-shear' specific viscosity vs. the product of concentration and intrinsic viscosity for amyloses of chain length 2800 (●) in water and chain lengths 2800 (◆), and 300 (▲) units in dimethyl sulphoxide.

A combination of NMR experiments, reported in detail elsewhere (14), show that amylose gels contain two distinct types of molecular structure:- 1. rigid B-type left-handed double helices and 2. more mobile amorphous (non-ordered) single chains. It is proposed that amylose gels contain cross-links (junction zones) due to inter-chain double helix formation over limited lengths of chain, with interconnecting single chain segments. For very long (>2000 units) amylose chains, the experimental evidence suggests that only limited aggregation of chains occurs around junction zones. With decreasing chain lengths, it appears that

Table 1 Comparison of critical gelling and overlap concentrations for aqueous amylose systems of various chain lengths.

unit chain length	initial overlap concentration	coil overlap concentration	critical gel concentration
300	7.2%	22%	1.1%
660	4.9%	15%	1.0%
1100	3.8%	11.6%	0.8%
2800	2.3%	7.0%	1.0%

increased aggregation of chain segments occurs until, for short chains (<~ 200 units) whole chain aggregation leads to precipitation predominating over cross-linking.

REFERENCES

1. Hizukuri, S., Takeda, Y., Yasuda, M. and Suzuki, A. (1981) Carbohydr. Res., 94, 205-213.
2. Takeda, T., Hizukuri, S., Juliano, B.O. (1989) Carbohydr. Res., 186, 163-166. and references cited therein.
3. Hizukuri, S. (1986) Carbohydr. Res., 147, 342-347.
4. Gidley, M.J. and Bulpin, P.V. (1987) Carbohydr. Res., 161, 291-300.
5. Pfannemüller, B. (1987) Int. J. Biol. Macromol., 9, 105-108.
6. Imberty, A. Chanzy, H., Perez, S., Buleon, A. and Tran, V. (1988) J. Mol. Biol., 201, 365-378.
7. Imberty, A. and Perez, S. (1988) Biopolymers, 27, 1205-1221.
8. Hizukuri, S. Kaneko, T. and Takeda, Y. (1983) Biochim. Biophys. Acta., 760, 188-191.
9. Gidley, M.J. (1987) Carbohydr. Res., 161, 301-304.
10. Pfannemüller, B. and Burchard, W. (1969) Makromol. Chem., 121, 1-17.
11. Pfannemüller, B., Mayerhöfer, H. and Schulz, R.C. (1971) Biopolymers, 10, 243-261.
12. Gidley, M.J. and Bulpin, P.V. (1989) Macromolecules, 22, 341-346.
13. Clark, A.H., Gidley, M.J., Richardson, R.K. and Ross-Murphy, S.B. (1989) Macromolecules, 22, 346-351.
14. Gidley, M..J. (1989) Macromolecules, 22, 351-358.
15. Gidley, M.J. and Bociek, S.M. (1985) J. Am. Chem.

Soc., 107, 7040-7044.
16. Ellis, H.S. and Ring, S.G. (1985) Carbohydr. Polym., 5, 201-213.
17. Clark. A.H. and Ross-Murphy, S.B. (1985) Br. Polym. J., 17, 164-168.
18. Miles, M.J., Morris, V.J. and Ring, S.G. (1985) Carbohydr. Res., 135, 257-269.
19. Sarko, A. and Wu, H.-C.H. (1978) Staerke, 30, 73-78.
20. Clark, A.H. and Ross-Murphy, S.B. (1987) Adv. Polym. Sci., 83, 57-192.
21. Morris, E.R., Cutler, A.N., Ross-Murphy, S.B., Rees, D.A. and Price, J. (1981) Carbohydr. Polym., 1, 5-21.

Uniaxial compressional testing of starch/hydrocolloid gels: effect of storage time and temperature

D.D.CHRISTIANSON and E.B.BAGLEY
USDA Northern Regional Research Center, 1815 N. University Ave, Peoria, IL 61604, USA

ABSTRACT
Corn and rice starch were gelatinized in water and in dilute solutions of xanthan, guar, carboxymethylcellulose (CMC) and carrageenan. After gelatinization the hot pastes were poured into ring-molds and held for 24 hours at room temperature. The rheological responses of the resultant gels were determined in lubricated uniaxial compression, and the data obtained were plotted as $f/2(\lambda-\lambda^{-2})$ verses λ^{-1}. λ is h/h_o, h being the sample height at time t and h_o the initial height. At low deformations the gels behaved as neo-Hookean materials with the plots being linear with zero slope. The intercept at $h/h_o = 1$ then gives a value which can be interpreted in terms of an apparent or effective cross-link density (CLD) for the neo-Hookean gel. At higher deformations extensive deviations from neo-Hookean responses were observed culminating in a maximum at which point fracture occurs. Detailed response depended both on the starch type and the hydrocolloid. Storage of the gel at room temperature had comparatively little effect on the gel behavior, contrary to that observed after storage at 4° C.

INTRODUCTION
Starch gels and pastes have been studied extensively to evaluate effects of added ingredients on gelatinization and gelation as well as the effect such ingredients have on stabilization of the gel during storage (1-6). Several common techniques such as X-ray diffraction, differential calorimetry (DSC) and dynamic mechanical analysis have been used to study specific effects such as retrogradation and staling of foods containing starch. A recent comparative study (7) of these techniques showed that wheat starch gels retrograded at a more rapid rate, as measured by X-ray and DSC techniques than when measured by mechanical procedures. These differences in

results suggest that in mechanical analysis only the intergrated macromolecular changes are observed whereas in other techniques specific molecular changes are more clearly defined.

Nevertheless, the physical and mechanical properties of foods in general and of food gels in particular are of value precisely because they measure the integrated effects of the system components. They measure the macroscopic properties that relate to such quality characteristics as texture, strength, rigidity, etc. Among the physical tests of interest are the uniaxial compression, simple shear and torsion tests. Earlier we have analyzed data from such experiments as if the materials were Mooney-Rivlin substances, while recognizing that this is an oversimplification. In the compression experiment, the magnitude of the force per unit cross-sectional area, measured in the unstrained state, is f. This is related for the Mooney-Rivlin material to the extent of compression, $\lambda = h/h_o$ (with h_o being the initial sample height) by

$$f/2(\lambda-\lambda^{-2}) = C_1 + \frac{C_2}{\lambda} \qquad \text{(Eq. 1)}$$

where C_1 is a quantity which, for vulcanized rubber, is proportional to the cross-link density. C_2 is a second parameter of uncertain physical meaning. While recognizing the limitations of the model when applied to starch gels it did prove convenient for correlating data. Simple shear and torsional data can also be treated by the appropriate modification of Eq.(1) for the same two parameters, C_1 and C_2 (8).

Gelatin behaves like a Mooney-Rivlin material in that not only do the three types of measurement give the same values of (C_1+C_2), but when C_1 and C_2 are evaluated separately, physically meaningful values of C_1 are obtained. It was found for gelatin that C_2 = zero, indicating neo-Hookean behavior. However, for starch gels, though (C_1+C_2) agreed among the measurements, the values of C_1 alone were found to be negative, a physically meaningless result. This was reported and the problem noted earlier (9).

One of the problems in applying Eq. (1) is computational. As λ^{-1} nears unity (small deformation) small errors in λ can be magnified as noted by Bagley et al. (2). Improved experimental technique and better procedures for determining h_o have allowed us to examine the initial portions of the plots of Eq. (1) for starch gels and it will be shown below that even starch gels are neo-Hookean at low deformations in compression, with $C_2 = 0$. This permits evaluation of C_1, which will be taken here as a measure of an apparent cross-link density (CLD) for these gels. The effect of added hydrocolloids on this apparent CLD will be examined. This method will also be used to investigate the effect of storage time and storage temperature on the CLD. In addition, the force/deformation data are obtained up to the gel fracture point.

MATERIALS AND METHODS

Materials

Starches and gums used in these experiments were obtained from commercial sources. Corn starch (34C1) was obtained from CPC Inc., Chicago, IL, and rice starch (No.S-7260) from Sigma Chemical Co., St. Louis, MO. Xanthan (No.G-1253) and kappa-carrageenan, Type III (No.C-1263) were obtained from Sigma Chemical

Co., St. Louis, MO.; cellulose gum (CMC) from Hercules, Inc., Wilmington, DE; and guar gum, Dycol HV4F, from National Starch and Chemical Corp., Bridgewater, NJ.

Gel Preparation and Rheological Testing

Starch was dispersed in distilled water and stirred for 15 min to wet the starch thoroughly before addition of the gum solution or water to obtain a dispersion of 10% solids. Gums were dispersed in water and gently stirred for 2 hours before mixing with starch dispersions. Final gum concentrations in the starch dispersions were 0.5% for guar, CMC, and kappa-carrageenan, and 0.25% for xanthan. The final mix was gently stirred for 20 minutes. At the gum and starch concentrations used in these experiments, the increases in viscosity contributed by the gums was minimal and would not significantly increase mechanical shearing of the granules during gelatinization.

Starch was gelatinized under carefully controlled processing conditions with minimal shearing at this dilution. Dispersions were heated in a Corn Industries Viscometer (CIV) to 94° C and held at 94° C for 30 min after the samples had reached the final temperature (24-27 min). The processed dispersions were poured into ring-molds (77mm diameter and 20mm height) and held at room temperature for 24 hours. Gels were also stored at room temperature (21° C or at 4° C) over 120-hour period and their properties measured at 24-hour intervals during the storage period. Compressional tests were carried out with an Instron Universal Testing Machine (Model 1122) with a 50 kg load cell and a crosshead speed of 0.5 cm/min. Demolded gels were liberally lubricated with paraffin oil on both upper and lower sides and compressed between teflon-coated plates. Two batches of gels were prepared with three or four gels from each batch being tested separately to determine batch-to-batch and sample-to-sample reproducibility.

RESULTS

Figure 1 shows stress (Pa) versus dh/h, where h is the height at time t and dh=(h_0-h). The results demonstrate typical reproducibility of the compressional data obtained in this study. These data represent the results for 3 replicates from each of two batches of gel prepared from corn starch and water. The reproducibility of the data is excellent (+/- 5%) until near the point of fracture. The observed stress at fracture varies from about 6 x 10^{-3} Pa to almost 8 x 10^{-3} Pa, with corresponding variation in the strain at fracture. Fracture data often show such variability, as noted by Mitchell (5). However, the data obtained at low deformations, where the cross-link density is measured, show good reproducibility.

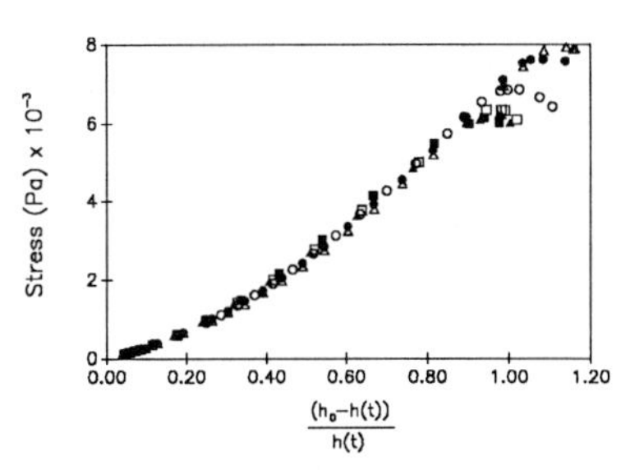

Figure 1. *Reproducibility Of Compressional Data for three replicates from two batches of Corn Starch Gels.*

The results of lubricated uniaxial compression for rice starch gels, gelatinized in water and in solutions of carrageenan, CMC, guar and xanthan, are shown in Figure 2 where $f/2(\lambda-\lambda^{-2})$ is plotted against λ^{-1}.

The linear portion shown at λ^{-1} up to 1.25 indicates that the gel is behaving as neo-Hookean material. This is particularly evident in the control gel (gelatinized in water). Here the initial linear region extends to values of λ^{-1} of 1.25, well beyond the region where serious computational problems can occur (2). For this 10% starch/water gel the value of C_1 is 0.42 x 10^{-3} Pa. For comparison purposes note that this is a factor of ten below the value for a 10% gelatin gel. A physically meaningful interpretation as to what constitutes these apparent "cross-links" in starch gels is not clear.

For a rice starch/water gel, deformations above 1.25 lead to an abrupt departure from the neo-Hookean behavior. Whatever the physical mechanism(s) giving the observed gel properties, it is evident that the deformation effectively "hardens" the gel. The curve rises rapidly, with a slight S-shape, through a maximum at which fracture occurs.

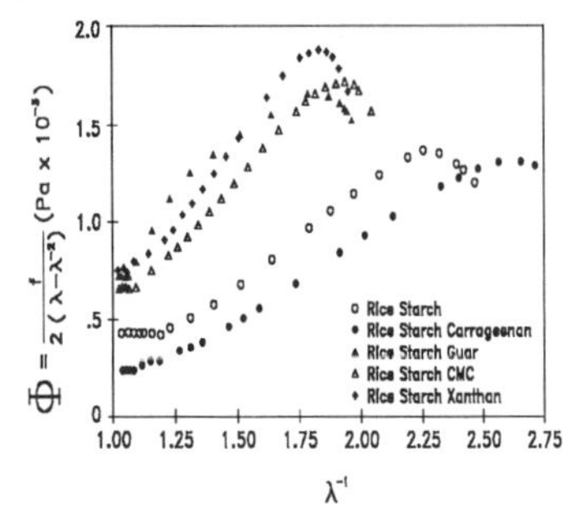

Figure 2. *Rice starch gelatinized in water and in hydrocolloid solutions where $f/2(\lambda-\lambda^{-2})$ is plotted against λ^{-1}.*

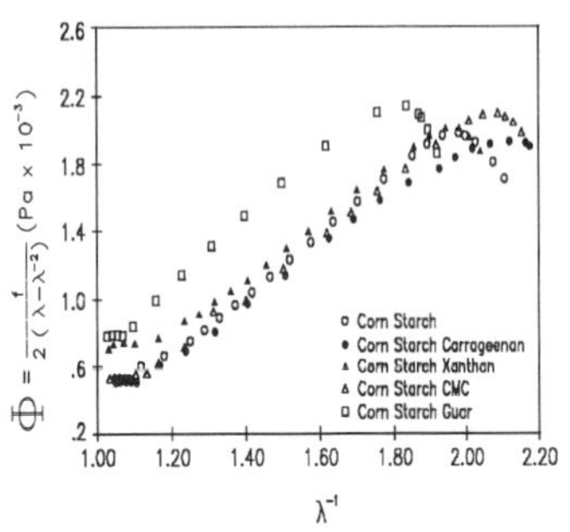

Figure 3. *Corn starch gelatinized in water and in hydrocolloid solutions where $f/2(\lambda-\lambda^{-2})$ is plotted against λ^{-1}.*

With gels prepared from gum solutions,the neo-Hookean region is shorter than with the rice starch/water,with the possible exception of the starch/carrageean gel, but in replicate experiments the same linear initial region always appeared (Table 1). The curves for guar, CMC and xanthan all are grouped above the control with an effective value of C_1 almost twice the control. The subsequent portions of all three curves are steeper than the control, with significantly higher stress and lower strain at fracture.

Figure 3 shows the equivalent data for corn starch gels. The control gel goes to a smaller strain for fracture and a higher stress than the rice gel control in agreement with earlier work (10). The neo-Hookean region is not as clearly delineated in these corn gels as in the rice gels, but an estimate of the initial cross-link density can still be made. For the control gel and the gels prepared with carrageenan and CMC the initial value of C_1 is about 0.52 x 10^{-3} Pa compared with the rice starch gel value of 0.42 x 10^{-3} Pa. The corn starch gels prepared with either xanthan or guar have a higher value of C_1, about 0.76 x 10^{-3} Pa. The overall appearance of the curves are close to the control except for the gel prepared with guar. The gums appear to have very little influence on the gel's elasticity or strength during this short-term structural development with the exception of guar. The higher CLD values of gels prepared with

Table I.
Apparent cross-link density and fracture properties of starch-hydrocolloid gels (24 hour room temperature gelation)

Additive	C_1 Average (Pa)	Stress at Fracture (Pa)	Strain at Fracture $(h_o - h) / h$
Corn Starch			
Control	5130 ± 130	6870 ± 910	1.03 ± .097
Carragean	5110 ± 200	7480 ± 960	1.14 ± .063
CMC	5690 ± 560	8180 ± 430	1.06 ± .069
Guar	7750 ± 590	6510 ± 210	0.87 ± .017
Xanthan	7500 ± 520	6740 ± 470	0.97 ± .048
Rice Starch			
Control	4330 ± 500	5940 ± 310	1.36 ± .041
Carragean	2450 ± 70	6670 ± 500	1.64 ± .058
CMC	6480 ± 270	5860 ± 300	0.98 ± .013
Guar	7250 ± 110	5340 ± 60	0.94 ± .011
Xanthan	7760 ± 300	6040 ± 560	0.87 ± .049

xanthan and guar would indicate that some association occurs with the fast retrograding amylose during gelation. Earlier work (11) indicated that both xanthan and guar associate with amylose during gelation of corn starch.

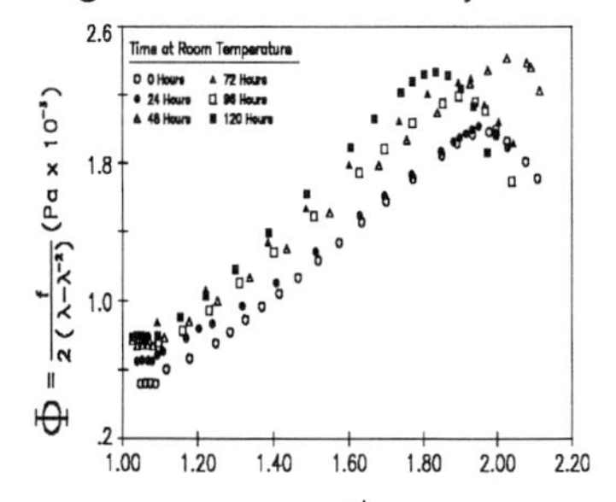

Figure 4. *Effect of room temperature storage on apparent CLD and gel fracture of corn starch gels sampled every 24 hours over a 120 hour period.*

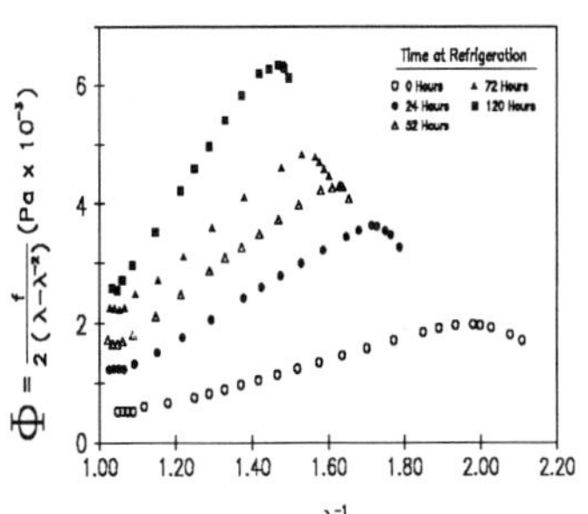

Figure 5. *Effect of refrigerated storage on apparent CLD and gel fracture of corn starch gels sampled every 24 hours over a 120 hour period.*

When corn starch gels were stored at room temperature over a period of 120 hours and measured at 24-hour intervals, there was a small, gradual, increase in the CLD and a relatively small change in the shape of the curves as the gels aged (Fig. 4). Similar results were obtained with rice starch controls subjected to room temperature storage.

To accelerate the effects shown in Figure 4 for corn starch gels, gels were stored at 4° C. The results shown in Figure 5 demonstrate very significant and systematic shifts in both the cross-link density and the rigidity modulus. First, for the 0, 24, 52 and 72-hour storage times, the neo-Hookean region is very evident and the intercept at $\lambda^{-1} = 1$ goes from .52 to 2.3×10^{-3} Pa. The 120-hour storage did not clearly show initial neo-Hookean behavior. Further, the gel rigidity increases progressively with storage time, with a systematic reduction in strain and an increase in stress at fracture. These progressive increases in both the CLD and gel strength are attributed to crystallization of amylopectin (AP) embedded in the amylose matrix. It appears that the hardening of the gels correlates with AP retrogradation well known to occur at refrigerated temperatures.

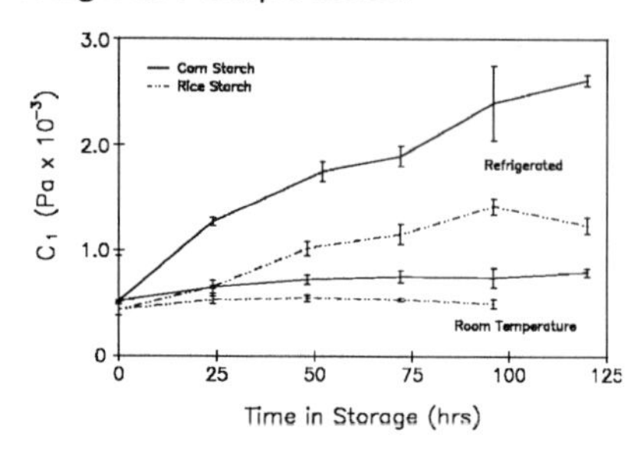

Figure 6. *Apparent cross-link density of corn and rice starch control gels measured at 24 hour intervals during room temperature and refrigerated storage.*

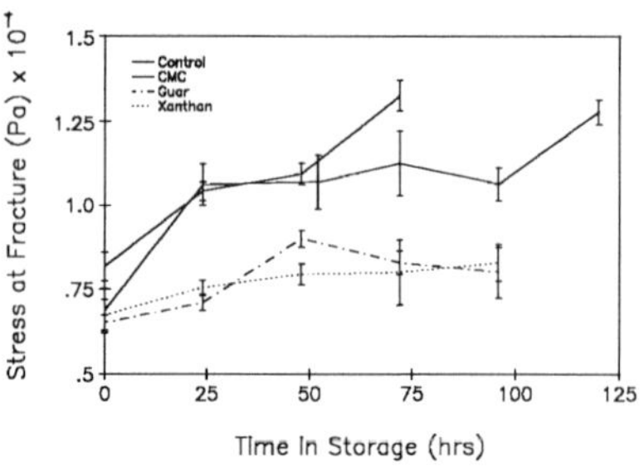

Figure 7. *Influence of hydrocolloids on the stress at fracture of corn starch gels during refrigerated storage.*

Comparative plots of storage time versus the C_1's of rice and corn starch control gels stored at room temperature and 4° C are shown in Figure 6. The CLD's of the control rice and corn starch gels observed after 24-hour gelation (noted here as zero time) did not change significantly during storage at room temperature, suggesting that only the rapid retrogradation of amylose plays a major role in the initial development of the CLD. Apparently the slow AP retrogradation has little influence on the initial development of the CLD's. Earlier studies by Miles et al. (12) have shown that the short-term development of gel structure is largely dependent on amylose aggregation.

The CLD's of both corn and rice starch/hydrocolloid gels after 24-hour gelation as shown in Figures 1 and 2 appear stable during storage at room temperature. However, under refrigerated storage when the amylopectin retrogrades more rapidly, the CLD's of starch/hydrocolloid gels increase with time of storage at a rate quite similar to the controls (Figure 6).

Figure 7 shows the change in stress at fracture with time of storage at 4° C of corn starch gels. For guar and xanthan the stress at fracture changes very slowly with storage time compared to the control and the starch/CMC gel.

CONCLUSIONS

Starch gels in lubricated uniaxial compression appear to behave as neo-Hookean materials at low strains. At higher strains, extensive deviations from this behavior occur up to fracture. A cross-link density level for starch gels, as measured at low compressional deformations, is reached within 24 hours of gelatinization and remains relatively stable during room temperature storage. However, CLD of gels progressively increase upon refrigerated storage, possibly due to the retrogradation of amylopectin both in the gelatinized granule fragments and in the exuded amylopectin in the amylose matrix. The stress at fracture increases with a concurrent reduction in strain at fracture with storage time of the starch gels, indicating that the gels begin to harden during 4° C storage. The latter is probably due to the retrogradation of amylopectin.

The apparent cross-link density (C_1) of rice starch gels is lower than that of corn starch gels; but, when rice starch is gelatinized in the presence of guar or xanthan, comparable CLD levels can be obtained within 24 hours at room temperature. Work is continuing on the measurement of the CLD of starch gels during processing under high temperature and mechanical work conditions.

References

1. Ring, S. G. and Stainsby, G. Prog. (1982) Food Nutr. Sci. 6, 323-329.
2. Bagley E. B., Christianson, D. D., and W. J. Wolf. (1985) J. Rheology. 29, 103-108.
3. Doublier, J. L. Llamas, G., and M. LaMeur. (1987) Carbohydrate Polymers 7(4), 251-175.
4. Sajjan, S. U. and M. R. Raghauendra Rao. (1987) Carbohydrate Polymers 7, 395-402.
5. Mitchell, J. R. Proc. Int. Workshop on Plant Polysaccharides-Structure and Function. (1984) Mercier, C. and Rinaudo, M. (eds.) INRA/CNRS, p. 93.
6. Russell, P. L. (1987) J. Cereal Sci. 6, 147-158.
7. Roulet Ph, MacInnes, W. M., Wursch, P., Sanchez, R. M. and A. Raemy. (1988) Food Hydrocolloids 2(5),381-396.
8. Treloar, L. R. G. (1958). The Physics of Rubber Elasticity, Oxford at the Clarendon Press.
9. Christianson, D. D., Navickis, L. L., Bagley. E. B., and W. J. Wolf. (1984) In "Gums and Stabilizers for the Food Industry 2" (Eds. G. O. Phillips, D. J. Wedlock and P. A. Williams) Pergamon Press, New York, pp. 123-134.
10. Christianson, D.D., Casiraghi, E.M., and Bagley, E.B. (1986) Carbohydrate Polymers. 6, 335-348.12. 513-517.
11. Christianson, D.D., Hodge, J.E., Osborne, D., and R.W. Detroy. (1981) Cereal Chem. 58(6), 513-517.
12 Miles, M.J., Morris, V.J., Orford, P.D., and Ring, S. G. (1985) Carbohydrate Res. 135, 271-281.

Rheology of starch-galactomannan gels

MARTINE ALLONCLE and JEAN-LOUIS DOUBLIER[+]
Sanofi-Bio-Industries, Baupte, 50500 Carentan, France
[+]Institut National de la Recherche Agronomique, LPCM, BP527 44026 Nantes Cedex 03, France

ABSTRACT

The gelation process of starch-galactomannan (guar gum and locust bean gum) mixtures has been described from oscillatory shear experiments using a Controlled Stress Rheometer (Carri-Med). It was found that the presence of galactomannans affected the kinetics of starch retrogradation. Moreover the mechanical spectrum shown by final systems was strongly modified when compared with starch alone. A starch gel may be described as a composite consisting of a dispersed phase (swollen granules mainly composed of amylopectin) embedded in a continuous network in which the main component is amylose. Starch-galactomannan mixtures can be described in a similar way, the continuous phase being in itself an admixture of amylose and galactomannan. Amylose gelation and hence starch retrogradation would be governed by the likely incompatibility between amylose and galactomannans.

INTRODUCTION

It is known that starch retrogradation is primarily influenced by amylose gelation. The characteristics of the gel is related to the degree of starch swelling and to the concentration of soluble amylose (1). A starch gel can be regarded as a composite in which swollen gelatinised granules are embedded in an amylose gel matrix (2,3,4). Three main factors influence the mechanical behaviour of such a system: the rigidity of the matrix gel, the volume occupied by the particles and their deformability. The incorporation of galactomannans into a starch paste is known to result in a dramatic increase of the paste viscosity. We recently demonstrated that this effect results from the artificial concentration of the hydrocolloid within the continuous phase (5). The object of the present investigation is to describe the rheological properties of starch-galactomannan gels on a similar basis.

MATERIALS AND METHODS

<u>Materials</u>

Commercial corn starch was obtained from Roquette Frères (France). Pure guar and locust bean gum were supplied by Sanofi-Bio-Industries (50500, Carentan, France). Gum solution was prepared by sprinkling the gum onto the sides of a water

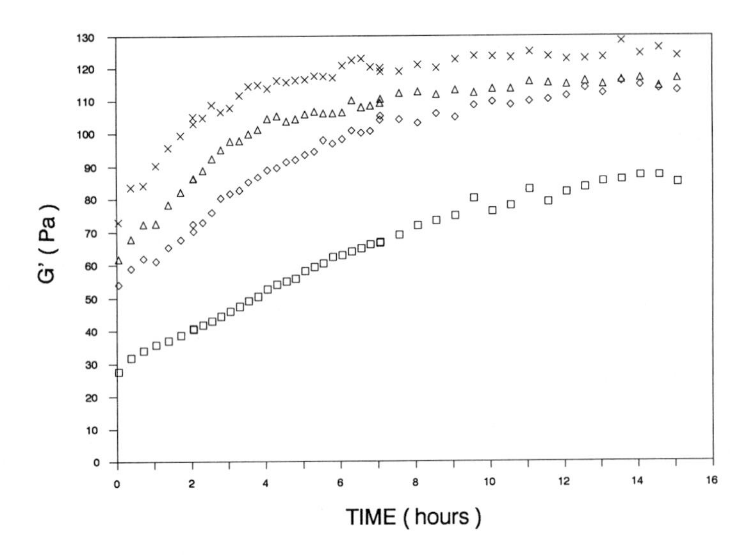

Figure 1. G' variation as a function of time. Starch concentration: 4%. Guar gum concentration: (□), 0%; (◇), 0.2%; (△), 0.32%; (×), 0.5%.

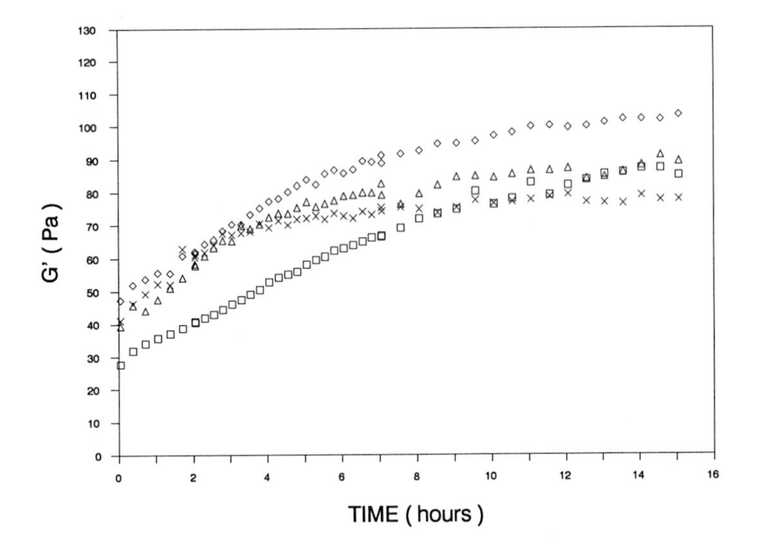

Figure 2. G' variation as a function of time. Starch concentration: 4%. Locust bean gum concentration: (□), 0%; (◇), 0.2%; (△), 0.32%; (×), 0.5%.

vortex. Stirring was continued at slow speed for 1 hr at room temperature; then the solution was heated to 80°C for 0.5 hr while stirring. Finally, the solution was filtered.

Preparation procedures
The mixtures contained 4% starch with galactomannan concentration ranging between 0.1 and 0.5%. Corn starch-galactomannan dispersions were heated in a double-walled, round bottom vessel. Stirring was achieved with an anchor-shaped blade rotating at 200 rpm. The maximum temperature (96°C) was reached in about 15 min and was held for 30 min before starting the rheological measurement. Concentration of the blends was determined by drying aliquots overnight at 103°C.

Rheological measurements
A Controlled-Stress Rheometer (Carri-Med) was used to perform oscillatory shear experiments (strain amplitude: 4%) with a cone and plate geometry (5cm diameter, 4° angle). Two measurements were carried out. The first one was the kinetics of gel formation (variation of G′ and G") of the mixture at 25°C, at 1 Hz for 15 hours. The second one was a mechanical spectrum of the final gel at 25°C from 0.01 Hz to 4 Hz.

RESULTS AND DISCUSSION
Figure 1 shows the kinetics of gel formation of starch-guar gum mixtures, where the galactomannan concentration varied from 0.1% to 0.5%. By comparison with starch alone at 4%, the storage modulus (G′) increased more rapidly in the case of mixtures than for starch alone. The beginning of the kinetics showed clear differences between mixtures when the galactomannan concentration increased (G′ increased more steeply when guar concentration increased). At the end of the experiment (15 hours), these differences were much less visible for all the mixtures (120 Pa). The storage modulus for all the mixtures was clearly superior to starch alone (85 Pa).

Figure 2 shows the kinetics of gel formation of starch-locust bean gum mixtures. It is seen that the effect of locust bean gum was less dramatic. The initial kinetics of gel formation was also accelerated but at a lesser extent and moduli of the final gels were of the same order as for starch alone (85 Pa) except for the 0.2% locust bean gum concentration which was slightly higher (100 Pa).

Figure 3 shows the mechanical spectra at 25°C of final systems. Starch alone at 4% and starch (4%) / guar gum (0.5%) mixtures are both represented. For starch alone, the mechanical spectrum was typical of a gel: G′ independent of frequency and the G"/G′ ratio of the order of 0.1. In contrast, the mixture showed a slight G′ increase as a function of frequency and the G"/G′ ratio was ~0.3. These two features show that the mixed systems had changed into a less elastic gel. Also, it should be mentioned that all the mixtures described in figures 1 and 2 displayed this type of behaviour.

Galactomannans being non gelling polysaccharides, the evolution of G′ corresponds to amylose gelation that takes place in the continuous phase. We show that the presence of galac-

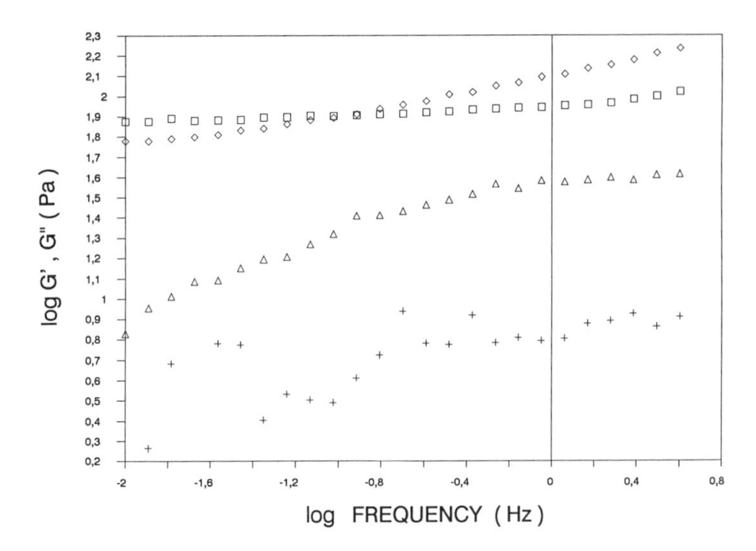

Figure 3. G′ and G" vs frequency. Starch 4% (□: G′, : G"). Starch 4% / guar gum 0.5% (◇: G′, △: G").

tomannan strongly affects the rheological evolution of the system. This being regarded as a composite with a dispersed phase embedded within a continuous phase mostly composed of amylose and galactomannan (5), such effect is primarily to be ascribed to amylose-galactomannan «interactions». From swelling-solubility experiments, we could estimate the volume fraction of the dispersed phase to be 0.68 and the amylose concentration in the continuous phase to be 3% (5). The actual galactomannan concentration in the continuous phase thus ranged between 0.31 and 1.53% when the total concentration of galactomannan varied from 0.1 to 0.5%. It was recently reported (6,7) on thermodynamic incompatibility of amylose with other α-glucans (dextran and amylopectin). It is likely that amylose-galactomannan «interactions» arise from a similar mechanism and that such incompatibility interferes with the gelation by accelerating the process in a first step and then slowing it more rapidly than with amylose alone. The difference between guar gum and locust bean gum may arise from their slightly different chemical structure.

REFERENCES

1. Ott, M. and Hester, E.E. (1965) Cereal Chem., 42, 476-484.
2. Ring, S. and Stainsby, G. (1982) Prog. Fd. Nutr. Sci., 6, 323-329.
3. Miles, M.J., Morris, V.J., Orford, P.D. and Ring, S.G. (1985) Carbohydr. Res., 135, 271-281.
4. Ring, S.G. (1985) Staerke, 37, 80-83.

5. Alloncle, M., Lefebvre, J., Llamas, G. and Doublier J.L. (1989) Cereal Chem. (in press).
6. Kalichevsky, M.T., Orford, P.D. and Ring, S.G. (1986) Carbohydr. Polym., 6, 145-154.
7. Kalichevsky, M.T. and Ring, S.G. (1987) Carbohydr. Res., 162, 323-328.

The fracture properties of starch gels

H.LUYTEN and T.VAN VLIET

Department of Food Science, Wageningen Agricultural University, Bomenweg 2, 6703 HD Wageningen, The Netherlands

ABSTRACT

Fracture properties of food and food-like materials can be studied in different ways. For potato starch gels it has been found that fracture stress and strain increased with increasing rate of deformation. These results were consistent with those found for creep and relaxation experiments: at lower stresses, fracture occurred at lower strains, but after longer periods of time. The reason for this behaviour is yet unknown.

INTRODUCTION

The fracture properties of food materials are important when handling or eating them. In spite of this only little basic research has been performed on foods. Most of the theory for fracture mechanics has been developed for elastic or predominantly elastic materials of which the fracture properties do not or hardly depend on the rate of deformation. However, for many foods they are strongly dependent on the rate of deformation, and this has several consequences for the usage properties. In this paper we will present some results of a study on the fracture properties of potato starch gels.

THEORY

A material fractures when a defect grows, new surfaces are formed and the material falls apart into pieces. Fracture starts (initiates) if the local stress in the material exceeds the fracture stress. Fracture propagates if there is enough strain energy available to form new surfaces (1). Inhomogeneities (defects) inside a material cause the stress and strain to be locally higher. Fracture therefore starts at such places. Defects thus lower the overall stress and strain at fracture. The sensitivity of a material to defects, i.e. the decrease of stress with increasing defect size, can differ among materials and is called the notch sensitivity of the material.

The fracture behaviour of a test-piece thus depends on

- the material of which the test-piece is made. The material properties determine the fracture stress near the point of fracture, the fracture energy and the notch-sensitivity.
- the size of the defects in the material.
- size and shape of the test-piece.
- the rate of deformation.

Dissipation of energy affects the fracture behaviour, because less energy is available to create new surfaces. This energy dissipation can be due to viscous flow of the material as well as to inhomogeneous displacement of material relative to each other (for example flow of liquid through the matrix) (2). Because energy dissipation often is time-dependent, fracture properties can be time-dependent as well. Shortage of energy often cause fracture to be slow.

MATERIAL AND METHODS

Dispersions of 10% potato starch (AVEBE, Foxhol, the Netherlands) in water were heated to about 65°C while gently stirring to prevent sedimentation. Long cylindrical moulds were filled with the dispersion and further heated in a microwave oven up to about 95°C. In this way samples without visible defects could be made. The material was stored at room temperature for 24 h before experiments were started. Cylinders with a height of 30 mm and a diameter of 34 mm were cut.
Uniaxial compression experiments were performed between perspex plates with an Overload Dynamics material testing machine, table model S 100. A 200 N load cell was used. Stress relaxation was followed at constant relative deformation (strain). A more extensive description of the experimental procedure followed has been given by Luyten (3). When testing at the same relative deformation rate, no effect of size and shape of the samples could be detected (Luyten, to be published). This means that we are almost sure that true material properties are determined (3), although this has to be checked further with different kinds of deformation.

Creep measurements in compression were done with the apparatus described by Mulder (4). Size and shape of the samples were equal to those used for the other experiments.

Figure 1 Compression curves (typical examples) of 10% potato starch gels at various initial relative deformation rates: 1.7 10^{-4} s^{-1}, 1.7 10^{-3} s^{-1}, 1.7 10^{-2} s^{-1}, 1.7 10^{-1} s^{-1}.

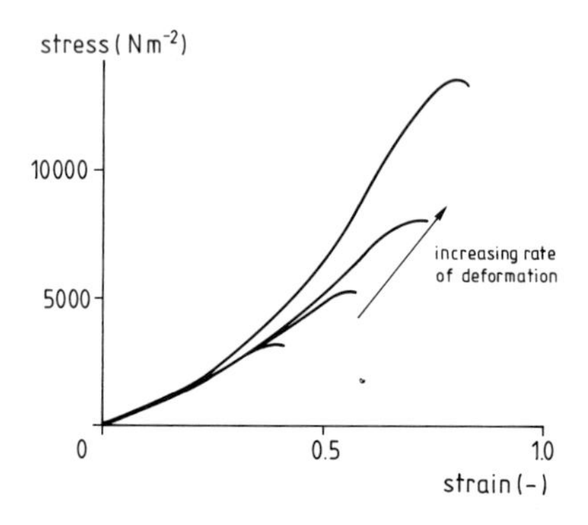

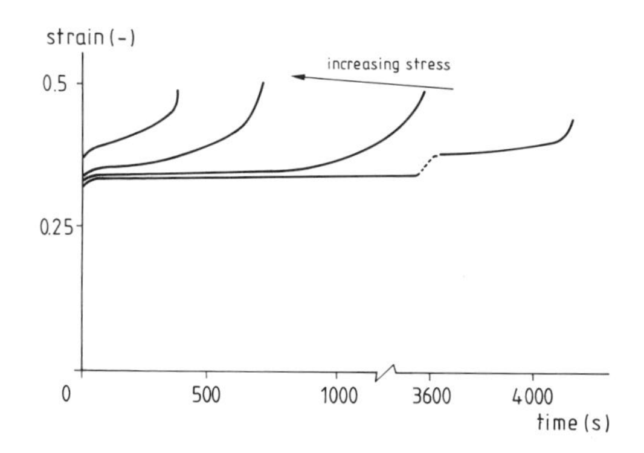

Figure 2 Creep curves for 10% potato starch gels measured in compression at various stresses: 1400 $N \cdot m^{-2}$ to 2800 $N \cdot m^{-2}$.

The specific fracture energy, R_s, was determined by cutting with a wire at a certain speed and measuring the force (5). A wire with a diameter of 0.1 mm was used for most experiments. Wires with diameters of 0.025 up to 0.5 mm gave the same results.

RESULTS AND DISCUSSION

The behaviour of starch gels at low stresses and short time-scales is elastic: the Young'smodulus E (initial slope of the compression curve) is independent of the rate of deformation (fig.1), the strain in creep experiments at low compression stresses is nearly constant in time (fig.2), the decrease in stress in relaxation experiments at low strains is almost negligible (fig.3). Therefore it may be concluded that the deformation behaviour at these small stresses and short times does not depend on time.

In contrast to the behaviour at low stress the fracture behaviour of starch gels is clearly time dependent. Both stress and strain at fracture increase with increasing rate of deformation (fig.1). In the creep experiments at lower stresses fracture occurred at lower strains but after longer periods of time. Obviously, during the period that the strain increased only very slowly with time, some restructuring processes occur in the gel which ultimately lead to fracture. The rate of these processes is slower at lower stresses. The relaxation experiments showed that if a sample is compressed further, implying higher stresses, it fractures earlier. At a low strain, there was for a long time only little stress relaxation, until the sample fractured relatively fast. This indicates that fracture of starch gels is a slow process and costs time.

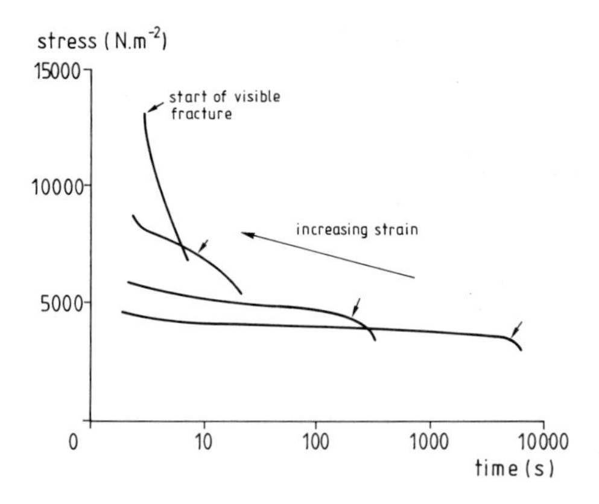

Figure 3 Stress during relaxation of 10% potato starch gels. The relative rate of deformation was 1.7 10^{-1} s^{-1}, the strain at relaxation was between 0.4 and 0.7.

Figure 4 Example of a force-time curve for a cutting experiment on a 10% potato starch gel. Cutting speed 5 mm min^{-1}. Wire diameter 0.1 mm.

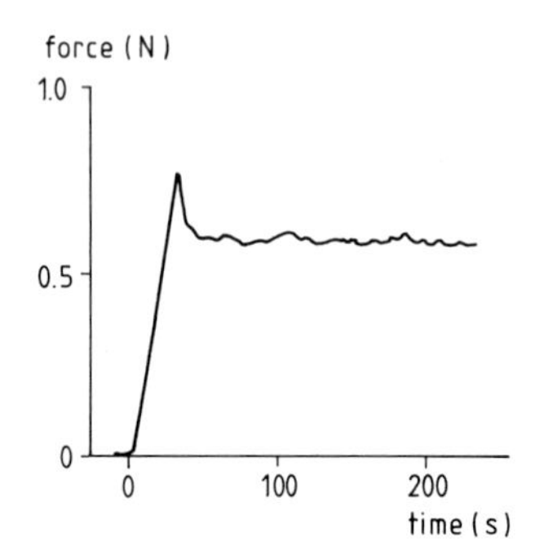

A possible reason for the slow fracture could be that not enough energy is available for crack propagation. The energy needed to fracture a starch gel was determined by cutting experiments (fig.4). From the results obtained it can be concluded that the specific fracture energy, R_s, is about 1-2 J m^{-2}. In compression experiments at comparable rates of deformation about 30 J m^{-2} is available, ten times as much as needed. Consequently, shortage of energy can not be the reason for the slow fracture observed. The explanation for this time dependent fracture in potato starch gels is the subject of present research.

CONCLUSIONS

The fracture behaviour of potato starch gels is time dependent, in contrast with the elastic behaviour observed for small stresses and short times. These time-dependent fracture properties are probably caused by slow fracture: fracture costs time. However, the reason for this slow fracture is unknown.

ACKNOWLEDGEMENT

This work is supported by AVEBE and by the Netherlands Programme for Innovation Oriented Research, Carbohydrates (IOP-K) with financial aid of the Ministry of Economic Affairs and the Ministry of Agriculture and Fisheries.

LITERATURE REFERENCES

(1) Atkins, A.G., Mai, Y-M.,'Elastic and plastic fracture',Horwood (1985)
(2) Vliet, T. van, Luyten, H., Walstra, P., Zoon, P.,'Time-dependent fracture and yielding behaviour of gels', to be published in J. Mat. Sci. (1990)
(3) Luyten, H.,'The rheological and fracture properties of Gouda cheese', Ph.D. thesis, Wageningen agricultural university (1988)
(4) Mulder, H.,'Het bepalen van reologische eigenschappen van kaas', Versl. Landbouwk. Onderz., 51 (1946)
(5) Atkins, A.G.,Vincent, J.F.V.,'An instrumented microtome for improved histological sections and the measurement of fracture toughness', J. Mat. Sci. Letters, 3 (1984) 310-312

The water and salt were heated to boiling then the macaroni was added; the water returned to boiling after 3 minutes.Heat was reduced (setting 2) and product simmered for 5.5 minutes.The macaroni was drained in a colander. The margarine was melted in another saucepan (setting 5). The flour was added and cooked with constant stirring; after removal from the heat the milk was gradually added with constant stirring to form a smooth sauce.The sauce was returned to the heat and stirred until thickened. Spices were added and mixed thoroughly.In the batches containing modified starch it was then sieved into the sauce and whisked to mix thoroughly. The grated cheese was added and the sauce heated until it had melted. All processes were timed and ingredients were from the same batch so that standard samples could be prepared.

Each batch was evenly divided into six aluminium foil containers (190mm x 90mm x 50mm) which were sealed with a cardboard lid prior to blast freezing (40/90 Mini Blast Freezer - Southern and Redfern Ltd ; air speed approximately 5m/sec) for 60 minutes or blast chilling (Foster QC 45 - Turbo - Blast Chiller; air speed approximately 4m/sec) for 40minutes. The packs were then stored at either -20°C or 0 to 3°C.

The packs were reheated prior to sensory evaluation to a centre temperature of 96°C in a stirred air Belling oven modified with electronic thermostats, air temperature 180°C + or - 1°C , for a time of 30 minutes (frozen) or 20 minutes (chilled).The samples were evaluated as soon as possible after reheating.

Sensory analysis

Twelve judges out of sixteen were selected as being sufficiently discriminating for use as permanent members of the panel. Sensory difference tests were done to determine that there were no detectable differences between identical batches. Each test batch was then compared with a freshly made batch of the same formulation. Table 1 shows the comparisons which were made by the judges. The declared control difference test was used to detect any differences between samples and a correlated t-test was done on the results to determine whether any detected differences were statistically significant (1).

Descriptive analysis with scaling (2,3) was then used on some significantly different results in an attempt to characterize any differences found.The samples tested thus are listed in Table 2.Nine judges were used and the results were statistically analysed by the analysis of variance and Tukey's Test.

RESULTS AND DISCUSSION

Statistical analysis of the declared control difference test data showed (Table 1) only one pair of results to be not significantly different; this was the fresh 50% wheat flour and 50% modified starch compared with the sample which had been frozen and stored for 1 week.It can be reasoned that the freezing and short storage time along with the modified starch prevented any detectable changes from occurring.

The differences which were detectable in the other samples

could have been due to starch retrogradation or to a change in one or more of the other characteristics.Some chilled samples were investigated further (see Table 2) using descriptive analysis with scaling; differences were found to be significant for creamy colour intensity (50:50 formulation) cheesy taste (50:50 formulation) and thickness (100% wheat flour formulation). With regard to the loss in cheesy taste.Tukey's test showed that the fresh batch(sample 1) was significantly different from the chilled 1 day stored batch(sample 2) and from the chilled 5day stored batch(sample 3) but the two batches which were chilled and stored were not significantly different from each other.It would seem the major loss of cheesy taste occurred during the chilling process and first day of chilled storage.This trend was also seen in the 100% wheat flour formulation but the differences were not significant; may be the wheat flour masked these changes.

With regard to thickness (100% wheat flour formulation)the chilled product stored for 5 days was rated the thickest and was significantly different from the fresh product but not from the chilled product stored for 1 day.Possibly retrogradation of the wheat flour occurred with time.

Tukey,s test showed that for creamy colour intensity the chilled 1 day stored batch(sample 2),50:50 formulation, was significantly different from the chilled 5 days stored batch(sample 3)- this score indicated the creamiest colour but the fresh batch was not signicantly different from either chilled batch. The main trend in the scores indicates the development of a more creamy colour on chilled storage.

Further work is required to elucidate these interactions.

Table 1 . STATISTICAL ANALYSIS OF DECLARED CONTROL DIFFERENCE TESTS

Test batch*	n	n-1	d	t	p
1 day chilled w. flour	44	43	-1.60	-9.25	<0.001
1 day chilled 50:50	40	39	-0.55	-4.11	<0.001
5 day chilled w. flour	52	51	-1.00	-3.89	<0.001
5 day chilled 50:50	52	51	-0.98	-6.16	<0.001
1 week frozen w.flour	40	39	-1.05	-4.30	<0.001
1 week frozen 50:50	40	39	-0.18	-1.00	NS
6 week frozen w.flour	48	47	-1.19	-9.52	<0.001
6 week frozen 50:50	40	39	-6.00	-3.59	<0.001

Key: * = Test batch compared to fresh batch of same formulation; w.flour = wheat flour; 50:50 = 50% wheat flour and 50% modified starch; n = number of responses; n-1 = degrees of freedom; d = mean of differences between hidden control and test sample; t = calculated result of correlated t-test; p = significance level.

Table 2 . DESCRIPTIVE ANALYSIS WITH SCALING

Characteristic	Thickening Agent	Sample Means 1	2	3	F value*
Creamy colour Intensity	100% W.flour	6.13	6.41	7.03	0.38
	50:50	6.84	6.00	9.03	4.99
Lustre	100% W.flour	10.44	10.53	9.78	0.72
	50:50	9.67	8.63	9.22	0.85
Cheesy aroma	100% W.flour	8.16	7.22	6.31	1.83
	50:50	8.47	6.69	7.94	2.11
Starchy aroma	100% W.flour	4.78	5.09	6.72	1.26
	50:50	5.72	5.47	7.31	2.64
Cheesy taste	100% W.flour	7.84	6.28	6.06	1.51
	50:50	9.19	6.59	6.31	12.07
Spicy taste	100% W.flour	4.47	4.47	3.44	0.34
	50:50	4.63	4.28	4.06	0.15
Thickness	100% W.flour	5.91	7.75	8.56	6.88
	50:50	8.09	9.19	8.22	2.29
Smoothness	100% W.flour	8.13	7.94	8.69	0.21
	50:50	8.22	8.94	7.75	0.48
Mouthfeel	100% W.flour	6.19	6.31	5.28	1.00
	50:50	5.56	5.31	6.13	0.41

Key:- 100% W.flour = 100% wheat flour; 50:50 = 50%wheat flour and 50% modified starch; * F value must exceed 3.74 to be significant at the 5% level and 6.51 to be significant at the 1% level (2); 1 = fresh sample; 2 = sample after 1day chilled storage; 3 = sample after 5 days chilled storage.

ACKNOWLEDGEMENTS

The authors thank Cerestar U.K. Ltd for supplying the starch samples and in particular Phillip Lawson for expert advice also Sylvie Kierzek and Jennifer Grubb for technical help as well as the Sensory Panel, without whom the project would not have been possible.

REFERENCES

1. Hill, M.A. and Glew, G. (1970) J. Fd. Technol. 5 ,187-191.
2. Larmond, E. (1977) Laboratory Methods for Sensory Evaluation of Food, 73 pages, Publication 1637/E, Agriculture Canada.
3. Stone, J., Sidel, J., Oliver, S. and Woolsey, A. (1974) Food Technol. 28 ,(11), 24-34.

Some characteristics of isolated starch granules and degradability of starch in the rumen of the cow

M.G.E.WOLTERS and J.W.CONE+

TNO-CIVO Institutes, Department of General Food Chemistry, Utrechtseweg 48, 3704 HE Zeist, The Netherlands

+State University Utrecht, Department of Large Animal Medicine and Nutrition, Yalelaan 16, 3584 CL Utrecht, The Netherlands

ABSTRACT

Concentrates with a high starch content fed to ruminants can result in a decreased content of milk fat and/or disturbances of ruminal fermentation. Differences in starch degradation can be attributed to the origin of the starch and may be related to its physicochemical properties. Starch granules have been isolated from raw materials. The isolated starch granules were studied by scanning electron microscopy. The contents of amylose and lipid-bound amylose were determined colorimetrically. The degradability of starch and of raw materials was determined in vitro both with rumen fluid and α-amylase.

INTRODUCTION

To meet the energy demands of dairy cows large amounts of concentrates are given. Rations high in readily fermentable concentrates can lead to ruminal acidosis and depression of feed intake (1).

Feed concentrates contain large amounts of starch. Starch-rich concentrates differ considerably in degradability by rumen fluid (2). The rate of starch breakdown, either by rumen fluid or by enzymes, depends on several characteristics of the starch in question (2, 3, 4, 5). The aim of this study was to obtain insight into the relationship between physical and chemical properties of starch and its degradability. Parts of the results obtained are presented here.

MATERIALS AND METHODS

Raw materials

Raw materials used were potato (cv. Bintje), maize silage, oat, maize, field bean (cv. Alfred), pea (cv. Finale), milocorn, barley, tapioca, rice and wheat.

Isolation of starch granules

Raw materials were soaked in water for 24 hours. The soaked samples were milled with an Ultraturrax. The resulting slurry was filtered over a 70 μm nylon filter to separate the starch from the protein. In the filtrate the starch granules deposited.

Amylose

Apparent amylose was determined colorimetrically according to Chrastil (6). Real amylose was determined after extraction with methanol to remove the lipids. Lipid-bound amylose was calculated as the difference between apparent and real amylose.

In vitro degradation of starch with rumen fluid

Rumen fluid was taken from a fistulated cow. The starch was degraded by incubating the sample with diluted rumen fluid for 6 hours at 39 °C (2). The remaining starch was hydrolysed with diluted HCl and the glucose was determined enzymatically according to Bergmeyer (7).

In vitro degradation of starch with α-amylase

The samples were incubated with α-amylase (Calbiochem) for 4 hours at 39 °C (2). The remaining starch was determined enzymatically according to Bergmeyer (7).

RESULTS AND DISCUSSION

Isolation of the starch

Starch was isolated from the raw materials. In the isolation procedure the starch granules were damaged as little as possible. All the starch was isolated from the raw materials. In the isolation procedure protein and fat were removed from the starch. Only rice starch had a considerable content of protein.

Apparent and real amounts of amylose

Lipids present in the starch are connected to amylose and form an amylose-lipid complex. The colorimetric determination of amylose as described by Chrastil is based on the complex formation between amylose and iodine (6). Apparent amylose was determined in the starch as such. Real amylose was determined after removal of the lipids. The difference between real and apparent amylose is a measure for the amount of amylose-lipid complex present in the starch.

Table 1 shows the variation in the content of real amylose. Especially bean and pea had a high content of real amylose. Only maize silage was rich in lipid-bound amylose.

Table 1. CONTENT OF APPARENT, REAL AND LIPID-BOUND AMYLOSE IN ISOLATED STARCHES (% of DRY WEIGHT)

	apparent amylose	real amylose	lipid-bound amylose
potato	33.0	31.9	-1.1
field bean	43.2	42.3	-0.9
milocorn	25.2	27.4	2.2
maize silage	23.2	30.1	6.9
wheat	24.5	25.7	1.2
oat	24.0	27.0	3.0
pea	43.6	43.5	-0.1
barley	25.2	26.0	0.8
maize	27.4	29.0	1.6
tapioca	18.0	20.1	2.1
rice	19.0	21.4	2.4

Lipid-bound amylose is calculated as the difference between real and apparent amylose.

In vitro degradation of isolated starch granules and starch in raw materials with rumen fluid and α-amylase

The degradability of starch in the raw materials and isolated starch granules was determined in vitro using rumen fluid and α-amylase (Table 2). Differences in degradability were much larger with α-amylase than with rumen fluid.

Table 2. IN VITRO DEGRADATION OF ISOLATED STARCH GRANULES AND STARCH IN RAW MATERIALS WITH RUMEN FLUID AND α-AMYLASE (% DEGRADED STARCH)

	rumen fluid		α-amylase	
	raw material	isolated starch granules	raw material	isolated starch granules
potato	28.4	15.2	3.0	2.6
field bean	23.9	15.8	12.0	8.1
milocorn	27.2	16.6	10.1	14.7
maize silage	29.7	15.7	19.3	15.1
wheat	20.7	20.8	20.0	17.3
oat	28.8	21.0	26.2	23.1
pea	20.9	5.6	27.0	26.6
barley	21.6	14.9	23.1	28.1
maize	24.0	19.7	18.7	37.3
tapioca	31.0	21.7	46.5	46.7
rice	40.5	31.0	61.6	60.4

For degradation with α-amylase there was a good linear correlation between the degradability of starch in the raw materials and isolated starch granules. With rumen fluid this correlation could not be seen. There seems to be a negative correlation between amylose content and degradability.

Scanning electron microscopy of isolated starch granules

The isolated starches were studied by scanning electron microscopy. Variations in size, shape and surface of the granules were observed. Figure 1 and 2 show starch granules of maize silage and of field bean respectively.

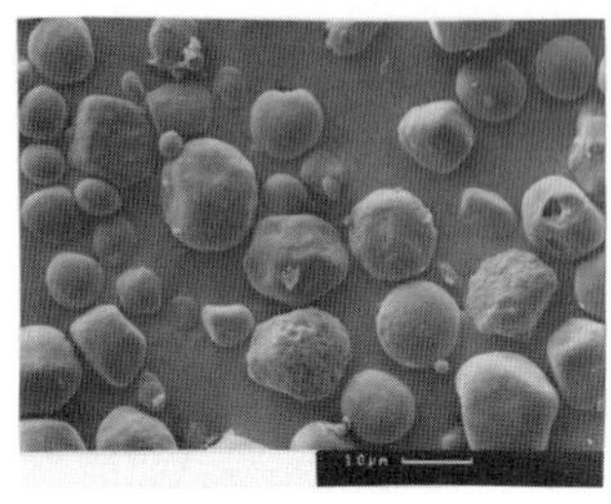

Figure 1. Starch granules of maize silage

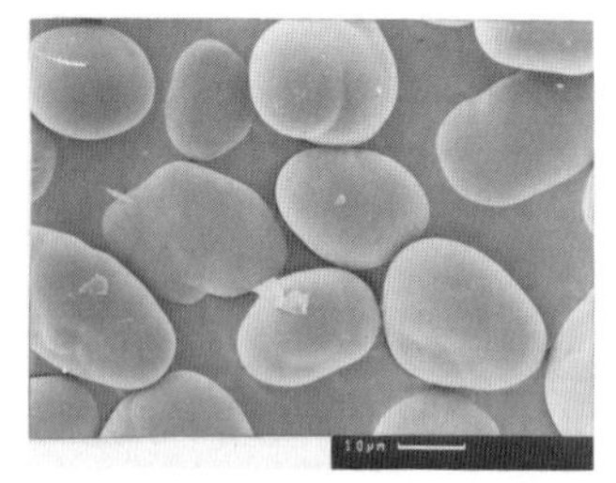

Figure 2. Starch granules of field bean

CONCLUSIONS

1. With the isolation procedure clean, undamaged starch granules were obtained.
2. Starch isolated from different raw materials shows a vast variation in degradability by rumen fluid and α-amylase.
3. Starch isolated from different raw materials differs in physical and chemical properties
4. There is a negative correlation between amylose content and degradability.

ACKNOWLEDGEMENT

This research was financially supported by the Produktschap voor Veevoeder, The Hague. The authors wish to thank Miss Riek Schreuder and Miss Marcella Vlot for technical support.

REFERENCES:

1. Malestein, A. and van 't Klooster, A.Th. (1986) J. Anim. Physiol. a. Anim. Nutr. 55, 1 - 13.
2. Cone, J.W., Clin-Theil, W., Malestein, A. and van 't Klooster, A. Th. J. Sci. Food Agric. 48, (in press).
3. Malestein, A., van 't Klooster, A.Th. and Cone, J.W. (1988) J. Anim. Physiol. a. Anim. Nutr. 59, 225 - 232.
4. Moran, E.T. (1982) Poultry Sci. 61, 1257 - 1267
5. Rooney, L.W. and Pflugfelder, R.L. (1986) J. Anim. Sci. 63, 1607 - 1623
6. Chrastil, J. (1987) Carbohydrate Research 159, 154 - 158.
7. Bergmeyer, H.U. (1970) Methoden der enzymatischen Analyse Vol. 2, 2nd edn., Verlag Chemie, Weinheim.

Part 3

GELATIN AND OTHER FOOD PROTEINS

Source and production of gelatin

G.STAINSBY
Procter Department of Food Science, University of Leeds, Leeds LS2 9JT, UK

ABSTRACT

Gelatins are polydisperse water-soluble products derived by denaturation and partial degradation from the insoluble, fibrous protein collagen. The chemical composition and structure of collagen, and the biochemical changes that are required in order to convert it into food grade gelatin, with minimal loss of useful functionality, are described. The commercial processes for making gelatin from the raw materials (mammalian skins and bones) are then related to these biochemical changes, and methods for assessing the quality of the gelatin are briefly outlined.

INTRODUCTION

Gelatin has been manufactured for at least 150 years, yet it remains an important industrial hydrocolloid, particularly in the food, pharmaceutical and photographic industries, as the thermo reversibility of its gelation is unique. At the turn of this century gelatin was one of the very few biopolymers available in abundance with high purity. These factors, together with its ease of dispersion, led to extensive studies, and it was soon regarded as typical of all proteins. In the light of more recent work we now know that, on the contrary, it is quite different in structure and behaviour from other proteins, and needs separate consideration.

The name gelatin really refers to a family of similar, water-soluble derivatives of the highly insoluble native protein called collagen. An appreciation of the structure of this protein is required before the conversion to gelatin, the properties of the gelatin and the differences between gelatins can be considered.

COLLAGEN

The structural organisation of the whole of the animal kingdom, from the most primitive invertebrate to man, depends to a major extent on the fibrous nature of collagen. In mammals, collagen accounts for about one third of the total body protein and is involved in an extraordinarily diverse range of physiological and mechanical functions. It is, therefore, found in many forms, including the flexible, interwoven fibres of skin and the mineralised network of bone, the two principal materials for gelatin-making. These forms are mainly composed of just one of the eleven genetically different collagens that have so far been identified, called type I collagen. Embryonic skin is largely a somewhat different type of collagen (type III), and this is gradually replaced during growth to physiological maturity by type I collagen.

Both types of collagen occur as highly organised aggregates of monomeric collagen, held together and strengthened by covalent cross bonds, and share with all collagens several common features. Monomeric collagen is always a highly asymmetric and stiff, but not rigid, protein containing three polypeptide chains, each with more than 1000 amino-acid residues. An essential feature is that almost the whole amino-acid sequence of each chain is in triplets of the form -gly-X-Y. Glycine thus occurs regularly along each chain and accounts for almost exactly one third of the residues. A further third is formed from proline, hydroxyproline and alanine, with all the usual amino acids (except tryptophan and, in type I collagen, the sulphur-containing residues) making up the rest of the X and Y positions. Collagen (and gelatin) is unusual in also containing hydroxy-lysine. This residue, and hydroxyproline, are virtually absent from all other proteins, except elastin. In analytical studies hydroxyproline is used as a convenient marker for collagen or gelatin.

The high proportion of imino residues, and their uneven distribution along the chain triplets, enables each chain to adopt a characteristic helix, very similar to that of synthetic polyproline and quite unlike the α-helix found in other proteins. The regular occurrence of glycine permits just three chains to wind around one another to form the rope-like collagen monomer. All collagens, in consequence, exhibit a characteristic wide angle x-ray diffraction pattern, with a meridional reflection at 0.29 nm that is diagnostic for the repeating pitch along the chains of this compound helix. Many collagens, including those from skin and bone, give a precisely banded structure when viewed in the electron microscope, the pattern repeating at 67 nm. It is the possession of all these chemical and physical characteristics that allow a protein to be defined as a collagen (1).

A reasonably detailed understanding of the somewhat unusual biosynthesis of monomeric collagen is available, together with a less certain description of its assembly into fibres (2). Briefly, each component chain of the monomer is synthesised separately, and contains some 1500 residues, i.e. it is almost half as long again as a collagen chain. The complete amino-acid sequences of these chains are known for the major types of collagen (3), and show that the chain consists in essence of five linked regions (Fig 1): a central region of triplets is linked at each end by a short disordered region, devoid of triplets, to a more organised 'globular' polypeptide.

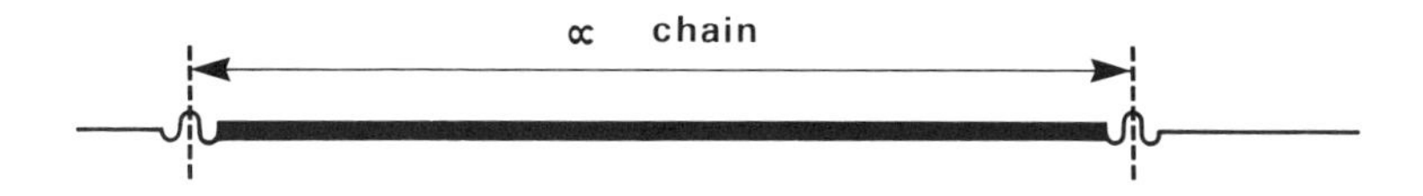

Fig 1. Schematic representation of a single chain of procollagen. Only the central region, between the vertical lines, is retained in collagen. The thickened line represents triplet sequences.

Whilst still inside the synthesising cell three completed chains assemble and become linked together by inter-chain disulphide bonds in the C-terminal globular peptides. This provides the correct chain registration for the chains to fold together from their carboxyl to amino ends. Specific proline and lysine residues are hydroxylated, and some of the hydroxylysine residues then glycosylated, to produce pro-collagen monomers which cannot self-aggregate whilst inside the cell. Subtle changes in these modifications relate to the eventual physiological function of the collagen.

After leaving the cell almost one third of the procollagen residues are discarded by hydrolytic cleavage of the globular regions, to leave only short (≈ 20 residue) disordered telopeptides on either side of the main helical region. What remains is monomeric collagen. Fibres are formed by the self assembly of these monomers in the extracellular fluids. Specific sequences of polarity and hydrophobicity, which are repeated along the helical section, initiate this assembly. The repeating unit contains 234 residues, and is some 67 nm long. Maximum electrostatic and hydrophobic interaction occurs when neighbouring monomers are in a staggered, parallel array (as in Fig 2, for a two-dimensional assembly). Such non-covalent interactions help covalent cross links to form between the monomers, as the four main linkage sites on each chain become suitably juxtapositioned. The first stage of the linking is a condensation between an aldehyde group and the ε-amino group of a lysine, or hydroxylysine. The aldehyde site arises by oxidative deamination of specific lysine and hydroxylysine residues in the telopeptides, whilst the required basic groups are located some 90 residues from each end of the helical domain. A collagen monomer of three identical chains thus has 12 potential sites for covalent linkage, but only a fraction of the telopeptide sites become oxidised.

These links, when first formed, are rather unstable to heat and to acidic conditions, but in time - and by processes that are still rather poorly understood - the links strengthen so that mature, polymerised collagen is very insoluble, difficult to swell, and quite stable to heat.

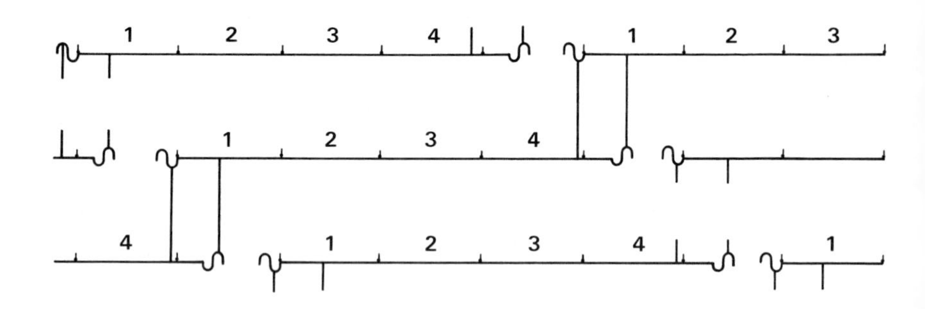

Fig 2. Diagram of a two-dimensional array of collagen monomers. Note the alignment of the repeating units and the space available for the disordered telopeptides. Just four crossbonds are shown, for clarity.

Superimposed on this maturation of the cross links between monomers there is the change, in skin, from type III to type I monomers during physiological maturation of the animal. The compositions of the chains in these monomer types are distinctly different (4). The three chains are identical in type III monomers but in type I monomers two compositionally different chains, called α1 and α2, are involved, the monomer containing one α2 and two α1 chains. Despite the differences, all the chains are homologous with respect to intra- and inter-molecular assembly, and the sites for cross bonding are strongly conserved (5). Type I collagen also contains an intra-chain cross bond, linking a pair of chains in each monomer in the N-terminal telopeptide region. The collagenous tissue for gelatin making is thus rather heterogenous as a polymer system, and the importance of the differences between the component monomers remains to be established. Gross and Rose (6) have suggested that α2 chains are more readily hydrolysed than the others, but Stainsby (7) has opined that the evidence for this is far from conclusive.

CONVERSION TO GELATIN

Though collagen is used to impart desirable mouthfeel to meat products and is the major component in manufactured edible containers for some foods, its dominant use in food is as gelatin. Here again it is the functionality which is all-important as the nutritional quality is low (8). As nutrients in a human diet, however, gelatin or collagen seem to be particularly beneficial during bone repair, burn recovery and general wound healing (9).

A tiny fraction of mature collagenous tissue - that which has been recently synthesised - is still monomeric or only lightly polymerised. This is easily extracted by cold, neutral salt or acidic buffers of low ionic strength (10), and the constituent intact chains separated by only mild heating to form a gelatin. Solubilisation of all the rest of the tissue, however, inevitably demands the breaking of covalent bonds. If only the cross bonds could be ruptured, at a sufficiently low temperature, then monomeric collagen would again result and intact chains be released on denaturation. At present, however, such a highly selective depolymerisation cannot be achieved. Endogenous collagenases attack just one weak bond, but this is in the helical region between residues 775 and 776 and the cross bonds remain intact (11). Even if some of these could then be broken by other means, it is unlikely that a good quality gelatin could be produced because gel-forming ability - the most prized characteristic - depends on the presence of intact α-chains (see page). Other collagenases are even less useful, as many bonds along the helix are attacked.

Hydrolysis in the disordered telopeptide region, however, has the same depolymerising effect as breaking cross bonds, and these regions contain bonds vulnerable to many proteases. In native collagen enzyme accessibility is very severely restricted by the tight cross bonding, though limited swelling occurs in mildly acidic and alkaline conditions. Optimal enzymatic activity near pH 3 or 10, is required, and then progressive swelling could take place as hydrolysis proceeds. Denaturation of the collagen must be avoided, or the released chains would be readily hydrolysed. It follows that low temperature activity is another essential feature for the enzyme, and that it must denature more readily than monomeric collagen. No enzyme is known which meets all these requirements, but pepsin and pronase have been used to produce good quality gelatins (12). To date, however, enzymatic conversion has not become a widespread commercial venture. Perhaps, in due course, modified proteases with the necessary thermal instability may be developed.

Commercial gelatin-making still relies on empirical experience and involves hydrolyses catalysed by acid or alkali, without the aid of enzymes. Studies of the end-groups in gelatins have shown that the splitting of peptide bonds has not been random, or all the constituent amino acids would appear in the same proportions as their occurrence in the primary sequence. Though glycine is found to this extent, the imino acids (proline and hydroxyproline), which occur as 2 residues in 9 in mammalian collagen, are found at most as 1 in 25 end groups (13). In terms of pH, temperature and time - the main variables - the conversion to gelatin has two stages. During the first of these the collagen is 'opened up' progressively using temperatures lower than the denaturation temperature of the monomer. (This falls quite sharply in acidic and alkaline conditions (14). Subsequently, and at much higher temperatures, the gelatin is 'melted out' and then recovered from the solution. The pH for this stage is either just mildly acidic or near-neutral, to minimise degradation. The number of end groups

After extracting to a protein concentration of about 5% the solution is clarified, until sparkling, by serial filtration and the ash reduced by ion-exchange. The liquor is then vacuum evaporated - to minimise hydrolysis and reduce odour - to about 30% solids, and then chilled, extruded to noodles, dried and coarsely ground. A customer's specification often requires further grinding, and the blending of these extracts.

Fresh pigskin, a raw material used extensively in the USA and increasingly in Europe, is a much younger collagen than most cattle hide. It requires only a short, mild acidic pretreatment before washing and extraction. Class A gelatins are produced. The normal equilibrium between the cis and trans forms of back-bone peptide bonds of prolyl residues can be disturbed by heating in acid (26). The commercial acidic pretreatment of collagen is unlikely to cause even a slight disturbance, however, as this would inhibit the folding needed for gelation, and very high quality pigskin gelatins are made. Good quality pigskin gelatin can also be made by lime pretreatment, but there are substantial losses into the lime (27).

The remaining major raw material, ossein, is usually pretreated in alkaline conditions. Though acidic treatment is possible it is rarely used commercially.

The compositions of the main raw materials, and details of the technological aspects of typical processes for converting them to gelatine, have been fully described (12). A valuable feature of this exposition is the information on yield and quality as functions of the severity of pretreatment and the extraction conditions. Johnston-Banks (28) summarises the main features of a modern, efficient process for the production of medium grade edible gelatine from tannery waste. The air-dry product of commerce is usually written as gelatine. It consists mainly of the protein gelatin, the balance being water with a little inorganic ash and an even smaller fraction of organic impurities, derived from lipids and mucoids which survive through the manufacturing processes. The water content varies with humidity. In the UK the gelatin content is usually some 85% w/w at the point of sale. Gelatine swells rapidly in cold water and dissolves completely in warm water, provided the moisture content is never reduced to about 1%. If it is, then permanent insolubility occurs.

QUALITY CONTROL

No account of the production of gelatin would be complete without a brief mention of the main aspects of routine quality control. In this regard, the criteria for good food-grade gelatines are not quite as demanding as those for photographic gelatin, as extremely minute quantities of some chemcials (originating in the raw material or in the processing aids) can seriously affect the sensitivity of the silver halide photo-graphic process.

The main physical properties, for grading any gelatine, are the Bloom strength and the viscosity, each under carefully standardised conditions. Comprehensive details for these, and for other physical properties, are available (20). The Bloom strength is essentially the weight required to make a carefully specified plunger depress the surface of a set gel by 4 mm. As the strength of a gel depends on concentration, and thermal history, these must be properly controlled. The jelly for this test is set in a bottle of specified dimensions and contains exactly 6.2/3% w/w air-dry gelatine (i.e. some 6% protein). It is conditioned for 16-18 hours at 10° before testing at 10°. [The strength of such a gel continues to rise with time, though at a decreasing rate. At this lifetime, the rise is some 1/2% per hour.] Under the test conditions the strength is approximately proportional to the square of the protein content. Although it is a large and highly complex deformation, in rheological terms, the Bloom strength has been shown to be linearly related to the absolute rigidity modulus, obtained using small, simple deformations. A medium grade gelatine has a Bloom value in the range 150-220 g, and a high grade a Bloom of up to about 300 g.

Although Bloom strength depends on pH, this is not controlled and the test is made without adjusting the pH given by the sample. This can be misleading for performance in particular uses, especially when these require the gel to be much nearer to the melting point, e.g. at a much lower concentration and/or at room temperature.

Viscosity is the second most important routine physical test. It is normally determined at 60°, again at the nominal concentration of 6.2/3% gelatine, and the results expressed in millipoise. pH is an important factor affecting viscosity, and should be stated.

Other physical properties which may be specified include setting time, melting point, pH and some agreed procedure for evaluating the particle size and hence the ease of dissolution of the powder.

Detailed methods for the chemical examination of gelatines are also available (30). Among the inorganic contaminants the main elements controlled by statute for food usage are copper (< 30 ppm), lead (< 10 ppm) and arsenic (< 1 ppm). Sulphite is sometimes used as a preservative, and in edible gelatine the statutory limit is 0.1%. Organic contaminants and calcium, one of the commonest ions in gelatine, must not contribute to a haze or to a visible colour, as gelatin solutions must be bright and colour-free.

CONCLUSIONS

Despite its structural complexity, and insolubility, collagen is readily transformed by quite simple processes to a water-soluble ingredient with very useful functionality for the food industry. This was quite satisfactorily achieved long before the structure of collagen was elucidated, and now that this is known in some detail it is potentially feasible for more effective conversions to be developed. Many vegetable proteins, particularly the cereal glutens, are as insoluble and intractable as collagen, and maybe in time they too will yield new ingredients for food use. Nor will such improvements be restricted to proteins: among the polysaccharides, pectin already parallels gelatin, and in due course there should be others.

REFERENCES

1. Ramachandran, G.N. (1967) in Treatise on Collagen, Vol.1. Chemistry of Collagen (ed. Ramachandran, G.N.) pp 103-183. Academic Press, London.
2. Fessler, J.H., Doege, K.J. and Fessler, L.I. (1987) in Recent Advances in Meat Research, Vol. 4. Collagen as a Food (eds. Pearson, A.M., Dutson, K.R. and Bailey, A.J.) pp 283-298. Van Nostrand Reinhold, New York.
3. Miller, E.J. (1985) Ann. N.Y. Acad. Sci., 460, 1-13.
4. Weiss, J.B. (1984) in Connective Tissue Matrix (ed. Hukins, D.W.L.) pp 17-53. Macmillan, London.
5. Hofmann, H., Feitzek, P.P. and Kuhn, K. (1980) J. Mol.Biol., 141, 293-314.
6. Gross, S. and Rose, P.I. (1975) J. Phot. Soc., 23, 33-43.
7. Stainsby, G. (1987) in Advances in Meat Research, Vol.4. Collagen as a Food (eds Pearson, A.M., Dutson, T.R. and Bailey, A.J.) pp 209-222. Van Nostrand Reinhold, New York.
8. Stainsby, G. and Ward, A.G. (1969) J. Soc. Leather Trades Chem., 53 2-11.
9. Pearson, A.M. (1987) in Recent Advances in Meat Research, Vol.4. Collagen as a Food (eds Pearson, A.M., Dutson, K.R. and Bailey, A.J.) pp 382-383. Van Nostrand Reinhold, New York.
10. Chandrakasen, G., Torchia, D.A. and Piez, K. (1976) J. Biol. Chem., 251, 6062-6067.
11. Bailey, A.J. and Etherington, D.J. (1987) in Advances in Meat Research, Vol.4. Collagen as a Food (eds Pearson, A.M., Dutson, T.R. and Bailey, A.J.) pp 305-323. Van Nostrand Reinhold, New York.
12. Hinterwaldner, R. (1977) in Science and Technology of Gelatin (eds Ward, A.G. and Courts, A.) pp 295-364. Academic Press, London.
13. Grand, R.J.A. and Stainsby, G. (1975) J. Phot. Soc., 23, 67-72.
14. Crosby, N.T. and Stainsby, G. (1962) Research, 15, 427-435.
15. Grand, R.J.A. and Stainsby, G. (1975) J. Sci. Fd Agric., 26, 295-302.
16. Saunders, P.R. and Ward, A.G. (1955) Nature, London, 176, 26-28.
17. Kemp, P.D. and Stainsby, G. (1981) J. Soc. Leather Tech. and Chem., 65, 85-90.
18. Ward, A.G. (1958) Leatherhead Food R.A. Sci. and Tech. Survey No. 31; Ward, A.G. (1977) in Science and Technology of Gelatin (eds. Ward, A.G. and Courts, A.) p xi. Academic Press, London.
19. Maxey, C.R. and Palmer, M.R. (1976) in Photographic Gelatin-2 (ed. Cox, R.J.) pp 27-36, Academic Press, London.
20. Frey, P. and Nitschmann, H. (1976) Helv. Chim. Acta, 59, 1401-1409
21. Tomka, I. (1982) U.S. Pat. No. 4, 360, 590.

22. Stainsby, G., Wooton, J.W. and Ward, A.G. (1969) in Proc. 1st Int. Congr. Food Sci. Technol., London, 1962. (ed. Leitch, J.M.) Vol.1. pp 753-764. Gordon and Breach, London.
23. Rose, P.I. (1977) in Theory of the Photographic Process, (ed. James, T.H.) 4th edition, pp 51-67. MacMillan, New York.
24. Marrs, W.M. (1984) Leatherhead Fd. Res. Report No. 461.
25. Shirai, K., Wada, K. and Kawamura, A. (1979) Agric. Biol. Chem., 43, 2045-2051.
26. Schmid, F.X. and Baldwin, R.L. (1978) Proc. Nat. Acad. Sci. U.S.A., 75, 4764-4768.
27. Takahashi, K., Shirai, K. and Wada, K. (1988) J. Food Sci., 53, 1920-1921.
28. Johnston-Banks, F.A. (1984) J. Soc. Leather Tech. and Chem., 68, 141-145.
29. Wainewright, F.W. (1977) in Science and Technology of Gelatin (eds. Ward, A.G. and Courts, A.) pp 508-534. Academic Press, London.
30. Leach, A.A. and Eastoe, J.E. (1977) in Science and Technology of Gelatin (eds. Ward, A.G. and Courts, A.) pp 476-506. Academic Press, London.
31. Hamilton, P.B. and Anderson, R.A. (1954) J. Biol. Chem., 211, 95-102.

Functional properties of gelatin

D.A.LEDWARD

School of Agriculture, University of Nottingham, Sutton Bonington, Loughborough, Leicestershire, UK

ABSTRACT

Gelatin is not a unique chemical entity but represents a range of polypeptides, the size, shape and charge distribution of which vary with the source and method of manufacture. The functional properties of one gelatin sample are unlikely to be the same as those prepared from either a different starting material or by different technologies. Although gelatin may serve many functions as a food ingredient, including acting as an emulsifying, stabilising, thickening, foaming or water binding agent its major use is as a gelling agent. The unique amino acid composition and sequence of gelatins enables high grade samples to gel at concentrations of 1% or lower. These gels are formed by the imino acid rich regions of the relatively flexible polypeptide chains adopting a helical conformation on cooling. These helices achieve stability by hydrogen bonding together to partially reform the collagen triple helix which can act as the junction zones of the gel; although not all such helices necessarily act as junction zones. The number and size of such zones affects the properties of the gel and depends on the nature of the solvent and thermal history of the system as well as the size and amino acid composition and sequence of the polypeptides.

INTRODUCTION

As can be seen from the previous chapter gelatin consists of a heterogeneous mixture of polypeptides which may be single chains of molecular weight 100,000 or less (α chains and their degradation products), dimers of molecular weight up to 200,000 (β -chains and their degradation products), or even higher molecular weight aggregates (Stainsby, this volume). In addition the amino acid composition and sequence is not unique and depends on the source and, to some extent, on the method of manufacture. Thus gelatins derived from cold water fish such as codskins contain only about 155 imino acid residues (proline and hydroxyproline) per 1000 whilst gelatins derived from the skins of warm blooded animals (pigs and cattle) contain about 221 imino acid residues per 1000. Important differences due to the method of manufacture include the observations that alkali extraction leads to loss of ammonia from glutamate and aspartate residues with the formation of the corresponding acids and consequent decrease in pK value (Stainsby, this volume); possible conversion of L-residues to the D-form may occur during alkaline, but not acid extraction, and alkaline extraction also appears to lead to more rupture of peptide bonds involving imino acid residues. As will be seen all these factors will affect the functional properties of the gelatin. Gelatin has many uses in food, as described in the following chapters, including use as a binder, thickener, emulsifier, fining agent, stabiliser and whipping agent. However its main use is undoubtedly as a gelling agent and it is in this respect that its properties are unique.

GELATION

Among the commonly used gelling agents gelatin is unique in that at concentrations as low as 1% it will form a thermoreversible gel that converts to a solution as the temperature rises to 30-40°C; thus gelatin gels tend to 'melt in the mouth'. This is obviously a very desirable property in a number of foods.

Current views regarding the gelation mechanism have recently been reviewed[1] and will merely be summarised here with special attention being paid to work reported since that date.

At temperatures above 40°C gelatins are assumed to exist as a random coils[2] but the high imino acid content must, to some extent, limit their flexibility[3,4] and it has been suggested that even at these temperatures the chains possess some structure, the actual amount depending on the imino acid content. It is now generally accepted that these pyrollidine rich regions of the chains act as potential junction zones in that, as the temperature decreases these regions, especially those with the sequence gly-pro-pro (or hypro)-gly-pro-pro (or hypro) tend to take up the poly-L-proline II helix, aggregates of three such helices, in a collagen fold formation, yielding the junction zones.

These junction zones are stabilised by interchain hydrogen bonds similar to those found in native collagen and some water molecules are also reorientated and bound within the structure[5]. It seems likely that at least some of these water molecules serve to stabilise the triple helix by hydrogen bonding in a similar manner to that proposed for native collagen[6,7]. Although proline is important, hydroxyproline in the third position of the triplet is believed to be the major determinant of stability due to its hydrogen bonding ability[8].

Although most theoretical treatments of a gelatin gel have assumed the junction zones are individual triple helices there is evidence, from electron microscopic studies carried out on a range of mature gelatin gels, that at least some of the junctions are composed of several triple helices aggregated together. This hypothesis though is not universally accepted[9].

Effect of Maturing Conditions on Gel Strength

The strength of a gelatin gel increases as the temperature decreases due to either an increase in the number of crosslinks[10], or the growth of existing junctions[11], or both.

Manipulation of the thermal history of the gel can markedly affect its strength[1]. Thus holding a solution at a temperature just below that required for nucleation gives rise to a weak gel with only a few junctions. These junctions though are very stable since the chains involved in their formation have sufficient flexibility to 'anneal' themselves into their most stable 'collagen like' conformation, other regions, of lower imino acid content, will not form junctions. However if this weak gel is subsequently chilled to a lower temperature the strength will increase as these junctions grow and new ones

Fig. 1. Increase in rigidity modulus of 2.0% gels of a commercial limed hide gelatin, at pH7, matured at the temperature shown for 74 hr prior to holding at 10°C for 90 hr. Control gels were chilled directly to 10°C. From ref. 24.

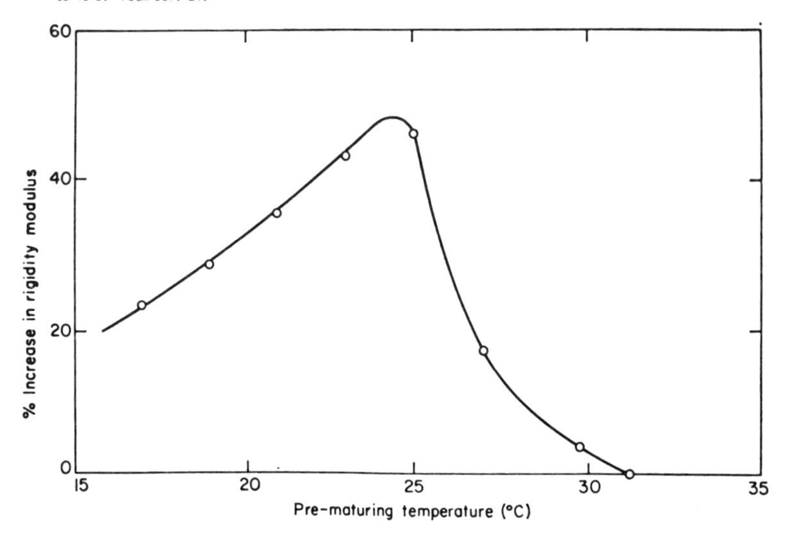

form. A solution of the same concentration snap chilled to this lower temperature will probably have a similar number of junction zones to those formed in the gel prematured at the higher temperature. The gel though will be weaker since the initial nucleation sites will not have been able to anneal themselves into their most stable conformation and encourage further growth of these zones on subsequent cooling[1] (Fig. 1). Oakenfull[12] has estimated that in dilute (1-2%) gels at high temperatures the average junction contains about 142 residues, i.e. 47 per chain, although at 1°C junctions can form involving only 30-50 residues[13]. Busnel <u>et al</u>[9] estimate that in gels matured at 10°C about 20 residues per chain are involved in the junction zones.

An interesting outcome of the above hypothesis is that a weak gel, with a few very stable linkages, matured at a high temperature, may well have a higher melting point than a strong gel of the same solution, with numerous weaker linkages, prepared by rapidly chilling to a lower temperature.

The strength of a gelatin gel increases with time when held at a fixed temperature, presumably due to the junction zones reorganising themselves or to new zones forming as the gels anneal. It has been claimed that on maturation the range of temperatures over which the junction zones 'melt' decreases as the average melting temperature increases[9].

Not unexpectedly the strength of a gelatin gel increases with concentration and when matured under the same conditions the rigidity modulus, G, over a small concentration range, is proportional to the concentration to the power n, where n is usually about 2 but may be as low as 1.27 for certain gelatins[14].

As with any gelling system the properties of the gel will depend, at least to some extent, on the nature of the solvent as well as the size, shape and chemical composition of the macromolecule. Gelatin is no exception. For example although the strength of a gelatin gel is little affected by pH in the range 4 to 10 outside these limits there is a sharp decrease in strength. This decrease may be due to the fact that at these extreme pH values the chains carry a high positive or negative charge and thus electrostatic repulsive forces inhibit their coming into suitable juxtapositions for the formation of junction sites. The junction zones are relatively non-polar in nature[1] and thus the effect of small changes in pH, and thus charge on the protein, would not be expected to be very marked.

The effect of pH on gel strength, (rigidity modulus) becomes far more pronounced in weak gels i.e. at low gelatin concentrations and high maturing temperatures[15] possibly reflecting the fact that electrostatic forces are of greater importance in influencing the aggregation of the more mobile chains found in these systems. Similarly the observation that high concentrations of neutral salts, at neutral pH, cause a lowering of rigidity modulus[16] indicates that electrostatic considerations are of some importance.

Non-electrolytes such as sugars and glycerol usually increase the gel strength of gelatin gels. NMR studies indicate that stabilisation by sugars is a structural effect and operates in the order sucrose > D-galactose $\geq$ D-glucose with fructose having no measurable stabilisation[17]. These workers suggest that the stabilisation is due to hydrogen bonding which operates in such a way as to involve water in a manner that is not found with sugars or gels alone. The addition of such compounds as dextran, methyl cellulose, polyethylene glycol & polyvinyl chloride may increase the rate of gelation by encouraging polymer - polymer contacts to the detriment of polymer-solvent ones.

Other chemicals (e.g. KSCN, LiBr, $CaCl_2$, urea, phenols) at high enough concentration totally prevent gelation. As such reagents are capable of breaking both hydrophobic and hydrogen bonds[18] they presumably act by preventing the stabilisation of the gel junction sites, either directly by preventing hydrogen bond formation and/or by modifying the structure of the liquid water in the vicinity of these sites.

<u>Rigidity Factor of Gelation Gels</u>

Although the above explains how the properties of a gelatin gel depend on maturing conditions, it is known that gelatins of the same weight average molecular weight from similar sources may have widely differing gel strengths when matured under identical conditions. This difference has been interpreted as an inherent structural feature of each gelatin, known as the 'rigidity factor'. Two factors which contribute to this phenomenon are undoubtedly the heterogeneity, with regard to size and shape, of the gelatin molecules and also the amount and distribution of the imino acids within the polypeptide chains. Gelatins of low imino acid contents, such as those derived from codskins invariably have poor gelling ability. It has been suggested[1, 19] that the pretreatment and method of preparation of the gelatin may be largely responsible for the wide range of rigidity factors found for gelatins from the same source. For example, if the gelatin is hydrolysed, in the apolar, pyrollidine-rich regions, which may result from alkaline treatment then one would expect the ability of the chains to take up the poly-L-proline II type configuration to be diminished as the imino residues at the chain ends would be unable to take part in this type of rearrangement. Thus hot alkaline hydrolysis causes a marked decrease in gel strength with decreasing molecular weight. If hydrolysis occurs in the polar regions, as occurs with tryptic and perhaps acid digestion, then spiralisation of the polypeptide chain should be "unaffected" and hence the gel strength will be independent of molecular weight, provided the change in molecular size does not affect the ability of the chains to link effectively in the gel network[1]. This explanation may be an oversimplification since alkali will also lead to some conversion of the isomers to the D-form which may inhibit helix (and gel) formation.

It is well established that the α_1 chains of collagen are richer in pyrollidine residues than the corresponding α_2 chains (220 and 205 respectively) and this is undoubtedly one reason why gels made from α_1 chains are more stable than those made exclusively from α_2 chains[20,21].

However the situation is more complex than this as the difference in pyrollidine content is not thought great enough to account for the large differences in stability found[20]. In addition it has been found that $(\alpha 1)_2\ \alpha_2$ renatured collagen is more stable than the other possible combinations[20]. It is reasonable to postulate that additional stabilisation between the different chain types occurs on gelation or renaturation. The observation that the setting time of a mixture of gelatins of different isoelectric points (5-6 and 6-9) is markedly decreased compared to the separate gelatins at the same concentration[1] suggests that electrostatic (charge) effects are important since in the mixed system the chains carry net charges of opposite sign, which should encourage attraction. Such considerations may explain, to some extent, why a gel made from a 2:1 mixture of α_1 and α_2 gelatins is at least as stable as one made exclusively of α_1 chains, since α_1 and α_2 chains possess different isoelectric points, they can be separated by ion-exchange chromatography.

The presence of a covalent linkage between α chains will obviously have some effect on gelation behaviour since at least two chains will be held in register. Their effect though is difficult to predict since they may contribute to the rigidity of the gel by increasing the number of crosslinks or they may encourage the formation of <u>intra</u>molecular (non-useful) junctions (see next section).

Takahashi, Shirai & Wade[21] found that gels made exclusively from β_{11} chains had similar melting points to those made from α_1 chains, at the same concentration and β_{12} chains had higher melting points than those prepared from α_2 chains. Thus the covalent linkages in these models do not appear to be detrimental to the melting points of the gelatins. Unfortunately the strength of the gels was not measured but the authors did observe that the enthalpy of melting was much lower for the dimers than the monomers. They suggest that the covalent linkage may inhibit helix formation and, if this is so, gels made from the dimers should be weaker. The melting points will presumably be independent of the presence of covalent bonds since such bonds are located at the chain ends. Thus the formation of more stable junctions, derived from imino acid rich regions away from the ends, and which dictate the melting point, should be unaffected. In gels matured at 10°C the enthalpy of melting for these purified gelatins was only 15-44% of the enthalpy of denaturation of acid soluble collagen[22] (Table 1).

<u>Table 1</u> Temperatures and enthalpy of melting of acid soluble collagen and gelatin gels prepared from its subfragments. Gels prepared from 2.4 to 3.1% solutions matured at 10°C for 20hr. Data from refs. 21 & 22.

Sample	Melting Temp. °C		ΔH
	onset	finish	mJ/mg
Acid soluble collagen (ASC)	36	42	65.2
ASC - gelatin	28.6	39.0	18.4
α_1 - gelatin	28.0	37.5	28.4
α_2 - gelatin	19.0	29.0	10.0
β_{11} - gelatin	28.2	38.0	20.1
β_{12} - gelatin	27.7	36.0	13.8

<u>Concept of Useful and Non-Useful Junctions</u>

So far in this discussion we have assumed all junction zones contribute to the gel network. This is obviously an oversimplification since long chains may form intramolecular bonds which will not contribute to the network and small fragments may take part in helix formation but not be of sufficient size to contribute to the network[1].

In solution of 40°C gelatins are laevorotatory and on gelation there is a marked increase in the laevorotatory dispersion of the system, due to the partial reformation of the poly-L-proline II helix. Ultimately, on complete renaturation to collagen the specific optical rotation is identical to that found for native tropocollagen. This change in specific optical rotation is therefore an index of the concentration of helices present in the gelled solution. Since the rigidity modulus is proportional to the square of the gelatin concentration (and thus the square of the helix content) it is not surprising that there is a linear relationship between the specific optical dispersion [- α] and the square root of the rigidity modulus, G, (Fig. 2). The relationship is apparently independent of pH (in the range 3.7 to 7), ionic strength and chemical modification of the gelatin but is concentration

FIG. 2. Relationship between the square root of the rigidity modulus, $G^{\frac{1}{2}}$ and specific optical rotation $(-\alpha)_{578}$ for 2.0% (o) and 5.5% (●) gels of a commercial limed hide gelatin matured under different conditions. From Ref. 24.

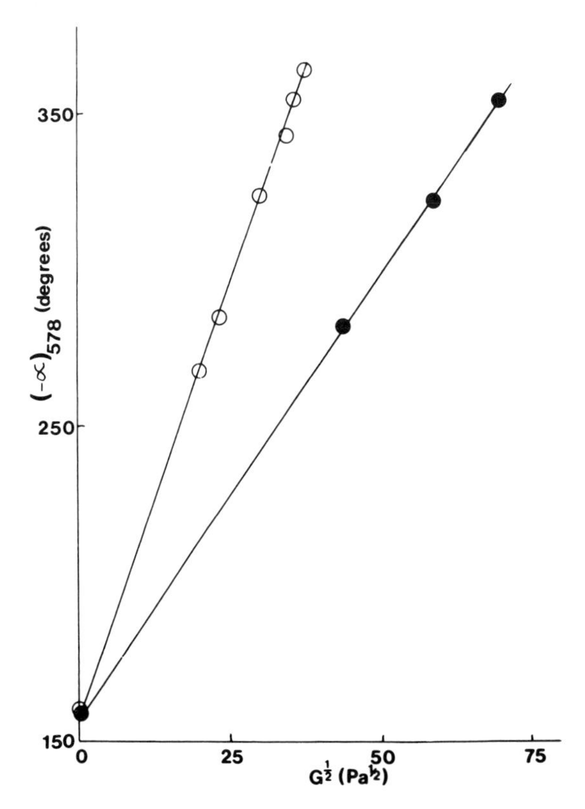

dependent[1]. It is also very dependent on the nature of the gelatin.

We have argued that the gradient of the plot of $[-\alpha]_\lambda$ against $G^{1/2}$, k, is an index of the of number useful to non-useful junction zones in the gel since $[-\alpha]_\lambda$ is a measure of the total poly-L-proline II helix content and G is related to the total number of network junctions. Thus if the helices aggregate to yield suitable junction zones for a gel network k will be small while if the helices associate intramolecularly to yield small cyclic aggregates which make little contribution to the gel network,k will be large. It is generally believed that at very low (<0.1%) concentration the mechanism is unimolecular with the triple helix being formed primarily by single chains folding back on themselves[9] i.e. k is infinite since no useful junctions form.

At higher concentrations the possibility of a single chain folding back on itself to form a non-useful cyclic aggregate must decrease due to the increased possibility of intermolecular stabilisation from other molecules i.e. k decreases with increasing gelatin concentration (Fig. 2). For a well characterised alkali processed gelatin Busnel et al.[9] found that at 0.025% concentration there was little increase in weight average molecular weight on maturation at 10°C even though helix formations was seen. However, the weight average molecular weight of a 0.2% solution doubled under similar conditions. These authors also claimed that the renaturation kinetics of gelatin solutions, in the concentration range 0.1 - 1.0%, suggest the mechanism is bimolecular i.e. the junction zones consist primarily of two chains, one looping back on itself to supply two segments of the helix whilst the second chain supplies the third segment. They also claim that the possibility of 3 separate chains forming the helix is remote. However Oakenfull and Scott[23] suggest that even at threshold concentrations, three chains are involved in the formation of the junction zone. The decreased possibility of non-useful junctions forming at higher concentrations may account for the observation that the relationship $G \propto C^2$ is only valid over the relatively small concentration range of 2 to 6%. The molecular weight distribution of the gelatin molecules would also be a key factor in determining the ratio of useful to non-useful linkages. Studies carried out on an acid degraded gelatin[24] clearly demonstrated that as the weight average molecular weight decreased from 190,000 to 35,000, the value of k, for 2.0% solutions,increased by 25% suggesting a marked increase in the ratio of non-useful to useful junctions with decreasing molecular size. Saunders & Ward[25] also determined both the rigidity modulus and specific optical rotation of 5.8% gels under different maturing conditions although no attempt was made to correlate these parameters. If such a correlation is made it becomes apparent that the slopes (k values) of the $[-\alpha]$ against $G^{1/2}$ curves decrease by about 20% i.e. the ratio of non-useful to useful junctions decreases as the weight average molecular weight decreases from 270,000 to 70,000. Thus it would appear that a minimum value of k exists at some specific weight average molecular weight i.e. at a given concentration there is an optimum molecular weight for the maximal formation of useful compared to non-useful junctions. Because of molecular size and shape heterogeneity is it is not surprising that the optimum weight average molecular weight is not unique but varies from gelatin to gelatin[1]. Long single chains in dilute solution will presumable be able to fold back on themselves and aggregate <u>intra</u>molecularly, leading to the formation of non-useful junctions. Samples containing chains of reduced size prepared by hydrolysis of the polar regions will still have the same ability to form the poly-L-proline II helix but stabilisation is more likely to occur primarily by <u>inter</u>molecular association and perhaps therefore give rise to gels of increased rigidity. Below a certain molecular size the possibility of forming an infinite 3-dimensional network must diminish, and although the ability to form the helix may also decrease, the effect will be less marked[24] thus k will increase. This simple analysis will be complicated by the presence of multichain gelatins since these links may contribute to the rigidity of the gel but not its optical rotation, thus decreasing k or, because they may hold the chains in register they may encourage the formation of non-useful <u>intra</u>molecular junctions and thus increase k. However the results of Takahashi <u>et al</u>.[21] suggest that, at least with dimers, no increase in non-useful junctions occurs.

<u>Summary of Factors Affecting Gel Strength</u>

From the above it can be seen that the strength of a gelatin gel will depend on many factors including

(a) Time:- it will only very slowly, if ever, reach a constant value when matured at a given temperature.
(b) Temperature:- it will be greater at lower temperatures. In addition, the value at any temperature will be increased by holding at a higher temperature but still below the nucleation temperature prior to chilling.
(c) Solvent:- differences due to modification of the solvent will be more marked at higher temperatures and lower gelatin concentrations.
(d) Amino acid composition and sequence.
(e) Weight average molecular weight:- the dependence though is complex being different for acid processed and alkali processed gelatins.
(f) Concentration:- over a small concentration range the strength is proportional to the square of the concentration i.e. $G \propto C^n$ where $n = 2$. However the relationship is not unique in that n may be as small as 1.27 for some gelatins.

Although it is generally accepted that poly-L-proline II type helices are the basis of the junction sites in gelatin gels formed by cooling, it has been shown that if a gelatin gel is formed by cooling a solution in the presence of known crosslinking agents, such as formaldehyde, helix formation is inhibited. However a gel may still form and it is assumed that in these cases most junction sites are covalent in nature and involve the crosslinking agent e.g. formaldehyde[26].

OTHER FUNCTIONAL PROPERTIES

Although it is the gelling ability of gelatin that has

received most attention by the Food Industry gelatin is, like virtually all proteins, surface active and will adsorb at air water or oil water interfaces and thus serve to stabilise foams and emulsions[27]. However it is not the most effective agent available and thus its use tends to be limited. Recently though it has been shown that mixtures of acidic and basic proteins are very effective foaming agents and foam stabilisers[28]. Since gelatins possess isoelectric points ranging from 4.7 to 9.6 the functionality of mixtures of different gelatins in such systems is worthy of study. The isoelectric points of the proteins can be extended by appropriate modification of the acidic or basic amino acid sidechains. Similar considerations may extend the usefulness of gelatins as stabilisers of frozen products where they may limit ice crystal growth. As gelatins are charged macromolecules, they may bind other charged molecules and thus aid in clarifying wines and beers. Such aspects of use are considered in the following chapters.

REFERENCES

1. Ledward, D.A. 1986. In 'Functional properties of food macromolecules' ed. J.R. Mitchell & D.A. Ledward pp. 171-201, Elsevier Applied Science, London.
2. Amis, E.J., Janmey, P.A., Ferry, J.D. & Yen, H. 1981. Polymer Bull. Berlin 6, 13-20.
3. Josse, J., Harrington, W.F. 1964. J. Mol. Biol. 9, 269-287.
4. Stainsby, G. (1977). In 'Science & technology of gelatin' ed. A.G. Ward & A. Courts pp. 179 - 247, Academic Press, London.
5. Naryshkina, E.P., Volkov, V. Ya, Dulinnyi, A-I Izmailova, V.N. 1982. Kolloidn. Zh. 44, 356 - 362.
6. Borstein, P. & Traub, W. 1979. In 'The proteins -IV' 3rd edn. ed. H. Neurath, R.L. Hill & C.L. Boeder, pp 411 - 632, Academic Press, London.
7. Glanville, R.W. & Kuhn, K. 1979. In 'The fibrous proteins: scientific industrial & medical aspects - 1' ed. D.A.D. Parry & L.K. Creamer pp. 133 - 150, Academic Press, London.
8. Privalov, P.L. 1982. Adv. Prot. Chem. 35, 1 - 104.
9. Busnel, J.P., Morris, E.R. & Ross-Murphy, S.B. 1989. Int. J. Biol. Macromol. 11, 119-125.
10. Bedborough, D.S. & Jackson, D.A. 1976. Polymer, 17, 573 - 584.
11. Nijenhuis, K. 1981. Colloid Polymer Sci. 257, 1017 - 1026.
12. Oakenfull, D. 1984. J. Food Sci. 49, 1103 - 1104, 1110.
13. Hauschka, P.V. & Harrington, W.F. 1970. Biochemistry, 9, 3734-3763.
14. Robinson, J.A.J., Kellaway, I.W. & Marriot, C. 1975. J. Pharm. Pharmacol. 27, 77 - 84 and 27, 653 - 658.
15. Bello, J., Bello, H.R. & Vinograd, J.R. 1962. Biochem. Biophys. Acta. 57, 214 -221.
16. Finch, C.A. & Jobling, A. 1977. In 'Science & technology of gelatin' ed. A.G. Ward & A. Courts, pp 249 - 294, Academic Press, London.
17. Naftalin, R.J. & Symons, M.C.R. 1974. Biochem. Biophys.

Acta. 352, 173 - 178.
18. Finch, A., Gardner, P.J., Ledward, D.A. & Menashi, S. 1974. Biochem. Biophys. Acta. 365, 400 - 404.
19. Grand, R.J.A. & Stainsby, G. 1976. In 'Photographic Gelatin - 2' ed. R.J. Cox, pp 76 - 90, Academic Press, London.
20. Tkocz, C. & Kuhn, K. 1969. Eur. J. Biochem. 7, 454 - 462.
21. Takahashi, K., Shirai, K. & Wade, K. 1988. J. Food Sci. 53, 1920 - 1921.
22. Menashi, S., Finch, A., Gardner, P.J. & Ledward, D.A. 1976. Biochem. Biophys. Acta. 444, 623 - 625.
23. Oakenfull, D & Scott, A. 1986. In Gums & Stabilisers for the Food Industry - 3' ed. by G.O. Phillips, D.J. Wedlock & P.A. Williams pp 465 - 475, Elsevier Applied Science, London.
24. Ledward, D.A. 1968. Ph.D. Thesis, University of Leeds.
25. Saunders, P.R. & Ward, A.G. 1958. In 'Recent advances in gelatin and glue research' ed. G. Stainsby, pp 197 - 203, Pergamon Press, London.
26. Rogovina, L.Z., Vasil'ev., V.G., Malkis, N.I., Slonimski, G.L., Titova, E.F. & Belavtseva, W.M. 1979. Vyskmol. Soedin Ser A. 21, 1235 - 1243.
27. Stainsby, G. 1986. In 'Functional properties of food macromolecules' ed. J.R. Mitchell & D.A. Ledward, pp 315 - 353. Elsevier Applied Science, London.
28. Poole, S. 1989. Int. J. Food Sci. & Technol., 24, (in press).

Interactions of gelatin with polysaccharides

V.B.TOLSTOGUZOV
Institute of Organoelement Compounds, USSR Academy of Sciences, Moscow, USSR

ABSTRACT

The functional properties of gelatin-polysaccharide mixtures are strongly dependent on two phenomena: complex formation and thermodynamic incompatibility of these macromolecular substances. Electrostatic interactions between gelatin and anionic polysaccharides lead to complex formation at low ionic strength (under 0.3) and pH values below the gelatin isoelectric point. Unlike globular proteins, gelatin forms electrically neutral insoluble complexes with anionic polysaccharides over a wide range of system compositions. Phase separation due to the incompatibility of gelatin and polysaccharides occurs where interpolymer complexing is inhibited and the bulk concentration of biopolymers exceeds a certain critical value -usually 1-4%. Thought is given to the effect of the two phenomena on the formation of structure and a set of physicochemical properties of model gelatin-polysaccharide systems, including complex, mixed, and filled gels.

INTRODUCTION

For twenty years now we have been investigating nonspecific interactions between the main classes of food proteins and polysaccharides. Here we will report some findings, using as example systems containing gelatin and polysaccharides. Then their applications will be briefly discussed. Most specialists admit definitely that the functional properties of food systems containing gelatin and polysaccharides is a paramount aspect of these biopolymers interaction. To study the relationships between the composition, structure, and properties of the systems containing these biopolymers is, therefore, one of the most important problems facing us today. Unfortunately, our understanding of these relationships is still rather poor, despite the fact that gelatin and polysaccharides are the historically longest studied polymers and the most common food hydrocolloids. What kind of interactions are of interest to food science and technology? We distinguish between two types of gelatin-polysaccharide interactions related either to attraction or repulsion of these macromolecules (Figure 1). Accordingly, they are responsible either for complexing or for immiscibility of these biopolymers in aquatic media (1-4). Since we are dealing here with polyelectrolyte interactions, the conditions for their complex formation or thermodynamic incompatibility are determined primarily by pH, ionic strength, charge density, and biopolymers concentration in a system. In the case of complexing, the Gibbs free energies of interaction between dissimilar macromolecule segments should exceed both the energy of thermal motion and the Gibbs energies of interaction of similar macromolecules with each other and with water. Besides the desolvation effects typical also of interactions between low molecular mass acids and bases, a change in the Gibbs free energy on gelatin-polysaccharide complexing should compensate the loss in the system configuration entropy. The latter seems to increase with chain flexibility. Interpolymer complexing occurs at pH values below the protein isoelectric point (IEP) and at low ionic strengths (usually under 0.3). Under these conditions, the polypeptide chains of gelatin have a net positive charge, i.e. they behave as polycations. The experimental data discussed below refer to alkali processed gelatin with an IEP of 4.75 as well as to carboxyl-containing anionic polysaccharides, primarily to alginates and pectins. These polysaccharides are polyanions. With pH varying from 4.75 to 2, an interaction of the positively charged macro-ions of gelatin with the negatively charged macro-ions of an anionic polysaccharide gives rise to interpolymer electrostatic complexes. These latter can be regarded as a new type of food biopolymers whose properties differ substantially from those of the macromolecular reagents. Complexes can be soluble (Figure 1A) and insoluble (Figure 1B). In the latter case, the system is separated into two phases. This phenomenon has been given the name of complex coacervation (5-7).

Thermodynamic incompatibility in solution mixtures of gelatin and polysaccharides shows up if electrostatic complexing is inhibited, i.e. at a sufficiently high ionic strength and at pH values above the gelatin IEP, where the macro-ions of the polymers have like charges. Generally, the conditions of a medium for biopolymer thermodynamic incompatibility should inhibit the interactions between dissimilar polymers and promote those of similar macromolecules. Of prime importance are also the excluded volume effects. In view of the unfavored interactions between segments of dissimilar polymers, each macromolecule shows preference for a surrounding of its own type. Solution mixtures with a low bulk concentration of polymers may be thermodynamically stable (Figure 1C). However, when the bulk polymer concentration is increased above a certain critical value, a system breaks down into two liquid phases (Figure 1D). Thus, on mixing a gelatin and

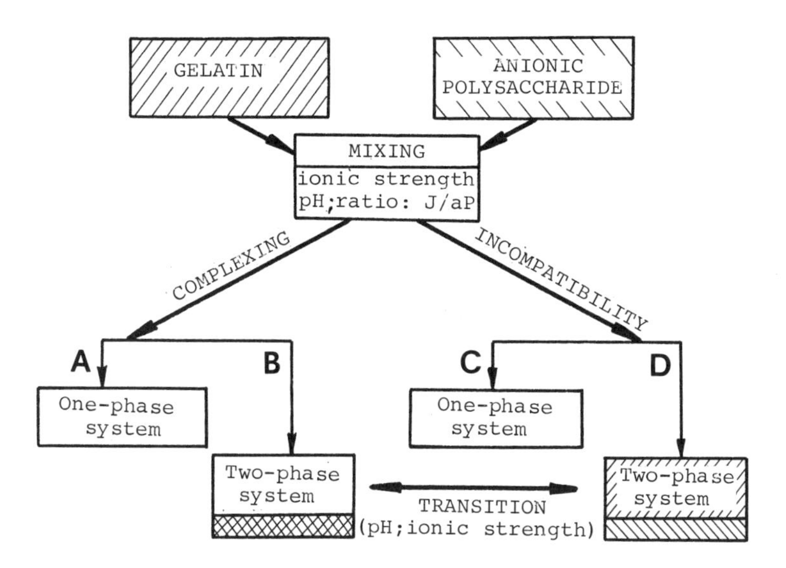

Figure 1. Types of interaction between gelatin and anionic polysaccharides.

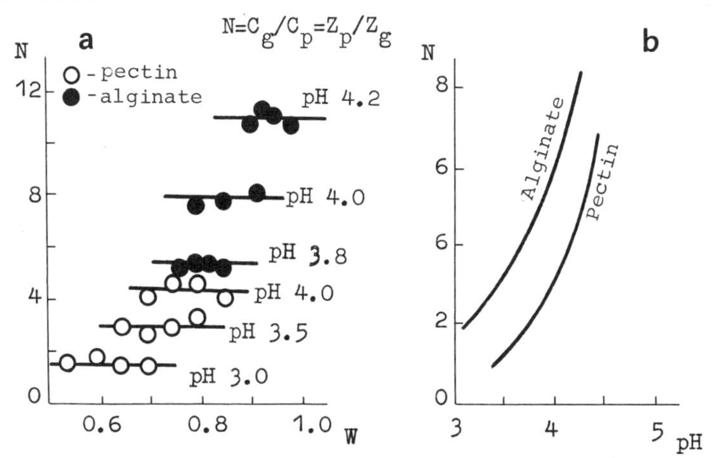

Figure 2. Ratio (N) of the concentrations of gelatin and polysaccharide (alginate or pectin) in the coacervate phase versus (a) weight fraction (W) of gelatin in the system $W = C_g/C_g+C_p$; (b) pH of the system.

and a polysaccharide solution, four different types of system are obtainable, namely two types of single-phase system (Figures 1A and 1C) and two types of two-phase system (Figures 1B and 1D). In the complex-formation conditions, the single-phase system is a solution of the complexes. The other type of a single-phase system is a solution of thermodynamically incompatible polymers, whose bulk concentration is below the phase separation threshold. Phase separation caused by the incompatibility of gelatin and polysaccharides (Figure 1D) leads to concentration of the biopolymers mostly in different phases of a system. By contrast, phase separation induced by complexing of the same polymers results in their concentration in one phase of a two-phase system, namely in the complex coacervate phase. This latter is formed by insoluble gelatin-anionic polysaccharide complexes(Figure 1B). Thus, depending upon the conditions of media and system composition, an interaction between gelatin and polysaccharides may result in four types of systems strongly differing in structure and properties. The discussion of some features of these systems starts with the complexing of gelatin and anionic polysaccharides.

GELATIN-ANIONIC POLYSACCHARIDE COMPLEXES

Formation of gelatin-anionic polysaccharide complexes is a particular case of interactions between oppositely charged polyelectrolytes in a solution. It is mainly electrostatic by nature. Unlike globular proteins, gelatin is notable for its random coil conformation at temperatures above 35°C. Its flexible chains are able to form a maximum number of contacts with the oppositely charged groups of an anionic polysaccharide. It is therefore typical of gelatin to form with oppositely charged anionic polysaccharides insoluble electrically neutral complexes. On the contrary, globular proteins interacting with anionic polysaccharides produce more readily, charged soluble complexes (3-9). A maximum yield of insoluble complexes is obtainable where the weight fraction ratio of gelatin and an anionic polysaccharide is equal to the ratio of their charges at a given pH of a system (Figure 2). The composition of insoluble complexes is determined by the charge ratio of polymer reactants (Figure 2A). It depends on the system pH value rather than on the initial ratio of polymer reactant concentrations (Figure 2A). The composition of insoluble gelatin-anionic polysaccharide complexes meets, as a rule, the requirement for complete mutual compensation of charges on the two polymer components. It is given by the equation $n=c(g)/c(p)=z(p)/z(g)$, where $c(g)$ and $c(p)$ are the concentrations of gelatin and anionic polysaccharide, while $z(g)$ and $z(p)$ are their charges, respectively. This means that for a given pH value, the weight fraction ratio of gelatin to an anionic polysaccharide in an insoluble complex is equal to the ratio of their charges. The compositions of complexes at various pH values calculated according to this equation from the titration curves are consistent with the data obtained by chemical analysis of the complex coacervate phase (Figure 2B). When the pH value is reduced from 4.5 to 2, the net charge of gelatin macro-ions increases, while that of ionic polysaccharide macro-ions decreases (Figure 2B). As a result (Figure 2b), the composition of electrically neutral complexes is changed, the insoluble complex is enriched with polysaccharide (8-10).

The complex coacervation process falls into two stages - complex formation and system separation (3,9). The former is formation of an electrically neutral complex due to the attraction between oppositely charged chains of polyelectrolytes. Presumably this process stage is stimulated by a drop in the free electrostatic energy of a system due to neutralization of oppositely charged groups, as well as by a possible increase in entropy due to dehydration. This stage occurs rapidly and is diffusion-controlled. At the molecular level, this stage can be regarded as a gradual attachment of the gelatin macro-ions (ligands) to an anionic polysaccharide (nucleus of a complex). i.e. as a process of mononuclear association. The anionic polysaccharide macro-ion is taken for the nucleus of a complex, because usually it has a higher charge density, compared to protein. The net charge of a polyanionic complex decreases with attachment of each successive ligand. The loss of charges on polymer reagents reduces hydrophilicity and solubility of a resultant complex. Electrostatic complexes exist at low ionic strengths and in urea solutions. Normally they dissociate where the pH is above the gelatin IEP or once the ionic strength of a system exceeds 0.2-0.3. On adding salts of calcium or some other polyvalent metals, the complex yield is reduced and their composition is changed. This is attributable to the binding of metal ions by anionic polysaccharides followed by reduction of their charge.

The second stage of the complex coacervation process is phase separation. It is not as fast as the first one and depends on the system concentration and composition. In electrically neutral complexes, aggregation and hydrogen bonding of their particles are promoted. Separation of the insoluble complex phase makes a system much more turbid, since the light scattering by the new phase particles is several orders of magnitude higher than that by the initial polymer reactant molecules. Insoluble complexes do form even where very dilute solutions (with a bulk concentration under 0.001%) of gelatin and an anionic polysaccharide are mixed. System turbidity increases very rapidly and it gets stabilized within 10-15 min. after the mixing of solutions. Stable dispersed systems of insoluble complexes are obtained up to bulk concentrations of 0.007%. In this range of concentrations, at a constant pH value, the system optical density can be used as a measure of the yield of an insoluble complex. As the bulk concentration of polymer reagents exceeds 0.01%, the dispersed particles of an electrically neutral complex are aggregated and precipitated. They form a phase of complex coacervate (8,9).

Soluble complexes may form in systems whose composition is far from that of an electrically neutral complex and that contain an excess of an anionic polysaccharide. According to free electrophoresis data, soluble complexes are polyanionic (8). The composition of soluble complexes depends on the system composition. As the relative content of gelatin increases, the composition of soluble complexes approaches that of an insoluble complex. On adding a small quantity (up to 0.05 M) of salt to a solution of anionic complexes, they aggregate and precipitate. This is also a complex coacervation phenomenon that seems to result from inhibited repulsion between similarly charged particles of a soluble complex. A critical salt concentration required for soluble comp

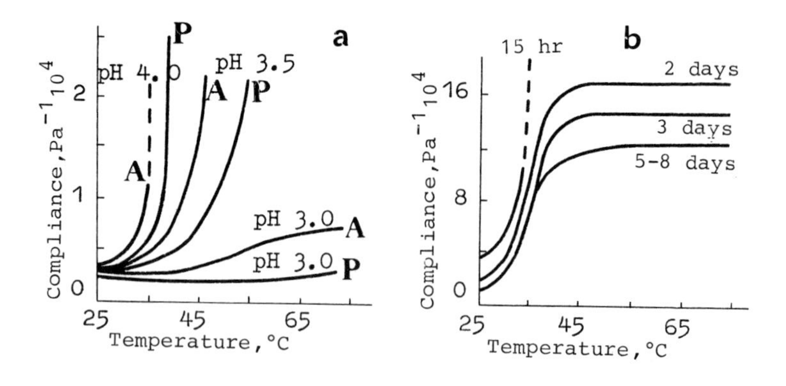

Figure 3. Thermomechanical curves of gels:
(a) the complex coacervate phase of systems: A- gelatin-alginate, P- gelatin-pectin (DE=35%) plotted for various pH values and 7-day ageing;
(b) gelatin-sodium alginate soluble complexes versus ageing time (pH 4.2; C_G=5%, C_A=2.5%).

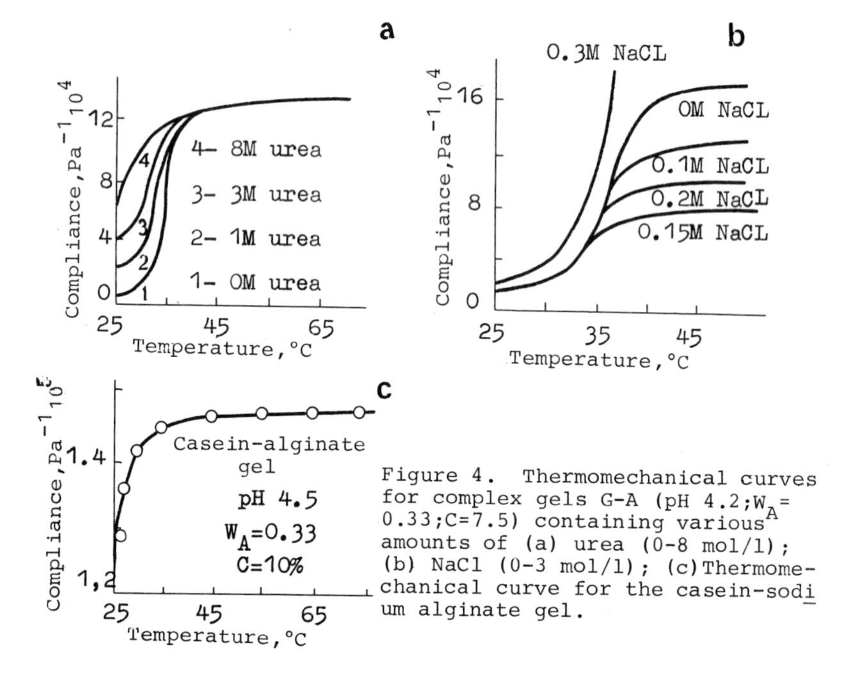

Figure 4. Thermomechanical curves for complex gels G-A (pH 4.2; W_A= 0.33; C=7.5) containing various amounts of (a) urea (0-8 mol/l); (b) NaCl (0-3 mol/l); (c) Thermomechanical curve for the casein-sodium alginate gel.

alginate. The soluble complex was obtained by mixing gelatin and alginate solutions at pH 4.2 and 40°C. Then the mixture of a 0.5% bulk concentration was lyophilized. To obtain a gel, a dry sample of the complex is dissolved in water at 40°C for 2 hours and cooled down to 20°C. A critical concentration for gelation of soluble complexes depends on their composition. For gelatin-alginate complexes prepared in the 2:1 proportion , it is 1.5% at 20°C. Complex gels with concentrations lower than 6% are thermoreversible. They are very similar to gelatin gels in their thermomechanical properties. As the concentration increases from 1.5 to 6%, the gel melting temperature rises slightly. The critical concentration for formation of the second network seems to be about 6%. The two types of network differ not only in critical concentrations but also in the rate of their formation. Therefore the properties of gels with concentrations over 6% are changed during storage. The fresh gels are similar to gelatin gels (Figure 3B). But after 1-day ageing they become thermoirreversible. The deformability of such gels increases sharply at 30-40°C but their elastic properties are preserved up to 70-80°C and higher. The increase of gel compliance at 30-40°C may be the result of melting of the gel network typical of gelatin. Rapid drying of gels during heating makes it difficult to study their thermomechanical properties at higher temperatures. That is why the thermomechanical curves are discontinued at 60-80°C. The elastic properties of gels at temperatures over 40°C (Figure 3B) increase with time and become constant in 5 days. Transition from thermoreversible to thermoirreversible gels may be due to the formation of the second network which is electrostatic in nature and does not melt on heating. The first network is destroyable on heating or addition of urea (Figure 4A) but is stable upon salt incorporation (Figure 4B). In contrast, the second network is resistant to heating and urea solutions but it dissociates at ionic strengths in excess of 0.3 and pH above the gelatin IEP. The ageing of complex gels seems to involve partial transition from the interligand non-electrostatic interaction of mononuclear complexes to electrostatic polynuclear complexes. Therefore, aged complex gels do not break down in concentrated urea solutions. Soluble complexes can form gels even in an 8-M solution of urea. Figure 4A shows the thermomechanical curves for complex gelatin-alginate gels obtained in solutions containing up to 8M urea. These gels do not melt, unlike gelatin gels or fresh complex gels without urea. Over 40°C their elastic properties are virtually independent of the urea content. To illustrate the general structure and properties of complex gels accounted for by the presence of an electrostatic network, Figure 4C shows the thermomechanical curve of a 10-% casein-alginate complex gel obtained in the 2:1 weight proportion of the components. To prepare the gel, sodium caseinate and sodium alginate solutions were mixed in the alkaline pH region, then the pH was set to 4.5. At this pH value and at low ionic strength, the obtained mixture undergoes gelation. The formed gel does not melt. Note that neither casein nor sodium alginate taken separately,will gel under these conditions (pH 4.5). It is also noteworthy that the thermomechanical properties of casein-alginate gels and of gelatin-alginate gels in urea solution are practically identical. It is a well-known fact that the interactions typical of gelatin gels fail in concentrated urea solutions. Thus, urea removes the specific features of gelatin gels and reveals the general features of complex gels. Formed by electrostatically held complexes, the 3-D network is easily disrupted by added NaCl. It is evident from Figure 4B that as NaCl concentration in a gel is raised from 0 to 0.15M, its elasticity increases at temperatures over 40°C. At higher salt concentrations (up to 0.25 M) the gel elastic properties are, on the contrary, reduced at this temperature, while below 40°C they virtually do not change. At NaCl concentrations above 0.3 M, thermoirreversible gels change over to thermoreversible gels due to dissociation of electrostatic complexes (Figure 4B). As a result, complex gels go over to gelatin gels containing sodium alginate. At 0.4 M and higher concentrations of NaCl, as well as at pH of a complex coacervate or of a concentrated solution of complexes exceeding the gelatin IEP, liquid systems break down into two phases. One of them is rich in gelatin, while the other is rich in polysaccharide. This implies a transition from complexing to thermodynamic incompatibility of biopolymers (Figure 1). This phenomenon occurs at a sufficiently high bulk concentration of biopolymers, provided their complexing is inhibited. The conditions for gelatin-anionic polysaccharide complex formation have been just considered. We should like only to note that at a low ionic strength gelatin is capable of forming weak complexes also with neutral polysaccharides. These complexes break down at ionic strengths in excess of 0.1-0.2 M. Under the same conditions, highly concentrated gelatin-neutral polysaccharide systems are phase-separated (22,24). Prior to taking up the problem of gelatin compatibility with polysaccharides, it must be mentioned that the phase-separation phenomenon shows up under the conditions typical of real food systems. In fact, concentrated as they are, many foods usually have a high ionic strength and pH above the protein IEP.

THERMODYNAMIC INCOMPATIBILITY OF GELATIN AND POLYSACCHARIDES

In food systems with many macromolecular components the problem of various polymer compatibility is of fundamental importance. Generally, biopolymers differing in chemical nature and conformation are incompatible in aqueous media (13,18). Gelatin and polysaccharides are typical examples of such polymers. Gelatin and most of food polysaccharides, being the degradation products of native biopolymers, are , as a rule, heterogeneous mixtures of biopolymers. Each contains macromolecular components strongly differing in chemical nature, molecular mass, chain rigidity, nature, number, location, and dissociation degree of ionizable groups. Such multicomponent mixtures containing unlike polymers may be incompatible in aqueous media. What is more, many food polysaccharides form colloidal dispersions rather than true solutions in aqueous media. In such systems, depletion flocculation can take place. Hence, in systems having one of the polymer components in the form of colloidal particles, phase separation is more pronounced than in solution mixtures of the same biopolymers. Thus, the problem of gelatin incompatibility with polysaccharides is very complicated and multiaspected. It has been recently discussed at length in several review papers (13,18,19). Here we will confine ourselves only to phase equilibria in gelatin-polysaccharide mixtures as well as to the properties of mix-

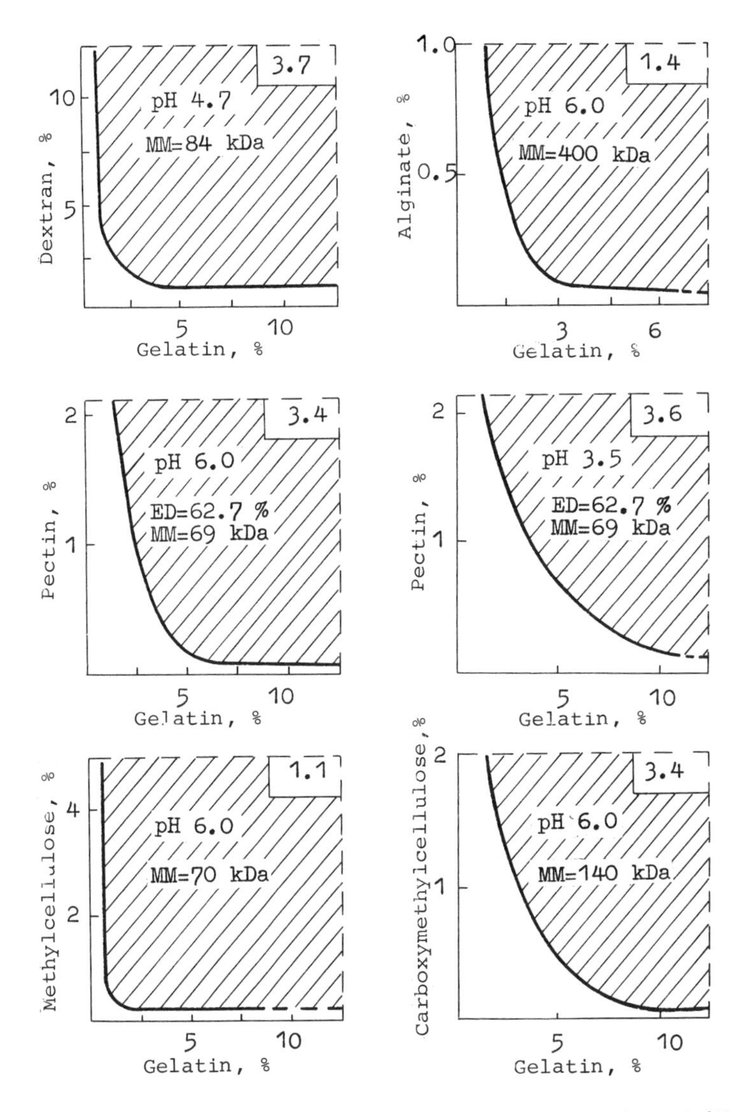

Figure 5. Phase diagrams for systems gelatin (MM=240kDa) - polysaccharide-water (at 40°C; 0.5 M NaCl). At the upper right there are indicated separation threshold values (%).

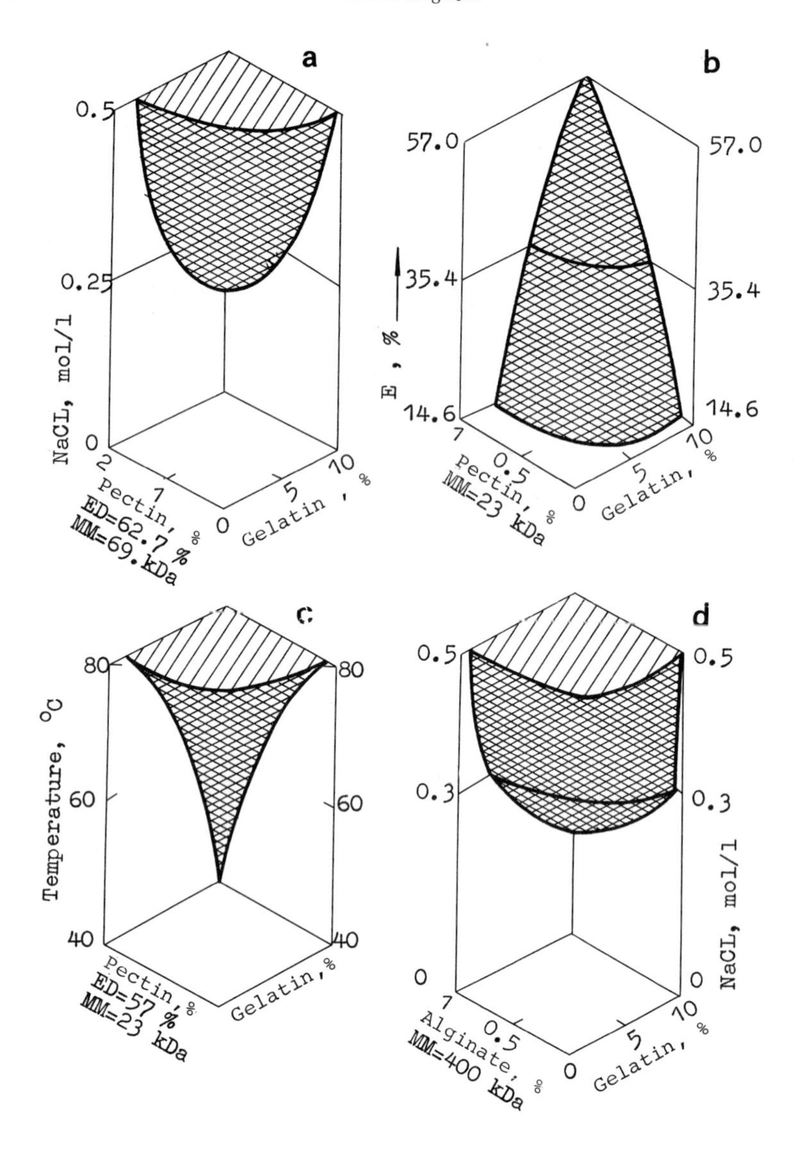

Figure 6. Phase diagrams for systems-gelatin (MM = 240kDa)-pectin-water for various (a) NaCl concentrations; (b) degrees of esterification; (c) temperatures;
(d) gelatin-sodium alginate-water systems for various NaCl concentrations.

ed and filled gels containing gelatin and anionic polysaccharides. The objective of the latter is to compare the properties of complex and mixed gels with the same anionic polysaccharides. The studied systems include gelatin mixtures with both anionic and neutral polysaccharides. Figure 5 shows phase diagrams for some gelatin-polysaccharide-water systems, namely those containing dextran, methyl- and carboxymethylcellulose, alginate, and pectin. The phase diagrams were produced at 40°C and 0.5 M concentration of NaCl. One can see the binodal curves separating the composition regions of one- and two-phase systems. The region under the binodal curve corresponds to one-phase mixtures of solutions,while the region above the curve represents compositions of two-phase systems. The figure also gives the values of phase separation thresholds, i.e. minimum bulk concentrations at which a given system breaks down into phases. For gelatin-polysaccharide systems, its value varies between 1 to 4%. For globular protein-polysaccharide mixtures it usually exceeds 4%. The lower compatibility of gelatin with polysaccharides is probably due to the statistical coil conformation of gelatin and its higher excluded volume, compared to globular proteins.

The higher the phase separation threshold and the larger the region under the binodal curve, the more compatible is gelatin with a given polysaccharide. For example, as the pH value of a gelatin-high-methoxyl pectin (DE=63%) system increases, the compatibility of these biopolymers decreases. Figure 6 shows the effects of salt concentration, temperature, and degree of pectin esterification (DE) on the incompatibility of gelatin with pectin and alginate. The incompatibility of these biopolymers increases at higher salt concentrations and pH values, as well as with lower DE of pectin. These results are typical of gelatin-polysaccharide solution mixtures (13,18,22). They normally have a lower critical point,related to a salt concentration (Figure 6A and 6D). The value of critical salt concentration for a gelatin - pectin and gelatin -alginate mixtures is about 0.2 M. Below this value gelatin and studied polysaccharides are completely compatible. Above it,the incompatibility of gelatin with polysaccharides increases, as the salt concentration is raised from 0.2 to 0.5 M. The higher the degree of pectin esterification, the better its compatibility with gelatin (Figure 6B). Once DE exceeds 50-60%, gelatin and pectin are completely compatible. Such systems do not separate into phases even at a 2-M salt concentration. Figure 6C shows the temperature effect on the compatibility of the same system. As the temperature rises, the system demixes and its incompatibility increases. These findings suggest that compatibility between gelatin and polysaccharides at pH above the protein IEP and at a high ionic strength is due to the formation of weak interpolymer complexes. As noted earlier, biopolymer incompatibility manifests itself in the inhibition of association of dissimilar macromolecules. Even at a low ionic strength, the repulsion of chains carrying like net charges is replaced by their association due to hydrogen bonding and electrostatic interaction of oppositely charged groups. At a high ionic strength, as a result of screening of electrostatic interactions between similarly charged macromolecules, their associates can be stabilized by hydrogen bonds and hydrophobic interactions. This may be favoured by an increase in the pectin DE and lowering of the system temperature. Hence, the predominant mechanism of gelatin-polysac-

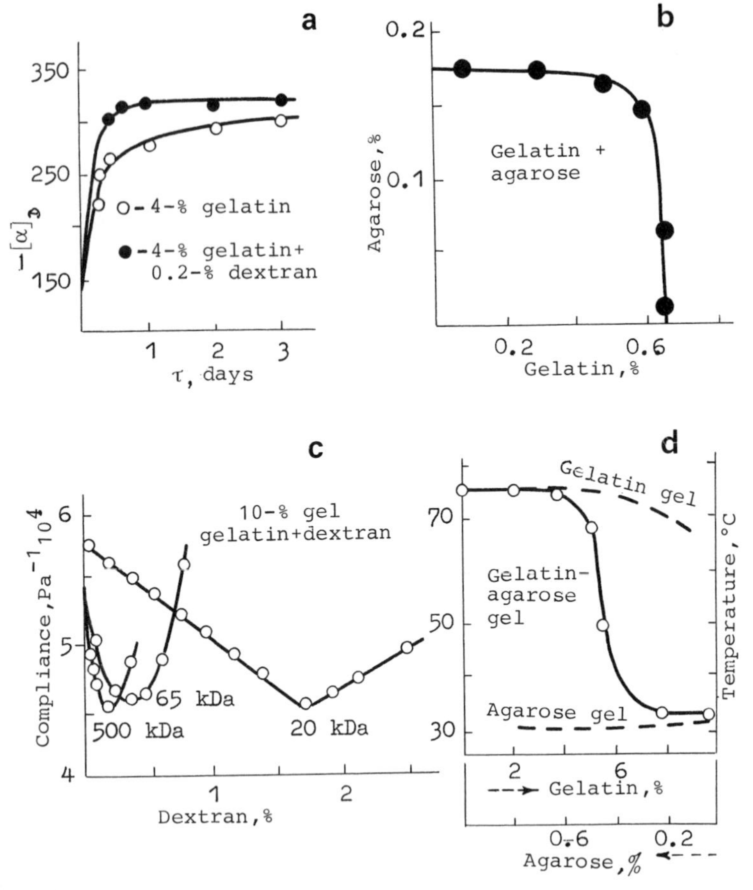

Figure 7.
(a) Effect of dextran on mutarotation of a 4-% gelatin solution at 6°C,
(b) Critical concentration for gelation versus composition of the gelatin-agarose-water system at pH 7.0 and ionic strength of 0.1 M
(c) Effect of dextran of various MM values on compliance of a 10-% gelatin gel;
(d) Temperature of gelation of the gelatin gel, agarose gel, and mixed gel versus system composition.

charide miscibility is the formation of their weak complexes,of various nature, that are destroyed by adding salt and raising the system temperature. Now let us consider some properties of mixed and filled gels.

MIXED AND FILLED GELS

Mixed gels containing two or more 3-D networks can be formed by mixing gelatin with various polysaccharides as gelling agents. In filled gels, one of the gelling agents, say, in the form of liquid or gel-like dispersed particles, serves as a filler of one or more 3-D networks of a gel that, in turn, may be either mixed or complex. Generally, mixed gels are obtainable if two or more gelling agents are co-soluble at concentrations exceeding their critical values of gelation and if they are able to form 3-D networks individually either in the whole system or in the volumes of its phases (1-3,11,25-31). Because of the excluded volume effects and the energy factors responsible for gelatin-polysaccharide incompatibility, the aqueous solution of a polysaccharide is a poorer solvent for gelatin than water. With the solvent quality reduced by an added polysaccharide in a gelatin solution, the gelatin molecules self-associate intensively and form more compact particles, thereby minimizing their contacts with the complex solvent components (25-29). This gives rise to several remarkable effects common to mixed gels (Figure 7). (1) In solution mixtures of gelling agents, the critical concentrations for gelation of each of them are lower than in their binary solutions. (2) The gelation of gelling agent mixtures is accelerated. (3) The elastic properties of gels of one-phase mixtures increase but they are decreased on transition to gels filled with liquid dispersed particles. And finally, the gel melting point is strongly dependent on gel composition. Below are examples of how these effects show up. Figure 7D indicates that the melting temperature of gelatin-agarose mixed gels changes discontinuously from the value typical of gelatin gels to that of agarose gels at mixture compositions corresponding to the liquid system phase inversion. Figure 7A shows the effect of added 0.2% dextran on the coil-helix transition in a 4-% gelatin solution at 6°C. It shows up in much faster rates of helix formation and gelatin gelation even with small quantities of added dextran (26). Similar results have been obtained for gelatin-methylcellulose and gelatin-agarose solution mixtures (19,24). The faster formation of a 3-D gel network from the aggregates of gelatin macromolecules in an incompatible biopolymer solution is attributable to the effect of depletion flocculation. The latter can intensify the interaction between the least compact and the most disordered end parts of gelatin macromolecular aggregates, thereby promoting their interpenetration and association. Intensified association of aggregates leads to faster formation of the gel network. Biopolymer incompatibility is also responsible for the lower critical concentration for gelation in gelling agent solution mixtures, compared to that in biopolymer-solvent binary systems. Data for a gelatin-agarose system are shown in Figure 7B (27,28). A similar result has been obtained for a gelatin-dextran system (26). Figure 7C shows the effect of dextran additives on gelatin gel compliance. With an increasing concentration of dextran, the gel compliance is first decreased and then increased. Its minimum value is related to the system phase separation, its position depending on the dextran molecular mass. It corresponds to the transition from one-phase gels to gelatin gels filled with liquid dispersed particles of a dextran solution. The elastic properties of such filled gels decrease with a higher volume fraction of the dispersed phase. This effect is due to the higher defectivity of the gel network (26). It also depends on the competition for water between the system phases containing biopolymers of different hydrophilicities (19,28). As a result of gelatin gelation, its concentration in the gel dispersion medium is lowered to a value below a critical concentration for gelation and, accordingly, the effects of gelatin excluded volume are less important. This increases markedly the free volume of a system that can be occupied by a polysaccharide or, in a more general case, by the second polymer molecules. This means that the dispersion medium of gelatin gel may be a better solvent for polysaccharides than the initial gelatin solution and that the properties of mixed gels should depend on the sequence of gelation of individual gelling agents in their mixtures. When a two-phase liquid system containing two or more thermodynamically incompatible gelling agents is converted into a gelled state in flow, this may yield filled gels with an anisotropic fibrous rather than isotropic structure. This effect stems from the deformation of dispersed particles of two-phase liquid systems in flow. In flowing water-in-water emulsions, the liquid droplets take the form of liquid filaments that are fixed by gelling one or the two phases. This phenomenon is basic to spinneretless fiber spinning studied on gelatin-containing systems (1-4,29,30).

Now we will briefly discuss the thermomechanical properties of mixed gels (2,11,31). Figure 8A shows the thermomechanical curves for a mixed gel containing 10% gelatin and 0.5% calcium alginate; the same gel containing also 0.5 M NaCl, as well as a 10-% gelatin gel and a 0.5-% calcium alginate gel. Figure 8B depicts the thermomechanical curves for the same mixed gel plotted during its heating and cooling. The mixed gel was prepared by cooling a solution mixture of gelatin and sodium alginate with subsequent immersion of the gel to a solution of calcium acetate. In a mixed gel, one of the two networks can be destroyed without gel destruction. For instance, the gelatin gel network is destroyed by heating or by hot water treatment. The calcium alginate network is destroyed by adding 0.5 M NaCl or by alkali treatment (31). The thermomechanical properties of mixed gels are extraordinary. Their thermomechanical curve falls into three segments. Below 30°C a mixed gel behaves just like a gelatin gel. Over 45-50°C, thermomechanically it looks like a calcium alginate gel. Within the intermediate temperature range (35-45°C), the gel is anomalously highly deformable, with a compliance maximum at 34-38°C. The phenomenon of anomalous gel compliance is reversible. It is reproducible under cooling (Figure 8B) and re-heating of a gel. The position and magnitude of the compliance maximum depend on the gel composition and on the rates of its heating and cooling. The anomalously high deformability of gels can be explained in terms of the steplike nature of the gelatin gel melting process. The gelatin gel network is formed by aggregates of macromolecules. In an early stage of mixed gel heating (up to 30°C) the networks break down into macromolecular aggregates. At intermediate temperatures (30-45°C) we deal

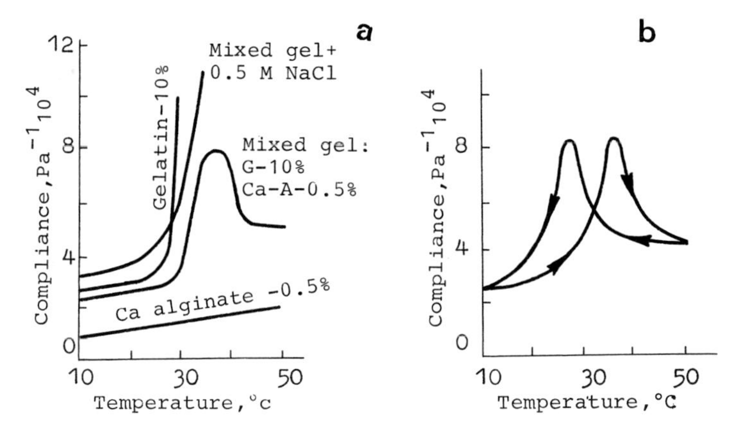

Figure 8. Thermomechanical curves (a): 10-% gelatin gel; 0.5-% calcium alginate gel: gelatin (10%)-calcium alginate (0.5%) mixed gel; the same mixed gel containing 0.5 M NaCl. (b) gelatin (10%)-calcium alginate (0.5%) mixed gel on heating and cooling.

with the calcium alginate gel whose dispersion medium is a relatively low-viscosity solution of aggregates of gelatin macromolecules. These aggregates are destroyed by further heating to 50°C and higher. The result is a solution of gelatin macromolecules. It has a higher concentration of solute particles, namely very flexible, strongly interacting macromolecules of gelatin; hence its viscosity is much higher than that of a solution of gelatin macromolecular aggregates. The increased viscosity of the dispersion medium of the calcium alginate gel reduces the deformability of the latter under the specified conditions of investigation. In thermomechanical studies, non-equilibrium values of gel compliance have been measured. They are governed by the dispersion medium viscosity: the higher the viscosity, the lower the values. The gel compliance is therefore maximal at 34-38°C where the cross-links are broken. Then it decreases with dissociation of macromolecular aggregates. On cooling the gel containing a gelatin solution, the processes take place in a reversed order, ending up again with the gelatin network. Similar results have been obtained for a gelatin-calcium pectinate system (31). Thus, the network of a thermoirreversible gel serves as a peculiar viscometer in the study of viscosity changes during gelatin gelation. A gel rheologically is likened to a sponge with very fine pores acting as a system of infinite capillaries. Now it becomes possible to measure changes in the viscosity of the gel dispersion medium under slight deformation, at very slow shear rates and obviously without destruction of macromolecular aggregates.

In conclusion, we should like to stress that the phenomena of thermodynamic incompatibility and complexing of gelatin with polysaccharides determine, to a large extent, the structural and mechanical properties of multicomponent food systems. We have en deavored to show, using some examples, that the problem of interaction between gelatin and polysaccharides both in liquid solutions and in gels is of critical importance for getting a better insight into the functional properties of food hydrocolloids.

REFERENCES

1. Tolstoguzov, V.B. (1974) Nahrung, 18, 525.
2. Tolstoguzov, V.B. (1978) Artificial Foodstuffs, Nauka (in Russian), Moscow.
3. Tolstoguzov, V.B. (1986) in Functional Properties of Food Macromolecules, (eds. Mitchell, J.R. and Ledward, D.A.) pp 385-415. Elsevier Applied Science, London.
4. Tolstoguzov, V.B. (1987) New Forms of Protein Foods, Agropromizdat (in Russian), Moscow.
5. Booly, H.L. and Bungenberg de Yong, H.G. (1956) Biocolloids and their Interaction, Springer Verlag, Wien.
6. Bugenberg de Yong, H.G. (1949) in Colloid Science (ed. Kruyt, H.R.), v.2 Elsevier, New York.
7. Evreinova, T.N. (1959) The Origin of Life on the Earth, Pergamon Press, London.
8. Vainerman, E.S., Grinberg, V.Ya. and Tolstoguzov, V.B. (1972, 1974) Colloid Polymer Sci., 250, 945; 252, 234.
9. Tolstoguzov, V.B. and Vainerman, E.S. (1975) Narhung, 19, 45.
10. Tschumak, G.Ya., Vainerman, E.S. and Tolstoguzov, V.B. (1976) Nahrung, 20, 321.
11. Tolstoguzov, V.B., Braudo, E.E. and Vainerman, E.S. (1975) Nahrung, 19, 973.
12. Tolstoguzov, V.B., Braudo, E.E. and Gurov, A.N. (1981) Nahrung, 25, 231, 317.
13. Tolstoguzov, V.B., Grinberg, V.Ya. and Gurov, A.N. (1985) J. Agric. Food Chem., 33, 151.
14. Imeson, A.P., Ledward, D.A. and Mitchell, J.R. (1977) J. Sci. Fd. Agric., 23, 661.
15. Imeson, A.P., Watson, P.R., Mitchell, J.R. and Ledward, D.A. (1973) J. Fd. Technol., 13, 329.
16. Stainsby, G. (1980) Food Chem., 6, 3.
17. Chilvers, G.R., Gunning, A.P. and Morris, V.J. (1988) Carbohyd. Polym., 8, 55.
18. Tolstoguzov, V.B.(1988) Food Hydrocolloids, 2, 195.
19. Tolstoguzov, V.B. (1988) Food Hydrocolloids, 2,339.
20. Muchin, M.A., Vainerman, E.S. and Tolstoguzov, V.B. (1976) Nahrung, 20, 313.
21. Muchin, M.A., Streltsova, Z.A., Vainerman, E.S. and Tolstoguzov, V.B. (1978) Nahrung, 22, 867.
22. Grinberg, V.Ya. and Tolstoguzov, V.B. (1972) Carbohyd. Res., 25, 313.
23. Varfolomeeva, E.R., Grinberg, V.Ya. and Tolstoguzov, V.B.(1980) Polym. Bull., 2, 613.
24. Grishchenkova, E. Antonov, Yu.A., Braudo, E.E. and Tolstoguzov, V.B. (1984) Nahrung, 28, 15.
25. Tolstoguzov, V.B. and Braudo, E.E. (1983) J. Texture Studies, 14, 183.
26. Tolstoguzov, V.B., Belkina, V., Gulov, V.Ya., Titova, E.F., Grinberg, V.Ya. and Belavtseva, E.M. (1974) Starke, 26, 130.

27. Braudo, E.E. Gotlieb, A.M., Plashchina, I.G. and Tolstoguzov, V.B. (1986) Nahrung, 30, 355.
28. Gotlieb, A.M., Plashchina, I.G., Braudo, E.E., Titova, E.F. Belavtseva, E.M. and Tolstoguzov, V.B. (1988), Nahrung, 32, 927.
29. Tolstoguzov, V.B., Mzhelsky, A.I. and Gurov, A.N. (1974) Colloid Polymer Sci., 252, 124.
30. Tolstoguzov, V.B. (1988) In Food Structure - Its Creation and Evaluation (eds. Blanshard, J. M. V. and Mitchell, J. R.) Butterworths, London.
31. Slonimsky, G.L., Tolstoguzov, V.B. and Izumov, D.B. (1970) Vysokomolek. Soed., B12, 403.

A study of the concentration-dependent conformations of α collagen using electron microscopy

S.J.MCBURNEY, D.M.GOODALL, P.D.BAILEY, A.WILSON[+] and M.HOPGOOD[+]

Chemistry Department, University of York, Heslington, York YO1 5DD, UK
[+]CCTR, Biology Department, University of York, Heslington, York YO1 5DD, UK

ABSTRACT

A fraction of pure α chains was separated from calf skin collagen, using gel filtration chromatography. The ultrastructure of the chains refolded at different concentrations (1.75 mg ml^{-1} and 0.05 mg ml^{-1}) in aqueous solution was studied using high-resolution transmission electron microscopy. The observed average length, 95±24 nm, of the structures formed at the low concentration is approximately one third the length of an α chain. This is consistent with intramolecular triple-helix structures formed by reverse folding.

INTRODUCTION

From kinetic studies of the folding of collagen and gelatin, using optical rotation to monitor the reaction, it has been found that conformational ordering exhibits concentration-independent kinetics below a concentration of approximately 0.1 mg ml^{-1}(1). This is consistent with a change of mechanism from inter- to intramolecular ordering as the concentration of macromolecule decreases. An intramolecular back-folded triple helix structure has been suggested to explain the kinetics in dilute solution(1). An acceptable space filling model of the model polypeptide $(PPG)_8$ back-folded has been proposed by Ramachandran(2).

In this study we are seeking direct evidence for intramolecular ordered structures using electron microscopy. Transmission electron microscopy (TEM) has been frequently used to obtain structural information about biological macromolecules. The preparative TEM technique used in this study has already been used to give information about collagen(3), myosin(4), spectrin(5) laminin, fibronectin and various types of native collagen(6), as well as gel-forming polysaccharides(7). The technique can reveal molecular dimensions of the molecules, conformational features, for example flexible sites or globular domains, and also time-resolved growth of polymer chains(8,9). The resolution of the technique is such that single collagen strands are not visualised, whereas triple-helix regions are. By inducing conformational ordering above and below the threshold value observed from kinetics, it was hoped that the different structures proposed for the inter- and intramolecular ordering regions would be seen using TEM.

EXPERIMENTAL

The α fraction from acid-soluble calf skin collagen (Sigma) was purified using gel filtration chromatography based on the procedure of Piez(10). Two consecutive separations were required before electrophoretic analysis (SDS PAGE) gave an α fraction free of cross-linked β and γ collagens.

Solutions of α collagen at either 0.05 mg ml^{-1} or 1.75 mg ml^{-1} were thermally denatured and then allowed to refold overnight in a cold room at $4^{o}C$. Time-dependent polarimetric measurements at $5^{o}C$ showed collagen to have substantially refolded under these conditions. The samples were diluted to a final concentration of 25 μg ml^{-1} in 50% (v/v) glycerol/water, and sprayed onto freshly cleaved mica using a low-volume nebuliser. The mica, kept on ice, was then transferred onto a cool stage and evacuated at 10^{-5} Torr for longer than 5 hours. The temperature during the drying was kept between 3-$8^{o}C$.

An electron beam evaporator (Bio-Rad) was used for low angle uni-directional shadowing of platinum, which was followed by carbon evaporation at 90°. The metal film replicas were floated off onto water and picked up on 400 mesh EM grids. The grids were studied using a transmission electron microscope (JEOL 1200 EX) and photographs taken at a magnification of x 25,000 - 50,000, followed by photographic enlargement of x 2.5.

RESULTS

Several grids of each sample were scanned and photographs taken from regions where the structures were free from entanglement or orientation. This was necessary since this technique is known to leave areas devoid of structure, as well as regions where the structures are entangled(6).

Typical regions of molecules refolded at the lower concentration of 0.05 mg ml^{-1} are shown in Figure 1. Width measurements of the molecules would be unreliable due to the shadow thickness(11). However, the average length of the rod-like structures, taken from 21 molecules, is 95±24 nm. The length of an intermolecular tropocollagen triple helix is known to be 300 nm(12).

Replicas of the control solution, which contained no protein, but had been prepared in the same way as the sample replicas showed no structural features. These replicas looked very similar to low-angle shadowed cleaved mica.

Figure 2 shows the refolded structures formed at the higher concentration of 1.75 mg ml^{-1}. At this concentration some molecular aggregation may have occured. At no other concentration nor on the control replicas were these longer fibrillar structures seen.

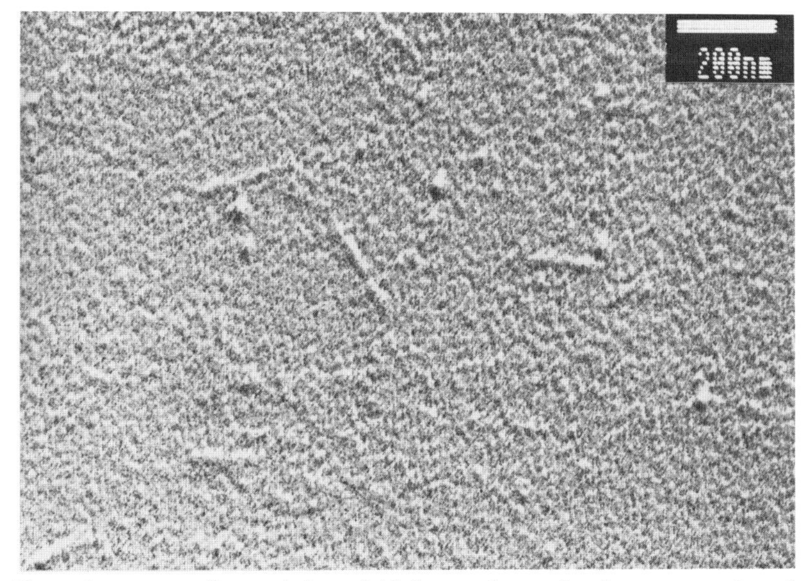

Figure 1. α collagen chains refolded at an intramolecular concentration of 0.05 mg ml^{-1}

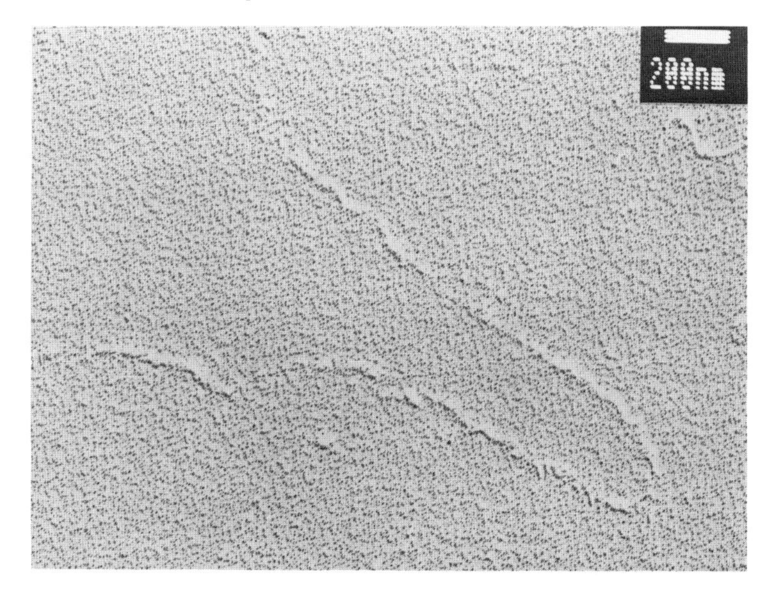

Figure 2. α collagen chains refolded at an intermolecular concentration of 1.75 mg ml^{-1}

DISCUSSION

A fraction of pure α chains of uniform length has been separated from a commercial source of calf skin collagen. This was necessary to minimise the polydispersity of structures formed on folding, which would not be achieved if the starting sample was heterogeneous with respect to chain length.

A perfectly back-folded triple-helix molecule would be anticipated to have a length of 100 nm. Using the purified α chains refolded at 0.05 mg ml^{-1}, a sufficiently low concentration to induce intramolecular reaction, an average length of 95±24 nm was found. Our observations are consistent with the formation of triple-helical and back-folded species, since the resolution of the technique permits only the triple strand to be seen and not a single chain. At this concentration, no longer structures are seen, whereas these are present in the sample refolded at a concentration of 1.75 mg ml^{-1}, where kinetic studies show intermolecular reaction occurs.

ACKNOWLEDGEMENT

We thank the A.F.R.C. and Rowntree Ltd for support through the Cooperative Studentship Award Scheme to S.J.McBurney.

REFERENCES

1 Harrington, W.F. and Rao, N.V. (1970) Biochemistry, 9, 3714 - 3724.

2 Ramachandran, G.N. (1968) Biopolymers, 6, 1771 - 1775.

3 Hall, C.E. and Doty, P. (1958) J.Amer.Chem.Soc., 80, 1269 - 1274.

4 Elliott, A. and Offer, G. (1976) Proc.R.Soc.Lond.B., 193, 45 - 53.

5 Shotton, D., Burke, B. and Branton, D. (1978) Biochimica et Biophysica Acta, 536, 313 - 317.

6 Engel, J. and Furthmayer, H. (1987) Methods in Enzymology, 145, 3 - 78.

7 Stokke, B.T., Elgsaeter, A., Skjak-Braek, G. and Smidsrød, O. (1987) Carbohydrate Research, 160, 13 - 28.

8 Dolz, R., Engel, J. and Kühn, K. (1988) Eur.J.Biochem., 178, 357 - 366.

9 Dolz, R. and Engel, J. (1989) Ann.N.Y. Acad.Sci., in press.

10 Piez, K.A., (1968) Analytical Biochemistry, 26, 305 - 312.

11 Shotton, D.M., Burke, B.E. and Branton, D. (1979) J.Mol.Biol., 131, 303 - 329.

12 Stryer, L. (1981) Biochemistry, 2nd edn., W.H. Freeman and Company, San Francisco.

The uses of gelatine

G.W.JORDAN

Gelatine Products Ltd, Sutton Weaver, Runcorn, Cheshire, UK

Thispaper can conveniently be divided into four areas:-

1) Food
2) Pharmaceutical
3) Photographic
4) Technical

FOOD

To consider gelatine for foods I think we should start by stating that the most important property of gelatine is that it forms a gel at almost any concentration above 1% and at any pH. However the protective colloid action and emulsifying properties of gelatine permit combination uses in products requiring stabilisation of foams, and also oil in water emulsions, with thickening or setting characteristics.

I will now list a number of food products explaining the role of gelatine in that particular product. Also, for the best results with each type of food product, the selection of a gelatine with the correct properties is essential and I hope that some of the following ideas will be useful.

Food Uses

1) Jelly Desserts
2) Confectionery
3) Marshmallow products, nougat etc
4) Dairy Products
5) Canned Meat Products
6) Meat Pies
7) Aspics
8) Wine fining and clarification of fruit juices

Jelly Desserts

Traditional jelly desserts are manufactured in two forms either as tablets in which the sugar, flavouring, colour and fruit acid have been dissolved and then gelled, or alternatively as jelly crystals, in which the same components are blended in the powdered form.

Jelly tablets are now only produced at a rate of roughly 50% of that produced in the mid Seventies. Sold in most retail outlets except Marks and Spencers but in fact sold in Marks and Spencers in Paris and Toronto.

It is basically a confectionery product, produced in a similar way to jelly babies (or the modern gummi bears) and then diluted with hot water before setting in a refrigerator. We used to think that you required a gelatine of excellent clarity at a pH 3.5 - this has changed since the use of real fruit juices and natural colours. Using modern methods, a gelatine of low electrolyte content particularly calcium is required. The prevailing low pH of these products would lead to some considerable reduction in the rigidity of the gel produced, but this factor to a large extent is offset by the co-operative effect of the sugar on the structure of the gelatine gel leading to a marked increase in the gel rigidity.

Traditionally a high Bloom gelatine is used for jelly tablets but I have a feeling that this stems from the old food laws which stated that jelly tablets had to have a minimum TS of 72% with sugar solid of 65% which left 7% of a 4 ounce or 112g as gelatine to set 1 pint or 586 mls of total volume. This was only possible with high Bloom.

For jelly crystals which are mainly sold in the catering field the grade of gelatine is comparatively unimportant. It is essential that the particular size of the gelatine is similar to that of caster sugar to ensure that the product dissolves quickly in the minimum amount of hot water before being adjusted to the correct volume with cold water. If a jelly contains sufficient gelatine of the correct Bloom it should form an acceptable gel in about two hours. In the case of tablets this can be reduced by dissolving it in a minimum of water in a micro wave oven before diluting with iced water and allowing to set in a refrigerator.

Confectionery

Certain types of confectionery such as gums and pastilles, utilise the ability of gelatine to form a thermally reversible gel. They usually require a chewy texture which permits the use of gelatine of considerably lower Bloom strength than in the case of jelly desserts. Some of these products also incorporate vegetable gums e.g. gum arabic, and starch in combination with gelatines to give a distinct change in texture. I must say, being a little prejudiced, of course, that the best confectionery is produced using solely gelatine as a jelly agent, witness the success of gummi bears. I am not an expert, but flavour release is apparently an important aspect of this type of product.

Other forms of confectionery which contain gelatine include certain types of what I would term lozenges or tablets and also confectionery pastes. The gelatine acts as a binder, preventing disintegration of the compressed sugar and flavour. The more the gelatine, the smoother the sweet.

Classic examples of confectionery pastes are liquorice allsorts where the gelatine is used in a similar way to tablets, as a binder. In the case of liquorice sweets it is boiled with the other ingredients to achieve the required texture. There is one exception to the rule, that one should never "overheat the gelatine" as this costs money. A low Bloom gelatine is sufficient for most tabletting and paste work.

Marshmallow Products

The important property of gelatine for effective use in marshmallows is its ability to support a foam, which is essentially a dispersion of air in a gelatine/syrup liquor, and also its ability to produce a rapid setting of the liquid medium in the form of a film around the entrapped air, so yielding a stable product.

The properties of gelatine which are important in this context are:-

a) Its influence on the liquid/air surface tension. Generally the lower the value, the greater the volume of foam produced.
b) The rate of setting of a gelled film to stabilise the foam produced.

Marshmallow products may be starch-deposited, extruded or deposited on biscuits. It is possible to use less gelatine in starch-deposited marshmallows than in other types, since the shape can be retained for a considerable time before removing from the starch. A high Bloom gelatine is normally used at a concentration of about 2%. Increased quantities of a lower Bloom gel strength may be used for marshmallow biscuit topping. As with extruded marshmallow, sufficient gelatine must be present to produce a rigid set, since the shape has to be defined immediately after aerating.

Nougat may also be included in this category, since it is a aerated product but of a much tougher texture than conventional marshmallows. This texture is achieved by the use of gelatine of low Bloom strength in a system of very high solids.

Dairy Products

In dairy products a number of different properties of gelatine are important. Some examples are:-

a) In fruit fool, milk jellies etc, the property of forming a reversible gel is important. A soft gel with a minimum amount of 'gelatine' background is necessary and gelatines of high Bloom strength are usually used.

b) In ice cream and deep frozen desserts the gelatine acts as a stabiliser, its presence ensuring a smooth texture free from large ice crystals. Bloom strength is not a significant factor in the choice of gelatine for these products.

c) The advantages of using gelatine in yoghurt are two fold:-
To avoid syneresis or whey separation to improve the texture of 'low fat' yoghurt.
When a carton of yoghurt is opened free liquid is often noticed on the surface of the product. This can be avoided by the addition of about 0.3% gelatine. The gelatine stabilises the milk gel, formed by the action of the lactobacillus and ensures a uniform product.
Gelatine also improves the eating characteristics of both low fat yoghurt and skimmed milk yoghurt. These two products have often been offered as low calorie health foods but, without gelatine, they are weak gels lacking the texture associated with a true yoghurt.

d) One developing use for gelatine is in the manufacture of low fat margarines. Gelatine acts as a stabiliser and texturiser and it is an increasing area of interest to the gelatine manufacturer. It is possible to achieve the traditional texture with a reduced fat content and hence achieve lower calorie counts.

Canned Meat Products

The meat packing industry still uses gelatine extensively in the canning of boned, cooked hams, brawns and jellied tongues. Gelatine serves to absorb the meat juices extruded during retortion, ensuring a pleasing appearance and a product that slices well. After retortion, the gelatine must set rapidly to a firm gel on cooling, so avoiding separation of the meat juices. Hence it must possess a sufficiently high gel strength to withstand the thermal degradation which inevitably accompanies the retortion process. It is also necessary that such gelatines have a very low content of calcium to avoid loss of clarity of the gel by precipitation of calcium phosphate derived from the soluble phosphates present in the meat juices. This applies particularly to such products as canned chicken, ham and tongue.

Meat Pies

The application of gelatine in meat pies utilises its ability to form a gel rapidly on cooling. Gelatine in the sol form is injected through the baked pie crust, filling in the areas between the meat and the pie casing. A rapid set is necessary to avoid excessive penetration of the liquid gelatine into the casing. A 5% gel is about right for this purpose.

A recent article in the Sunday Telegraph had some very unfair and unpleasant things to say about this injection method and I am glad to say that the gelatine industry has extracted an apology!

Aspics

A gelatine for aspic must be of excellent clarity at a pH of about 4.5 and a concentration of 2%. Alkali processed gelatines have to be handled carefully to avoid iso-electric point haze as the pH of these products is often at the iso-electric point and the concentration of about 2% also encourages the development of this haze.

Wine Fining and the Clarification of Fruit Juice

This process uses the amphoteric character of the gelatine molecule. The charge, +ve or -ve, varying with the pH of the system and the iso-electric point of the gelatine.

As a result of this polyelectrolyte behaviour gelatine is used extensively in the clarification of wines and fruit juices in order to remove the unstable and undesired levels of polyphenolic compounds which cause cloudiness and sedimentation in storage and an unacceptable astringency in taste. The positively charged gelatine molecule co-acervates with the negatively charged tannin molecule in dilute solution to form a floc, which can be removed by decanting and further filtration.

PHARMACEUTICAL

1) Hard Shell Capsules
2) Soft Shell Capsules
3) Tablets
4) Surgical Sponges and Blood Plasma expanders

Hard Capsules

The capsules are formed on mould pins, the surface of which carry a lubricant to facilitate the subsequent removal of the capsule.

On a fully automatic machine the mould pins are dipped in the gelatine solution, withdrawn and rotated to provide adequate film distribution during the sol-gel transformation. The formed capsules pass through a series of drying chambers and the dried capsules are subsequently stripped from the pins which are re-lubricated for return to the dipping cycle.

The gelatine employed in the hard shell capsule process must be capable of uniform film formation on the pins and rapid setting to a gel. The distribution of the gelatine film on the lubricated pin surface must be such as to provide an adequate 'stripping edge' to enable effective automatic removal of the capsules after completion of the drying operation with the minimum of applied force. It is also essential that the nature of the gelatine does not contribute to defects in the capsule walls such as oil holes and 'bubbles'.

A very high Bloom gelatine prepared to a very strict specification is important for hard shell capsule manufacture. I sometimes wonder if the very tight specifications for gelatine were developed to hide shortcomings in the engineering of the capsule machines.

Soft Shell Capsules

It is surprising that the principle of the gelatine based soft capsule was discovered in Paris in 1833 and registered as a patent in 1834. It consists of plunging a small sack containing mercury into a solution of gelatine and then demoulding the capsule formed on the outside of the bag. After drying this was filled with medicament and sealed with a drop of gelatine.

Today there are two main manufacturing processes mainly injection of medicament without a noticeable seal and the most widely know method which consists in the sealing of two separate halves.

The first method employs two concentrate nozzles. The centre column is the fill or medicament and the outer column a mixture of gelatine and glycerine in aqueous solution. This thin stream, which must have a slightly greater density than the oil, dissociates into drops. The speed and size of the drops are regulated by an intermittently opening diaphragm.

The drops fall through the column of oil and become round in shape almost perfectly spherical. The recovered capsules from the bottom of the oil column can then be washed and dried. These capsules are always of one colour, seamless and round.

On the other hand the rotary die encapsulating machine produces a capsule with a longitudinal seam which is formed and filled simultaneously.

The capsule wall is a mix of gelatine, glycerine and water, in an approximate concentration of 40% gelatine, 20% glycerine and 40% water. This ratio is varied dependent on the characteristics required from the finished capsule. The mix is prepared by mixing and de-aeration often under vacuum.

Two spreader boxes,with automatic level control,supply the liquid mix to two air cooled cylindrical drums. Films of the required uniform thickness are formed on the surface of the drums; the uniformity of this film is of the utmost importance. Capsules are manufactured by passing these films over two identical revolving die rolls with cavities on the periphery. A pump forces pre-measured charges of fill material through a heated feeder head, placed between the two films into matching die roll cavities. The filled capsules are formed by pressure between the die rolls. The remaining film, often referred to as 'net', is collected for reprocessing by addition to fresh capsule walls mix.

The machines are precision engineered and fill weights are accurate to within ± 2%. Capsules are produced at rates of up to 60,000 per hour.

On leaving the encapsulating machine, the capsules are automatically washed in a solvent and then undergo multistage drying for up to 48 hours.

This method has the considerable advantage of manufacturing capsules of almost any size from 0.1 ml to considerably more than 4 ml in oval, round oblong or even bottle shapes.

The specification for Bloom and viscosity for gelatines for soft shell capsules has changed very little in 50 years. We would usually suggest 150 Bloom at 35mPa.s viscosity. Electrolyte content has to be minimised and obviously physical characteristics e.g. colour and clarity are also important, low sulphur dioxide content is obviously very important due to possible bleaching of capsule wall colour.

The gelatine has to dissolve in the glycerine/water solution, as quickly as possible and a minimum of aeration is required in the capsule base and hence the capsule wall. Air bubbles cause leakages in the capsule walls and further contamination of other capsules during manufacture, and, of course, waste of a very expensive medicament fill.

Tablets

A medical dosage in the form of a tablet is produced by compressing granular or powdered materials into specific shapes and sizes by mechanical dies. Non-compressible materials may be tabletted by the use of gelatine as a binding agent. The pre-compression procedure involves the dispersion of the constituents of the tablet as a suspension in a gelatine sol, this then being dried and milled to uniform particle size, before being compressed to shape.

Hydrolysed gelatines or cold water dispersible gelatines can be used without pre-dissolving in hot water to form a sol. In fact many tablets with a high dosage, active ingredient, of very unpleasant taste are often manufactured using this technique.

Since the gelatine is to be used as an adhesive, the Bloom strength is not important, however, due to the fact that for many years an acceptably pure gelatine could not be produced below 150 Bloom, the minimum Bloom strength for pharmaceutical purposes is still specified at 150, although it is not necessarily the best grade for this purpose

Problems arise in this area to differentiate shapes of tablets and confectionery to avoid accidental consumption by young children. Great care is taken in this area with extra precautions in outer packaging and also the stamping of codes and trade marks on each individual tablet.

Surgical Sponges and Blood Plasma Expanders

Surgical sponges may be formed from gelatine by incorporating filtered air into a sterile aqueous solution of the gelatine. The resultant foam is dried after the addition of a cross linking agent to reduce the solubility of the gelatine. The sponges after further sterilisation and soaking in a blood clotting medium are applied at the site of the bleeding.

A high bloom gelatine is required for this purpose, having physical properties sufficiently high to withstand the degradation accompanying the sterilisation process.

Gelatine can also be used as blood plasma expanders and are designed to maintain the circulatory blood volume during shock. Gelatine may be modified to render it suitable as a transfusion medium in place of blood serum or human plasma and as a constituent of intravenous injectables.

PHOTOGRAPHIC

There is conjecture about the future of gelatine in the photographic industry with the introduction of electronic photography but the evidence indicates that the development of simple very sophisticated cameras mean that more films are used and not less. Nowadays not 1 but 4 or 5 exposures are made every time an interesting subject appears because of the ease of use of modern cameras.

It is possible that in 20 or 30 years time, electronic photography will have developed sufficiently to replace what I would term chemical photography and the photographic gelatine manufacturer will wish to sell in other fields.

TECHNICAL

1) Microencapsulation

The coating of microencapsulation dyes on paper is the basis of the production of N C R type Paper. The formation of the wall material of these microcapsules is based on the ability of an acid processed gelatine to form co-acervate complexes with polysaccharides such as Gum

Arabic. This co-acervate phase is produced by strict pH control around minute droplets of the dyes in a solvent which is immisable with water. The capsules are used to coat paper and when the microcapsules are broken the ink is released.
The classical method involves the use of Gum Arabic and an acid pigskin gelatine with a iso - electric point around pH8, and a high Bloom strength, to form a firm gel quickly on cooling.

2) Paper Sizing
The use of technical gelatin in this field is largely confined to bank notes and is responsible for the rather nice crackle produced when handling them.
The method consists of dipping the paper into a sizing solution removing the excess sizing solution by squeeze rolls and then drying the paper.

3) Electro-deposition of Metals
Gelatine is used in electroplating processes to produce smooth metallic deposits on the cathode from acid solutions of metallic salts. In the absence of gelatine, electro-deposition generally occurs as a rough crystalline coating on the electrode. The presence in the electrolyte of about 0.5% by weight of gelatine, however, appears to exert a powerful protective colloid action, inhibiting crystallisation and permitting deposition of the metal in the colloid form so providing a smooth hard surface.

4) Other uses which are of interest are photogravure, screen printing, collotype printing, the manufacture of abrasive papers and even masonry moulding.

Since the early seventies, cold water dispersable gelatines have been available and are sold in reasonable quantities throughout Europe and the rest of the world

I think it is safe to say that no-one has yet succeeded in preparing a practical cold water table jelly without five pages of instruction. However, they are used in mousse type products and extensively in creams and bakery fillings.

The next speaker has the onerous task of reviewing the future of gelatine. I am convinced that the special properties of gelatine lend themselves to increasing use in the manufacture of low fat health foods, both sweet and savoury

Ironically, in the last twenty years or so, the biggest increase has been in the use of hydrolysed gelatine in the pharmaceutical and health food sector. This, of course' means removing the one property that makes gelatine unique amongst proteins.

With the aid of the confectionery companies a little marketing would not go amiss. Why, for instance did Marks & Spencer give up selling table jellies, and would they recommence if more publicity was given to the ease of preparation of table jellies using microwave techniques. Do housewives realise that a table jelly can be used to prepare five desserts for the price of one "ready to eat" product.

The gelatine industry will, I am sure, continue to experience the periodic feasts and famine that have occurred regularly for many years, maybe we should all try harder to have less serious famines and better feasts in the future.

The future of gelatine

C.-G.HAGERMAN
R & D Department, Extraco AB, S-264 00 Klippan, Sweden

INTRODUCTION

Since prehistoric times products from connective tissues have been used for their gelling properties. The industrial convertion of the raw material into gelatine also came at a very early stage of the industrial revolution.

In other words gelatine is an old product and it has had a long time to be tested into many different fields before being established in its major strongholds of today, namely
Photography
Pharmaceuticals
Food.

Roughly 200 000 tons of gelatine were produced in the world in 1988 and of that amount about 25 000 tons were used in the photographic field, about 15 000 tons in the pharmaceutical field, mainly for capsule production, and the rest in principle for food applications.

In the future it is most likely that the bulk of gelatine will be used in the same applications as today, even if its use in some of them will be threatened by new techniques or other ingredients.

PHOTOGRAPHY

In the photographic industry you could for instance say that there is on one hand an ambition to get a deeper knowledge of the gelatine molecule and thus become aware of the factors that affect it in the photographic process. The aim is, of course, to be able to produce emulsion films with even greater divergence in performance than today. But on the other hand the camera industry has for many years tried to develop cameras that will work entirely without any emulsion film. So far there is no real sign of success, but a discovery in that direction is likely to come in the future.

PHARMACEUTICALS

The capsule production has today production equipment that can produce capsules with an enormous speed. This development has forced gelatine producers to tighten the gelatine specifications in order to produce gelatine capable of keeping track with the machinery. In the future, if the development continues, the gelatine industry will perhaps have to meet other problems besides setting times, gliding and bubbling to be able to satisfy the production standards.

There will certainly be a need in the future to find a better understanding of how different parameters within the gelatine molecule affect the different aspects of the gelatine capsule.

FOOD

Like today most gelatine produced in the future will be used in food production. The major fields are today

Confectionery

Jelly desserts

Dairy products.

The best estimate for the future is that it is among those groups gelatine will be used also in the future. However, people in the industrialized world are today very much aware of the fact that they are generally eating too much, so a change in consumption will come and that will, of course, also affect gelatine foodstuff.

The gelatine confectionery is of tradition a European product. The major challenge from cheaper raw materials, like starches, has already happened some 15-20 years ago. The products which survived that attack and are still based on gelatine will survive also in the future.

It is also more likely that our European habit of eating gelatine confectionery will be spread to the rest of the world rather than to imagine a decline in this field. Imagine that every Chinese takes only one jelly baby of 4 grammes weight a day with 6 % gelatine in it, close to 90 000 tons of gelatine will be needed for that production only.

The traditional jelly desserts, however, will probably see a declining demand on behalf of new desserts with a lower sugar content. In the future there will be more and more demand for convenience desserts, since more women are today working outside their homes. This will, of course, cause some changes in the gelatine field but also create possibilities for the development of new products of ready-to-eat desserts and fast-setting desserts. Already today you can see this development. The younger generation are not eating the same desserts as the older generation and that trend will most certainly continue.

In all the industrialized part of the world people are eating too much. At the same time as this is going on, a huge part of the same population are involved in some type of slimming operation from fasting to just temporarily cutting their daily intake of sweets or whatever they think that they can be,without for the moment.

But slowly people are getting more and more aware of what they are eating. In the future low-calorie food and foodstuffs with a lot of fibres will be of even greater interest than today. WHO recommends for example that the total fat content of food shall not exceed 35 %.

People are also starting to show a greater interest in the list of ingredients than they used to do. Products with an E-number are in some countries looked upon with suspicion and foodstuffs containing them are avoided. Gelatine is in most countries regarded as a foodstuff, without any limitations, and has no E-number. When it comes to stabilisation, viscosity adjustment and jelling, gelatine therefore is an excellent choice, compared with most other functional ingredients in the same field.

Many food emulsions with a high fat content, like margarine and mayonnaise, are stable without adding stabilisers.

Products which attempt to look and taste like those fatty emulsions, but with a much reduced fat content, need stabilisers to be stable. Today there is gelatine for example in low-fat spreads and fruit yoghurts for that reason.

Low-fat margarine types are produced with a lower and lower fat content and consequently a higher and higher gelatine content.

80 % fat   No gelatine
60 % fat   Approx. 0.5 % gelatine
40 % fat   Approx. 3 % gelatine.

Some years ago there was not much evidence that a 40 % low-fat margarine of the same texture and taste as normal margarine could be produced, but today trials and test launchings of products containing as low as 10-25 % fat are going on.

"The Dairy Market at a Glance" says about the yellow fats market in Great Britain for the 1990's, that the total market will show a further decline. Butter will continue to be the main loser but also the margarine consumption will decline. Instead the low-fat spreads will expand rapidly to as much as 25 % of the total yellow fats market at the end of this period.

The use of gelatine in fruit yoghurts once started with the gelatine as an aid for a longer shelf life of the product. The gelatine prevents syneresis in the product and has made it easier to rationalize the yoghurt production. Today it is needed not only for that reason but also for giving the low-calorie yoghurts their texture.

The general idea of all those changes of the calorie content of the products is, of course, that the new products shall resemble the old fatty ones as much as possible.

The future will show if other products in the same field will be as successful as the low-fat spreads and the yoghurts.

It is, however, not sure that the growth of gelatine products will come entirely from the jelling type of gelatine. Hydrolysed gelatine is a name for all gelatines without jelling behaviour. From being products made from the last extractions of the gelatine production, the gelatine hydrolysates have found their own niche among the gelatines. Today the raw material is as vital for this type of products as it is for normal gelatine, since a low taste level is needed for success.

By use of different types of enzymes, the taste level can

be kept low and it is also possible to adjust the viscosity of the product to a narrow range.

There is probably just as many functional products among the hydrolysates ranging from what is regarded as zero bloom strength down to free amino-acids as there is among the gelatines ranging from zero bloom to the highest gel strength.

Gelatine hydrolysates are today found in many products where the gelling behaviour of gelatine is a disadvantage.

In wine clarification the smaller hydrolysate molecule has proved as effective as the bigger gelatine molecule. And the product is much easier to handle.

In sport drinks and products like dietary food supplement it is the fact that gelatine hydrolysate is a pure tasteless protein that is important. By adding tryptophan and methionine to the hydrolysate, a product with a very high nutritional value is obtained.

Sport drinks have come with the younger generation. It is one of those fast food articles that, if composed right, with harmless and nutritional valuable ingredients, will grow in the future. Dietary food supplement on the other hand, is a type of well-composed dry food that can be given to elderly people or hospitalized persons having difficulties eating solid food. By dissolving one package of the article in water the patient gets his daily need of protein, vitamins and other essential foodstuffs.

With the development towards a faster and faster way of living, it is most likely that these types of products will grow in the future.

By use of different techniques the length of the peptide chains in the gelatine molecule can be controlled. This can be made by use of different enzymes and, if necessary, in combination with fractionation of the product. In that way many new functional products can probably be produced. Already there are products produced in such a way that they can replace gum arabic in both confectionery and emulsions in a proportion close to 1:1.

Gum arabic has always been a unique carbohydrate giving low viscosities at very high concentrations. The fact that a gelatine hydrolysate can replace it in that close proportion gives rise to a question: Perhaps it is possible to find other gelatine hydrolysates with functional behaviours that can act instead of the normal stabilisers used today. Since gelatine, as said before, is a product with no E-number, it could certainly become a popular alternative.

CONCLUSION

The future of gelatine looks bright. It will continue to be strong, especially in the food sector. With the change of living and eating that will come in the future, more ready-to-eat products will be needed. There will also be a call for more low-calorie and nutritionally well-balanced food. Both these demands are in favour of the use of gelatine and gelatine hydrolysates.

Gelations of soymilk and 11S-globulin separated from soybeans – Tofu-making qualities of Japanese domestic soybeans

M.YOSHIDA, K.KOHYAMA[+] and K.NISHINARI[+]

Kanagawa Prefectural Institute of Agriculture, Hiratsuka, Kanagawa, 259-12, Japan

[+]National Food Research Institute, Tsukuba, Ibaraki 305, Japan

ABSTRACT

A texturometer or a curdmeter has been practically used for evaluation of tofu (soybean gel) qualities. Dynamic viscoelasticity measurements were carried out in order to examine the gelation process of soymilk and 11S globulin in the presence of glucono-δ-lactone (GDL) in the present work. The storage and loss moduli were observed as a function of time after the addition of GDL. Gelation curves fitted well to a first order reaction equation. The saturated value of the storage modulus (G') was mainly dependent on concentration of 11S globulin. The rate constant (k) was an increasing function of the concentration of GDL.

1. INTRODUCTION

In recent years, soybeans are recognised as an important crop as well as rice in Japan[1-3]. Under these circumstances, soybean qualities and tofu-making qualities (TMQ) have attracted much attention from many investigators. In previous studies TMQ were evaluated as hardness by a texturometer or a curdmeter. Since there have been very few studies on gelation of soymilk, dynamic viscoelasticity measurement of tofu gel was performed in the present work.

It is known that 11S globulin mainly governs the hardness of tofu gel here we also examined gelation of 11S globulin-GDL systems.

2. MATERIALS AND METHODS

2.1 Sample Soybeans

Five varieties of soybeans, *ENREI, TACHISUZUNARI, OKUSHIROME, TAMAHOMARE, TSUKUIZAIRAI* produced in Japan in 1987 were used as samples.

2.2 Preparation of Soymilk

5g of soybeans were dipped in 50 ml of distilled water for 18 hours at 20°C and homogenized by a Waring blender and polytron-homogenizer. It was filtered through a two-fold gauze. The filtrate was used as soymilk.

2.3 Preparation of 11S Globulin

11S globulin was isolated from *ENREI* by a method of Thanh *et al*, as follows[4]. Soybeans were ground and defatted by n-hexane. The whole buffer extract extracted with tris-(hydroxymethyl)aminomethane-HCl (Tris-HCl) buffer pH 7.8 from the defatted soybean meal, and centrifuged for 15 min. at 10,000 rpm at 20°C to whole buffer extract 2N HCl was added so that pH might be 6.6. The extract was dialysed against Tris-HCl buffer at pH 6.6 for 3 hours at 2~3°C, and centrifuged for 20 min. at 10,000 rpm at 2°C. The precipitate was freeze-dried. The powder obtained was dissolved in Tris-HCl buffer (A) or Na-phosphate buffer (B). The solution (A) was submitted to ultracentrifuge sedimentation analysis and the solution (B) was separated by SDS-polyacrylamide gel electrophoresis[5-6]. This globulin showed peaks in the molecular weight range from 17,000 ~ 40,000 by electrophoresis and showed only one peak in the ultracentrifuge sedimentation analysis. Therefore, this globulin was confirmed as 11S globulin[1 4 7 8].

2.4 Measurement of Gelation Process

Soymilk and 11S globulin solution were heated in boiling water for 5 min. and cooled to room temperature. The solution was heated at 80°C and GDL was added. The storage and loss moduli of the solution were observed by a Rheolograph Sol (Toyo Seiki Seisakusho Ltd.) as a function of time[9].

2.5 Analysis of Gelation Curves

Observed gelation curves were fitted well to the equation (1) of the first order reaction by the least square method. The origin of the time t=0 was taken as the point where G' began to deviate from the base-line.

$$G = G_{sat}[1-\exp(-kt)] \quad (1)$$

where G (t) = storage (G') or loss (G") modulus
Gsat = saturated value of G' or G"
k = rate constant of the gelation
t = time

3. RESULTS AND DISCUSSION

3.1 Gelation of Soymilk

Figure 1 shows the gelation curve of soymilk prepared from *ENREI*. Curves for G" were not so different among soybean varieties we examined. G' increased exponentially, and tended to the saturated value after about one hour. Table 1 shows tofu hardness measured by a curdmeter, the storage modulus at 60 min and the saturated value of G' for soymilk samples prepared from five varieties of soybeans. G' after 60 min. and Gsat agree fairly well with hardness, and give a good index for TMQ. TMQ will be predicted before the gelation is completed.

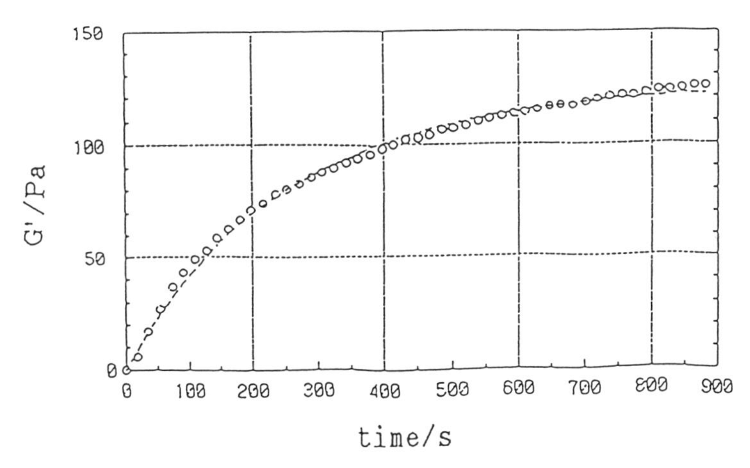

Figure 1. A typical gelation curve for 5%.soybean protein of soymilk prepared from ENREI in the presence of 0.4% GDL.
ooo : experimental,. --- : calculated.

varieties	curdmeter(g/cm^2)	G'(60)(Pa)	G'sat(Pa)
ENREI	47	260	251
TACHISUZUNARI	46	231	225
OKUSHIRIME	40	204	195
TAMAHOMARE	34	166	164
TSUKUIZAIRAI	30	169	161

Table 1. Tofu hardness by a curdmeter, storage modulus G' at 60 min by a Rheolograph Sol, and saturated storage modulus G'sat (calculated by curve fitting) for about 5% soymilk in the presence of 0.4%GDL.

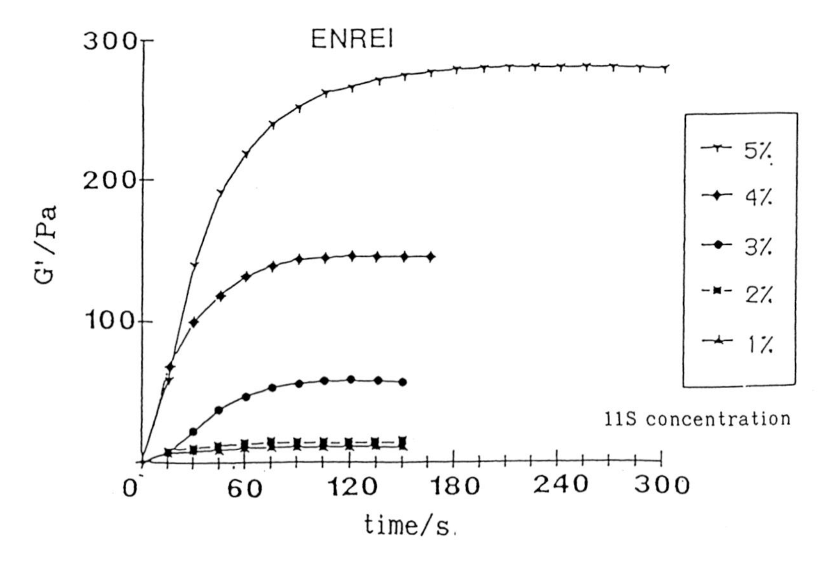

Figure 2. Gelation curves of 11S globulin from ENREI in the presence of 0.4% GDL. Concentration of 11S globulin: 1 ~ 5%.

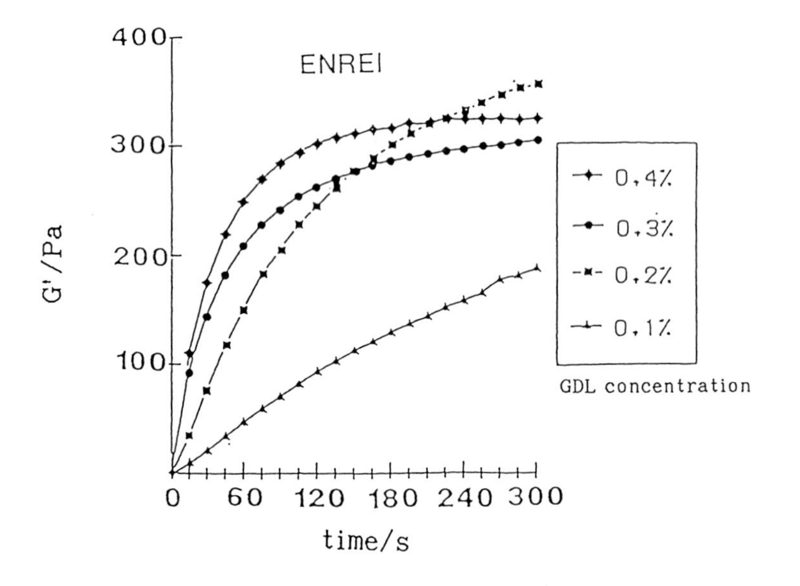

Figure 3. Gelation curves of 5% 11S globulin from ENREI in the presence of GDL of various concentrations.

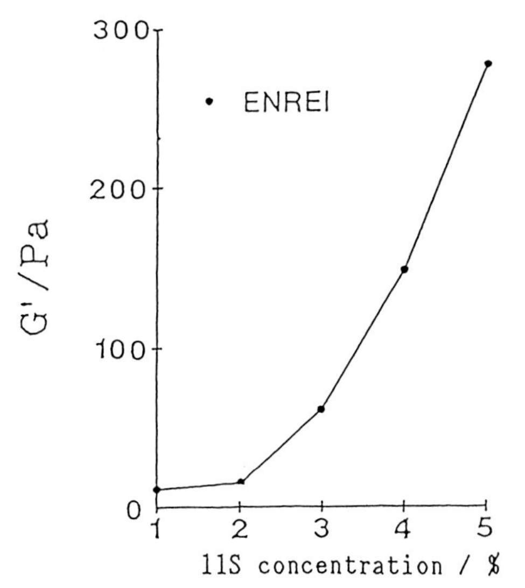

Figure 4. Dependence of saturated storage modulus G' on the concentration of 11S globulin in the presence of 0.4% GDL.

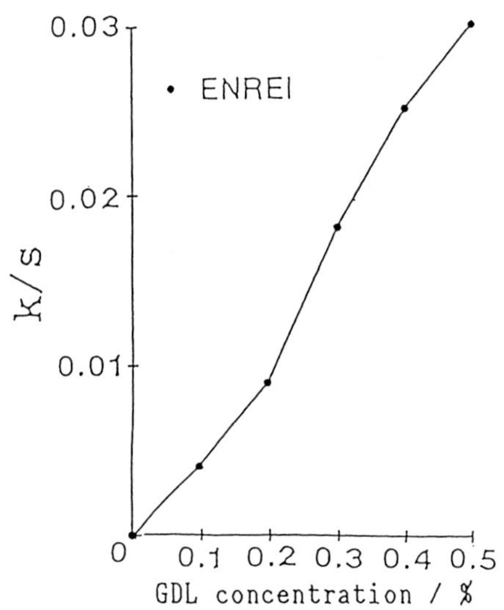

Figure 5. Dependence of rate constant k on the concentration of GDL for 5% 11S globulin solution.

3.2 Dynamic Viscoelasticity of 11S Globulin

Figure 2 shows gelation curves of 11S globulin of various concentrations in the presence of 0.4% GDL. Figure 3 shows gelation curves of 5% 11S globulin in the presence of GDL of various concentrations. G'sat of 11S solution with a constant GDL concentration increased with increasing concentration of 11S as shown in Figure 4, and the rate constant k of the gelation of 5% 11S globulin solution increased with increasing concentration of GDL as shown in Figure 5.
Consequently the saturated value of G' was mainly dependent on concentration of 11S globulin (Figure 4), and the rate constant k was governed by the concentration of GDL (Figure 5). A similar tendency was observed for the clotting of casein in the presence of rennet[10]. The concentration dependence of the storage modulus G' of biopolymer gels is described by an exponential function (2)

$$G' \propto C^n \qquad (2)$$

where G' = storage modulus
C = concentration of the biopolymer (n ~ 2 in many cases)

In the case of heated soybean in globulin gels it was found that n = 5 in the concentration range of 7.5-58.4%[11], but in our 11S globulin gels, n = 3. The difference in this exponent n may be attributed to the difference of polymers and also to the presence of GDL in our case.

We found that TMQ is evaluated by the storage and loss moduli. The dynamic viscoelastic method predicts TMQ at the stage of gelation before tofu is made, and it requires only a small amount of soymilk sample.

4. REFERENCES

1. K.Hara, Bulletin of the Agricultural Research Inst.of Kanagawa Prefecture, **129**, 58-68 (1987), **130**, 85-90 (1988).
2. K.Saio *et al,* Agr.Biol.Chem., **33**, 1301 (1969).
3. K.Saio *et al,* J.Food Science, **38**, 1139-1144 (1978).
4. V.H.Thanh *et al,* Plant Physiol., **56**, 19-22 (1975).
5. K.Kitamura *et al,* Theor.Appl.Genet., **68**, 253-257 (1984).
6. T.Sugimoto *et al,* Agr.Biol.Chem., **51**, 1231-1238 (1987).
7. T.Mori, Agr.Biol.Chem.,**43**, 577 (1979).
8. H.Yamauchi, New Food Industry, **24(8)**, 43-58 (l982).
9. K.Nishinari in "Shokuhingaku Souron (An Introduction to Food Science)", S.Masusige and T.Noguchi ed., pp.120-138, Asakura, Tokyo (1988).
10. M.Tokita, Hikichi, R.Niki and S.Arima, Biorheology, **19**, 209 (1982).
11. T.M.Bikbow, V.Ya.Grinberg, Yu.A.Antonov, V.B.Tolstoguzov and H.Schmande, Polymer bulletin, **1**, 865-869 (1979).

The effects of pH and temperature on the physicochemical properties of globin solutions and the rheological properties of the gels

K.AUTIO, M.SAITO[+], K.KOHYAMA[+] and K.NISHINARI[+]
Technical Research Centre of Finland, Food Research Laboratory, Biologinkuja 1, SF-02150, Espoo, Finland
[+]National Food Research Institute, Tsukuba, Ibaraki 305, Japan

ABSTRACT

In contrast to other globular proteins, bovine globin forms a gel at low concentration. Although this property is of practical importance, there have been no studies on the relationship between the physicochemical properties of globin and the viscoelastic properties of the gels.
The physicochemical properties of globin solutions were studied by circular dichroism (CD) and ultracentrifugal analysis, and the structure formation of the gels was investigated by dynamic viscoelasticity measurements.
Two different kinds of network structures are formed: a less ordered random structure above 60 ^{o}C and a more ordered structure above 85 ^{o}C. The former is typical for high-pH gels (pH 5.5) and the latter for low-pH gels. CD analysis suggests that globin is more unfolded at low pH-value. The α-helix content decreases from 39% at pH 5.6 to 24% at pH 3.6.

INTRODUCTION

The gelation of globular proteins usually requires unfolding of the protein as a first step, typically achieved by heat treatment. In some cases the subunit dissociation is also a prerequisite. The second step is the aggregation of the unfolded proteins (3).
Spray-dried bovine globin has been shown to consist of unsymmetrical aggregates $\alpha_3\beta_2$ (Ip=8.2) and $\alpha\beta_2$ (Ip=7.9) which are dissociated by acid and urea treatment (5). No studies however have been made on the aggregation mechanism of globin molecules or the aggregates.
pH is one of the most important parameters for globin gels; a

difference of even 0.1 pH unit can change the gel properties dramatically (2). At higher pH values an aggregated gel structure is formed, which is rigid but breaks down at low deformations. At low pH values a more ordered gel structure with low rigidity is obtained.
The extent of unfolding and the aggregation state of bovine globin molecules at different pH values are studied. Relationships between the viscoelastic properties of the gels and structure at subgelling concentrations are discussed.

EXPERIMENTAL

Preparation of globin

Bovine globin was prepared by the method of Autio *et al.* (1). After separation the globin was concentrated and spray-dried. The protein content of the powder determined by the Kjeldahl method (N x 6.25) was 90%.

Globin solutions

Globin powder was dissolved in slightly acidic water (pH 3.0) at 4.5% concentration and, after dissolving was complete the pH was adjusted with 1N NaOH.

Rheological measurements

The dynamic viscoelasticity measurement was carried out with a Rheolograph sol (Toyo Seiki Seisaku-Sho Ltd, Tokyo, Japan) equipped with a programmable thermocontrol unit. Sinusoidal strain with a frequency of 2 Hz and an amplitude of 0.02 strain units was applied to the sample and the stress was detected by the load cell. The two sine waves will have a phase difference, δ , and this is used to give the storage and loss components G'(storage modulus) and G" (loss modulus). The mechanical loss tangent, tan δ was calculated from the storage and loss moduli (G"/G').
The dynamic viscoelasticity measurements were also done with a Bohlin Rheometer VOR (Bohlin Reologi AB, Lund, Sweden). A concentric cylinder measuring system was used at the heating rate 1 ^{o}C/min.

Measurement of circular dichroism (CD)

The CD analyses were done at 20 ^{o}C using a Jasco A-500 spectropolarimeter (Japan Spectroscopic Co, Ltd, Tokyo Japan) with a scanning wavelength from 200 to 260 nm. The protein dispersion (in 0.025 M acetic acid - acetate buffer, pH 3.6, 4.3, 4.8 and 5.6) was centrifuged at 39,000 x g at + 5 ^{o}C for 30 minutes, before and after heat treatment, then diluted to protein concentration of 0.034% with the buffer. The helix content was calculated as molar ellipticity residue of α-helix at 222 nm.

Ultracentrifugal analysis

Sedimentation coefficient was determined at 20 °C by a Hitachi 282 ultracentrifuge, operating at 55,000 rpm and using UV-absorption system with a scanner. Protein powder was dissolved in 0.1 M acetic acid - acetate buffer, pH 3.7; 4.9; 5.3 and 5.6 and in 8M urea. The solutions were centrifuged as before. The protein concentration varied from 0.04 to 0.2 %. The sedimentation coefficient was calculated and extrapolated to zero protein concentration.

RESULTS AND DISCUSSION

Rheological measurements of globin gels

Figure 1 shows the storage modulus as a function of temperature for globin solutions with pHs 4.7, 5.2 and 5.7. The higher the pH value, the lower is the gelation temperature. At pH 5.5 the globin solution is opaque and the aggregation is random; at pH 4.5 the gel is transparent and the aggregation more ordered. It is suggested that different kinds of network structures are formed at different temperatures: the insoluble particles aggregate at lower temperature and the soluble protein forms a "real gel" at higher temperature. At pH 4.7 globin does not form a gel if heated to 80 °C, but only if heated to 85 °C, as can be seen from Fig. 2.

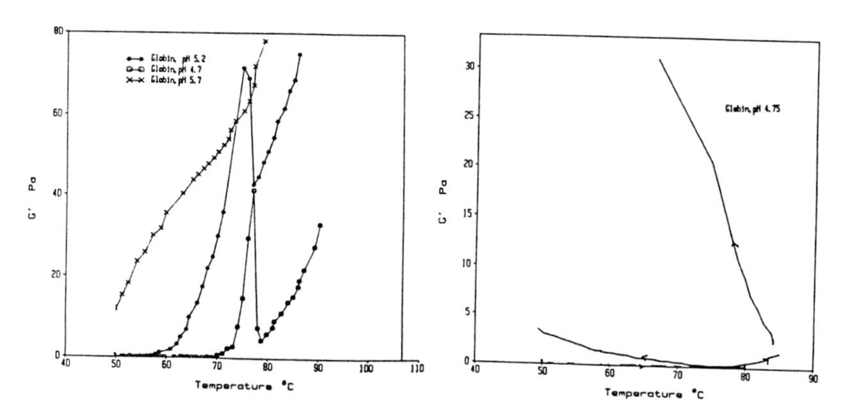

Figure 1. Heat-induced gelation of globin solutions at different pH values.

Figure 2. The effect of heating temperature on the gel formation during cooling.

CD Analysis

The α-helix content of native bovine haemoglobin is approx.73% (4). When the haeme is removed from the protein, the segments near the haeme pocket unfold, lowering the α-helix content in globin preparates (Table 1). CD analysis showed that globin molecules are more unfolded in acidic conditions. Approx. 15% of the α-helix content is lost between pH 5.6 and 3.6. Heating caused a decrease in the α-helix content of globin at pH values 4.8 and 5.6. The denaturation temperature of globin was reported to be between 45 and 50 ^{o}C (4).

Sedimentation coefficient

The sedimentation coefficient ($S^{o}_{20,w}$) for globin (Table 2) varied from 1.7 to 2.1 x 10^{-13} at different pH-values (3.7-5.6) and was 1.9 x 10^{-13} in the presence of 8M urea suggesting that globin molecules exist in the subunit. It has been shown with titration curves that spray-dried globin dissociates around pH 6.0 to α and β subunits (5).
The effect of heat treatment on the tertiary structure is probably quite an important factor in the gelling of globin solutions and needs to be further studied.

Table 1. α-Helix Content (%) in Globin Solutions.

pH	Non heated	Heated
3.6	26	24
4.3	24	26
4.8	29	21
5.6	39	23

Table 2. Sedimentation Coefficients for Globin Solutions.

pH	S_{20w} x 10^{13}
3.7	1.7
4.9	1.9
5.3	1.9
5.6	2.1
8M urea (pH 5.3)	1.9

REFERENCES

1. Autio, K., M.Kiesvaara and Y.Mälkki, U.S. Pat. 518525. Appl. 552032. May 21, 1985.
2.Autio, K. Xth International Congress on Rheology, August 14-19, 1988. Australia, Sydney, p.164.
3.Clark, A.H. and S.B. Ross-Murphy, Adv. Polym. Sci. 83 (1987) 57-192.
4.Hayakawa, S., Y.Suzuki, R. Nakamura and Y. Sato, Agric.Biol. Chem. 47 (1983) 395-402.
5.Kanko, S. and K. Autio, J. Chromatography 324 (1985) 395-406.

Evidence for protein crosslinks in mung-bean starch

C.G.OATES

Department of Biochemistry, National University of Singapore,
10 Kent Bridge Crescent, Singapore 0511

ABSTRACT

Rheological and physicochemical measurement confirm that mung-bean starch exhibits restricted swelling and high granule stability. The fine structure of amylopectin fractionated from mung-bean starch was examined by enzymic hydrolysis. Limited debranching and reduced β-amylase activity was shown, such a phenomenon is typical of chemically cross-bonded starch. Further study using a proteolytic enzyme suggests that the amylopectin fraction of mung-bean starch is stabilised by the presence of protein crosslinks.

INTRODUCTION

The granules of starches derived from legumes show limited swelling and solubility. Furthermore, these granules are resistant to breakdown at high temperature and shear, as shown by typical Brabender viscosity curves (1). Tolmasquim (2) proposed that such properties are the result of natural cross-links within the starch granule. Starch phospho-diester bonds have been suggested as possible cross-links, but the low phosphorous levels of 0.010 to 0.016% commonly encountered in legume starches would preclude any possibility of extensive cross-linking of this nature. Lipids associated with starch granules can restrict the absorption of water and release of soluble material by the starch granule. The lipid content of legume starches of 0.12-0.20% (3) would be too low to result in the extent of restriction exhibited by these starches. Further explanation, proposed by Hoover & Sosulski (4) suggests that covalent bonds within the starch granule lead to the reduced swelling. The purpose of this work was to investigate the existence of peptide bridges in mung bean starch.

MATERIALS & METHODS

Starch isolation and preparation

Starch was isolated from Thai mung beans by the procedure of Schoch & Maywald (1) with minor modifications (Oates & Wong, in preparation).

10% suspensions of starch were incubated at 45°C for 48 hrs. To some of the suspensions was added 0.1% bromelain (Sigma chemicals). Samples were again dried after washing with ethanol and diethyl ether.

Starch swelling power and solubility

Swelling power and solubility determinations were carried out for the temperature range 70-95°C by the procedure of Ring (5).

Viscosity measurement

Suspension of 7% starch were incubated (45-55°C) for 24 hrs, some of the suspensions also contained 0.1% bromelain. The viscosity of the suspensions was measured at regular time intervals at a shear rate of 0.45 S^{-1}, using a Brookfield DVII RVT viscometer and small sample cell adaptor (Cell SC4-13R, Spindle SC4-21). The cell was maintained at a constant temperature at 80°C or 95°C. In all cases readings were taken after a 10 minute equilibration period.

Starch fractionation

Mung bean starch was fractionated according to the method of Takeda & Hizukuri (6). The amylopectin was repeatedly, washed with water and freeze dried.

Enzymatic digestion

Enzymatic studies were performed using crystalline suspensions of sweet potato β-amylase and pullulanase (Sigma chemicals). Some samples prior to enzymatic hydrolysis were incubated with bromelain (0.1 mg/10 mg amylopectin) at pH 5.2, for 48 hrs at 45°C. Prior to hydrolysis with β-amylase/pullanase, these samples were heated at 100°C for 15 minutes to denature the protease and analysed for reducing sugar activity.

Enzyme hydrolysis followed the procedure of Hood & Mercier (7). Gel permeation chromatography using Sephacryl S200 was carried out in a 100 x 25 mm column flow rate 8-9 ml/hr. Total sugar was by the phenol-sulphuric acid method (8) and reducing sugar by the neocuproine method (9).

RESULTS AND DISCUSSION

Swelling power and solubility

Mung bean starch showed a limited two stage swelling and solubility pattern typical of legume starches and cross-bonded starches. Following reaction with bromelain, a non-specific protease, both swelling and solubility were significantly increased (Table 1).

Table 1. Solubility and swelling pattern of mung bean starch, with and without bromelain pre-treatment.

Temperature °C	Swelling Power		Solubility/%	
	Non-treated	Bromelain-treated	Non-treated	Bromelain-treated
70	6.8	7.0	6.8	7.8
75	7.9	8.8	8.8	9.6
80	11.8	14.0	12.0	19.3
85	13.0	15.8	16.8	24.4
90	14.2	17.8	22.0	27.8
93	16.0	18.2	23.7	29.5

Measurement of the mean granule size of starches in solution heated at 70-90°C, revealed that granules pre-treated with bromelain were some 36.5 - 37.5% bigger than the non-treated material (Table 2). Both granule populations exhibit d the same mean size prior to heating in water. If mung bean starch possessed natural peptide cross-links, these would be degraded by the action of bromelain, thereby allowing an increase in both swelling and solubility. Bromelain treated granules still showed lower swelling and solubility as compared with non-legume starches. This may reflect incomplete hydrolysis of the peptide bridges, due to limited penetration by the bromelain into the starch granule.

Table 2. Granule size of native and swollen starches

Temperature ^{o}C	Granule size/μm Starch	Bromelain treated starch
native	2.64	2.64
70	3.63	2.64
80	4.95	3.63
90	6.27	4.62

Viscosity measurement

Solution viscosity of bromelain pre-treated mung bean starch, heated to 80 or 95°C, increased with extended periods of incubation at 45 or 50°C (Table 3). This trend is particularly interesting as with time the viscosity of non-treated suspensions decreases. The drop in viscosity is thought to be the result of annealing effects or changes in lipid-amylose interactions (Oates & Wong, in preparation).

Table 3. Viscosity (Cpx10^{3}) at a shear rate of 0.45S^{-1} of 7% starch suspensions heated to 80 and 95°C.

Sample	Incubation Temperature	Cell Temperature	Time of incubation/Hr 0	1	2	4	8
Bromelain	50	80	7.4	8.1	9.2	7.1	6.4
Digested	45	80	6.1	-	8.5	6.8	6.5
Starch	45	95	6.1	-	8.4	5.9	5.4
Starch	50	80	14.0	7.0	6.3	5.6	2.6
"	45	80	9.6	-	3.8	2.6	1.8
"	45	95	7.8	-	6.8	5.8	4.5

Amylopectin Analysis

Amylopectin extracted from mung bean starch is more resistant to β-amylase hydrolysis (Table 4) than other non-legume starches. Hoover et al (4) using α-amylase also found similarly low levels of hydrolysis in legume starches. Hydrolysis by β-amylase of the non-bromelain treated amylopectin was 30.0% (Table 4). This constituted only 60% of the amount of hydrolysis obtained after pre-digestion with bromelain. Furthermore, complete hydrolysis by the combined action of β-amylase and pullanase was only shown by bromelain treated samples. The non-treated material was only 49.5% hydrolysed. Incubation with bromelain alone did not lead to any increase in measured reducing power.

Table 4. Extent of hydrolysis by β-amylase and β-amylase and pullanase, values are % ± S.D.

Sample	β-amylase	β-amylase and Pullanase
Starch	30.0 ± 4.85	59.5 ± 3.32
Starch & Bromelain	53.6 ± 2.92	102.0 ± 6.77

The elution profile of the pullanase debranced amylopectin (Fig. 1) shows a broad peak centered at an elution volume of 414 ml and a much smaller peak immediately after V_o. The elution profile for the bromelain pre-treated samples exhibited the double peak typically associated with pullanase hydrolysed amylopectin (7). Such results would suggest that in non-bromelain hydrolysed amylopectin some of the A or B chains are not accessible to pullanase. If this is the case, it can be expected that only the outer chains will be hydrolysed leaving a high molecular weight component (peak at V_o). As the result of extensive incubation with bromelain, many of the 1,6 bonds are made accessible and a more complete de-branching can take place.

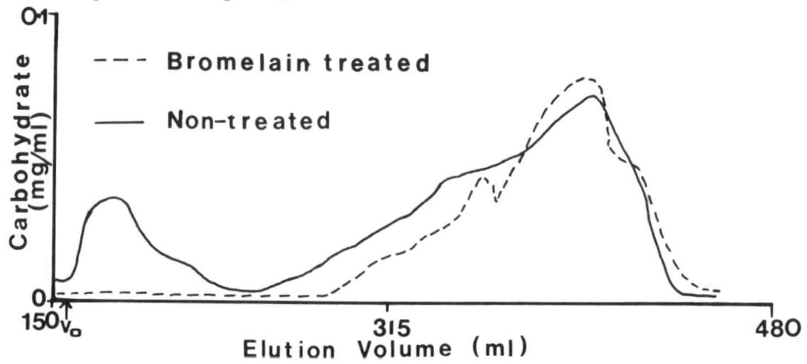

Fig. 1. Elution profile of pullanase debranched mung bean amylopectin.

This work would therefore suggest the possible existence of peptide cross-links within the amylopectin fraction of mung bean starch, which are responsible for maintaining the high level of granular integrity observed by this material.

ACKNOWLEDGEMENTS

The authors thank Miss Lee Woan Peng for technical assistance and gratefully acknowledge the support of the National University of Singapore.

REFERENCES

1. Schoch, T.J. & Maywald, E.C. Cereal Chem. 45 (1968), 564.
2. Tolmasquim, E., Correa, A.M.N. & Tolmasquim, S.T. Cereal Chem. 48 (1972), 132.
3. Biliaderis, C.G., Maurice, J.J. & Vose, J.R. J. Fd. Sci. 45 (1980), 1669.
4. Hoover, R. & Sosulski, F. Starch/Starke 37 (1985), 181.
5. Ring, S.G. Starch/Starke 37 (1985), 80.
6. Takeda & Hizukuri. Carb. Res. 148 (1986), 299.
7. Hood, L.F. & Mercier, C. Carb. Res. 61 (1978), 53.
8. Dubois, M., Gilles, K.A., Hamilton, J.K., Rebers, P.A. & Smith, F. Anal. Chem. 28 (1956), 350.
9. Dygert, S., Li, L.M., Florida, D. & Thoma, J.A. Anal. Biochem. 13 (1965), 367.

Part 4

PECTIN

Pectin – a many splendoured thing

W.PILNIK
Agricultural University, Department of Food Science, PO Box 8129, 6700 EV Wageningen, The Netherlands

ABSTRACT

The chemistry and enzymology of pectin is discussed. Pectin is present as a structural polysaccharide in the middle lamella and the primary cell wall of higher plants. It is therefore part of the diet of man and responsible for many technology and quality aspects in the processing of fruit and vegetables. Pectin is also industrially extracted from apple and citrus pomace and traded internationally in purified and standardized forms. Some aspects of pectin raw materials and of pectin standardization are considered. The manufacturing process comprises almost all unit operations of chemical engineering. Uses of pectin as a gelling agent in food gels and some newer uses as a stabilizer of acid milk products and as a clarifying and decontamination agent are discussed. The intensive research efforts on gel formation mechanisms and on gel properties are considered and research needs in this field are presented. Pectin attracts great attention as dietary fibre in purified form or as a constituent of dietary fibre preparations for which sometimes non-dietary functional properties are claimed. Regulatory problems in connection with pectin manufacture, pectin uses and the application of pectin enzyme technology are considered.

In 1824 the French chemist Henry Braconnot made a communication to the Société Royale Académique in Nancy where he was director of the public gardens (1). He reported about a substance which he had found present in all fruits and vegetables which he had so far investigated. The compound was obtained as a viscous extract, it had the characteristics of an acid and gave a translucent gel when the aqueous extract was mixed with alcohol or sugar. In his communication Braconnot said (in free translation): "I propose for this substance the name of pectin, from Greek word PECTIS which means coagulum. .... I am quite convinced that it will have many applications in the art of the confiseur ...". Braconnot also commented on the fact that his pectin seems to be present in all plants and he expressed his belief that it would be seen to have important functions in them. These visions of Braconnot have become absolutely true: Pectin to-day is an important food additive, mainly for "the confiseur" (2). Pectin has been identified in the edible parts of higher land plants as important constituents of the primary cell wall and the middle lamella and as such is part of our daily food intake. Scientists concerned with growth, consistency and pathology of plants are studying it and food technologists engaged in fruit and vegetable processing are constantly concerned with pectic changes by chemical or enzymatic action which they endeavour to prevent or to promote in view of the quality of their products. Industrial manufacture of pectin is a challenge to the chemical engineering side of food technology and the dream of food technologists in developing countries (3,4) conscious of the high pectin content of some of their waste materials. Much interest is expressed in the dietary fibre properties of pectin and its use as such.

Figure 1. Structural elements of pectin: Homogalacturonan chain and the kink produced by 1,2 linked α-L-rhamnose in the chain.

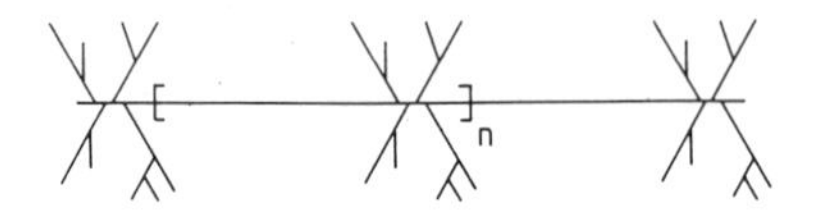

Figure 2. Model of pectin with repeating units of homogalact-uronan and hairy regions (5).

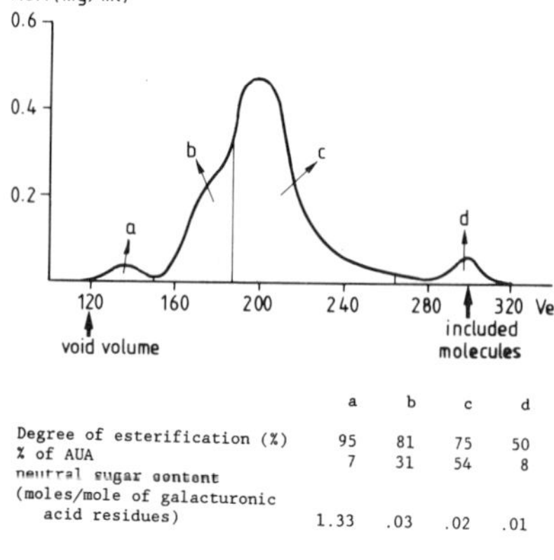

	a	b	c	d
Degree of esterification (%)	95	81	75	50
% of AUA	7	31	54	8
neutral sugar content (moles/mole of galacturonic acid residues)	1.33	.03	.02	.01

Figure 3. Gel filtration of a pectate lyase degraded pectin. AUA = anhydro-uronic acid content, VE = elution volume, the eluent was water. Fraction α is the least degraded because of its high degree of esterification. It contains only 7% of AUA but most of the neutral sugars; obviously a hairy region (5).

Chemically pectin appears as a polyuronide, a straight chain of a few hundred α-D-galacturonic acid molecules linked by 1,4-glycosidic linkages which are all diequatorial due to the Cl conformation. In carefully extracted pectins from most plants 70 to 80% of the galacturonic acids are methyl-esterified. Pectins are not pure poly-uronides; there are 1,2 linked α-L-rhamnose molecules interspersed in the galacturonan chain interrupting its conformational regularity by kinks (Fig. 1). In carefully extracted pectins from many fruits and vegetables one finds 1 to 4 molecules rhamnose and 10 to 15 molecules arabinose and galactose per 100 molecules of galacturonate. The L-arabinose and D-galactose molecules are covalently bound to rhamnose molecules as complicated side chains. In our laboratory we have established a model of apple pectin (Fig. 2) with repeating units of smooth homogalacturonan regions and hairy regions in which rhamnose and the other neutral sugars are concentrated (5). This model has been seen to apply also to pectin extracted from other sources. Other structural elements are acetyl groups and xylose side groups. The importance of these varies with the source of pectin. These findings have been made possible by applying techniques of specific break down of the pectin molecule by ß-elimination or specific enzymes, chromatographic separation of the fractions and their analysis. An illustration is given in Fig. 3 (5).

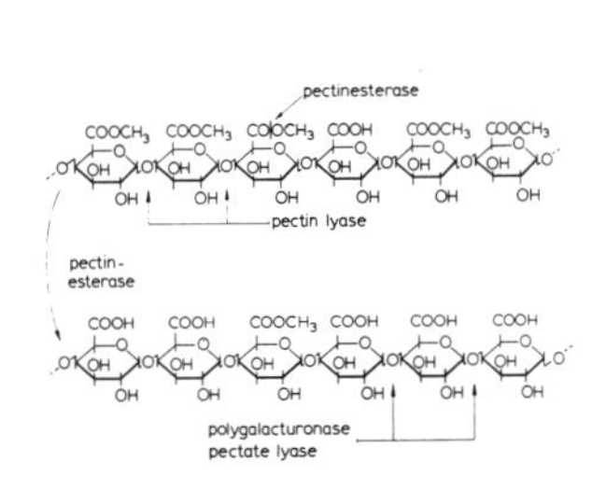

Figure 4. Points of attack of pectic enzymes.

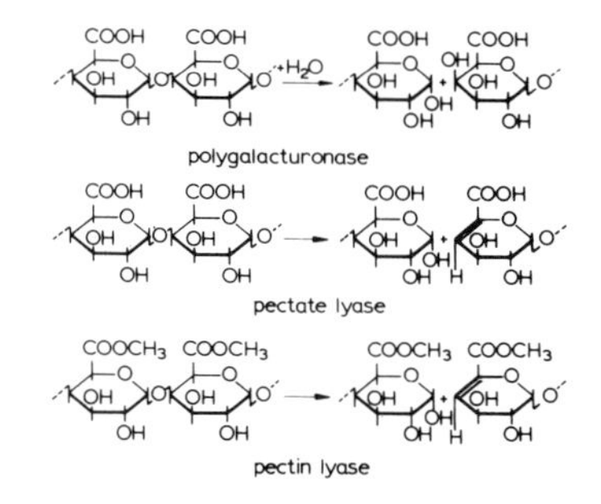

Figure 5. Mechanisms of depolymerization: Polygalacturonase by hydrolysis, lyases by ß-elimination.

Until just recently pectic enzymes have been defined entirely by their action on the galacturonan backbone of pectin (Fig.4) (6). Pectin methylesterases (PE) split off methanol from pectin, transforming it gradually into low ester pectins and pectic acid. Plant PE attacks pectin at the non-reducing end or next to a free carboxyl group and then proceeds along the molecule by a single chain mechanism creating blocks of free polygalacturonic acid which are extremely calcium sensitive. Fungal PE has a random de-esterifying action. There are three groups of depolymerases: Polygalacturonases (PG) hydrolyzing glycosidic linkages next to free carboxyl groups, pectate lyases (PAL) splitting glycosidic linkages next to free carboxyl groups by ß-elimination and pectin lyases (PL), splitting glycosidic linkages next to esterified carboxyl groups (Fig. 5). PL therefore can degrade highly esterified pectin alone, PG and PAL are able to do this in conjunction with PE action. PE and PG occur in plants; they are also produced by moulds together with PL. Commercial pectinases which are mostly derived from <u>Aspergillus niger</u> are therefore a mixture of these activities. PAL is a bacterial enzyme. De-esterification and depolymerisation are also achieved by chemical means. Under acid conditions at low temperatures, de-esterification occurs. If the temperature is raised above 40°C depolymerisation starts and with increasing temperature it proceeds at higher rates than the de-esterification. However, even at high temperatures the α-1,4-linkage between galacturonic acids is quite stable under acid conditions. If de-esterification without depolymerisation is desired under acidic conditions, e.g. pH 0.3 and 40°C, the half-time value for de-esterification is about 60 hours. Starting from a high methoxyl pectin, a random distribution of the methoxyl groups is obtained and this gives calcium sensitive low methoxyl pectins with good use properties.

Alkaline de-esterification, even at room temperature, at pH 10 proceeds with about 100 times that rate. However, under these conditions the pectin will also be rapidly depolymerised by ß-elimination; according to the mechanism which we have already seen for the lyases. This hydroxide ion catalysed reaction starts already at pH values near 4.5. This explains the unique heat stability - pH curve of pectins (Fig. 6) (7).

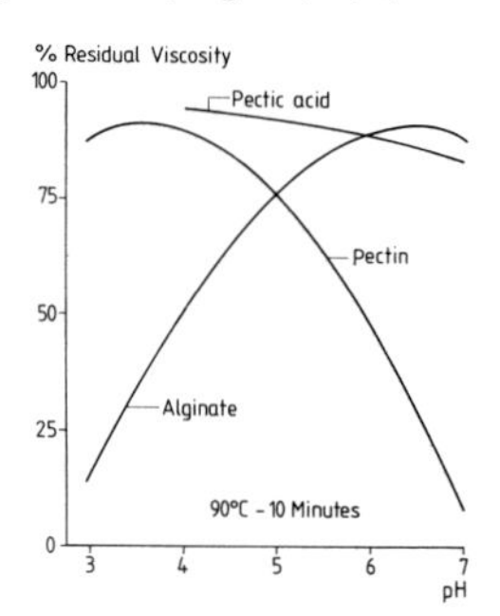

Figure 6. Stability of hydrocolloids measured by residual viscosity after heating (7).

These basic facts of pectin chemistry and pectin enzymology help to understand and solve food technology problems and contribute to process and product development. At the same time there is a feedback to polysaccharide science. Our purification procedures for enzymes deliver preparations which are useful for structure work. Recently one commercial pectinase preparation was found to contain an enzyme able to reduce the molecular weight of hairy region preparations(8). In purified form it allowed us to obtain a range of oligomers consisting of galacturonic acid, rhamnose and galactose with rhamnose at the non-reducing end. We have proposed the name rhamnogalacturonase for this enzyme that obviously splits the glycosidic linkage between galacturonic acid and rhamnose and must be considered a new pectolytic enzyme. Separation of oligomers with high rhamnose content and the use of highly sophisticated NMR techniques allowed us to establish that the L-rhamnose is present in the α-form (9).

In vegetable processing chemical ß-eliminative break-down of pectin is responsible for the soft consistency obtained by heat processing. Undesirable consistency loss can be controlled by activating endogenous PE by a long time - low temperature blanching process. The de-esterification of the cell wall pectin increases cell cohesion by calcium bridges and reduces ß-elimination by reducing the points of attack (10). In many fruit juices endogenous pectin has a nuisance value when endogenous PE causes calcium pectate flocculation resulting in cloud loss. If clear juice is desired clarification can be sped-up by addition of commercial pectinases. Press yield of many juices is improved by breaking down pectin in the mash. Many fruit juice technologists see the future of their industry in complete liquefaction achieved by joint action of cellulases and pectinase. No complicated press systems are necessary. The quality of the product is good but in many countries it may not be called a juice because in their legislation fruit juices are defined by technology (11). Obviously apple pectin manufacturers are not delighted with such developments.

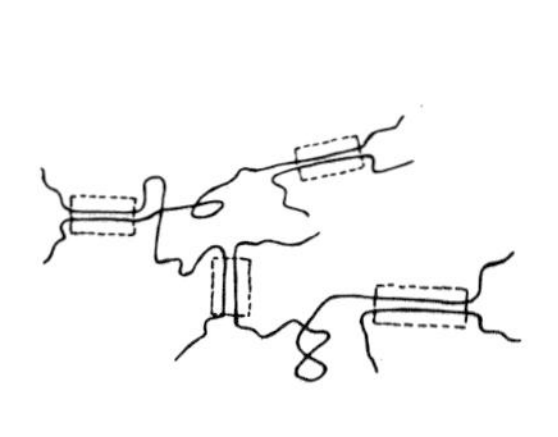

Figure 7. Formation of a gelnet by junction zones.

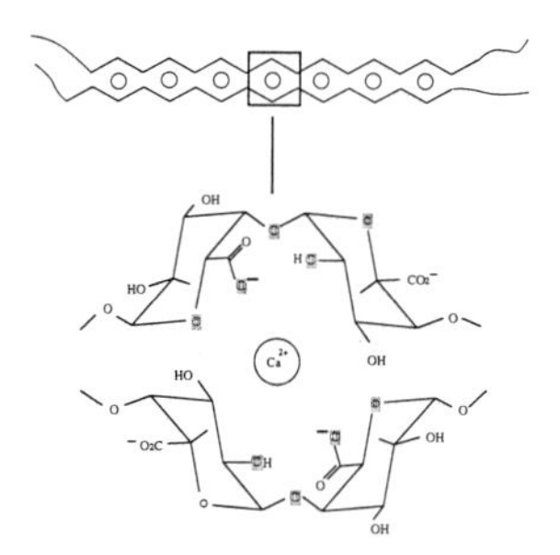

Figure 8. Egg-box structure of junction zones in calcium pectate gels (17).

Pectin is of course mainly known as a gelling agent for industrial and household preparation of jellies, jams and marmalades. In the EEC it has the number E440 which makes it a recognized additive with specifications of purity, a chemical description and in some cases rules for conditions of use. It is made in modern factories from the residues of citrus and apple juice manufacture (12) and the annual production is estimated near 20.000 tons (13). The commercial product is expected to be a light coloured, bland tasting powder, water soluble, having a constant and standardized gel forming capacity and setting behaviour. The product without standardizing diluents and/or buffer salts should have minimum 65% galacturonic acid (EEC and FAO specifications). The gel formation mechanism of pectin gels is similar to that of other gelling polysaccharides (Fig. 7) (14,15): Solvent, eventually with co-solutes, is trapped in the interstices of a 3-dimensional network held together by junction zones, associations of regions of polymer chains. In order to be able to form such associations the chains must have a regular conformational structure. There must also be structural elements causing the termination of a junction zone to prevent the formation of insoluble micelles (retrogradation). Obviously for pectin molecules the junction zones must be formed within the smooth regions. The terminating elements can be supposed to be a rhamnose kink or a hairy region or acetylated units or a change in the methoxylation pattern. Commercial pectins are able to form two different types of gels. In the more common type the pectin must be at least 55% esterified (high methoxyl pectin) and the binding forces between the chains are hydrogen bridges and hydrophobic forces between methoxyl groups (16). Both are promoted by low water activity and low pH values to reduce the electric charge and the electrostatic repulsion between chains. In this way the traditional jam recipe has come about: 65% sugar, pH 3, 0.3% high methoxyl pectin. Such a gel is not heat reversible. The second type of gel is obtained with pectin with less than 50% esterification. The smooth regions of such low methoxyl pectin contain sequences of de-esterified polygalacturonate which can align to junction zones, composed of pairs of chains between which calcium ions are co-operatively bound (Fig. 8) (17). The di-axial linkages provide the spatial order between chain segments to accommodate the calcium ions like eggs in a box; one speaks about egg box structure. Terminating structural elements are the same as in the low water activity gels. The formation of such gels is not dependent on water activity or pH; fruit juices can be gelled with the addition of some calcium and milk gels are formed with their natural calcium content. The gels are heat reversible. It is interesting to note that pectins suitable for producing low sugar gels have been developed in the USA during world war II with regard to the shortage of sugar. At present low sugar products are considered important because they help to reduce sugar intake. They are still minor products for the pectin and the jam boilers' industries and it is somewhat surprising to see most of the research efforts on gelling of pectins being devoted to calcium pectate gels. Methods of mathematical modelling, elegant experiments with calcium binding and with competitive inhibition of gelling, measurements involving optical rotation, light scattering, viscosimetry, potentiometry and conductometry, the mechanical properties of gels are used to characterize junction zones and their distribution in the pectin molecules (18,19,20). Some of these studies (18) are model studies comparing low methoxyl pectins made by alkaline or acid de-esterification (random distribution of free carboxyl groups) and by plant PE (blockwise distribution of free carboxyl groups). It is by no means obvious which system gives gels of better properties, i.e. smooth, resiliant structure with no syneresis as compared with grainy and brittle gels showing syneresis. The food technologist knows that greater calcium sensitivity and therefore presumably blockwise distribution of free carboxyl groups expresses itself in difficulties during the preparation of gels for which a strong tendency for coagulation must be overcome. An intermediary binding of the calcium ions to a chelating agent is very helpful but is linked to regulatory problems. An almost ideal solution is given by the use of amidated pectins (Fig. 9) (21). So far there is little known why the replacement of about one third of the free carboxyl groups of a low methoxyl pectin by amide groups removes all difficulties of coagulation, dissolution, syneresis and helps to obtain smooth, firm and resiliant gels. A "no fail" method would be the use of PE. Food technologists working with PE rich fruits have noticed strong gelation to occur in pulps if no counter measures are taken. From this experience trials have been made for a PE-pectin dessert gel: Dissolve high methoxyl pectin in milk, add some PE and eventually sugar, flavour and colour. One can predetermine the time of gelation but continuing PE action contracts the pectin molecules and soon after setting the gel will show strong syneresis. A slow release PE inhibitor still needs to be developed. If one examines the pectin in such a gel just after its formation one is surprised to see that the degree of esterification shows less than a 10% decrease. Will this gel still fit the egg box juntion zones model?

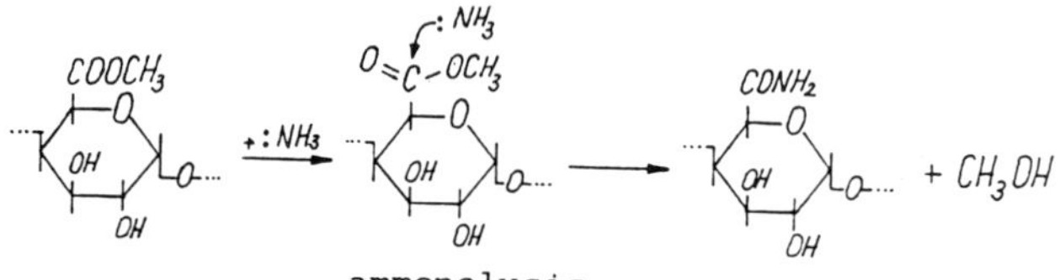

ammonolysis

Figure 9. Ammonolysis of ester groups to amide groups.

Commercial low methoxyl pectins are made by acid de-esterification in alcoholic suspension. Non randomness in distribution of acid groups may come from varying exposure of molecules to reagents. In citrus pectins there is the possibility of endogenous PE having acted in the raw material. This has its main effect on solution properties and gelling behaviour of high methoxyl pectins. In cases of pectins having the same average degree of esterification but different setting temperatures the "off" pectin is usually seen to have blocks of free galacturonate or at any rate a different charge distribution. Methods to check this next to the classical methods of calcium binding are conductometry, specific enzyme degradation and analysis of the oligomers, ion exchange chromatography (22) and quick and convenient - if one has the necessary equipment and the funds to buy the expensive columns - high performance ion exchange chromatography (Fig. 10) (23). The elution patterns of the citrus PE treated samples in Fig. 10 explain why even high ester pectins are calcium sensitive. Fig. 10 also confirms earlier findings (8) that fungal PE attacks the ester groups in a random manner and a patent (53) exists to replace the acid de-esterification process for low methoxyl pectin by an enzyme process. This is an attractive perspective; unfortunately moulds produce PE together with PL and PG and the pure PE market is too small to be interesting for enzyme manufacturers.

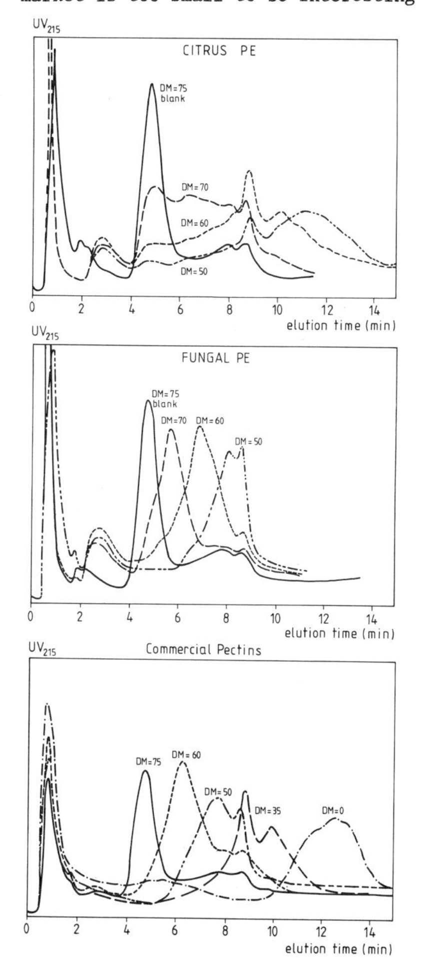

Figure 10. High performance ion (23) exchange chromatography.
Commercial pectins are acid de-esterified apple pectin production samples. Fungal PE are samples obtained from the DM 75 apple pectin by de-esterification with fungal PE to the calculated average DM.
Citrus PE are samples obtained from the DM 75 apple pectin by de-esterification with citrus PE to the calculated average DM.
DM = degree of esterification.

In view of the many factors influencing the preparation of calcium pectate gels and their properties it is not surprising that such pectins are traded on performance tests and no standard for gel strength exists. Computer programs for test plans to evaluate specified gel properties (rigidity and tendency for syneresis) in relation to specified conditions (Brix, pectin concentration, buffer salts, pH, calcium concentration) have been published (24).

The situation is much simpler for the sugar-acid-pectin gels obtained with high methoxyl pectins (25). Producers and users of pectin have been benefitting from an understanding to trade pectins according to so-called SAG grades, calculated from the sagging of a standard gel under its own weight, measured by a simple microscrew arrangement. Due to a surprisingly simple almost linear relation between pectin concentration and SAG gel strength pectins can be standardized to a certain SAG strength by mixing with sugar. Obviously there is an economic advantage in obtaining high molecular raw pectin. This is mainly a question of raw material. Average molecular weights are distributed round 75,000 for apple pomace, 90,000 for lemon pomace and 120,000 for lime pomace; these figures combined with the pectin content of these raw materials (about 15, 30 and 45% resp.) make one understand the trend for pectin factories to be established in citrus regions. SAG measures elastic properties of a gel. Gel strength measured by a breaking strength measurement is more sensitive to molecular weight (Fig. 11) (26). The food technologist can therefore decide whether he wants a gel of a certain stand-up quality to be less or more resistant to breaking. This is helpful in the case of clear jellies. For jams and marmalades with fruit pieces or pulp these measurements, if possible at all, will give different results. Unfortunately there is little progress in the development of rheology to psychorheology (27). There is even a semantic barrier between the food technologist and the physical rheologist: they cannot even agree on a definition of a gel (28). The jam industry seems to be moving away from the gel consistency for which pectins are graded. The advantage of heat stability under acid conditions is true only in respect of other gelling agents, agar, carrageenan, alginates. If thickening rather than gelling is desired, there are acid stable hydrocolloids like galactomannans and cellulose derivatives. Microbiological gums like xanthan and gellan also threaten the position of pectin. A valid measuring system which enables the food technologist to assess objectively sensory properties will also enable him to evaluate and make use of the many properties inherent in the pectin molecule. This is also true for the second largest use of pectin: the stabilization of acid milk products (29). Pectin addition to drink yoghurt keeps the casein micelles in suspension. The effect itself is easily measured but there is a lack of knowledge about the mechanism. An evaluation of the findings on properties of pectins in solution made in connection with gelling properties must still be made in respect of protein reactivity and might give inspiration for new directions in this important research. The possibilities opened by preparative chromatography to fractionate pectins according to size and charge (distribution) and make determinations of properties of these groups rather than to obtain average values mean a new dimension in pectin research. Molecular weights can be reliably determined and the properties of pectin preparation in view of applications, raw materials and extraction processes evaluated (30,31).

Pectin extraction is the subject of a special contribution. For a food technologist a pectin factory is sheer delight. A flow sheet reads like the contents section of a book on chemical engineering and each step of the

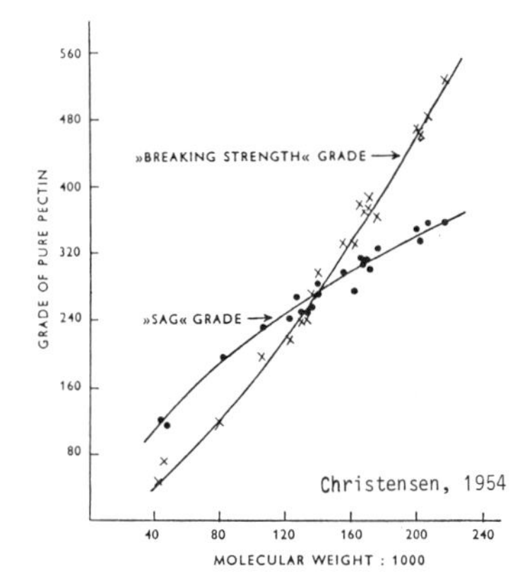

Figure 11. Breaking strength and SAG grades of various pectin samples (26).

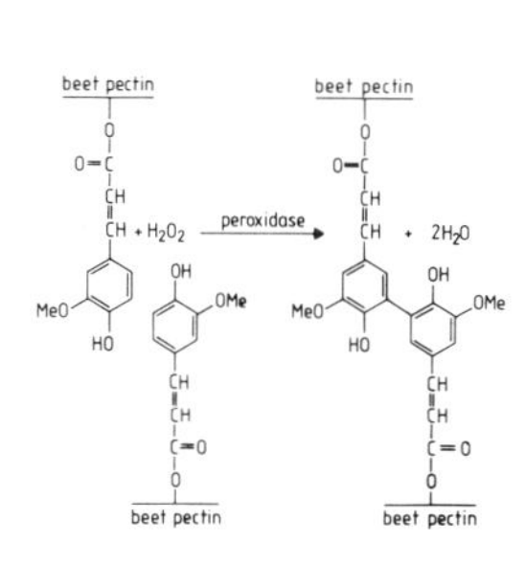

Figure 12. Crosslinking reaction between sugar beet pectin chains by phenolic coupling of ferulic acid substituents (35).

production process is a combination of the knowledge of pectin chemistry and engineering ingenuity. Restraints from environmental problems must be overcome and may well determine processes and materials. In Eastern Europe such restrictions are less stringent. Contrarily to the starch and cellulose industry, the pectin industry is not actively persuing development of chemical derivatives. It still remembers the high costs involved to obtain full clearance for food use of amidated pectins. The relatively small volume of business and the practical absence of profitable non-food uses are not encouraging for such ventures, quite apart from the fact that one does not want to lose the sympathy of the chemophobic consumers towards a "natural" additive. It is therefore interesting to note special features in pectins from certain raw materials. Sugar beet pectin has always been of interest because of the availability of large quantities of sugar beet pulp. However it has poor gelling ability due to its content of acetyl groups (32). However, due to these hydrophobic substituents, sugar beet pectin is surface active (33) which may make it useful as a natural emulsifying agent. Non gelling pectin ethylesters are also considered for this purpose (34). Recently feruloyl substitution of side chains of sugar beet pectins has been identified (35). This opens the possibility for increasing molecular weight and finally obtaining a gel by phenolic coupling (Fig. 12), a different principle than junction zones, namely chemical covalent binding between molecules. This is achieved by hydrogen peroxide with peroxidase which will not remain active in the final gel or pectin preparation. Again the food technologist is confronted with regulatory questions. Is sugar beet pectin really pectin? It does not gel but will upon removal of acetyl groups, e.g. when preparing low methoxyl pectin from it. It does not reach the 65% minimum content of galacturonic acid stipulated in FAO and EEC specifications but these are man made and not laws of nature. Sugar beet pectin is produced in the Soviet Union and was produced in Germany and Sweden during and after the war. It is interesting to reflect that pectin is the only polysaccharide gum which we insist making from waste material and also our prospecting for new raw materials is directed towards waste. It is striking that nobody has ever developed any plan for an agricultural or horticultural crop with pectin as primary product as starch in potatoes or sugar in sugar beet.

Man takes in daily 4 to 5 g pectin from the vegetables and fruits he eats in a normal western diet. To this quantity about 1%, 40 mg, may be added from pectin used as additive. In the last 15 years there have been recommendations from nutritionalists for an additional intake of at least 10 g a day. The reason is the wide acceptance in the mid-seventies of the dietary fibre hypothesis of Burkitt and Trowell (36) according to which the low fibre diets of industrialized societies are causative factors in the development of diseases like constipation, obesity, haemorrhoids, diverticulosis, cardiovascular disease, diabetes, colon rectal cancer and others, whereas the better bowel activity achieved by high fibre diets protects against these diseases or can even cure some of them. The acceptance of the fibre hypothesis brought a flood of activities, writings, symposia, promotion of fibre preparations, sometimes bordering on mass hysteria, but also serious research, chemical and physiological, and for the food technologist the not to be overheard warning of the interest of the consumer in nutrition which has not diminished even if the fibre craze has abated somewhat. Dietary fibre is defined as the sum of lignin and non starch polysaccharides. A good average value for fibre content of fruits and vegetables would be between 2 and 3% of the fresh edible material, pectin being about one third of this (37). Cereal preparations have much higher contents, e.g. wheat bran 40%. Of course, the dry matter content of cereals is also much higher. Fibre analysis goes much further (38), the various polysaccharides are quantitatively identified and their structure is elucidated. Clinical research has shown that they have different effects, pectin belonging with guar gum to the class which has no colon activity but a clearly established hypocholesterolemic action for which in human studies a daily uptake of 15 g is required (39). Other beneficial effects of pectin and guar are the prevention of postprandial hypoglycemic attacks after gastric surgery (dumping syndrome) (40) and a reduction of postprandial glucose and insulin concentration in the blood which can benefit diabetics (41). In view of the well known calcium binding capacity of pectin in vitro calcium balance was studied in human volunteers (42) taking 36 g pectin per day for 6 weeks. There was no effect. This can be explained by the complete bacterial digestion of pectin in the gut. The recommended and sometimes effected high fibre intake has in Britain prompted government sponsored research on eventual toxic effects. In the caecum of rats fed pectin as sole fibre material, increased activities of ß-glucuronidase, nitrate reductase and nitroreductase were found, all having implications of toxicity and carcinogenicity (43). However more recent studies (44) showed that these effects are absent in humans and in rats with human gut flora. In the meantime the adverse results had inevitably been reported in review articles and caught the attention of popular science writers, e.g. of the Economist (June 27, 1987) where such toxicity aspects were pointed out under the title "Fibre moral". The pectin industry has exercised restraint in propagating pectin for dietary purposes. Pectin technologists feel that their business is to produce and sell pectin to be used as additive under GMP conditions. Others must take the responsibility for high level intake as dietary product. It is very interesting to note that fibre material is reaching the market, being in fact cell wall material from sugar beet, potatoes, apple pomace, citrus peels, pea pods etc. with pectin contents up to 30%. Depending on pre-treatments such preparations may have functional properties and their use is recommended as thickening and gelling agents instead of the extracted, purified pectin (45). Already in the early twenties, long before the concept of dietary fibre was known, so-called fibre pectin was patented (46,47) and produced. The reason was the still existing difficulty to separate extracted pomace from the pectin liquor and the legitimate question "why bother?". A present day parallel development is fibre carrageenan.

In the Soviet Union there has always been an interest in health qualities of pectin. There have been reports of pectin preparations for workers exposed to lead to reduce lead uptake (48). There are reports now about pectin drinks to bind radionuclides. These are chemically near to calcium and it is doubtful whether they are removed. For lead however experiments are reported from the German Democratic Republic (49) in which the intake of 8 g pectin per day reduces significantly the lead level in the blood. The increased lead excretion is mainly through the urine and there are some theories that the hairy regions of pectin may play a rôle. It is further interesting that the pectin is given in the form of a "dietetic bar" which the Central Institute for Nutrition in Potsdam-Rehbrücke has developed. In Western Europe the possibilities of incorporating such quantities of pectin in a diet have been studied with little success.

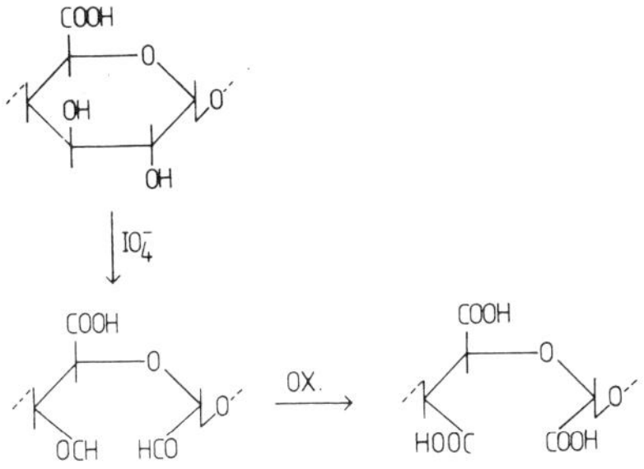

COOH
O
OH
O
OH
IO_4^-
COOH
O
O
OCH HCO
OX.
COOH
O
O
HOOC COOH

Figure 13. Tricarboxyl pectin (51).

Recently pectin has found some uses in food as a processing aid. Pectic acid added to wine will reduce heavy metal content and can be filtered off quantitatively (50). For such ion exchange applications Russian scientists have developed a tricarboxyl pectin (51) by treating with periodate and oxidizing the aldehyde groups (Fig. 13). It is highly questionable whether such a material would be admitted in Western Europe even as a processing aid without extensive testing. Pectic acid can also be used for the fining of fruit juices (52) replacing the "juice foreign" traditional fining agents gelatine or sodium silicate. The economics of these uses still needs to be examined. Such uses, for which the molecular weight is less important, would be highly welcome.

For the food technologist pectin means vegetable processing and fruit juice manufacture, functionality and consistency of foods, application technology and nutrition, raw material and environmental problems, toxicology and regulatory aspects. It means biochemistry, microbiology, rheology, sensory analysis, instrumental analysis, chemical engineering. It is indeed a many splendoured thing.

REFERENCES

1. Braconnot, N. (1825) Ann. chim. phys. 28 (2), 173-178.
2. Nelson, D.B., Smit, C.J.B. and Wiles, R.R. (1977) in Food Colloids, (ed. Graham, H.D.) Avi Publ. Co. Inc. Westport Conn.
3. Francis, B.J. and Bell, J.H. (1975) Trop. Sci. 17, 25-42.
4. Simpson, B.K., Egyankor, K.B. and Martin, A.M. (1984) J. of Food Processing and Preservation 2, 63-72.
5. Vries, J.A. de, Voragen, A.G.J., Rombouts, F.M. and Pilnik, W. (1986) in Advances in the Chemistry and Functions of Pectins (eds. Fisman, M. and Jen, J.) ACS Symposium Series No. 310, American Chemical Society, Washington DC, pp. 38.
6. Pilnik, W. and Rombouts, F.M.R. (1981) in Enzymes and Food Processing (Eds. Birch, G.G., Blakebrough, N. and Parker, K.J) pp. 105-128. Londen.
7. Pilnik, W. and MacDonald R.A. (1968) Gordian 68 (12), 531-535.
8. Voragen, A.G.J. and Pilnik, W. (1989) in Biocatalysis in Agricultural Biotechnology (Eds. Whitaker, J.R. and Sonnet, Ph.E.) ACS Symposium Series No. 389, Am. Chem. Soc., Washington DC, pp. 93-115.
9. Colguhoun I.J. et al. submitted for publication.
10. Steinbuch, E. (1976) J. Food Technol. 11 313-315.
11. Pilnik, W. (1988) in X. Intern. Congress of Fruit Juices, Orlando/Florida, U.S.A., pp. 159-179.
12. Pilnik, W. (1988) in Lebensmitteltechnologie, 2nd ed. (Ed. Heiss, R.) Springer-Verlag, Berlin, pp. 228-234.
13. Vincent, A. (1986) Proceedings IFST (U.K.) pp. 107-111.
14. Rees, D.A. (1969) Adv. Carbohydrate Chem. Biochem. 24, 267.
15. Morris, V.J. (1986) in Functional Properties of Food Macromolecules (Eds. Mitchell, J.R. and Ledward, D.A.) Elsevier London.
16. Oakenfull, J.G. and Scott, A.G. (1985) Food Technology in Australia 37 (4), 156-158.
17. Rees, D.A. Carbohydrate Polymers 2, 254.
18. Thibault J.-F. and Rinaudo, Margeurite (1985) British Polymer Journal 17 (2), 181-184.
19. Thom, D., Dea, I.C.M., Morris, E.R. and Powell, D.A. (1982) Prog. Fd. Nutr. Sci. 6, 97-108.
20. Powell, D.A., Morris, E.R., Gidley, M.J. and Rees, D.A. (1982) J. Mol. Biol. 155, 517-531.
21. Kim, W.J., Smit, C.J.B. and Rao, V.N.M. (1978) J. of Food Science 43, 74-78.
22. Taylor, A.J. (1982) Carbohydrate Polymers 2, 9-17.
23. Schols, H.A., Reitsma, J.C.E., Voragen, A.G.J. and Pilnik, W. (1989) Food Hydrocolloids 3, 115-121.
24. Erhardt, V., Krause, M., Seppelt, B. and Bock, W. (1980) Lebensmittelindustrie 27, 107-110.
25. Crandall, P.G. and Wicker, Louise in Advances in the Chemistry and Functions of Pectins (eds. Fisman, M. and Jen, J.) ACS Symposium Series No. 310, American Chemical Society, Washington DC.
26. Christensen, P.E. (1954) Food Res. 19, 163.
27. Wood, F.W. in Sherman, P. (1979) Food Texture and Rheology Academic Press London.
28. Peleg, M. (1983) Food Technology A 54-61 (November).
29. Iversen, E.K. (1984) Nordisk Mejeriindustri 11, (8) 67-71.
30. Lecacheux, D. and Brigand, G. (1988) Carbohydrate Polymers 8, 119-130.
31. Fishman, M.L., Gillespie, D.T., Sondey, S.M. and Barford, R.A. (1989) J. Agric. Food Chem. 37, 584-591.
32. Pippen, E.L., McCready, R.M. and Owens, H.S. (1950) J. Am. Chem. Soc.,

72, 813-816.
33. Smolenski, K. and Pardo, W. (1932) Chem. Listy 25, 446.
34. Popova, M., Stamov, St., Pancheva, T., Krachanov, Chr. and Berova, N. (1989) Proc. International Symposium, Food Additives of Natural Origin, Plovdiv. FECS Nr. 149.
35. Rombouts, F.M. and Thibault, J.F. (1986) in ACS Symposium Series 310 "Chemistry Function of Pectins" (Eds. Fishman, M.L. and Jen, J.J.) Am. Chem. Soc., Washington, D.C., pp. 49-60.
36. Birch, G.G. and Parker. eds. K.J. (1983) Dietary Fibre, Applied Science Publishers London.
37. Bock, W. and Krause, M. Ernährungsforschung 23 (4) 100-105.
38. Olsen, A., Gray, G.M. and Chiu, M. (1987) Food Technol. 41 (2) 71-80.
39. Kay, R.M. and Touswell, A.S. (1977) J. Clin. Nutr. 30, 171-175.
40. Jenkins, D.J.A., Gassull, M.A., Leeds, A.R., Metz, G., Dilawari, J.B., Slavin, B. and Blendis, L.M. (1977) Gastroenterology 73, 215-217.
41. Jenkins, D.J.A., Leeds, A.R., Gassull, M.A., Cochet, B. and Alberti, K.G.M.M. (1977) Annals of Internal Medicine 86 (1), 20-23.
42. Cummings, J.H., Southgate, D.A.T., Branch, W.J., Wiggins, H.S., Houston, H., Jenkins, D.J.A., Jivraj, T. and Hill, M.J. (1979) Br. J. Nutr.41, 477-485.
43. Conning, D.M., Mallett, A.K. and Nicklin, S. (1984) in Gums and Stabilisers for the Food Industry 2, Applications of Hydrocolloids (Eds. Phillips, G.O., Wedlock, D.J. and Williams, P.A.) Pergamon Press Oxford.
44. Rowland, I.R. and Mallett, I.R. (1986) in New Concepts and Developments in Toxicology (Eds. Chambers, P.L., Gehring, P. and Sakai, F.) Elsevier Science Publishers B.V. (Biomedical Division) pp. 125-138.
45. Speirs, C.I., Blackwood, G.C. and Mitchell, J.R. (1980) J. Sci. Food Agric. 31, 1287-1294.
46. Beylink, F.G. (1921) USP 1393660.
47. Leo, H.T. (1936) USP 2038582.
48. Obodovskaja, D.A., Tepper, R.Y.A. and Beznosik, A.K. (1978) Konservnaya i ovoshchesushil'naya promyshlennost 5, 21-22.
49. Walzel, E., Bock, W., Kujawa, M., Macholz, R., Raab, M. and Woggon, H. (1987) in Mengen- und Spurenelemente (Ed. Anke, M.) Leipzig.
50. Wucherpfennig, K., Otto, K., and Wittenschläger, L. (1984) Die Weinwissenschaft 38 (April), 132-139.
51. Aimukhamedova, G.B., Ashubaeva, Z.D. and Umaralicv, Z.A. (1974) in Chemical modification of pectinuous substances, Academy of Sciences of the Kirgiz Sovjet Republic, Institute of Organic Chemistry.
52. Wucherpfennig, K., Otto, K., Strohm, G. and Weber, K. (1986) Flüssiges Obst 581-590.
53. Ishii, S., Kiho, K., Sugiyama, K., Sugimoto, H. (1979) Ger. P. Appln. 2843351.

Commercial sources and production of pectins

C.D.MAY
H.P.Bulmer Ltd, The Cider Mills, Plough Lane, Hereford HR4 0LE, UK

ABSTRACT

Commercial pectins have been produced from fruit wastes, normally from the juice production industry, for over 50 years. The earliest product, still made today, was a concentrated liquid extract from apple pomace, and this was followed by precipitated pectins from both apple and citrus, preferably lime or lemon. Alternative raw materials such as sugar beet residues and sunflower heads are being considered in some areas. Pectin is extracted under hot acid conditions which also cause degradation of the material, and a compromise has always to be achieved between maximum yield and quality. For many purposes the degree of esterification must be reduced and methods are discussed. Solid pectin can be precipitated either by alcohols or by metallic salts followed by acid leaching. The process is completed by drying, grinding, and blending to a standard functionality, often a standard gelling strength.

INTRODUCTION

The production of commercial pectin started in order to meet a need - the need for a consistent material to improve the set texture of jams made from certain fruits, especially strawberry. It is not certain just when the use of jellying juices and other crude pectin extracts started - it may be as old or older than commercial jam making. Certainly by early in this century jam makers were buying dried apple pomace from which they could extract the pectin they required. The first pectin producers, by making a concentrated pectin extract which could be stored until needed, gave the jam maker a more convenient and consistent gelling agent, and by developing a more controlled process, made better use of the pomace. It was soon realised that although a liquid pectin was convenient, it was bulky and expensive to transport over long distances. Powder pectins were therefore introduced and this enabled citrus peel to be used as an alternative raw material.

PECTIN PRODUCTION

Raw Materials

Pectin production depends on a number of factors, including access to raw material, water, energy, and effluent disposal at reasonable prices. Fruit is never grown or even processed solely in order to produce pectin. This is in contrast to the manufacture of many other food hydrocolloids which have either been based on gathering a wild raw material which is now cultivated specifically for hydrocolloid production (gum arabic, carrageenan, etc) farmed specifically (eg guar) or produced by fermentation. Like the manufacture of gelatine, pectin is essentially a by-product industry, and depends on waste from a primary process for its raw material. The availability of raw material, and its quality, are entirely dependent on the market for juice and, in the case of citrus, oils. These markets determine in the longer term what fruits are grown, and in the short term which fruits are purchased for processing. In some areas, the fresh fruit trade is the major market for the fruit, and only substandard fruit or the surplus reaches the juice factory - in other areas, fruit is grown specifically for processing, as with cider apples, and with citrus fruit in Brazil and to a degree in Florida. Apple pomace of good quality is obtained from processing fresh rather than cold stored fruit, and is therefore only available for a short season. For this reason it is always dried and stored before use, as a viable pectin plant needs to run all year. The citrus season can be much longer, and in some areas such as Brazil, citrus fruit of one type or another is available all the year. Under these circumstances the peel can be used directly, but it can equally be dried and if necessary transported larger distances.

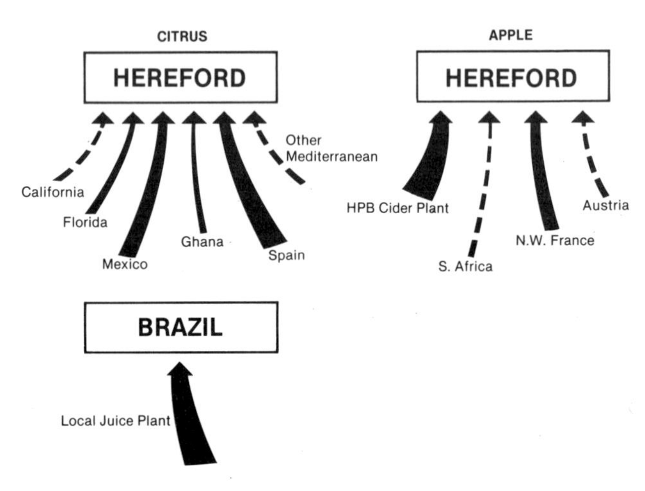

Figure 1 H P Bulmer Pectin - Major raw material sources.

Thus, pectin companies in areas such as northern Europe, which were founded to make apple pectin from locally processed apple pomace, have been able to expand by processing imported citrus peel. In my company, we have not only done this, but also have a joint venture plant in Brazil which takes peel from an adjoining juice operation (Fig 1). Clearly, for long distance transport, it is most economical to transport the peel with the best pectin value, not just on weight basis, but in terms of overall quality. There is therefore considerable competition to purchase the best raw materials, such as dried lime peel. If peel is to be dried, it is important to preserve the pectin contained in it in the best possible condition. Washing to remove most of the acidity is essential, and the drying temperature needs to be carefully controlled as pectin is quite readily degraded at high temperatures. It is also vital to ensure that the moisture content is low enough to prevent bacteriological action which will not only damage the pectin but also can cause spontaneous combustion in store.

Extraction and isolation

The only commercial process for extracting pectin involves heating the raw material with heat and acid - a process which also tends to degrade the pectin. Indeed, it would seem from recent studies from Pilnik's group (2) that the structure of plant materials may make it essential to break some of the pectin chains in order to free pectin from the cellulose skeleton of the tissue, even when specific enzymes are available. The reactions involved must therefore be very similar to those involved in the acid degradation of pectin, and show a similar dependence on temperature (Fig 2). However, extraction is more dependent on pH and this offers the chance of some selectivity. One function of the lower pH is to

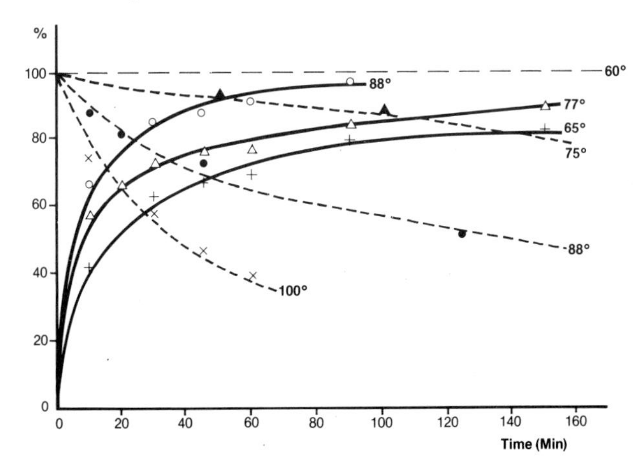

Figure 2. Pectin yield and molecular weight at various temperatures - (---) weight yield,(- - -) relative molecular weight.

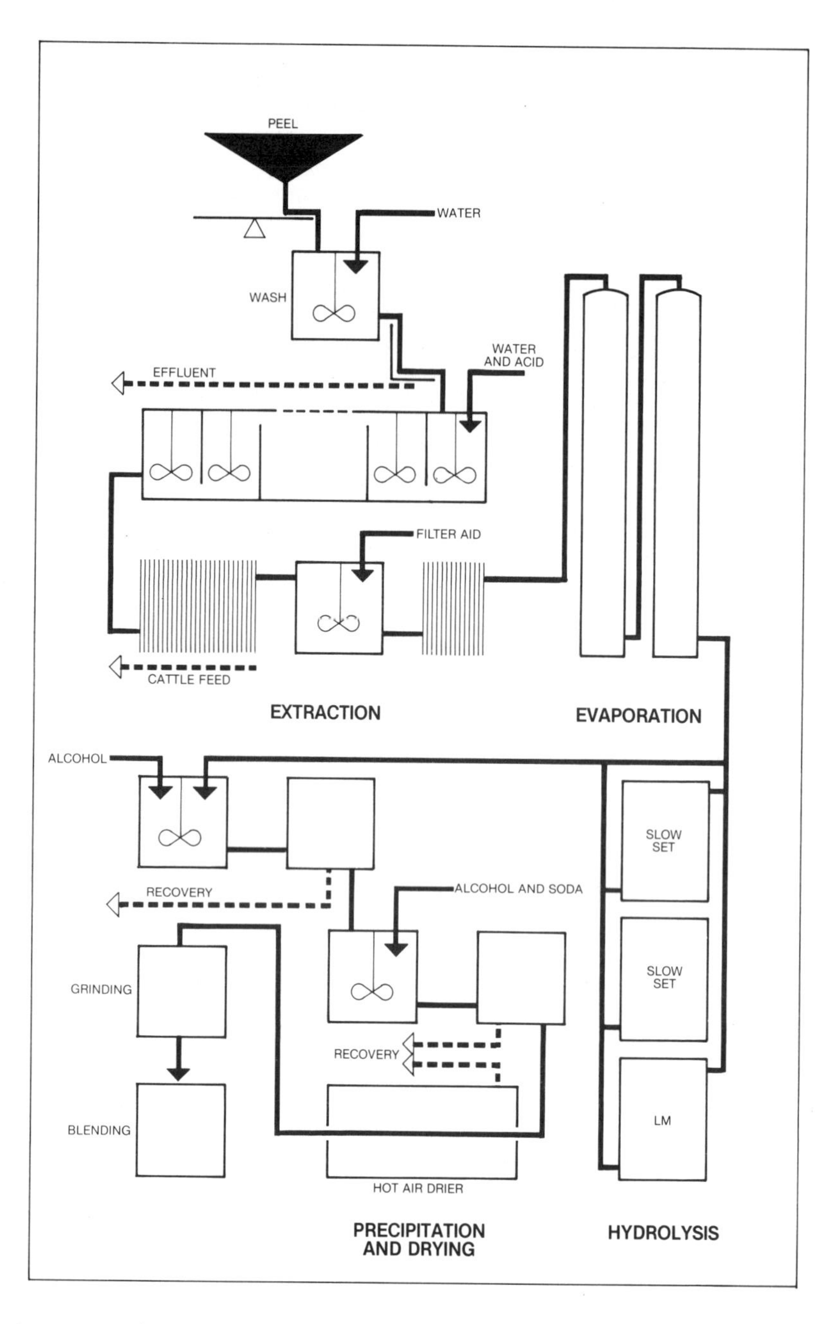

Figure 3. The Bulmer Pectin Process. (Reproduced with permission from Elsevier Scientific Publishers (4)).

supress ionisation of the carboxyl groups in the pectin, and thus reduce calcium binding, but this is not sufficient to release a good yield of high molecular weight pectin.

The physical problems of extracting the viscous pectin from the cellular residue can be reduced by increasing the rate of stirring and hence the shear forces during extraction, but this causes breakdown of the solids and can lead to problems in separating the solids from the extract. An alternative approach involving a special reactor where cyclic variations in pressure effectively pump the pectin out of apple pomace during extraction has been suggested by a Bulgarian group (3). It is essential to balance the physical and chemical factors very carefully in order to get the best out of a particular raw material. These conflicting requirements immediately set the pectin producer a problem; to what extent does he try to obtain the maximum weight yield of pectin, when this will compromise the quality of his product. Typical conditions of extraction are a pH of about 2.0 and temperatures between 70 and 100°C for one or two hours, or longer if it is intended to reduce the degree of esterification concurrently. Pectin makers have evolved their own particular compromises in the light of the raw material available, the type of equipment used, especially for separation of the residual solids, and the range of products demanded in the market at the time their plant was designed.

The Bulmer process, indicated in outline in Figure 3, evolved from the production firstly of rapid setting liquid pectin, followed by slow set liquid, and then by powdered pectins, apple and later citrus. Consequently our extraction process is designed to produce rapid set pectin, and slow set is produced by hydrolysis in solution. In the production of citrus pectin, peel is washed to remove some of the colour, then dropped into hot dilute nitric acid. The slurry passes through a series of stirred compartments, and the solid is separated on filter presses. The extract is clarified by keiselguhr filtration. The solution is then concentrated about four fold.

It may then be passed almost immediately to precipitation, or held at a lower pH for several days to make either slow set or low methoxyl pectin (Fig 4). Once more, the correct conditions depend on obtaining the right balance between competing reactions. Hydrolysis of ester groups is increased markedly by reducing the pH, but rather less by increasing temperature. Degradation reactions are, in contrast, very sensitive to temperature increase. The best way to achieve a controlled acid hydrolysis with minimum degradation is to use a low pH, a low temperature, and to accept a reaction time measured in days rather than hours.

This slow hydrolysis process lends itself to control by daily sampling and chemical analysis for the degree of esterification. Some experience is needed to relate this chemical analysis to the rate of set in a standard jelly, as the relationship varies with different source materials. The hydrolysis reaction is normally stopped by precipitation (Fig 4) at the appropriate time. After separation, the pectin is washed with aqueous alcohol, the pH adjusted by adding alkali, and dried. A low temperature is important to retain a high

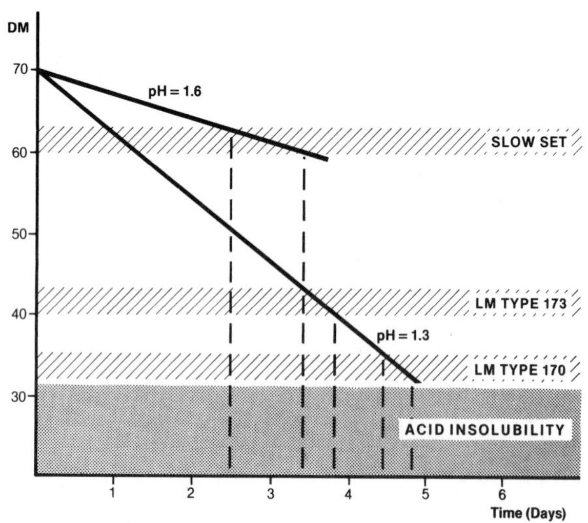

Figure 4. Solution Hydrolysis - Reaction rates and precipitation times for products of different degrees of methylation (DM)>

and dried. A low temperature is important to retain a high molecular weight, ensuring a good grade yield and maintaining the properties of the pectin for specific applications such as stabilisation of protein dispersions, and mouthfeel in drinks.

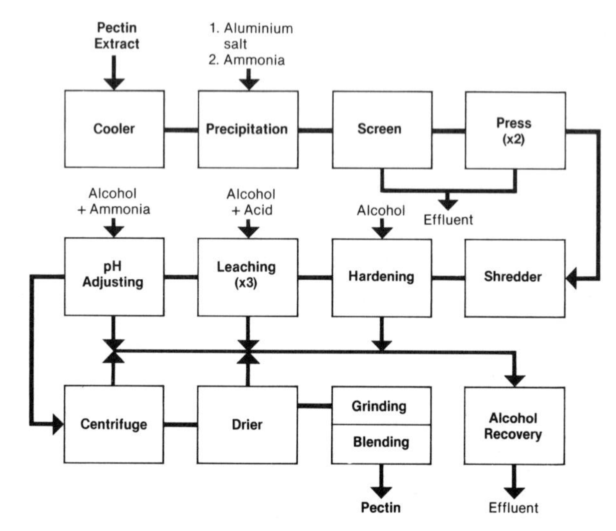

Figure 5. Aluminium Precipitation Process for pectin isolation.

Aluminium Precipitation

The alternative process (Fig 5) for isolating pectin from solution makes use of the well characterised interaction between negatively charged pectin and a positively charged colloidal metal hydroxide, normally aluminium (5). The basic chemistry has been explored by Joslyn and de Luca (6). If an acidic pectin extract is mixed with an aluminium salt, no precipitate is visible, but there is already a change in the pH of the system. On adding an alkali, such as ammonia or sodium carbonate to raise the pH to around 4, a bulky gelatinous precipitate is produced. Because citrates complex aluminium strongly, it is important that the citrate level of the extract is low, in other words the peel must have been thoroughly washed. The precipitate produced from a typical pectin extract is a bright slightly greenish yellow, although this colour is due to impurities - if a purer powder pectin solution is precipitated with an aluminium salt, the precipitate is white. The precipitate needs to age for a while to become firmer in texture, and can then be separated either by flotation or by using some kind of screen. At this stage the water content is very high, and it is usual to follow the screening with a pressing operation.

It is now necessary to remove the aluminium from the pectin without the pectin going back into solution.The first stage is to replace the water from which the yellow mass has been recovered with a non- solvent for the pectin, normally an aqueous alcohol mixture. Once the mass is suspended in alcohol, acid may be added, and the aluminium starts to go into solution. Several acid washing stages are necessary to remove sufficient of the aluminium to give a product which dissolves readily to give a solution of low viscosity. Excess acid is then washed away and the final slurry treated with ammonia or other base to partially neutralise the pectinic acid groups.

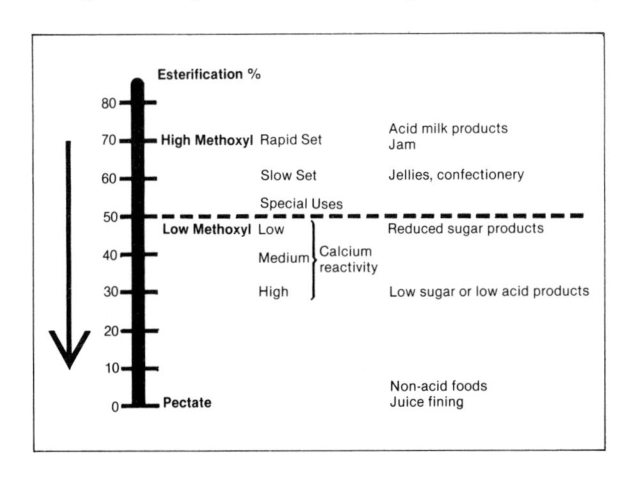

Figure 6. Deesterified pectins and their applications.

Options for Deesterification

Hydrolysis of a controlled proportion of the methyl ester groups is the major way of controlling the detailed gelation properties of a pectin. The chart (Fig 6) shows the products which can be made with different degrees of esterification. Most pectin manufacturers supply a wide range of types from rapid setting high methoxyl pectins which gel with acid and high concentrations of sugar, to low methoxyl types with a range of reactivities to calcium. The ester hydrolysis reaction can be carried out in different ways and at different stages in the pectin process (Table 1)

Table 1. Methods of acid Deesterification Compared

Method	Advantages	Disadvantages
Before extraction	Easier Extraction	Large volumes of difficult slurry
	Buffer stock of peel (wet)	Not easy to control
During extraction	Single stage process	Variable residence time make continuous
Concentrated extract	Easy to sample and control	Long reaction times - large tanks
		Preservative or aseptic process needed
Alcohol Suspension	Smaller volume	More flammables in plant
	Can improve colour and clarity of pectin	Limited amount of hydrolysis possible

The first option is to treat the peel or pomace before extraction, usually with acid. This means handling large volumes of slurry and controlling the time and temperature during which the solid is held. It can however make the subsequent extraction much easier. An alternative is to combine this treatment with extraction by extracting the pectin at moderate temperatures over a long residence time when a slow set pectin is required. It is easier to change the extraction process in this way if extraction is conducted batchwise, but chemical engineering considerations make continuous processes desirable. Other alternatives are to hydrolyse the ester groups in the concentrated pectin extract as mentioned above, or to hold the pectin in a warm acidic aqueous alcohol slurry. This last process has sometimes been combined with the aluminium removal process after aluminium salt precipitation.

Amidated pectins

Apart from this range of degrees of esterification, there is one other chemical modification of pectin which has gained acceptance. An alternative method of deesterification is to treat pectin with ammonia (7). To avoid excessive alkaline degradation, the reaction is usually conducted heterogeneously using a suspension of pectin in an aqueous alcohol. Under these conditions a significant number of ester groups are converted to primary amide groups. The amounts of deesterification and amidation depend on the temperature, and also on the nature of the alcohol in the suspending medium. It is also possible to control the composition of the product by selecting the degree of esterification of the starting pectin, so a wide range of amidated pectins, mostly but not exclusively of the low methoxyl type, but with a range of useful properties, can be prepared.

ALTERNATIVE RAW MATERIALS - BEET, SUNFLOWER, ONION

Because pectin is derived from waste products, there is never control of the availability of the raw material. The scale of the citrus juice industry is such that there will never be a complete shortage of citrus peel, although the more desirable types such as lime peel are somewhat limited. Some of the possible peel sources are not economically available because there is a minimum economic size for either a pectin plant based on fresh peel or for a peel washing and drying plant. Peel from a small isolated juice plant is therefore not effectively available. The apple juice industry is considerably smaller, and recent growth has been accompanied by the introduction of enzyme treatment techniques which rely on degrading the pectin in the fruit to release a greater yield of juice. This type of process is particularly valuable in processing dessert and culinary apple varieties, and gives less benefit with cider fruit, and has therefore had less impact in northern France and in England than elsewhere.

Beet Residue from Sugar Production

One alternative pectin containing material which is available in large quantity in temperate countries is the residue from beet sugar production (8). As a conventional gelling pectin beet pectin has two serious defects. Firstly, it is low in molecular weight, and secondly it is significantly acetylated. Acetylation tends in general to hinder gel formation, presumably by hindering the approach of pectin chains to form junction zones. It is possible to deacetylate the pectin by acid, but this also leads to the removal of some of the methyl ester groups, leading to a low methoxyl pectin. The process also reduces molecular weight even further, so the quality of the pectin/calcium gels which can be formed is very poor. If acidified methanol is used, the degree of methyl esterification can be maintained, but the molecular weight is still inadequate. One frequent observation is that beet pectin extracts are very troublesome to process because of foaming and it may be that beet pectins may find a use as a foaming or emulsifying agent, rather than in gelling applications.

Sunflower Heads

It is also known that the tissue of the seed head of sunflowers has a high pectin content (9). Selected samples can produce pectin of a very high molecular weight, although this too is acetylated. The higher molecular weight gives the opportunity to modify the molecule and retain sufficient chain length to produce an acceptable product. The major problem with this material is that the highest pectin yield and quality occur at an earlier stage than commercial maturity for oil yield and for easy removal of the oil seeds. Contamination of the pectin with small quantities of oil which can easily be oxidised is another potential problem. There is however some evidence that sunflower pectin is being produced commercially in China.

Other Potential Pectin Sources

Pectin is also present in many other fruits and vegetables. Most of the fruits are either not processed on a sufficiently large scale, or, like blackcurrant, would be difficult to purify because of colour or flavour. One source which might have some potential if flavour problems can be overcome is onion (10) which has been shown to contain high methoxyl pectin with good gelling power.

CONCLUSIONS

The pectin industry is now a sophisticated operation which can take variable natural raw materials and convert them to a whole range of well-characterised products. The development of the process to meet new requirements, both from customers and environmental and energy constraints will provide an ongoing challenge to scientists and engineers in the industry.

REFERENCES

1. Kertesz, Z.I. (1951) The Pectic Substances, Interscience Publishers, New York, USA.
2. Reynard, C.M.G.C., Voragen, A.G.J., Schols, H.A., Searle - v. Leeuwen, M.F., Thibaut, J.F., and Pilnik, W. (1989) in "Food Science - Basic Research for Technological Progress" Ed. Roozen,J.P., Rombouts, F.M., and Voragen, A.G.J., Pudoc, Wageningen, The Netherlands, 163-170.
3. Kratchanov, C., Marev, K., Kirchev, N., and Bratanoff, A. (1986) J.Food Technol., 21, 75-761.
4. May, C.D. (1989) Carbohydrate Polymers (in press).
5. Joseph, G.H., and Havighorst, C.R. (1952) Food Engng., 24, 87-89, 134-137, 160-162.
6. Joslyn, M.A., and de Luca, G. (1957) J. Colloid Sci., 12, 108-130.
7. Joseph, G.H., Keiser, A.H., and Bryant, E.F. (1949) Food Technol., 3, 85.
8. Phalak, I., Chang, K.C., and Brown, G. (1988) J Food Sci., 53, 830.
9. Lim, M.J.Y., Humbert, E.S., and Sosulski, F.W. (1976) Can. Inst. Food Sci. Technol. J., 9, 70.
10. Fatah, A.F.A. (1987) quoted in Indian Food Packer, 41, 120.

Gelation of very low DE pectin

STEEN BUHL
The Copenhagen Factory Ltd, Denmark

ABSTRACT

Pectin with very low degree of esterification, DE, is often referred to as pectate. Here pectate is defined as a pectin with detectable amounts of methylester groups and not more than DE = 5.

Pectate is, due to the low DE, extremely Ca^{++}-reactive and gelation will take place instantaneously, when pectate comes in contact with a Ca^{++}- solution.

Therefore it has been necessary to apply a special technique using EDTA and Glucono-delta-lactone (GDL). A controlled gelation of pectate is obtained by a slow reduction of pH caused by hydrolysis of GDL whereby Ca^{++} is gradually released from EDTA.

Pectate of both sodium- and potassium forms has been examined.

It has been revealed that pectate of both salt forms created very firm and brittle gels with Ca^{++}. The gelstrength or breaking-strength is very sensitive to pH and Ca^{++}-concentration.

It was also observed that pectate of potassium form gave much higher gelstrength than the sodium form. Generally, potassium ions seem to have a synergistic effect on Ca^{++}-gelation of pectate.

The effect of potassium on gelation of pectate has shown that pectate can jellify in presence of potassium ions as the only cations. This potassium induced gelation is very dependent on pH and potassium concentration. Two types of gels have been observed.

One type is formed at pH=3.6-4.5. This gel has a opaque appearance and high gelstrength.

The other type is formed at pH=3.1 or lower. This gel is transparent and has a rather low gelstrength.

An interesting point is that K-induced gelation does not take place in the pH-range of 3.1-3.6. This lack of gelation may be due to conformational change of pectate at pH=3.1-3.6.

INTRODUCTION

Ultra low DE pectin - pectate is a pectin with 0 < DE < 5. In order to conform with the pectin definition, there must be a detectable amount of methylester groups.

Ultra low DE pectin is produced in either sodium- or potassium form. Pectin of either form is hot water soluble when neutralized to pH=3.0-4.0, measured in a 1% solution. If the pectin has been neutralized to pH=7 (1% solution) it will be cold water soluble.

A pectin with DE below 5 is, of course, extremely Ca^{++}-reactive. Due to this reactivity ultra low DE pectin is difficult, if not impossible, to dissolve in hard water.

Dissolution shall take place in demineralized or at least Ca^{++}-free water.

In order to bring the pectin to gelation certain factors must be fulfilled: 1) sufficient pectin, 2) presence of cation, 3) suitable pH.

Ultra low DE pectin can jellify in water provided a suitable combination of pectin concentration, presence of Ca^{++} and pH.

Experiments have shown that 0.5% w/w pectin is too little to form a gel in water no matter the Ca^{++}-level and pH. As a rule-of-thumb, 1% pectin or higher can form a gel in water given adequate Ca^{++}-levels and pH.

At pH=7 ultra low DE pectin will practically speaking be fully dissociated. At this pH there will be maximum Coulombic repulsion between the pectin molecules. In order to form a gel at this pH a certain amount of Ca^{++} must be present to overcome the repulsive forces. As pH decreases less Ca^{++} will, consequently, be needed to create a gel.

When all gelling factors are within the correct range, gelation will occur spontaneously. It is therefore necessary to apply a special technique in order to obtain a controlled Ca^{++}-release.

EXPERIMENTAL

EDTA is used as sequestering agent. EDTA binds Ca^{++} stronger than pectate at pH=7.0-7.5. As pH is reduced EDTA looses its complexing power, and pectate starts to react with Ca^{++}, and a gel is gradually formed. This method is also described in 1).

Procedure for Gel Formation.

1. Dissolve the pectate (normally 5.00 g) in 400 ml demineralized water and adjust pH=7.0-7.5 with 1.0 N and 0.1 N NaOH.
2. Add calculated amount of a $CaCl_2$-solution containing 27.56 g $CaCl_2$, $2H_2O$/l and 46.00 g EDTA/l. 10 ml. of this solution contains 75 mg Ca^{++}. The $CaCl_2$ solution is adjusted to pH=7.0-7.5 with 20% NaOH-solution.
3. Add the desired amount (normally 1-20 g) of Glucono delta-lactone (GDL) and stir for 2 minutes.
4. Adjust to 500 g with demineralized water.
5. Transfer liquid to Bloom glass with tape along the brim.
6. Let sample stand at room temperature for about 24 hours.
7. Remove the tape and cut the gel along the top with a cheese cutter and measure gel strength (break) on Stevens texture analyser, LFRA.

LFRA-measurement.

1. Plunger, normally 24 mm.
2. Plunger speed 20 cm/min.
3. Chart speed 0.5 mm/second.

In cases where plunger (24 mm) exceeds maximum load, we have used 12 mm or even 6 mm plunger. For comparison reasons we have used a factor 4 using plunger 12 mm, instead of 24 mm, and factor 16 for plunger 6 mm.

Materials

As ultra low DE pectins we have used
GENU Pectin type LM 1912 CS Z (sodium form)
and
GENU Pectin type LM 111 CS Z (potassium form).

Both types are unstandardized i.e. no sugar has been added.

RESULTS

The basic parameters have been assessed based on results shown in figure 1.

FIG. 1

EFFECT OF PECTATE CONCENTRATION - 15 mg Ca^{++}/g PECTATE

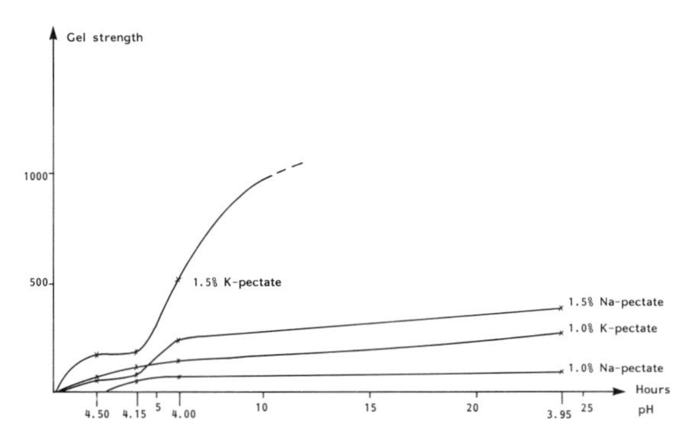

After about 24 hours GDL can be considered fully hydrolyzed. Also the gel has reached a reasonable stable condition.

We have chosen 1% pectin concentration as a basis for these examinations. As indicated 1.5% pectin will provide very firm gels especially when Ca^{++} is increased and pH lowered.

The curves also indicate a substantial difference in gelling performance between sodium- and potassium forms of the ultra low DE pectin.

The potassium form gives much higher gel strength.

pH is adjusted by addition of GDL. Table 1 shows the typical pH-values in the pectin solution after 24 hours. The pH will, of course, to some extent depend on the amount of salt added, e.g. $CaCl_2$.

Table 1. RELATION BETWEEN g GDL ADDED TO THE PECTIN SOLUTION (TOTAL 500 g) AND FINAL pH (AFTER 24 HOURS)

GDL (g)	pH
0.5	5.2
1.0	4.7
1.5	4.3
2.0	4.1
3.0	3.9
5.0	3.6
10.0	3.3
20.0	2.9
40.0	2.5

Ca++-gelation

The effect of Ca^{++} and pH is shown in figures 2 and 3.

FIG. 2

FIG. 3

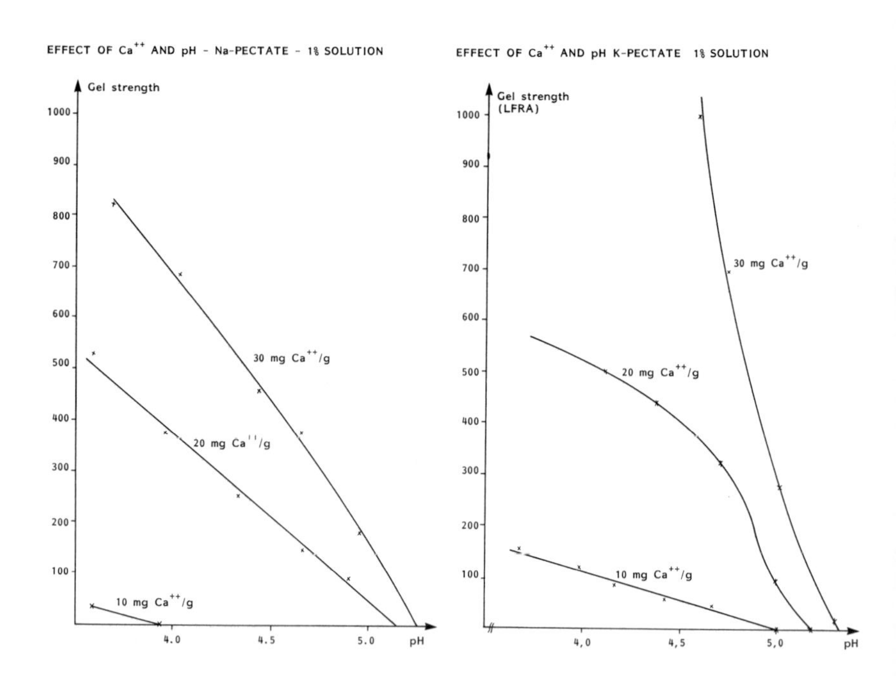

The general pattern is clear. Low Ca^{++} content can only provide gel at low pH. It is also obvious that final gel strength is very pH dependent.

The curves could give the impression that pectate gels cannot be formed at pH=5.5 or higher. That is, of course, not true. The zero gel strength at pH=5.5 or higher is due to the Ca^{++} - EDTA complex.

Gelation of pectate at pH=7 will require another sequestering agent than EDTA.

By comparing figure 2 and 3 it is again revealed that pectate in potassium form responds much stronger to Ca^{++} and to a decrease in pH than the sodium form.

The gels appear transparent,though gels formed from potassium pectate tended to contain some haziness.

None of the examined Ca^{++}-gels could be melted when heated to 100°C. This indicates that the gelling temperature under these conditions is above 100°C.

Effect of Ca^{++}, K^{+} and pH on gelstrength of pectate

The previous results indicate an impact of K^{+} on gelstrength of pectate. Both types of pectate were therefore examined for the effect of added potassium. Potassium was added as KCl. The results are shown in figures 4 and 5.

FIG. 4

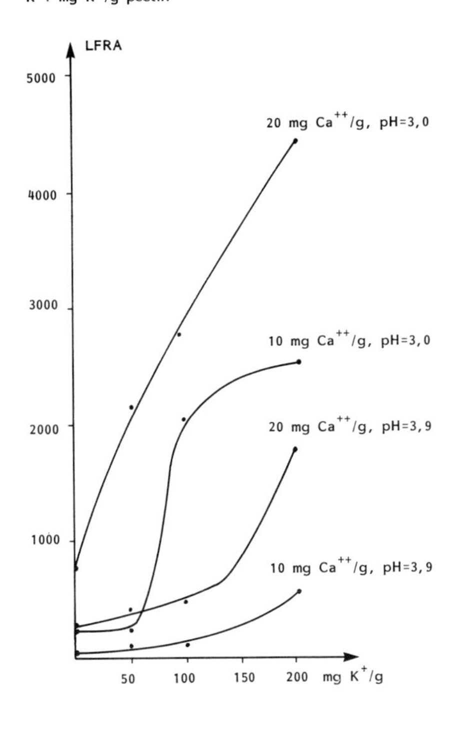

FIG. 5

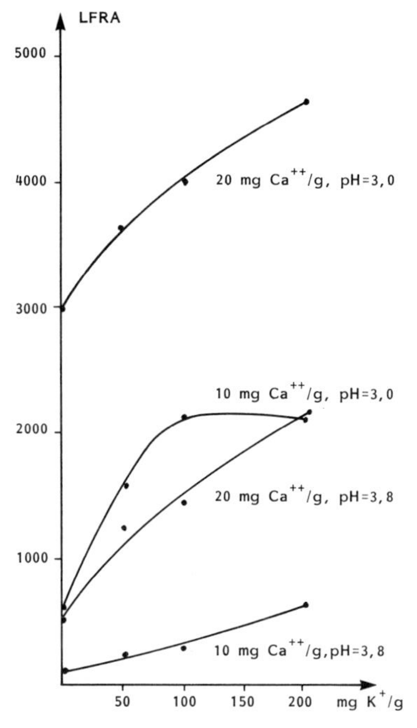

The figures confirm the effect of K^{+}. It is clear that even relatively small amounts of K^{+} will increase gelstrength of pectate. K-pectate obtains higher gelstrength than the sodium form at lower K^{+}-concentration, but at about 200 mg K^{+}/g pectate the two pectate forms seem to reach the same level of gelstrengths. The results indicate that potassium ions have a synergistic effect on Ca^{++}-gelation of pectates.

Potassium induced gelation of pectate

The test system was changed in such a way that Ca^{++} and EDTA were omitted and instead various amounts of KCl was added. pH was adjusted, as previously, with variable amounts of GDL.

The intention of this change was to examine whether K^+ can cause gelation or not. The results are shown in figures 6 and 7.

FIG. 6

Na - PECTATE 1% SOLUTION EFFECT OF K^+ AND pH ON GEL STRENGTH

K^+ is given in mg K^+/g pectin

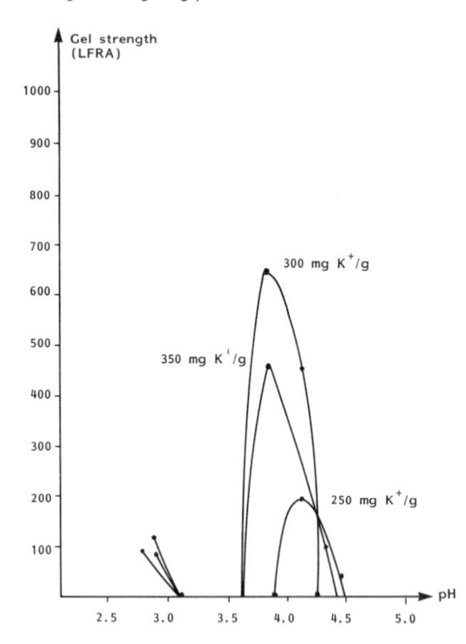

FIG. 7

K-PECTATE 1% SOLUTION EFFECT OF K^+ ON GEL STRENGTH VS pH

K^+ level in mg K^+/g pectin

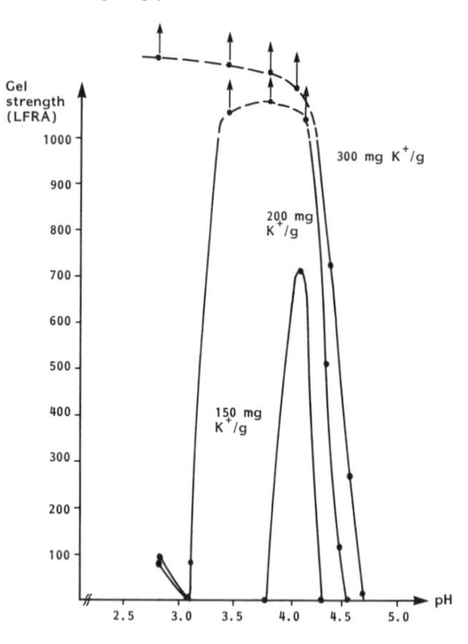

The figures show that both pectate forms start to gel at about pH=4.5. The sodium form requires about 250 mg K^+/g pectate and the potassium form about 150 mg K^+/g pectate.

Sodium pectate forms two types of gels. One is formed at pH=3.6-4.5. This type of gel is opaque and brittle. The gel shows heavy syneresis after breaking. The second type of gel is formed at pH=3.1 or lower. This gel is transparent and weaker than gel formed at pH=3.6-4.5. Gels at pH=3.1 or lower have little tendency for syneresis. The two different gels are separated by a pH-zone where gels are not formed.

Potassium pectate can also form the two different types of gels - an opaque gel, which is formed at about pH=4.5 and the transparent gel is, like for sodium pectate, formed at pH=3.1 or lower. However, the pH-range in which the opaque gel forms is very dependent on K^+-concentration. The opaque gel can be formed at lower pH-values if K^+-concentration is increased. At the extreme, which corresponds to about 250 mg K^+/g pectate, the transparent gel is not formed at all - only the opague gel is observed.

The impact of K^+ at low pH is shown in figure 8.

FIG. 8

EFFECT OF K^+ ON PECTATE ON Na- AND K- FORM AT LOW pH
Pectin concentration: 1% w/w

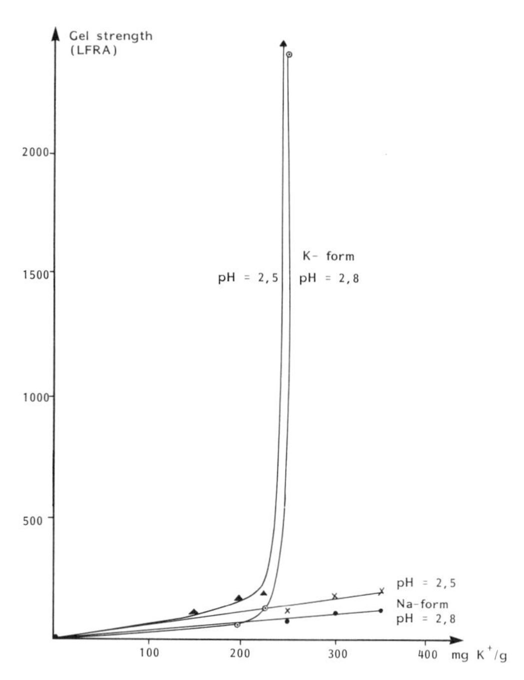

It was noticed that neither Na- nor K-pectate can form gels in the absence of K^+ and the gelstrength increases proportionally to the K^+-addition. All the Na-pectate gels were transparent.

K-pectate also forms gels in presence of K^+. Transparent gels are formed as long as K^+-level is below a certain level - but at 250 mg K^+/g pectate gelstrength suddenly rises and the gel turns into the opaque type. This observation corresponds fairly well with results obtained in figure 7. Here was shown that 200-300 mg K^+/g pectate generate only the opaque type of gel.

All the transparant gels at pH=2.5-2.8 were heat reversible. This applies to Na-pectate as well as K-pectate.

DISCUSSION

Ultra low DE pectin, pectate, is extremely Ca^{++}-reactive. It has been necessary to apply special methods to achieve a controlled gelation. The method used here is a Ca-EDTA pectate system. At high pH Ca^{++} is complexed to EDTA. Addition of GDL will gradually lower pH and cause a slow release of Ca^{++}.

This method has a draw-back that it is not possible to examine Ca^{++}-gelation at pH=7.0, where gelation of course can take place.

Both Na-pectate and K-pectate have been examined. Both types of pectate can form Ca^{++}-gels in water at 1% use level. Generally, the gels are very hard and brittle. Gelstrength or breaking-strength is very sensitive to Ca^{++} level and pH: None of the examined Ca^{++}-gels could be melted, when heated to 100°C. Thermoreversibility can be obtained by an increase of pH, due to presence of EDTA.

It has been demonstrated that potassium plays a role in Ca^{++}-gelation of pectate. Pectate in potassium form gains higher gelstrength than the sodium form. Also, addition of a potassium salt like KCl causes an increase in gel-strength for both Na- and K-pectate. The source of K^+ is not restricted to KCl. K_3-citrate or similar K-containing salts may also be used to obtain this K-effect.

It was revealed that gelation of pectate is not restricted to divalent cations. Potassium ions can under certain circumstances cause gelation of pectate. The K-induced gelation appears in two forms. One type forms gels at pH=3.6-4.5. These gels are opaque, very hard and brittle. There is observed heavy syneresis in these gels. The other type of gel is formed at pH=3.1 or lower. This type of gel is transparent and much softer than at pH=3.6-4.5 and there is only little tendency for syneresis.

These observations lead to the assumption that pectate gels appear in two different conformations dependent on pH and on K^+-concentration. It is known from literature (2, 3, 4) that pectin can form HM-gels or LM-gels. The HM-gels are believed to have a 3_1-conformation in the junction zones, whereas LM-gels correspondingly have a 2_1-conformation. It may therefore be reasonable to assume that pectate can appear in the same two conformations.

Potassium ions have in hydrated form smaller radius than sodium ions. This may suggest that K^+ can provide a more efficient shielding of dissociated carboxylic groups. Better shielding of the carboxylic groups means reduced Coulombic repulsion between pectate molecules. At a certain K^+-concentration repulsion will be so low that hydrogen bonds may be formed. However, hydrogen bonds are established between undissociated galacturonic acid molecules, which may explain why pH must be reduced to about 4.5 before K-induced gelation can take place. The conformation of pectate in relation to pH and in presence of K^+ is not known, but it may be the 2_1-form at pH=3.6-4.5 and 3_1-form at pH < 3.1.

As pH is reduced pectate becomes less dissociated which is believed to induce a gradual change from 2_1-conformation to 3_1-form. This transition from 2_1- to 3_1-form prevents pectate from forming hydrogen bonds, mainly because of sterical hindrance. This transition seems to take place at pH=3.1-3.6. Below pH=3.1 pectate will probably be mainly on the 3_1-form, and hydrogen bonds can now be established in this new conformation.

Na- and K-pectate respond somewhat different to addition of potassium ions. Na-pectate forms two different types of gels independent of K^+-level (up to 350 mg K^+/g pectate).

K-pectate, on the other hand, seems more sensitive to high K-levels. K-pectate forms only one type of gel, the opaque type, when K^+-level is sufficiently high e.g. 250 mg K^+/g pectate. If K^+-concentration is below this value,the K-pectate can form both the opaque and the transparent type of gel like Na-pectate.

The fact that sodium pectate does not provide K-induced gelation at pH=3.1-3.6 in absence of Ca^{++} opens interesting possibilities for handling of pectate as jellifying agent.

A pectate jelly can be prepared in the following way. Sodium hexa metaphosphate is added to the fruit juice as a sequestering agent (Ca^{++}-complexing agent). K_3-citrate is added to bring up K^+-concentration to about 150–200 mg K^+/g pectate in the final product. The juice is then adjusted to pH=3.2 (3.1–3.6). Pectate solution at pH=3.2 is mixed with fruit juice. Finally is GDL added to reduce pH to about 2.8–2.9. The whole process takes place at room temperature. The gelation time is normally between 30-60 minutes.

The texture of the jelly may be adjusted by K^+-dosage, pH and pectate concentration.

Finally, it shall be mentioned that the impact of K^+ on gelation has also been observed for other commercially available LM-pectins. Addition of K-ions can boost up the performance of a LM-pectin though the effect is not so pronounced as for pectate.

REFERENCES

1) Toft, K.
   Gel formation in Alginate-Pectin Systems Using Slow Acidifier.
   Workshop at the European Science Foundation in Uppsala, Sweden, April 26-27, 1983.
2) Walkinshaw, M.D., Arnott S.; Conformations and interactions of pectins, II Models for junctions zones in pectinic acid and calcium pectate gels., J. Mol. Biol., 153, 1075, 1981.
3) Rees, D.A., Polysaccharide conformation in solutions and gels recent results on pectins. Carbohydrate Polymers, 2, 254, 1982.
4) Rees, D.A., Polysaccharide gels. Chem. and Ind., 630, 1972.

The mechanism of formation of mixed gels by high methoxyl pectins and alginates

DAVID OAKENFULL, ALAN SCOTT and EUGENE CHAI

Food Research Laboratory, CSIRO Division of Food Processing, PO Box 52, North Ryde, NSW 2113, Australia

ABSTRACT

High methoxyl pectins can interact with alginates to form gels under conditions at which neither polysaccharide forms gels independently - without the added sugar required for gelation of the pectin and without the calcium required for gelation of alginate. Gelation is known to occur through a direct and specific interaction between the two polysaccharides, but the process is not yet understood in detail. Measurements have been made of the effects on the shear modulus of the mixed gels of, (i) varying the ratio of alginate to pectin at a fixed total concentration of polysaccharide and (ii), varying the total concentration of polysaccharide at a fixed ratio of alginate to pectin. Measurements were also made of viscosity of alginate-pectin solutions at concentrations too low to gel but otherwise under gelling conditions. The viscosity data indicate aggregation as a step preceeding gelation. Analysis of the shear modulus data suggests that the junction zones in the gel network are formed by a 1:1 interaction involving extended segments of the two polysaccharide chains.

INTRODUCTION

The number of gelling agents available for food use is limited. Consequently, mixed systems have great potential for producing new textures or for new product applications (1). High methoxyl pectins can interact with alginates to form gels under conditions at which neither polysaccharide forms gels independently - without the added sugar required for gelation of the pectin and without the calcium required for gelation of alginate (2-5). The gel strength and the melting temperature both depend on the ratio of alginate to pectin and this system would appear to offer opportunities to adjust the textural characteristics of the gel to suit particular product applications (5).

To study this synergistic interaction further, we have determined the effects on the shear modulus of the mixed gels of, (i) varying the ratio of alginate to pectin at a fixed total concentration of polysaccharide and (ii), varying the total concentration of polysaccharide at a fixed ratio of alginate to pectin. The concentration dependence of shear modulus can be used to derive information about the gel network, in particular the size and thermodynamic stability of the junction zones where the polysaccharide molecules are crosslinked (6,7). The results suggest that the junction zones involving extended segments of the two polysaccharide chains in a 1:1 interaction. In addition, there appears to be significant aggregation in a step preceeding gelation.

MATERIALS AND METHODS

Materials and Gel Preparation

Sodium alginate ((Type VII; Sigma) and citrus pectin (Davis Germantown) were purified by dialysis against distilled water. The degree of esterification of the pectin was 69.7%. The alginate had a ratio of mannuronic to guluronic acid of 1.5. Appropriate volumes of aqueous solutions (2%) of the individual polysaccharides were mixed and heated on a boiling water bath. Citrate buffer (210 g/l citric acid; 20 g/l sodium hydroxide; pH 3.02) was added to the hot solution (1 ml buffer to 9 ml polysaccharide solution) which then gelled on cooling. The gels were equilibrated at 15°C for 16 h before measurment of shear modulus or melting point.

Shear Modulus

For measurements of shear modulus, the gels were prepared in small cylindrical tubes (radius 1.25 cm). The shear modulus was measured by the method of Oakenfull, Parker and Tanner (8).

Melting Temperature

Gels were formed in capped tubes. A small glass 'marble' was placed on the top surface and the tube immersed in a water bath. The bath was heated slowly (*ca* 5 deg/min.) and the temperature at which the 'marble' fell to the bottom was noted.

Viscosity

Intrinsic viscosities were determined by standard procedures (9) using a suspended level viscometer.

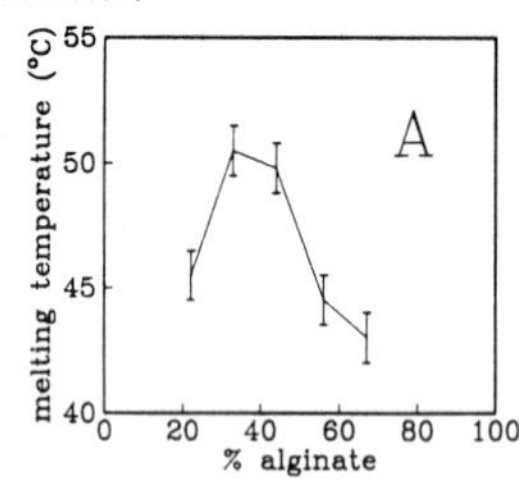

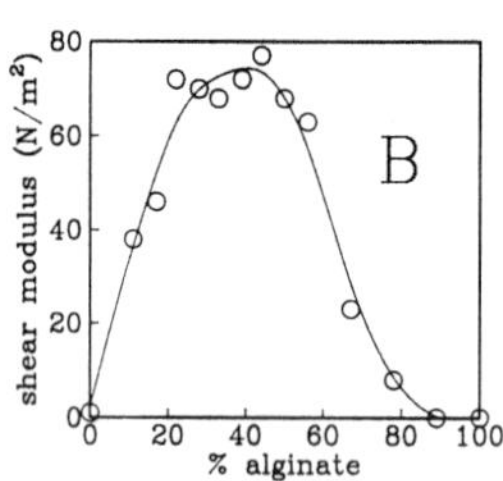

Figure 1. Effect of increasing percentage of sodium alginate on gelation of an alginate-pectin mixture at a fixed concentration of total polysaccharide (9 g/l) in citrate buffer (0.1 M) at pH 3.02. A: *melting temperature.* B: *shear modulus at 15°C.*

RESULTS

(1) Effect of Alginate-Pectin Ratio

The properties of the mixed gel are sensitive to the ratio of alginate to pectin, as shown in Fig. 1. Shear modulus and melting temperature both increase initially with increasing percentage of alginate, reaching a maximum at about 50%. Above 50% alginate, the melting temperature and gel strength both decrease. [Similar results were reported by Toft (2).]

(2) Effect of Added Sucrose

The effect of added sucrose on gelation is complex. In Fig. 2 we show the change in shear modulus with increasing concentration of sucrose, for gels prepared with a fixed concentration of polysacharide (13.4 g/l; 55% alginate). The curve is sigmoidal. Sucrose initially increases the gel strength, a maximum is reached at about 30% sucrose, the gel strength then declines until the sucrose concentration reaches 50% and then increases again.

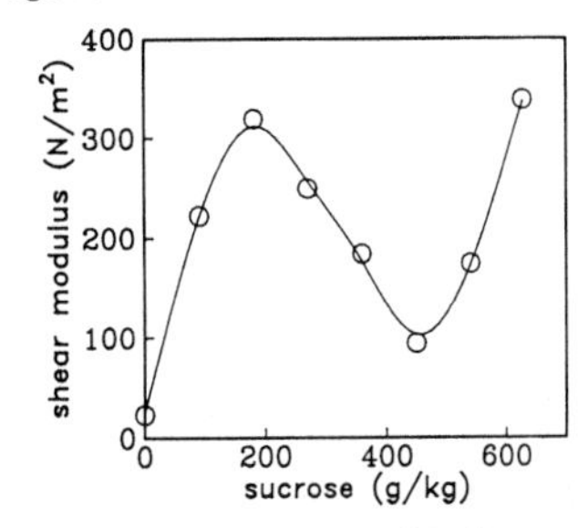

Figure 2. Effect of sucrose on the shear modulus of gels formed from a fixed concentration of alginate and pectin (13.4 g/l total polysaccharide; 55% alginate) at 15°C and pH 3.02).

(3) Effect of Polysaccharide Concentration

At a fixed ratio of alginate to pectin, the shear modulus (G) increases with the total concentration of polysaccharide (c) as shown in Fig.3.

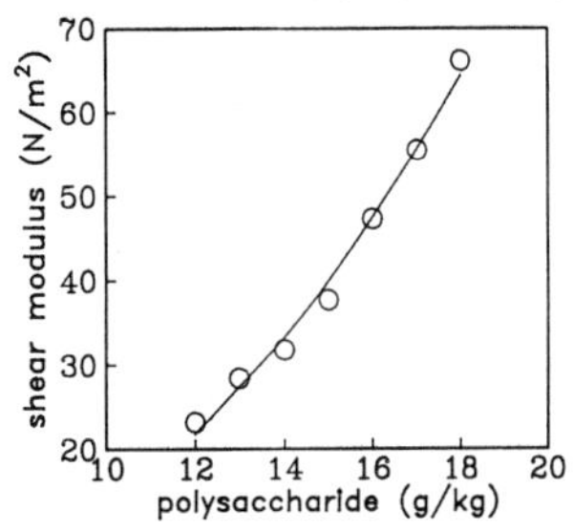

Figure 3. Shear modulus (G) vs total concentration of polysaccharide (c) for a fixed ratio of alginate to pectin (55% alginate) at 15°C and pH 3.02.

(4) Intrinsic Viscosity and Huggins Coefficient of the Mixture and Individual Components

The concentration dependence of viscosity was measured, at 25° and pH 3.02, for the sodium alginate, the pectin and a 55% alginate-pectin mixture. From these data, values were calculated for the intrinsic viscosity ([μ]) and Huggins coefficient (H_c). These results are given in Table I.

Table I. *Intrinsic viscosity ([μ]) and Huggins coefficient (H_c) for sodium alginate, pectin and a 55% alginate-pectin mixture at 25° and pH 3.02.*

	Alginate	Pectin	Pectin/Alginate
[μ] (ml/g)	114	279	310
H_c	0.035	0.13	0.57

DISCUSSION

The results presented in Fig. 1 confirm the previous observations of Toft (2) that the gel strength and melting temperature are sensitive to the ratio of alginate to pectin, with maxima close to a ratio of 1:1. The effect of added sucrose, shown in Fig. 2, also confirms Toft's observations but extends them over a wider range of concentration. Our results reveal an initial increase in gel strength with added sucrose, reaching a maximum at about 20%. Higher concentrations of sucrose then cause the gel strength to decreases, as found by Toft, but at about 50% sucrose the trend reverses and the gel strength increases again. This behaviour suggests that esterification of the pectin is required simply to reduce electrostatic repulsion. In the gelation of high methoxyl pectins alone, the ester methyl groups appear to play a more active role. High concentrations of sucrose, or similar additive, are required to increase the magnitude of the hydrophobic interaction between the ester methyl groups to give the junction zones thermodynamic stability (10). In the mixed system, it seems that at high sucrose concentrations, association of pectin with pectin becomes thermodynamically more favourable than association of pectin with alginate. Additional information can be derived (6,7) from the concentration dependence of the shear modulus (Fig. 3). Estimates of the size and thermodynamic stability of the junction zones are given in Table II, with equivalent results for high methoxyl pectin alone for comparison. The results suggest the same 1:1 interaction of the polysaccharide chains and similar thermodynamic stability (K_j) but the junction zones appear much larger in the mixed system. In addition, there appears to be aggregation in a pre-gelation step as the theoretical curve can be made to fit the experimental data points only when the molecular weight of the network forming polymer is larger than would be expected for either component singly. The viscosity data for solutions too dilute to gel (Table I) also indicate greater association in the mixture than occurs with either of the components singly. The values for the intrinsic viscosity ([μ]) and Huggins coefficient (H_c) both point in this direction (9), being greater for the mixed polysaccharide than for either of the components singly.

Table II. *Junction zone parameters calculated from the model of Oakenfull (6) from the concentration dependence of shear modulus at 25° for a fixed ratio of alginate to pectin (55% alginate).*

	M^a	M_j^b	n^c	K_j^d
Alginate/Pectin	468000	227000	1.91	2.3×10^5
HM Pectin	113000	10600	2.03	2.6×10^4

[a]*Number average molecular weight of the polysaccharide.* [b]*Number average molecular weight of the junction zones.* [c]*Number of associating units per junction zone.* [d]*Association constant for junction zone formation (in units of mole fraction).*

REFERENCES

1. Oakenfull, D. (1987) CRC Crit. Rev. Food Sci. Nutr., 26, 1-25.
2. Toft, K. (1982) Prog. Food Nutr. Sci., 6, 89-96.
3. Thom, D., Dea, I.C.M., Morris, E.R. and Powell, D.A. (1982) Prog. Food Nutr. Sci., 6, 97-108.
5. Morris, V.J. and Chilvers, G.R. (1984) J. Sci. Food Agric., 35, 1370-1376.
6. Oakenfull, D. (1984) J. Food Sci., 49, 1103-1104 & 1110.
7. Clark, A.H. and Ross-Murphy, S.B. (1985) Br. Polymer J., 17, 164-168.
8. Oakenfull, D.G., Parker, N.S. and Tanner, R.I. (1989) J. Texture Stud., 19, 407-417.
9. Bohdanecky, M. and Kovar, J. (1982) Viscosity of Polymer Solutions, pp 166-213, Elsevier, Amsterdam.
10. Oakenfull, D. and Scott, A. (1984) J. Food Sci., 49, 1093-1098.

Pectin application – some practical problems

L.BOTTGER
Department of Stabilisers, Grindsted Products, 8220 Brabrand, Denmark

ABSTRACT

This paper deals with some of the most common errors in connection with the production of jelly. Using a slow-set pectin as an example, the effect of an incomplete dissolution of pectin, an incorrect filling temperature and an inaccurate dosage of acid on the internal strength of jelly is illustrated, and the thermal breakdown of pectin in pectin solution as well as in boil is also mentioned.

1. INTRODUCTION

The jam and jelly products on the market today are usually characterised by being of a superior and uniform quality, one of the reasons being that the production process involved is well-known, but also because the equipment needed to ensure proper production control is relatively simple.

However, occasionally product defects do occur, caused by changed process parameters, although inappropriate handling of the pectin used might also cause defects. The most common error is insufficient internal strength, possibly combined with syneresis in the finished product.

Several of the factors which should be avoided in order to prevent product defects are discussed below.

2. MATERIALS AND METHODS

In the trials fruit was replaced by an artificial juice consisting of calcium citrate (0.003 M), potassium citrate (0.01 M) and citric acid (0.06 M). The pH in the juice was 3.4-3.5.

The pectin used was a 150° US SAG slow-set pectin (DM 66) produced from lime peels. The pectin was used in a dosage of 0.4%.

Procedure

Initially a 4% pectin solution is made. Pectin/sugar (1:4) is dissolved in water at a temperature of 80°C.

The juice, water and the required amount of sugar are boiled, after which the pectin solution is added, and finally citric acid is added to pH 3.1 (as standard), and the product is filled.

Internal strength is measured with a Stevens Texture Analyser (body ½", speed 0.5 mm/sec, depth 4 mm).

3. RESULTS AND DISCUSSION

Pectin dissolution

A proper dissolution of pectin is an essential condition for making a good-quality product. This seems obvious, but experience has demonstrated that insufficient dissolution of pectin is a frequent cause of product defects.

Most plants today are equipped with pectin dissolution tanks. These are often of the high-speed mixer type, which makes it possible to produce up to 8% pectin solutions (but typically 4% solutions) by adding the pectin directly to the water phase under vigorous agitation. This seldom causes any problems, but care should be taken when dimensioning the system as the viscosity of pectins is not equal, although they are all 150° US SAG. If the viscosity gets too high during dissolution (agitation intensitivity too low), the pectin will tend to form lumps which will be extremely difficult to dissolve.

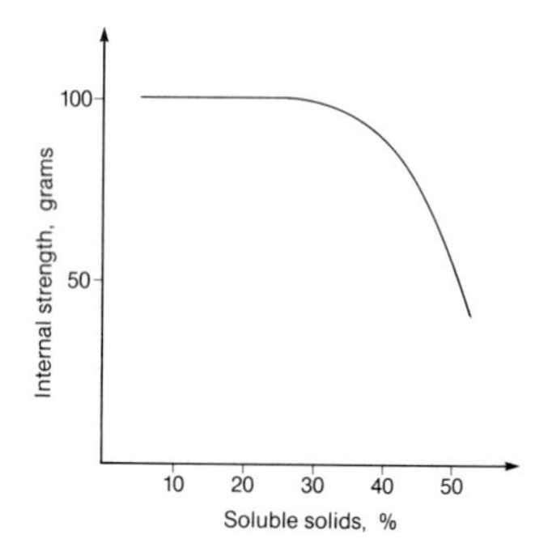

Fig. 1: Dissolution of slow-set pectin - the influence of the soluble solids content on the internal strength.

Provided the pectin dissolution tank is equipped with a slow-speed propeller agitator, it will be necessary to dry-blend the pectin with sugar before the pectin is dissolved in water. A typical blending ratio is 1 part of pectin to 4 parts of sugar. When working with 4% pectin solutions, it is important to control the blending ratio pectin:sugar, since an excess solids content at the time of dissolution will influence the internal strength in the finished product, cf. figure 1, where internal strength is drawn as a function of total solids content at the time of dissolution. It appears that around 35% soluble solids, corresponding to 7.7 parts of sugar to 1 part of pectin, decreases internal strength considerably. This is due to the fact that the pectin has not been completely dissolved, which besides the decrease in internal strength may result in a grainy or slightly flour-like texture in the product. Furthermore, this situation requires an increased consumption of raw materials in production.

It is worth mentioning that the above applies to slow-set pectin. When working with a rapid-set pectin, the above drop in internal strength can be seen as early as at a soluble solids content of approx. 25%.

The risk of an incomplete dissolution of pectin is present of course, if the system does not feature a pectin dissolution tank, but dissolves pectin directly in the boiler.

Another factor which may cause problems if the plant is not equipped with a pectin dissolution tank is the content of free calcium in the fruit which is used, if the pectin/sugar blend is added after the fruit has been added to the boiler. In figure 2, which illustrates internal strength as a function of mg calcium per kg jelly at various final solids contents, a distinct decrease in internal strength can be seen when working with fruit of a high calcium content.

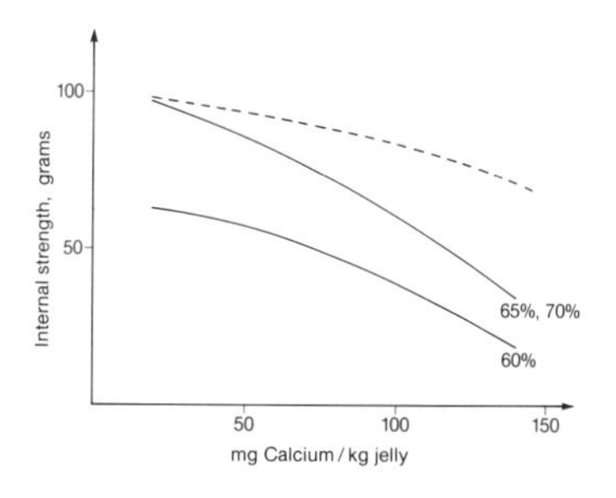

Fig. 2: Dissolution of slow-set pectin - the influence of the calcium content of the fruit on the internal strength.

For example, in an apple jelly containing 45% fruit, in which the calcium content is typically 40-50 mg per kg jelly, the drop in internal strength will be rather small (up to 10%) at a soluble solids content of 65-70%, whereas in a black currant jelly, in which the calcium content may vary between 100 and 150 mg per kg jelly, the drop in internal strength will be more than 50%. This undesired drop in internal strength can partly be prevented by adding a sequestrant agent such as sodium hexametaphosphate before the pectin is added, or by making a presolution of pectin and adding this presolution to the fruit, cf. the dotted curve.

Stability of pectin solutions

Pectin solutions are normally made by using hot water at around 80-90°C, at which temperature the pectin will dissolve rapidly. Some jelly manufacturers make a pectin solution in connection with each single batch, and in such cases the holding time of the pectin solution will be relatively short. Others make a larger amount of pectin solution in the morning, and use it during the day for the whole day's production. In such cases, the holding temperature of the pectin solution is important for the internal strength in the finished product, since a breakdown of the pectin as a function of temperature and time takes place (1). Figure 3 makes it clear that if you wish to let a pectin solution stand for a long time, the temperature should be 60°C or even below. After only two hours standing at 70°C, the drop in internal strength will be just under 10%, and at 82°C approx. 20%.

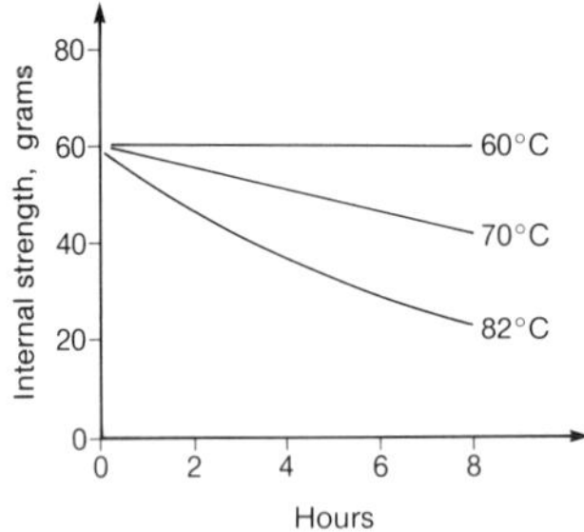

Fig. 3: Stability of 4% solutions of slow-set pectin. Influence of time and temperature on internal strength.

Stability of pectin in the boil

Boiling during the process is necessary for several reasons:

- To preserve the product by stopping enzyme activity and killing any mould and yeast fungi that may be contained in the raw material.
- To concentrate the product by evaporation of water.
- To remove air from the product, which partly improves the appearance of the product, and partly reduces the risk of the oxidation of flavour and colouring agents.
- To obtain sugar equilibrium in jams containing fruit pieces.
- To invert part of the sugar added, if required.
- To avoid pregelation phenomena when adding the acid and/or pectin solution.

Prolonged heat treatment should be avoided, as this would affect the product in a negative way and result in the break-down of pectin, loss of colour, loss of flavour and perhaps undesired sugar inversion.

It has previously been demonstrated that pectin solutions should be kept below 60°C due to breakdown of pectin, but the question is how much pectin is broken down during boiling and what happens if a production stop occurs, e.g. the filling line stops after pectin has been added to the boil.

The answer depends on soluble solids content, pH, temperature and time. If acid has not been added and the fruit is not too acidic, which means that pH in the boil is approx. 3.5, it is possible to cool the batch, especially at low solids contents. At high solids contents, i.e. close to 70%, this procedure should not be used, since the setting temperature is still around 80°C, cf. fig. 7. If production stop is expected only to last a very short time, it would probably be advantageous to maintain the temperature with a view to subsequent heating before the addition of acid. In figure 4 the boils have been kept at 80°C for up to two hours. It appears that the drop in internal strength in the finished product from a solids content of 65 and 70% is less than 10%. This drop in internal strength is due to a breakdown of the pectin that is added. It also appears that the drop is only half of that of a pure pectin solution, cf. figure 3. The fact that the drop at 60% solids is approx. 20% is probably not due to a stronger breakdown of the pectin at the lower sugar content, but rather the fact that the pH is a little higher than 3.1, since the internal strength at 60% solids content in this pH-area is very sensitive to pH, cf. fig. 10.

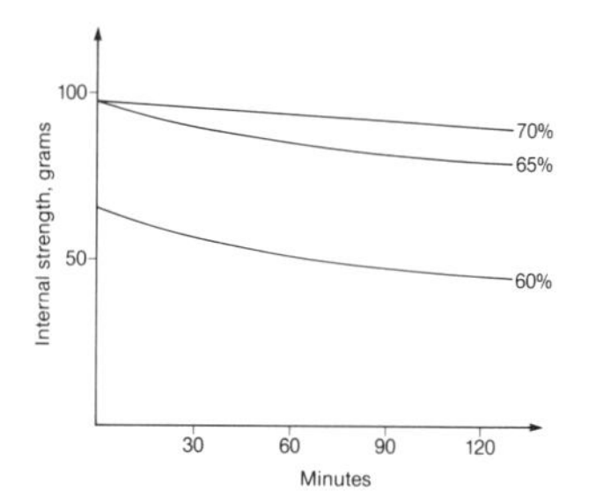

Fig. 4: The stability of pectin in boil (80°C) at a pH of 3.5. The influence of time.

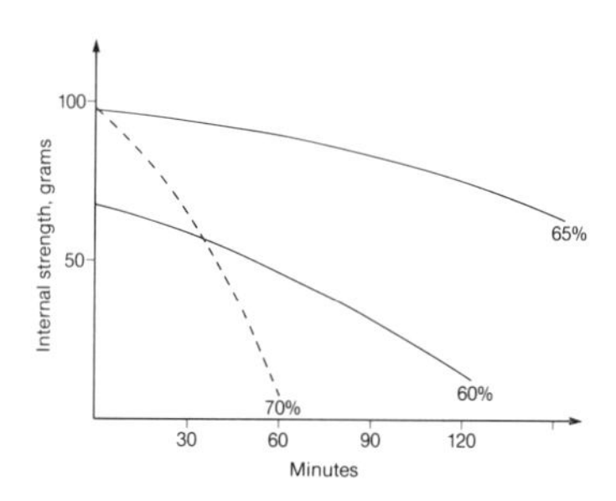

Fig. 5: Stability of pectin in boil (80°C) at a pH of 3.1. The influence of time.

Provided that the acid has already been added to the boil, the possibilities will be limited, since it will not be possible to cool the boil due to pregelation, at any rate not at a soluble solids content of approx. 70%, cf. figure 5. For gels with 60 and 65% solids respectively, there is still the possibility of cooling the product, since the setting temperature for these boils, for most practical purposes, is below approx. 60°C. Comparing figure 5 with figure 4, the drop in internal strength for boils containing 65% solids is almost identical at pH-values of 3.5 and 3.1. At a pH of 3.1 the drop at 60% solids is even more noticeable than at a pH of 3.5, but as already mentioned this is primarily due to a pH in the boils that is too high.

Pectin is broken down during boiling, but this breakdown, thanks to the protecting effect of sugar, is considerably less than the breakdown which may take place during the dissolution of pectin. If cooling of the product is required, e.g. in connection with production stop, in order to counteract the breakdown of pectin (as well as the loss of colour and flavour), the risk of destroying the product is much bigger due to pregelation at high solids contents, and especially at a low pH, compared to the quality deterioration caused by the breakdown of pectin.

Setting temperature

The pectin that is sold today is either rapid-set, medium-rapid-set, slow-set or extra-slow-set pectin. These types have normally been standardised to 150° US SAG, and the setting time indicated is determined according to Joseph and Baier's method (2), using the SAG boil.

However, we think that more relevant information will be obtained by specifying the setting temperature. Today there is no generally accepted method for setting temperature determination, which is probably due to the complexity of the problem as regards rate of cooling, soluble solids content, pH, etc. Hinton (3) carried out some research in the 1940's, and Olliver a.o. (4) developed a special modification of the SAG boil in the 1950's. The principle of both methods is to evaluate setting temperature visually. Ehrlich (5) used a viscosimeter to determine the setting temperature in connection with determining of the setting temperature's dependence on DM-values.

Neither of the methods included an investigation of the lowest temperature at which a test gel could stand without gelling for a given period of time (in this case one hour). This temperature is of practical importance, since it tells the jelly producer the temperature at which gelling conditions are present. To examine this we have used a Bohlin Rheometer, in which the test gel is exposed to such small oscillating movements that the risk of destroying any gel that is formed gel is insignificant. The instrument measures when the test system changes from the viscous stage to the elastic stage by measuring the phase angle between the deformation and the response of the sample. In the viscous area the phase angle difference is 90°, and in the elastic area the difference is 0°.

The instrument can also be used to determine the setting temperature at different rates of cooling of the test gel (6).

The results of our tests appear in figs. 6 and 7, which confirm previous reports (7), i.e. that the setting temperature decreases with decreasing amounts of soluble solids and increasing pH. The fact that the setting temperature depends on the rate of cooling is also confirmed (3,5).

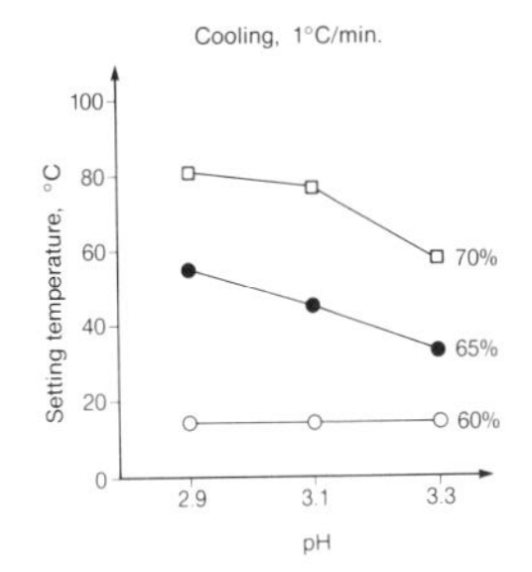

Fig. 6: Setting temperature - slow-set pectin. Dependence on pH and soluble solids content.

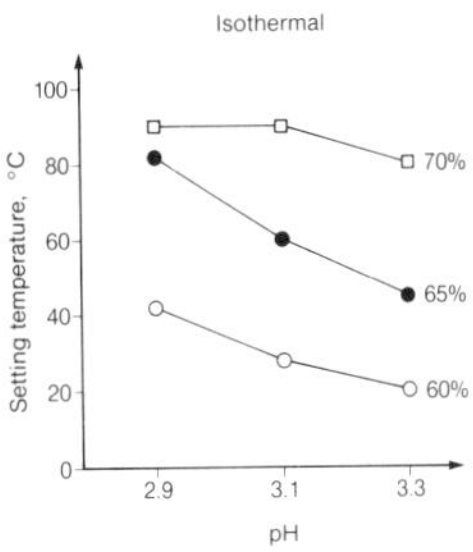

Fig. 7: Setting temperature - slow-set pectin. Dependence on pH and soluble solids content.

The results from fig. 6 might be of interest when a heat exchanger is inserted between the boiler and the filling line, or for those who use vacuum boilers, whereas the results in fig. 7 are of special interest if production stop occurs, or if products are filled directly from the boiler.

Fig. 7 is also of interest if vacuum boilers are used, since the setting temperatures found indicate what the temperature of the boiler should be when the pectin solution is added to the boil. At soluble solids contents of up to 65% this will seldom cause any problem, but at higher solids contents, i.e. 70% or more, the risk of pregelation will be present. Figure 8 illustrates the influence of the boiler temperature on internal strength at different solids contents before the pectin solution (60°C) is added. It is clear that the boil should be heated before pectin is added at high soluble solids contents.

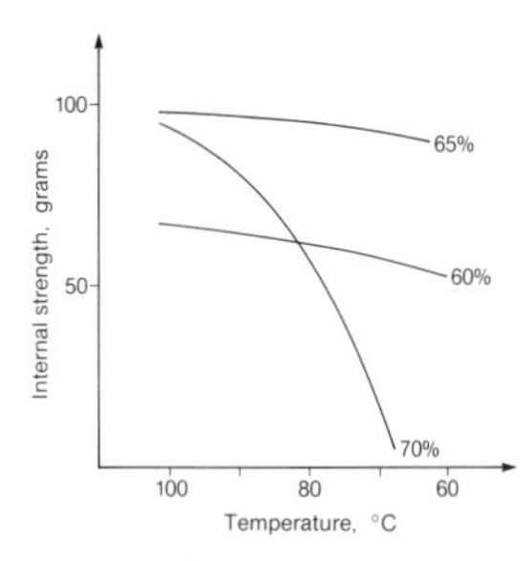

Fig. 8: Addition of pectin solution. The influence of the boiler temperature on the internal strength.

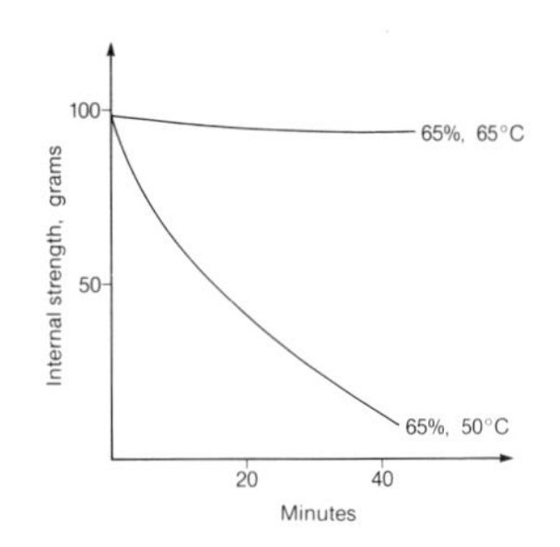

Fig. 9: Effect of filling temperature and time - slow-set pectin.

On the other hand, the setting temperatures from fig. 7 should not be used in order to determine the time at which acid can be added to the boil, since the citric acid (50% w/v) normally used will lead to a very low local pH resulting in pregelation of the boil, provided the agitation is not very effective.

If it is necessary to fill directly from the boiler into large containers (e.g. 10 or 25 kg), or if the stacking of products on pallets immediately after filling is required, it is tempting to cool the total boil in order to reduce the thermal load of the product (breakdown of pectin, as well as loss of colour and flavour of the fruit). Attention should be paid to this point, since the risk of destroying the product completely is present.

At a soluble solids content of 70% and a setting temperature of more than 90°C, a cooling to only 80°C will make it impossible to measure internal strength, i.e. the jelly will be almost liquid if the filling time is around 30 minutes. At a soluble solids content of 65%, the setting temperature, according to fig. 7, is 60°C. The filling temperature, however, should preferably be a little higher, e.g. 65°C, as a little too high soluble solids content and/or a little too low pH will make the setting temperature increase considerably. If filled at 50°C, which corresponds to the setting temperature at pH = 3.20-3.25, the product will be ruined and almost liquid in consistency, cf. fig. 9.

At a soluble solids content of 60%, the problem is less important due to the low setting temperature.

When the setting temperature of the boil exceeds the desired filling temperature, the safest solution would be to insert a heat exchanger between the boiler and the filling line, or to dose acid into the pipeline immediately before filling.

The influence of pH

One of the most important factors influencing the quality of the finished product which has not yet been mentioned, is the final pH of the product. In fig. 10 the correlation between internal strength and pH can be seen at soluble solids contents of 60, 65 and 70%. It appears that with increasing amounts of soluble solids, the internal strength optimum is displaced towards higher pH-values, which, although with a slightly different result, has been reported earlier (8).

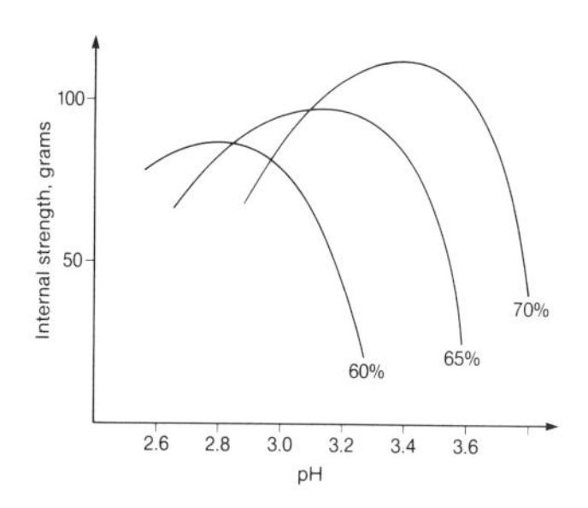

Fig. 10: Internal strength as a function of pH at different soluble solids contents - slow-set pectin.

Fig. 10 makes it clear that products with a soluble solids content of 60% are very sensitive to variations in pH around 3.1, since internal strength decreases considerably from 3.0 to 3.2, resulting in absolutely no gelling at around 3.3.

At a soluble solids content of 65% there is more flexibility, and pH control in this area is not quite as necessary. However, pH control is necessary when reaching high soluble solids contents around 70%, as there is a pronounced tendency to pregelation when pH drops. Besides, the curve (70%) illustrates that rather strong gels can be made at rather high pH-values, which is an additional factor indicating that caution should be exercised with pregelation when the pectin solution is added to boils with high soluble solids contents, as gelling conditions may already be present.

The alternative to an inefficient pH control is often to increase the dosage of pectin resulting in considerably increased raw material costs.

4. CONCLUSION

We have discussed some of the factors influencing the quality of jellies which often lead to product defects such as the lack of internal strength or syneresis.

The most frequent errors are

- insufficiently dissolved pectin,
- incorrect filling temperature, and
- incorrect pH.

Due to variations in different types of fruits, change to other types of sugars, and the natural variations within each type of pectin (variation in for example DM for slow-set pectin), the results obtained will only be indicative. For example, the setting temperature of a slow-set pectin may vary more than 5°C.

When changing to other types of pectin, e.g. rapid-set pectin, the setting temperature will increase to such an extent that the use of rapid-set pectin in products with high soluble solids contents is impossible due to pregelation in the boiler.

REFERENCES

1. Ikkala, P., Paper presented at IJPA, Bi-Annual Meeting, Atlanta, Oct. 1985.
2. Joseph, G.H. & W.E. Baier, Food Technology, 18, 18-21, Jan. (1949).
3. Hinton, C.L. Journal of the Science of Food and Agriculture, 1, 300-7, (1950).
4. Olliver, M., P. Wade & K.P. Dent, The Analyse, 82, 127-8, (1957).
5. Ehrlich, R.M., Paper presented at IFT Annual Meeting 1977 in Philadelphia, Pa.
6. Ikkala,P., "Gums and Stabilisers for the Food Industry 3". Ed. Phillips, G.O., Wedlock, D.J. and Williams, P.A. Elsevier Applied Science Publ. 1986.
7. Doesburg, J.J. & G. Grevers, Fd. Res., 25 (5), 634
8. Stuewer, R., N.M. Beach & A.G. Olsen, Ind. Eng. Chem. Anal. Ed. 6, 143 (1943).

Pectins in high and low sugar preserves

P.REARDON
Premier Brands UK Ltd, The Orchard, Chivers Way, Histon, Cambridge CB4 4NR, UK

1. What Are Preserves

Preserves are one of the few foods whose formulations are governed by an EC Directive. The Jam & Similar Products Regulations 1981 provide the UK version of this and specify both which mandatory and optional ingredients are to be employed.

The UK Regulations define the various categories of preserves as:-

Extra Jam
Jam
Extra Jelly
Jelly
Marmalade
UK Standard Jelly
Fruit Curd
Fruit Flavour Curd
Mincemeat
Reduced Sugar Jam
Reduced Sugar Jelly
Reduced Sugar Marmalade

For each category both the minimum fruit content and minimum sugar solids are established and any other formulation constraints applied.

Fruit Preserves MUST contain both Fruit and one or more sweetening agents.

Jams can employ fruit in either pulp or pureè form or a combination of both. Normally jam contains a minimum of 35% fruit. Extra Jams require a higher fruit inclusion (usually at least 45%) and purées may not be used.

All types of Jellies are based exclusively on fibre-free juices and extracts while Marmalades are sourced only from citrus fruits in a wide range of derivative forms - fruit pulps, centre pureè, citrus peel, fruit comminute or juice. Jelly fruit contents follow the appropriate Jam categories while Marmalades require a minimum of 20% citrus fruit to be used.

There are specific regulations for the remaining specialised products which will not concern us here.

Sweetening agents are also specified in the Regulations and these include:-

White Granulated Sugar
Sugar Solution
Invert Sugar Syrup
Glucose Syrup
Dextrose Monohydrate
Fructose
Cane Molasses
Honey

Of these Refined Sugar and Glucose Syrup are the most important commercially.

Now as well as the required ingredients - Fruit and Sugar there are additional ingredients which can be employed on an optional basis:-

These include:-

Pectins
Fruit Acids
Buffers
Colours
Antioxidants
Antifoam Agents
Preservatives)
Gums) Reduced Sugar Product
Carageenans) only

There is also provision in the regulations for other ingredients to be added to Jams designed specifically for manufacturing outlets such as Bakeries. Humectants, texture modifiers, sequestrants and flavours are the key ones used.

We have examined so far the components of preserves but we must also address the question:

" What are the characteristics which accurately describe a typical jam, jelly or marmalade?"

Well, preserves are sugar - acid - pectin gels which are designed to be readily spreadable. They should also have an attractive appearance with a distinctive colour, and in some instances noticeable chunks of fruit suspended within the gel matrix. Preserves must also have strong fruity and tangy flavours characteristic of the named variety and an acceptable level of sweetness.

The prime characteristic of any preserve however is that it is gelled and the key gelling agent in conventional preserves is pectin.

2. PECTIN USAGE AND CHOICE

Commercial pectins available to UK manufacturers are obtained from two sources, citrus peel and apple pomace. Citrus pectin is typically supplied in the form of a standardised dry powder which can be reconstituted with warm water in a high speed blender immediately prior to jam making. For ease of handling the strength of the solution is usually kept fairly dilute.

Apple pectin is normally available as a viscous solution preserved with sulphur dioxide to extend its shelf life. It is used directly.

In high speed Preserves Manufacturing operations, whether continuous or batch, pectin solution is normally metered in

CHOICE OF PECTIN - HIGH TSS SYSTEMS

PRODUCT	TSS	DEGREE OF METHYLATION	CHOICE OF PECTIN
Jams for Confectionery or Biscuit Applications	75 - 85%	59 - 63%	Extra Slow Set
Jellies (or Jams for Bakery Applications)	65 - 69% (70 - 78%)	63 - 67%	Slow Set
Jelly Marmalades	65 - 69%	69 - 71%	Medium Rapid Set
Jams with suspended fruit pieces	65 - 69%	71 - 74%	Rapid Set
Jams with suspended fruit pieces	60 - 65%	74 - 78%	Ultra Rapid Set

[NB Filling pH and Filling temperature must also be considered] volumetrically, after passing through a fine filter to remove any undissolved matter. Made-up pectin solutions should be used on the day of preparation as enzymic degradation can result from microbiological attack and this will lead to a major reduction in gel power. Pectin solutions should also not be held at high temperatures for prolonged periods of time if steady loss of gelation properties is to be avoided.

Before finalising his jam recipe the manufacturer must make a choice from a wide range of available pectins.

The first consideration must be the soluble solids of the product. If it is above 60% then the choice will be limited to one or more high ester pectins. If the TSS is between 25% and 55% then either an amidated or non-amidated low ester pectin will be chosen, possibly with a gum as a texture modifier and syneresis controller. Finally if the TSS is below 25% then carrageenan alone or carrageenan in combination with an amidated LM pectin is advised. The next consideration is more complex - the nature of the product must be identified. Are there fruit pieces to be uniformly suspended within the gel structure? Is clarity (i.e. absence of air bubbles) a major factor? What is the product pH? What is the filling temperature? What sort of texture is desired? What about the carbohydrate spectrum and its influence on the rate of gelation?

The answer to these questions will identify in the case of high sugar systems whether a rapid or slow set pectin is to be used and even whether a blend is desirable to target on a particular gelling rate.

The two major properties which characterise individual pectins and which must be known and understood by the preserves manufacturer are:

Gel Strength
Rate of Gelation

Gel Strength or Gelling Power of a pectin is in reality a measure of 'Value for Money'. It is determined by the USA-SAG method in which the pectin to be assessed is blended as part of a standard recipe, to produce an acidified sucrose jelly of known TSS and pH, following an established procedure. After de-moulding on to a glass plate a micrometer is used to determine the distortion under gravity (i.e. the SAG) at the centre of the jelly surface. The instrument used is called a Ridgelimeter. The degree of distortion is a measure of the gel strength of the pectin. Most commercial HM pectins are now made available standardised to 150°grade.

The second parameter - Rate of gelation - provides guidance on the optimum choice of a particular pectin or pecting blend. The rate of gelation can be measured either by determining the setting temperature (the temperature at which gelation starts to occur) but more usually by the setting time (the time before gelation commences). Setting time is assessed by placing the liquid jelly (at a given temperature) in a transparent container, and then into a water bath controlled at 30°C. By gently rotating the container the point at which the jelly can be seen to begin to form, is duly noted. The time interval is the setting time.

CHOICE OF AMIDATED PECTIN - LOW TSS SYSTEMS

PRODUCT	TSS %	TYPICAL DEGREE OF METHYLATION %	TYPICAL DEGREE OF AMIDA-TION %	CHOICE OF PECTIN
Retail Jams	45 - 55	35	15	SLOW
Retail Jams	40 - 50	32	17	MEDIUM
Retail Jams & Fruit Desserts	20 - 50	30	20	RAPID

CHOICE OF NON-AMIDATED PECTIN - LOW TSS SYSTEMS

PRODUCT	TSS %	TYPICAL DEGREE OF METHYLATION %	TYPICAL DEGREE OF AMIDA-TION	CHOICE OF PECTIN
Retail Jams	50 - 55	48	N\A	SLOW
Retail Jams	50 - 55	40	N\A	MEDIUM
Retail Jams	35 - 50	35	N\A	RAPID

So, from knowledge of the recipe and the required end product characteristics and given the necessary raw data on individual pectins, a meaningful choice of gelling system can be made.

It must be remembered however that certain jam formulations are naturally high in pectin from the fruit and predicting the contribution of this 'natural' pectin is much more difficult.

We have seen that differentiating between various high methoxyl pectins, all of which are manufactured and standardised to the same grade, is simply a question of the preferred degree of esterification. This in turn relates to the rate of gelation and provides an indication of whether a rapid, medium or slow set pectin is required for a particular application.

The situation is more complex with low methoxyl pectins, as not only varying degrees of esterification are encountered but also varying degrees of amidation. Both these factors influence the calcium sensitivity, particularly the latter.

Amidated rapid set pectins can have a degree of methylation between 27 and 30% (20 - 22% amidation), medium set pectins correspondingly 31 - 33% (17 - 18%) and slow set pectins 34 - 36% (14 - 16%). The figures for non-amidated pectin classes are typically:- 34 - 38% (rapid) 38 - 40% (medium) and 40 - 48% (slow).

The final choice however must bear in mind the TSS targeted, the calcium content, operating pH and the desired product texture (soft or firm, spreadible or short), whether a high degree of clarity is required and whether fruit pieces are to be suspended. Vegetable gums such as guar or carob gum and xanthan gum of microbial origin, also have a part to play in the texture modification of Reduced Sugar Jams and the control of syneresis. Many products currently on sale in the UK contain such additives.

The data sheets provided by the pectin manufacturer enable an initial choice of all systems to be selected but there is no substitute for bench-scale experimentation followed by plant trials to confirm this choice.

3. HIGH SUGAR PRESERVES

Jams, Jellies and Marmalades are manufactured by blending together most of the ingredients, raising the temperature and then evaporating to the required soluble solids either atmospherically in batches of up to 100Kg or by vacuum boiling in batches up to 2000 Kg. Both batch and continuous systems are available for either option.

Vacuum boiling is employed at temperatures down to 60°C, in order to avoid heat damage and caramelisation and to ensure full retention of fruit volatiles. The preserve must then be gently heated to 80 - 85°C, prior to filling, to pasteurise the product and ensure shelf stability. Filling temperatures of the same order are also employed, regardless of whether residual preservative is present in the end product.

Since the mechanism of gelation involving high ester pectins is pH dependent it is necessary to ensure that jams are prepared within an acceptable pH band.

The addition of fruit acids such as citric or tartaric can

VACUUM BOILING V OPEN PAN BOILING

	OPEN PAN	VACUUM PAN
B A T C H S I Z E	70 - 100 Kg	1000 - 2000Kg
COOKING TEMPERATURES	>100°C	60 - 70°C [80 - 85°C for sulphited jams]
C O O K I N G pH	2.8 - 3.2	3.5 - 3.7
F I N A L Ph	2.8 - 3.2	3.0 - 3.4
PASTEURISATION STAGE	NO	YES
FILLING TEMPERATURE (Retail Products)	80 - 90°C	80 - 90°C
ADVANTAGES	Flexibility, Low Capital Investment, Useful for tonnages	Large Batch Size Can easily be automated. Low temperatures lead to enhanced quality. Low laboured requirement.
DISADVANTAGES	Labour intensive Operator dependent Loss of fresh fruit volatiles. Dangers of Caramelisation particularly on high TSS products. Small batch sizes	Relatively inflexible. High capital cost. Not recommended for low tonnage lines.

supplement the acid contribution from the fruit. In vacuum boiling operations where lower temperatures apply, the use of buffer salts such as sodium citrate may be employed to raise the pH temporarily, thus avoiding the dangers of premature gelation. On vacuum boiling operations the bulk of any acid is added therefore after evaporation but prior to pasteurisation.

Most jams are made to a final pH of 2.8 - 3.4, the lower end of the range favouring an increased gelling rate and hence favoured primarily by open pan boiling systems.

Unlike amidated LM pectin systems, high sugar preserves gels are not thermally reversible and the phenomenon of broken gels' cannot be reversed by re-heating the product.

Typical retail preserves in the UK have soluble solids in the 65 - 69% region. As the TSS is raised the gelling rate is increased and it may be necessary to either adjust the pH or to move to a slower set pectin.

The role of the carbohydrate sweeteners is of great importance. Collectively they largely determine the product final solids, water activity and preserving power, contribute in a major degree to the 'body' and 'mouthfeel' characteristics of the end product and play a crucial role in the gelation process.

Individually they assist in varying degrees with the creation or prevention of crystallisation, to the modification of sweetness and of texture, in the influencing of pectin setting temperatures and even to gel strength itself.

Contrary to the conclusions in some published literature our experience suggests that 63DE glucose syrup increases the pectin setting temperature and decreases gel strength when replacing sucrose weight for weight on a dry solids basis.

Conversely the inclusion of fructose in the formulation decreases the setting temperature and increases the gel strength.

Syneresis (or gel shrinkage) is not normally a problem with high TSS preserves, provided that the gel has been correctly formed and not mechanically damaged by stirring or pumping. High glucose levels however can aggravate the situation, presumably by increasing the setting temperature and also modifying the gel texture to one less resistant to external rupturing.

Two more serious situations however relate to the production of crystal clear jellies free of bubbles and the suspension of fruit within the gel matrix.

In the former situation slow set pectins are the norm so that sufficient time is provided for all air contained within the batch to be expelled naturally. Control of the process and avoidance of entrapping excess air within the batch are also essential in this regard.

In the latter case rapid pectins are recommended so that early gelation is induced before the fruit pieces have time to float (or in the case of citrus peel, to sink).

The most critical products to manufacture are jelly marmalades, which as jellies require slow set pectins for reasons of gel clarity, while as products containing suspended peel rapid set pectins are the norm.

Clearly a compromise is needed and so we work to a precise setting temperature which is achieved by a blend of pectins operating under carefully controlled product TSS and pH conditions.

Finally the problem of soft gels or absence of gelation can be traced to either lack of sufficient pectin, to low TSS, to incorrect pH (too high does not favour gelation, too low can lead to broken gels) or to poor process control (high temperatures, combined with high pH and long holding times can lead to pectin destruction).

4. REDUCED SUGAR PRESERVES

The manufacturing process for low sugar jams, jellies and marmalades is very similar to that employed for high sugar preserves. The mechanism of gelation however is very different. Within limits pH is not the critical factor but the presence of calcium ions is.

The required calcium content of the formulation is contributed by the fruit, by the water (particularly in hard water areas) and by the addition of calcium salts. This latter addition is essential if the low ester pectin used is non-amidated but sometimes it proves necessary even with amidated pectins.

The problem of syneresis is much greater with reduced sugar preserves. This is because the high sugar content, which is an essential part of the composition of traditional jams and marmalades, stabilises the gel structure and prevents the bound water from being exuded.

With reduced sugar products this stabilisation cannot be provided and much greater care must be taken to avoid syneresis, particularly at solids below 40%. Choice of the correct pectin at the appropriate level and avoidance of premature gelation and mechanical damage are essential. In addition excessive calcium levels can promote syneresis and must be avoided.

Controlling the suspension of fruit and liberating air bubbles also require careful control of the setting temperature of the pectin and this in turn requires the correct choice of degree of esterification and degree of amidation for a product at a particular TSS and of particular product characteristics. Again the calcium content of the system must also be known as enhanced levels can lead to increased setting temperatures, broken gels, syneresis and retention of air bubbles.

Conversely the choice of a pectin with a too low setting temperature or a precise recipe lacking sufficient calcium can lead to fruit separation, slack gels or even absence of gelation.

Finally the overall texture must be considered. Is the gel too short and not spreadible enough? The answer to this question is likely to relate to TSS, calcium level and choice of pectin.

Hydrodynamic evidence for an extended conformation for citrus pectins in dilute solution

STEPHEN E. HARDING, GISELA BERTH[+], ABIGAIL BALL and JOHN R. MITCHELL

University of Nottingham, Department of Applied Biochemistry & Food Science, Sutton Bonington LE12 5RD, UK

[+]*Akademie der Wissenschaften der DDR, Zentralinstitut für Ernahrung, Potsdam-Rehbrücke, German Democratic Republic*

ABSTRACT

Data from hydrodynamic measurements (sedimentation velocity, sedimentation equilibrium and viscometry) are suggestive of an extended rather than a spheroidal conformation for citrus pectins in dilute solution. Evidence is presented from 1. the Wales-van Holde parameter $k_s/[\eta]$ for a series of pectin fractions and unfractionated material, and 2. the slopes of Mark-Houwink sedimentation and viscosity plots for the same series of fractions.

INTRODUCTION

Pectins have a wide variety of uses as gelling and thickening agents in the food industry. Their properties will be influenced by their conformation under the particular conditions in which they are used. A study using sedimentation equilibrium, sedimentation velocity and viscosity measurements on fractionated citrus pectin was undertaken to investigate the gross conformation of citrus pectin (sphere, rod, random coil etc.) under dilute aqueous solution conditions.

EXPERIMENTAL

The initial fractionation and viscometric characterisation had been carried out earlier by Berth and co-workers (1,2). The sample was a citrus pectin (Koch-Light Ltd.) with a degree of esterification of ~60% and a galacturonate content of ~70%. Concentrations (w/v) were determined by dry weight of pectin. Chromatographic separations had been carried out on GPC columns of Sepharose 2B/Sepharose 4B, in a phosphate buffer at pH 6.5 with 1mM Na_2EDTA and 0.3mM NaN_3 added. 10ml fractions were taken from the column and their intrinsic viscosities determined in a Viscomatic (FICA, France) viscometer at 25.0°C (1,2).

For the present study, consecutive fractions were taken,

combined in pairs, and the resultant 20ml fractions freeze dried prior to analysis. The solvent used for sedimentation analysis of the 20ml fractions was a phosphate chloride buffer of I=0.3, pH=6.5). For sedimentation equilibrium, 1mM Na_2EDTA and 2mM NaN_3 were addded, and the sample was exhaustively dialysed against the solvent before use.

Sedimentation equilibrium was performed using a Beckman Model E analytical ultracentrifuge equipped with Rayleigh interference optics, a 5 mW He-Ne laser light source and an RTIC temperature control system, following low speed sedimentation equilibrium procedures (3) to determine weight average molecular weights (M_w). Initial concentrations were of the order of 0.7 mg/ml.

Sedimentation velocity experiments were carried out using an MSE Centriscan 75 analytical ultracentrifuge at 20.0^oC, with scanning schlieren optics set at 546nm. Sedimentation coefficients at a series of concentrations (s_c) were determined, corrected for solvent density to conditions of water at 20.0^oC and extrapolated to ´infinite dilution´ against concentration (corrected for radial dilution in the ultracentrifuge cell), to determine the infinite dilution sedimentation coefficient (s) using the expression:

$$1/s_c = (1/s)(1 + k'_s c)$$

where k'_s is the concentration dependence parameter of the sedimentation coefficient for s_c values corrected for solvent density. This is further corrected for solution density using (4):

$$k_s = k'_s - \bar{v}$$

where $\bar{v}$ is the solute partial specific volume [determined here by precision densimetry using an Anton Paar (Graz) DMA 02C precision density meter as (0.57 ± 0.01) ml/g].

In the paper of Berth et al (2), values for the intrinsic viscosity are given for fractions of 10ml bandwidth. The equivalent intrinsic viscosity values for combined adjacent fractions (i.e. of 20ml bandwidth) were calculated from the weighted average of each set of two 10ml values.

CONFORMATIONAL PROBES

Knowing the molecular weights, sedimentation coefficients and intrinsic viscosities of the pectin fractions, it is possible to examine their possible conformations by two approaches:

1. Use of the "Wales-van Holde" parameter. The value of the ratio $k_s/[\eta]$ is characteristically ~1.6 for spheres and random coils, and substantially lower for asymmetric molecules (5-7).

2. Mark-Houwink approach. Two of the Mark-Houwink type equations are relevant here:

$$[\eta] = K'M^a \quad ; \qquad s = K''M^b$$

The values of a and b can be estimated from the slopes of log-log plots, and have characteristic values for the three main conformational groups (Table 1).

Table 1. Values of a and b for different classes of macromolecular shape (from Ref. 8)

	a	b
sphere	0	0.67
random coil	0.5-0.8	0.4-0.5
rod	1.8	0.15

RESULTS

Table 2. Results for fractionated pectins.

Fraction	k_s(ml/g)	$[\eta]$ (ml/g)	$k_s/[\eta]$
9-10	81.6±21.9	809.3	0.10
13-14	92.8±14.0	624.5	0.15
15-16	70.6±10.2	479.4	0.15
17-18	62.2±32.8	341.3	0.18
19-20	27.0±5.8	217.5	0.12
21-22	119.3±52.0	160.5	0.74

The values of $k_s/[\eta]$ for all the pectin fractions (Table 2) can be seen to be substantially lower than 1.6, indicating an extended, rod like conformation in all cases. This is confirmed by the plot of log s versus log M_w (Figure 1a), which has a slope from linear regression of 0.17±0.10, very close to the value of 0.15 for rods, but clearly not consistent with a spherical or random coil conformation.

The initial slope of (2.0±0.3), [estimated by linear regression of the first five data points] of the log $[\eta]$ versus log M_w plot (Figure 1b) appears to support this rod like classification. It is however significant to note that there is a decrease in slope with increasing molecular weight which could be an indication of a change in shape to a more flexible form as the molecular weight increases. Not shown in Fig 1b are two points of higher molecular weight which have anomalously low values for the intrinsic viscosity. We feel this is due to the presence of appreciable amounts of compact high molecular weight spherical aggregates contaminating the sample (1,2). If these species are also present (in proportionally lesser quantities) in the lower molecular weight fractions, this could provide an alternative explanation for the observed change in slope as molecular weight increases. Such effects would not be observed

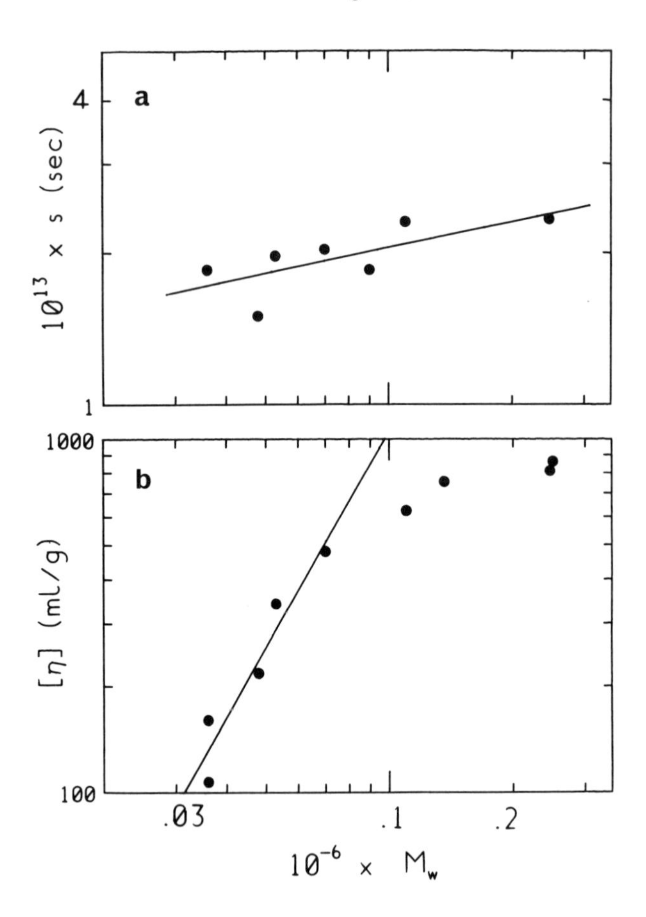

Figure 1. "Double Log Plots" of (a) Sedimentation Coefficient and (b) Intrinsic Viscosity Data versus Molecular Weight for Citrus Pectin Fractions

in the log s versus log M_w plot (Fig 1a), since the sedimentation coefficients will not be affected (higher molecular weight species removed by the centrifugal field).

DISCUSSION

Our results appear in all cases to support an extended rod-like conformation for citrus pectins in dilute aqueous solution. This is in agreement with evidence from theoretical calculations on polygalacturonate chains (9). Independent light scattering work on pectins has also indicated an extended conformation (10), although the wormlike chain model was considered more appropriate.

If true, the extended conformation found for citrus pectin, in comparison with many other polysaccharides, is probably due to: (i) The severe steric hindrance of the α-1-4 galacturonate linkage, which is biaxial, and (ii) The relatively low molecular weight.

Further, the results we have found suggest that it is inappropriate to estimate pectin molecular weights from measurement of intrinsic viscosity using a single Mark-Houwink relation. This, and other related phenomena will be considered in more detail in a future publication.

REFERENCES

1. Berth, G. (1988) Carbohydr. Polymers 8, 105-117
2. Berth, G., Dautzenberg, H., Lexow, D. and Rother, G. (1989) Carbohydr. Polymers (in press)
3. Creeth, J.M. and Harding, S.E. (1982) J. Biochem. Biophys. Meth. 7, 25-34
4. Harding, S.E. & Johnson, P. (1985) Biochem. J. 231, 543-547
5. Wales, M. and van Holde, K.E. (1954) J. Polymer Sci. 14, 81-86
6. Creeth, J.M. and Knight, C.G. (1965) Biochim. Biophys. Acta 102, 549-558
7. Cheng, P.Y. & Schachman, H.K. (1955) J. Polym. Sci. 16, 19-30
8. Smidsrød, O. and Andresen, I.L. (1979) Biopolymerkjemi, Tapir, Trondheim.
9. Bailey, E., Mitchell, J.R. and Blanshard, J.M.V. (1977) Colloid & Polymer Sci. 255, 856-860
10. Jordan, R.C. and Brant, D.A. (1978) Biopolymers 17 2885-2895

Improvements in the methanolysis of pectins by enzymic prehydrolysis

BERNARD QUEMENER and JEAN-FRANCOIS THIBAULT
Laboratoire de Biochimie et Technologie des Glucides, INRA, BP 527, Nantes 44026 Cedex 03, France

ABSTRACT

An improved methanolysis procedure has been developed for the simultaneous determination of galacturonic acid and main neutral sugars in pectins. The procedure involves enzymic hydrolysis with a purified commercial liquid preparation (SP249) followed by methanolysis with M methanolic HCl. The methyl glycosides released are analyzed by high performance liquid chromatography on C18 phase. The usefulness of this method is demonstrated by comparison of results to those obtained by methanolysis with M HCl in methanol (methanolysis method 1), by methanolysis with H_2SO_4 in methanol after 72% H_2SO_4 pretreatment for 3h (methanolysis method 2), by colorimetric determination of galacturonic content using metahydroxydiphenyl, as well as by gas liquid chromatography of the alditol acetate derivatives of the neutral sugars released by hydrolysis with sulfuric and trifluoroacetic acid.

INTRODUCTION

Determination of the constituents of polysaccharides generally involves hydrolysis with sulfuric or trifluoroacetic acid. The various stabilities of the free monosaccharides necessitates different optimum conditions of hydrolysis. Methanolysis causes less destruction than does aqueous acid. However, though methanolysis of polysaccharides yields more reliable results, the release of uronic acids from anionic polymers such pectins or alginates is not complete. In order to improve this alternative method of cleavage, we have examined the benefit of an enzymic prehydrolysis in case of pectins.

Table I. NEUTRAL SUGARS AND GALACTURONIC ACID CONTENTS (%) OF PECTINS

Pectin	Method	GalA	Rha	Fuc	Ara	Xyl	Gal	Glc
Sugar beet	M H_2SO_4, 100°, 2 h		1.9	0.10	3.3	0.15	7.8	0.10
	2M CF_3COOH, 120°, 2 h		4.8	0.07	3.1	0.20	7.3	0.11
	Methanolysis method 1	38.5	5.1	nd[a]	3.1	0.15	7.2	nd
	Methanolysis method 2	42.6	4.0	nd	nd	nd	nd	nd
	Proposed method	57.5	5.1	nd	3.1	0.15	7.5	nd
Apple	M H_2SO_4, 100°, 2 h		1.0	0.05	0.4	1.50	2.8	3.10
	2 M CF_3COOH, 120°, 2 h		2.0	0.07	0.5	1.40	2.6	2.60
	Methanolysis method 1	51.5	2.1	nd	1.0	1.20	2.5	3.50
	Proposed method	74.2	2.2	nd	1.0	1.40	2.6	3.60
Citrus	M H_2SO_4, 100°, 2 h		0.7	0.06	1.1	0.09	2.3	0.09
	2 M CF_3COOH, 120°, 2 h		1.2	0.04	1.0	0.16	2.2	0.13
	Methanolysis method 1	47.4	1.4	nd	1.0	0.08	2.2	nd
	Methanolysis method 2	47.2	1.5	nd	nd	nd	nd	nd
	Proposed method	77.1	1.4	nd	1.1	0.20	2.4	nd

a Not determined

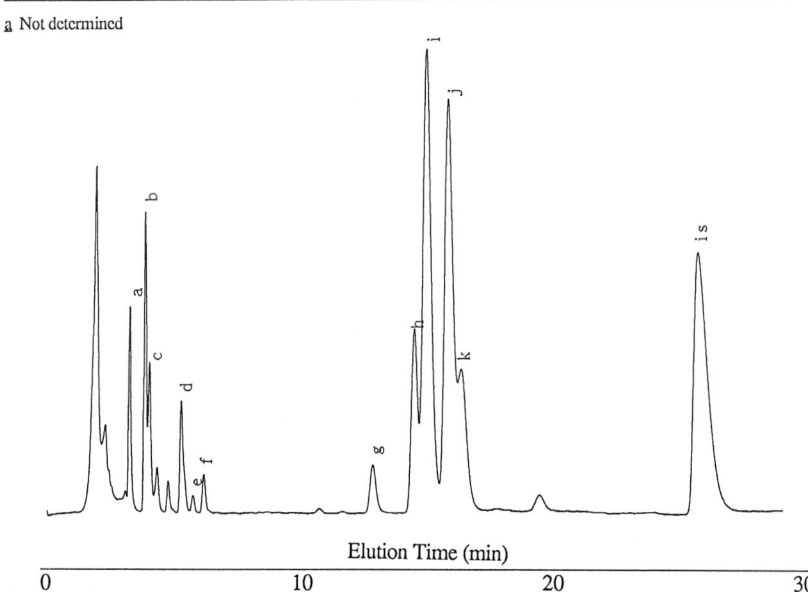

Figure 1. H. p. l. c. of methanolysis products from apple pectin on a Merck Superspher C 18 (2
x 0.4 cm) cartridge eluted at 0.9 mL/min by water (the eluate was monitored by differential refrac
trometry); is : internal standard (dimethyl L-tartrate).

MATERIALS AND METHODS

The sugar beet pectin (57.8% of galacturonic acid content (1)) and the citrus pectin (72% of degree of methylation, 79.2% of galacturonic acid content) were obtained from Kobenhavns Pectinfabrik (Copenhagen, Denmark); the apple pectin (28% of degree of methylation, 76.8% of galacturonic acid content) was from Unipectine (Redon, France). The polysaccharides were purified by precipitation of their 2% aqueous solution with four volumes of absolute ethanol or by precipitation with cupric ions (apple pectin only). The sugar beet pectin was dehydrated by solvent exchange and ground (particle size < 0.5mm). The citrus and apple pectins were freeze-dried. All the samples were dried at 40° over P_2O_5, under vacuum, for 24h at least before treatment.The enzyme cocktail used was a SP249 commercial liquid preparation obtained from NOVO (Denmark) and mainly containing pectolytic activities. This preparation was purified by precipitation with $(NH_4)_2SO_4$ (90% of saturation) in order to eliminate the low molecular weight sugars. This purified enzymic solution was diluted in ultrapure water to have a protein concentration of approximatively 1mg / mL.

The improved methanolysis procedure was as follows : the dried pectin (10mg) was transferred to a 10mL Reacti-Vial and was incubated at 45° for 1h with 400µL of purified SP249 liquid preparation. The hydrolysate was evaporated three times to dryness (with 500µL of anhydrous methanol every time) at 45° in a stream of air using the Reacti-Vap Evaporator (Pierce and Warriner, U.K.). 2mL of methanolic M HCl (2) containing the dimethyl L-tartrate as internal standard (2mg / mL) (3) was added, the tube sealed with Teflon-lined septa in the closed screw cap and methanolysis was then carried out at 85° for 16h. Following neutralisation by Ag_2CO_3 (120mg) and centrifugation, the supernatant was evaporated (40°, under vacuum) and the residue was solubilized in water (2mL). Separation of the methyl glycosides was carried out on a Merck Superspher C18 (25x0.4cm) cartridge, eluted at 0.9 mL / min by water. The eluate was monitored by differential refractometry. For quantification,relative response factors of standard sugars to internal standard were used and were calculated from the main glycoside peak,namely,that which was well separated from those of other derivative sugars (a, b+c, d, e, f, g, h+i+j+k, for galactose, glucose, arabinose, mannose, xylose, rhamnose, and galacturonic acid respectively, Fig.1). For g.l.c. analysis, the neutral sugars were derivatized to alditol acetates (4) and analyzed at 220° on fused silica capillary column (30mx0.32mm i.d. with DB225, 0.15µm film thickness, J &W Scientific). Hydrogen was the carrier gas and myo-inositol was used as internal standard.

RESULTS

The results are summarized in Table I. Methanolysis with method 1 and method 2 (5) gave low amounts of GalA, regardless of the origin of the pectin. These values were significantly increased after the enzymic prehydrolysis and became close to those obtained by colorimetric determinations (1). Method 1 and g.l.c. after acid hydrolysis gave comparable determinations of arabinose, xylose, galactose and glucose. The content in rhamnose, which is higher after TFA hydrolysis than after H_2SO_4 hydrolysis, is slighly increased by the proposed method. It can therefore be concluded that pectins may be analyzed for their main neutral sugars as well as GalA by methanolysis coupled to enzymic prehydrolysis with subsequent h.p.l.c. analysis.

REFERENCES

1. Thibault, J.F. (1979) Lebensm.-Wiss. U.-Technol., 12, 247-251.
2. Chambers, R.E. and Clamp J.R. (1971) Biochem. J., 125, 1009-1018.
3. Hjerpe, A., Antonopoulos, C.A., Classon, B., Engfeldt, B.and Nurminen, M. (1982) J. Chromatogr., 235, 221-227.
4. Sawardeker, J.S., Sloneker, J.H. and Jeanes A. (1965) Anal. Chem., 37, 1602- 1604.
5. Roberts, E.J., Godshall, M.A., Clarke, M.A., Tsang, W.S.C. and Parrish, F.W. (1987) Carbohydr. Res., 168, 103-109.

Effect of enzymes derived from orange peel on citrus and sugar beet pectins

CRAIG B. FAULDS and GARY WILLIAMSON

Department of Molecular Science, AFRC Institute of Food Research, Colney Lane, Norwich NR4 7UA, UK

ABSTRACT

Treatment of sugar beet pectins with a partially purified orange peel extract resulted in formation of a Ca^{2+}-dependent gel, at sugar beet pectin concentrations as low as 1%. The enzymes responsible are pectinmethylesterase(PME), which demethoxylates sugar beet pectin, and pectinacetylesterase (PAE), which is able to release acetate from sugar beet pectin. Commercial PME preparations contain enough PAE impurity to form sugar beet pectin gels. However, gel strength is slightly increased by addition of further PAE. The gels are compared to citrus pectin gels formed in the same way. The treatment requires no heat and involves no extremes of pH.

INTRODUCTION

In contrast to citrus pectin, sugar beet pectin has poor gelling properties in the presence of either sugar or calcium(1). This has been attributed to the presence of O-acetyl groups attached to C2 and C3 of galacturonic acid residues(2), and to a relatively low molecular weight(3).

We report here on an enzymic method for producing good calcium-dependent sugar beet pectin gels by limited deacetylation and demethoxylation using orange peel enzymes. The method should be suitable for inducing gel formation of the pectin after incorporation into a food material at ambient temperatures.

MATERIALS AND METHODS

Pectins

Orange peel pectin was type 105 'rapid set' citrus pectin (H.P. Bulmer Ltd., Hereford, England) with a degree of methoxylation of 69-71%. Sugar beet pectin was prepared using an unpublished method of H.P. Bulmer Ltd. (HNO_3, 38mM, pH1.7, 45°C, 16hr, then 70°C for 1hr). It contained 5.6% (w/w) of methoxyl groups, as determined by a method based on Hills et al(4), and 1.8% of acetyl groups as determined by ^{13}C-NMR(2).

Enzymes

Pectinmethylesterase (EC3.1.1.11) and triacetin acetylesterase (AE) (3.1.1.6) were obtained from Sigma Chem. Co. Ltd. PAE activity was detected by adding enzyme to 35mM succinate pH6.0/0.5% sugar beet pectin/15mM NaCl at 30°C. The reaction was stopped by heating to 100°C for 1min. After cooling, released acetate was measured using an enzyme-based kit (Boehringer-Mannheim Ltd.). Blanks were performed using a boiled (2min) enzyme solution. PME was assayed using a pH stat(5). One unit of enzyme activity (U) corresponds to the release of 1μmol of methanol (for PME) or acetate (PAE) per minute at 30°C.

Formation of gels

Pectin gels were prepared by mixing together pectin, water and $CaCl_2$ before adding enzyme, to a final volume of 5ml. Gel strength was assessed after 3 hr using an Instron(6). The gel strength is expressed as log [shear modulus] = log G^1.

RESULTS AND DISCUSSION

On mixing orange peel PME with various concentrations of citrus pectin in the presence of $CaCl_2$, the viscous solution rapidly formed a gel above 0.5% pectin (figure 1). In the absence of calcium or enzyme, no gel was produced.

A similar experiment was performed on sugar beet pectin (figure 1). This also formed a gel, although a higher concentration of pectin (1%) was required. This demonstrated that sugar beet pectin is a substrate for orange peel PME. The demethoxylation reaction presumably occurs by a similar mechanism to citrus pectin, resulting in blocks of unsubstituted uronic acid residues which associate with other blocks via calcium. However, sugar beet pectin contains 2-9% (w/w) acetyl groups(7), and it has been reported(8) that the close association of pectic molecules required for gelling is prevented by steric hindrance if the main chain carries substituents, such as an acetyl group at C(2) or C(3) of a galacturonic acid residue.

We therefore examined our PME preparation to see if any PAE activity could be detected. Using sugar beet pectin as a substrate, approximately 0.3mg of acetate was released per gram of pectin/hr/mg protein. Orange peel is reported to contain an AE(9). This enzyme has been shown to deacetylate triacetin and related compounds (9). No PAE activity of this enzyme was reported on acetic anhydride-acetylated citrus pectin(3), although the method of detection of this activity may have been unsuitable.

A commercial preparation of AE was therefore tested for PAE activity. A time dependent release of acetic acid is observed using 0.5% sugar beet pectin in succinate buffer (figure 2). After 60 min, 4mg of acetate was released from 1g of sugar beet pectin in the absence of $CaCl_2$. This is equivalent to ~20% of the total acetyl groups present. This demonstrated the presence of PAE activity in orange peel.

We therefore propose that the mechanism of formation of a

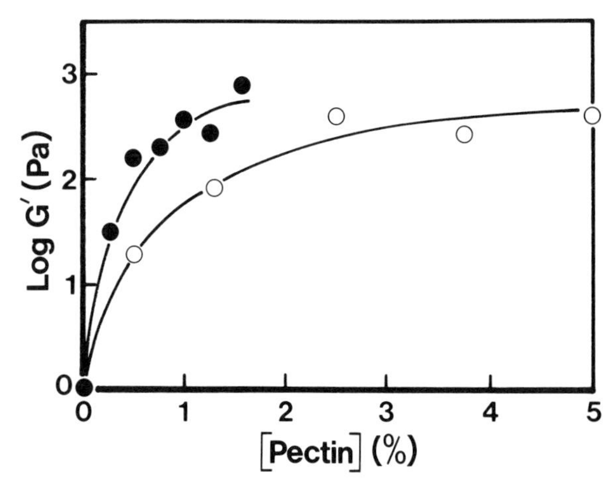

Figure 1. The effect of pectin concentration on gel strength. $CaCl_2$ (40mM) and PME (45U) were added to unbuffered solutions of citrus pectin (●) and sugar beet pectin (○).

good sugar beet pectin gel using orange peel extract requires two enzymes, PME and PAE. The former produces blocks of demethyoxylated galacturonic acid residues whereas the latter deacetylates galacturonic acid residues, thereby reducing the steric hindrance caused by O-acetyl substitution. In order to see if the sugar beet pectin gels formed using the PME preparation could be improved by supplementary PAE, we made gels with equal PME concentrations and added PAE to some. The additional PAE produced only a slight increase in log G^1. Possibly, only a low [PAE] is required, and most of this is already present in PME preparations. Sufficient PAE would be required to deacetylate only enough blocks of galacturonic acid residues to overcome steric hindrance in these regions. In the future, PAE needs to be purified free of PME so that the mechanism, specificity and relative importance to gel formation can be ascertained.

In summary, therefore, a commercial orange peel pectinesterase preparation induces gel formation in sugar beet pectin solutions at room temperature in the presence of $CaCl_2$. We propose that the gelation requires the action of two enzymes in the orange peel extract, PME and PAE.

ACKNOWLEDGEMENTS

We thank Jenny Matthew for preparing the sugar beet pectin, Vic Morris for helpful discussions, and Geoff Brownsey and Mike Ridout for running samples on the Instron.

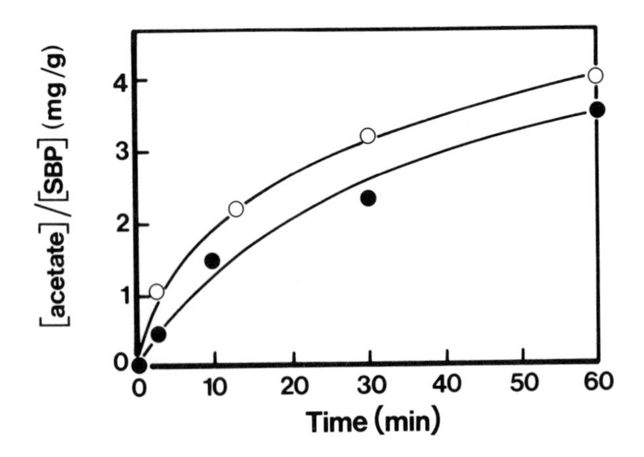

Figure 2. The release of acetic acid from sugar beet pectin by AE. Pectin (0.5%) in 25mM sodium succinate pH6.0 was incubated with 0.4mg/ml crude AE (4.3 triacetin hydrolysis U/ml) in the presence (●) and absence (○) of 5mM $CaCl_2$ at 30°C.

REFERENCES

1. Michel, F., Thibault, J-F., Mercier, C., Heitz, F. and Pouillaude, F. (1985) J. Food Sci., 50, 1499-1502.

2. Keenan, M.H.J., Belton, P.S., Matthew, J.A. and Howson, S.J. (1985) Carbohydrate Res., 138, 168-170.

3. Pippen, E.L., McCready, R.M. and Owens, H.S. (1950) J. Am. Chem. Soc., 72, 813-816.

5. Kertesz, Z.I. (1955) Methods Enzymol., 1, 159-162.

6. Johnson, R.M. and Breene, W.M. (1988) Food Technology, 42, 87-93.

7. Dea, I.C.M. and Madden, J.K. (1986) Food Hydrocolloids, 1, 71-88.

8. Thibault, J-F. (1986) Carbohydrate Res., 155, 183-192.

9. Jansen, E.F., Jang, R. and MacDowell, L R. (1947) Arch. Biochem., 15, 415-431.

A simplified method for the determination of the intrinsic viscosity of pectin solutions by classical viscosimetry

T.P.KRAVTCHENKO and W.PILNIK

Agricultural University, Department of Food Science, Biotechnion, Bomenweg 2, 6703 HD Wageningen, The Netherlands

ABSTRACT

Viscosimetric measurements were performed on various commercial pectin solutions with an Ubbelohde viscosimeter. Intrinsic viscosities were determined by extrapolating to zero concentration:

- the reduced viscosity: η_{sp}/C
- the logarithm of the reduced viscosity: $\ln(\eta_{sp}/C)$
- the inherent viscosity: $(\ln\eta_{rel})/C$
- the combined relation: $\frac{1}{C}\left[2\left(\frac{\eta}{\eta_0}-1-\ln\frac{\eta}{\eta_0}\right)\right]^{1/2}$

The classical η_{sp}/C-C relation exhibits a non linear behaviour. Thus, the determination of intrinsic viscosity by linear extrapolation to zero concentration of η_{sp}/C leads to a 5-10% subevaluation. The empirical $\ln(\eta_{sp}/C)$-C relation provides intrinsic viscosity values very close to the ones given by the inherent viscosity and combined relation. Moreover, the values provided by the combined relation are fairly constant with concentration. This behaviour allows determination of the intrinsic viscosity from a single-concentration measurement without loss of accuracy.

INTRODUCTION

Many practical applications of pectins such as formation of gels are directly related to their molar mass. Thus the determination of viscosity of pectin solutions has often been used to calculate their molar mass according to the principles initiated by Staudinger in the 30s.

Solution viscosity is basically a measure of the size or extension in space of polymer molecules (1). It is therefore empirically related to the molar mass. Mark (2) and Houwink (3) proposed the relation:

$$[\eta] = k\ M^{\alpha}$$

where $[\eta]$ is the limiting viscosity number, more commonly called intrinsic viscosity, and k and α are two constants dependent on the couple molecular shape-solvent. The simplicity of measurement and the usefulness of the viscosity-molar mass relation make of viscosity measurement an extremely valuable tool for the molecular characterization of pectins.

Part of the problem is then the accurate determination of the intrinsic viscosity. Several methods were proposed to extrapolate the intrinsic viscosity from viscosity measurements performed at different concentrations. Some of the available methods for the intrinsic viscosity determination of polymers were compared in order to improve the accuracy as well as the ease of the current intrinsic viscosity determination of pectins.

THEORY

The concentration dependence of the specific viscosity in the very dilute region may usually be expressed as a power series of the concentration (4):

$$\eta_{sp} = [\eta]C + k[\eta]^2C^2 + k'[\eta]^3C^3 + \ldots \quad (1)$$

where k is a dimensionless constant known as the Huggins factor (5).

From this expression one can easily recognize the relations currently used to determine the intrinsic viscosity:

$$[\eta] = \lim_{C\to 0} \frac{\eta_{sp}}{C} \quad (2)$$

η_{sp}/C being called the reduced viscosity

$$[\eta] = \lim_{C\to 0} \frac{1}{C} \ln \frac{\eta}{\eta_0} \quad (3)$$

$\ln(\eta/\eta_0)/C$ being called the inherent viscosity

In order to correct the imperfect linearity of the η_{sp}/C-C relation of polystyrene solutions, Staudinger and Heuer (6) established the empirical relation:

$$[\eta] = \lim_{C\to 0} \ln \frac{\eta_{sp}}{C} \quad (4)$$

In their study on pectin viscosity, Owens et al (7) used the relation 4 by plotting η_{sp}/C on a semi-logarithmic scale.

Moreover, combination of the former relations using the reduced viscosity (relation 2) and the inherent viscosity (relation 3) provides the following combined relation which has already been reported (8,9):

$$[\eta] = \lim_{C\to 0} \frac{1}{C} \left[2\left(\frac{\eta}{\eta_0} - 1 - \ln\frac{\eta}{\eta_0}\right)\right]^{1/2} \quad (5)$$

MATERIAL AND METHODS

Various unstandardized commercial pectins were taken in our own collection. Their brief analytical characteristics are given in table 1.

Viscosities were measured in an Ubbelohde capillary viscosimeter as described by Van Deventer-Schriemer and Pilnik (10). Pectins were dissolved in a 0.1 M tris-succinate buffer pH 6.0 (molarity refers to succinic acid) with 0.01 % thiomersal. Temperature was set at 30°C. Flow times were recorded with a stopwatch with a precision of 0.1 second. All concentrations are expressed in galacturonic acid content.

Table 1: ANALYTICAL CHARACTERISTICS OF PECTIN SAMPLES DETERMINED BY THE TITRATION PROCEDURE.

	Galacturonan content % raw pectin	Degree of Methylation (%)
Apple A	68.8	50.4
Apple B	72.3	74.3
Apple C	69.1	60.3
Apple D	63.7	62.1
Apple E	60.9	73.6
Lemon A	75.1	71.8
Lemon B	74.4	72.8
Lemon C	69.5	61.1
Lemon D	70.1	71.5
Lime A	71.3	62.6
Lime B	70.2	69.5

RESULTS AND DISCUSSION

Flow times through a capillary viscosimeter were recorded for 11 different commercial pectin samples at different concentrations in order to determine the intrinsic viscosity by means of various available procedures. Table 2 gives the intrinsic viscosities calculated with the use of relations given in the theoretical part of this paper.

Table 2: INTRINSIC VISCOSITIES OF PECTIN SAMPLES DETERMINED BY EXTRAPOLATION TO ZERO CONCENTRATION USING 4 DIFFERENT PROCEDURES.

	Reduced viscosity	ln of red.visc.	Inherent viscosity	Combined relation
Apple A	5.48	5.75	5.74	5.75
Apple B	6.88	7.85	7.78	7.75
Apple C	6.47	7.10	6.96	6.96
Apple D	4.72	4.76	4.78	4.78
Apple E	5.83	5.93	5.94	5.92
Lemon A	4.30	4.48	4.47	4.48
Lemon B	5.23	5.64	5.63	5.62
Lemon C	4.72	4.76	4.73	4.73
Lemon D	6.20	6.36	6.34	6.34
Lime A	5.87	5.99	5.95	5.99
Lime B	6.26	6.49	6.52	6.50

The values obtained by different procedures are very close to each other (SD $< 10^{-2}$) except for the reduced viscosity method which provides values 5-10 % lower. While using the relation given by Owens et al (7), this difference leads to a sub-evaluation of the weight average molar mass of 2-10 %.

This difference may be explained by the non-linear behaviour of the η_{sp}/C-C relation as it can be seen from figure 1. Logarithmic regression of the reduced viscosity against C provides a better extrapolation to zero concentration of the intrinsic viscosity than the linear regression.

Inherent viscosity and the combined relation also allow a very good extrapolation of the intrinsic viscosity. This may be explained by the fact that both relations are better mathematical simplifications of the general concentration dependence of the reduced viscosity (relation 1) than the reduced viscosity.

In order to improve the accuracy of the intrinsic viscosity determination of pectin solutions, we must therefore recommend the use of either the logarithmic extrapolation of the reduced viscosity, the extrapolation of the inherent viscosity or the extrapolation of the combined relation instead of the currently used linear reduced viscosity extrapolation.

As it can be seen from figure 1, the slope of the combined relation is very small. This suggests that a single measurement at low concentration allows to estimate the intrinsic viscosity.

Figure 1: INTRINSIC VISCOSITY DETERMINATION OF LEMON PECTIN B BY EXTRAPOLATION TO ZERO CONCENTRATION USING 4 DIFFERENT RELATIONS

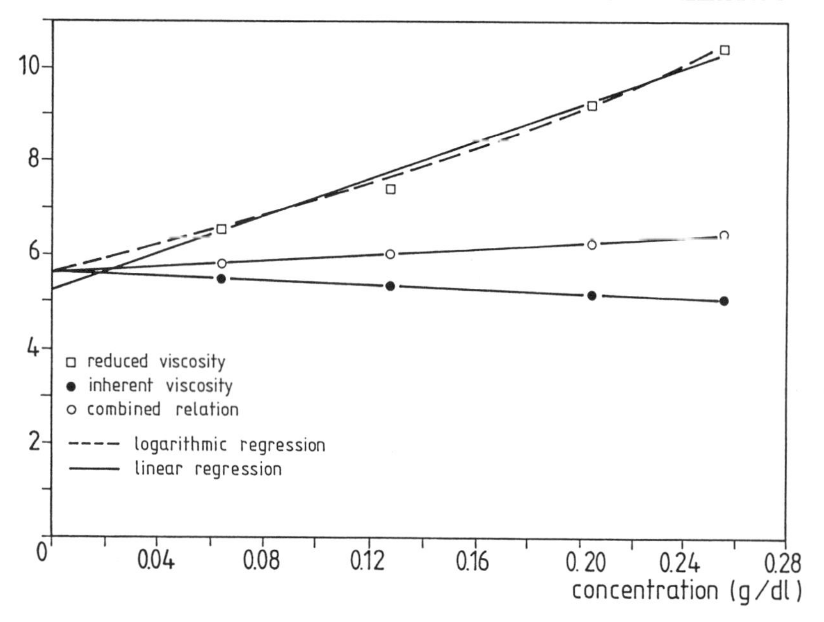

Table 3 compares the values of intrinsic viscosity obtained by either multi-point or single-point (0.03-0.06 g/dl concentration) extrapolation to zero concentration of the combined relation (relation 5). The latest method provides an estimation of the intrinsic viscosity with an error which is not higher than 5%. It is still possible to reduce the concentration of the solution to be measured to 0.01 g/dl. Moreover, accuracy might be improved by using an automated system for flow time

Table 3: COMPARISON OF MULTI-POINT AND SINGLE-POINT EXTRAPOLATION FOR THE DETERMINATION OF THE INTRINSIC VISCOSITY OF PECTINS USING THE COMBINED RELATION (relation 5).

	Multi-point Extrapolation	Single-point Extrapolation	*
Apple A	5.75	5.89	(0.055)
Apple B	7.75	8.13	(0.057)
Apple C	6.95	7.24	(0.057)
Apple D	4.78	4.86	(0.032)
Apple E	5.92	6.05	(0.031)
Lemon A	4.48	4.60	(0.065)
Lemon B	5.62	5.82	(0.064)
Lemon C	4.73	4.80	(0.035)
Lemon D	6.34	6.49	(0.035)
Lime A	5.95	6.08	(0.036)
Lime B	6.50	6.67	(0.035)

* pectin concentration (g/dl)

measurements. In such conditions, the intrinsic viscosity of pectin solutions can be accurately estimated (error < 1%) by a single measurement at low concentration by using the combined relation. This makes the viscosimetric procedure much faster without loss of accuracy.

ACKNOWLEDGEMENTS

The authors thank Doctor Lecacheux from Groupement de Recherche de Lacq, Elf Aquitaine (France) for his helpful comments and discussions and Hanneke Reitsma for her technical assistance in viscosity measurements.

This work was supported by Sanofi Bio Industries (France).

REFERENCES

1. Billmeyer F.W.Jr., (1971), Textbook of Polymer Science, 2nd ed., Wiley Interscience.
2. Mark H., (1938), in "Der feste Köper", Sänger R. ed., Hirzel, Leipzig.
3. Houwink R., (1940), J. Prakt. Chem. 157, 15.
4. Eisenberg H., (1976), Biological Macromolecules and Polyelectrolytes in solution, Clarendon Press, Oxford.
5. Huggins M.L., (1942), J. Amer. Chem. Soc. 64, 2716
6. Staudinger H., Heuer W., (1934), Z. Phys. Chem. A 171, 129.
7. Owens H.S., Lotzkar H., Schultz T.H., Maclay W.D., (1946), J. Amer. Soc. 68, 1628.
8. Lecacheux D., (1982), PhD thesis, Université Pierre et Marie Curie, Paris VI.
9. Morris E.R., (1984), in "Gums and Stabilizers for the Food Industry 2. Application of Hydrocolloids", Phillips G.O., Wedlock D.J., Williams P.A. eds, Pergamon Press.
10. Van Deventer-Schriemer, Pilnik W., (1987), Acta Alimentaria 16(2), 143.

In situ pectin de-esterification in bramley apples – a comparison of the properties of pectinesterase *in situ* and when extracted

K.KING

Department of Agriculture (NI) and Queen's University of Belfast, Food and Agricultural Chemistry Research Division, Newforge Lane, Belfast, Northern Ireland

ABSTRACT

Pectinesterase activity was compared in situ and when extracted from bramley apple waste (peel, cores and offcuts). Optimum activity was obtained at 60 °C for both the in situ and extracted enzymes but the latter was less heat stable. Both enzyme activities decreased above 0.3 M sodium chloride with an optimum at 0.2 M for the in situ enzyme. The extracted enzyme had a lower and broader pH optimum (pH 7.0 to 9.0) compared to the in situ enzyme which had a sharp optimum at pH 10.0.

INTRODUCTION

Pectinesterase (pectin pectoyl hydrolase; EC 3.1.1.11) catalyses the release of methyl ester groups from pectin. The enzyme has been found in numerous higher plants and is generally thought to have an important role in the softening of fruit tissues during ripening. Pectinesterase also has a number of important uses in food processing. Extracted enzyme can be added during maceration of fruits and vegetables to increase juice extraction and also directly to the juice, as a clarifying agent. Some fruits and vegetables show an improved final product texture when held at elevated temperatures, ie, 60-70° C, prior to further processing. A firming of texture occurs due to activation of pectinesterase resulting in calcium bridging of the pectin. Another application[1] uses pH to stimulate pectinesterase in situ producing low methoxyl pectin for use as a gelling or thickening agent[2].

Due to the nature of the applications of pectinesterase in the food industry, the value of characterising its properties using extracted enzyme preparations is questionable. The work reported illustrates some of the apparent differences in pectinesterase characteristics and properties when assayed using an extracted crude enzyme preparation and an in situ method.

MATERIALS AND METHODS

The apple waste material (peel, cores and offcuts) was obtained from a local processor, transported in a cool box at between 8-10° C, frozen and stored at -20° C.

All chemicals were Analar grade.

Determination of dry matter content

Weighed samples of defrosted apple material were freeze dried and the dry matter calculated. Dried samples were stored at -18° C.

Extraction of pectinesterase

Pectinesterase was extracted overnight from 500 g frozen apple waste using 1000 ml 0.2 M borate buffer (sodium tetraborate and boric acid) at pH 8.5. The filtrate obtained through muslin was fractionated using ammonium sulphate and the fraction between 30 and 85% saturation retained, resuspended in 0.01 M borate buffer and dialysed overnight. The enzyme extract was frozen in 10 ml aliquots.

Preparation of in situ pectinesterase suspension

In situ pectinesterase activity was assayed in an apple waste suspension. Dried apple waste was suspended in 0.1 sodium chloride pH 8.5 overnight, and made up to 100 ml.

Measurement of pectinesterase activity

Pectinesterase activity was measured in 0.5% apple pectin (BDH 250 grade), 0.1 M NaCl, pH 8.5 and 30° C using an autotitrator (Radiometer). The activity was calculated as μ mol COOH groups released per minute, per gramme dry matter from the gradient of volume of NaOH against time.

RESULTS AND DISCUSSION

An optimum temperature of 60° C was found for both the in situ and extracted pectinesterase (Figure 1) although few assays were carried out above this temperature due to hysteresis effects on the pH electrode of the titrator. This optimum is similar to 63° C found for the variety Ralls[3] but higher than 53° C reported for Golden Delicious and York Imperial[4]. The response of pectinesterase to increasing temperature was much greater in situ than when extracted with 6 and 4 fold increases, respectively, from 20° C to 60° C. The temperature stability, however, was found to be considerably different (Figure 2). In situ pectinesterase activity showed little change between 20 and 40° C, but above

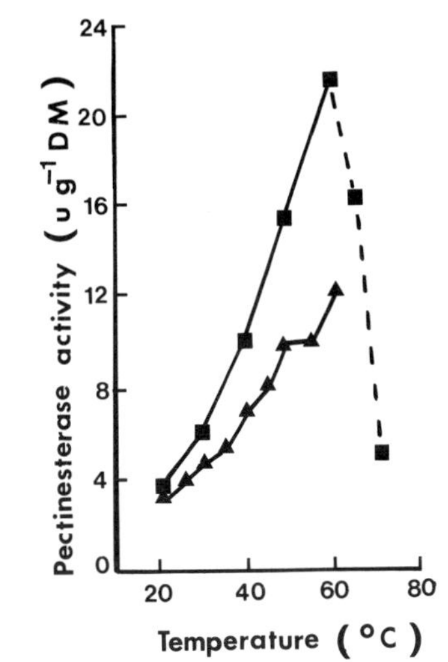

Figure 1. Effect of temperature. ■ <u>in situ</u>; ▲ extracted.

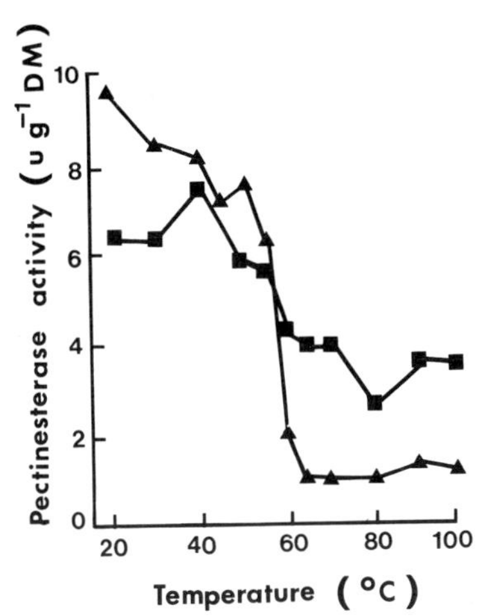

Figure 2. Temperature stability. Samples held for 5 min at temperature. ■ <u>in situ</u>; ▲ extracted.

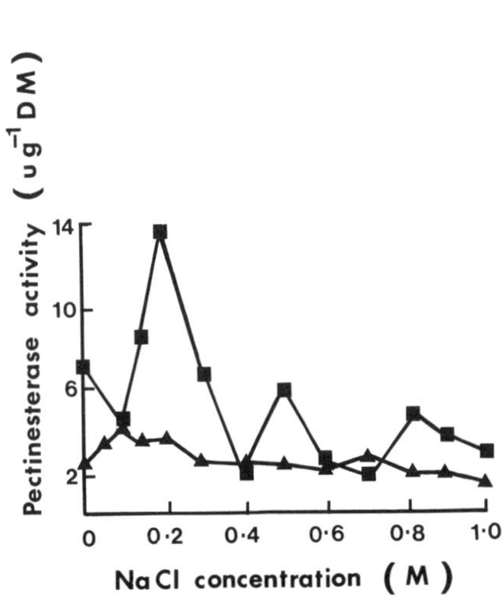

Figure 3. Effect of NaCl concentration. ■ <u>in situ</u>; ▲ extracted.

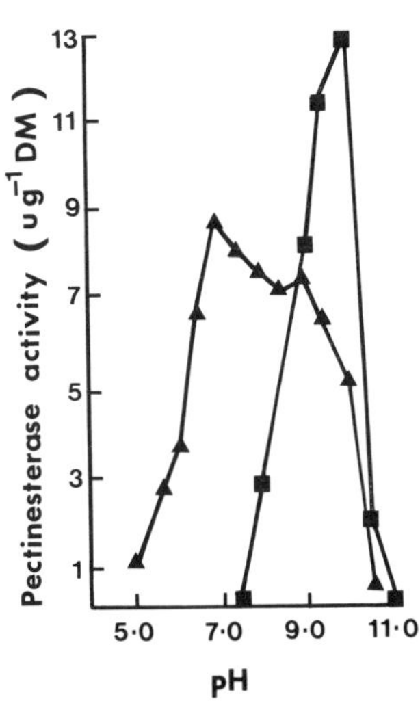

Figure 4. Effect of pH. ■ <u>in situ</u>; ▲ extracted.

40° C activity decreased to about 60% at 60° C. Above 60° C there was little change in activity. The extracted enzyme, however, declined steadily in activity (rate average 9% 10° C^{-1}) from 20° C to 55° C and then rapidly to a constant level above 65° C, of 14% of that at 20° C. Pollard and Kieser (1957)[5] found a similar enhancement of pectinesterase stability in apple juice compared with the isolated enzyme. Sodium chloride concentrations above 0.3 M reduced pectinesterase activity both _in situ_ and when extracted (Figure 3). However, at lower sodium chloride concentrations _in situ_ pectinesterase activity showed a peak optimum at 0.2 M whereas no optimum was found with the extracted enzyme. There was also a difference in response to pH (Figure 4). _In situ_ activity was only detected between pH 7.5 and 10.5 with a sharp optimum at pH 10.0, whereas the extracted enzyme had activity over a much broader pH range of 5.0 to 11.0 with an optimum between 7.0 and 9.0. These latter values agree with pH optima reported in the literature of 6.6[5], 6.5 to 7.5[4] and 7.3[3]. However, _in situ_ pectinesterase activity would be very low or undetectable at these pH values. The pH curve obtained for _in situ_ pectinesterase activity reflects not only the activity of the enzyme, but also the 'extraction' of the enzyme and its stability over a short time (assay time of 5 min). A pH value higher than that for optimum activity is generally used for extraction[5], but as extraction usually takes place over several hours, pH stability of the enzyme is also important.

For each characteristic examined, differences in response were found between the _in situ_ and extracted enzyme. It is probable that the _in situ_ enzyme requires activation, possibly by solubilisation before activity can be determined and therefore careful consideration is required before extrapolating extracted enzyme data to the _in situ_ enzyme.

ACKNOWLEDGEMENT

The assistance of Ms H. Bustard in carrying out the experimental work is gratefully acknowledged.

REFERENCES

1. Buckley, K., Mitchell, J.R. and Burrows, I.E. (1978) British patent No 1508 993. Food product and method.
2. Mitchell, J.R., Buckley, K. and Burrows, I.E. (1978) British patent 1525 123. Food binding agent.
3. Miyain, K., Okuno, T. and Sawai, K. (1975) (Purification and physico-chemical properties of pectinesterase extracted from apple fruits (Var. Ralls)). Bulletin of the Fac. of Agriculture, Hirosaki University. _24_, 22-30.
4. Lee, Y.S and Wiley, R.C. (1970) Measurement and partial characterisation of pectinesterase in apple fruits. J. Am. Soc. Hort. Sci. _95_, 465-468.
5. Pollard, A and Kieser, M.E. (1957) The pectase activity of apples. J. Sci. Food Agric. _2_, 30-36.

Novel and selective substrates for the assay of endo-arabinanase

BARRY V.MCCLEARY

Megazyme (Australia) Pty Ltd, 6 Altona Place, North Rocks, New South Wales, 2151, Australia

ABSTRACT

Substrates and assay procedures for the measurement of endo-1,5-α-L-arabinanase in crude, technical pectinase preparations have been developed. The method of choice employs carboxymethyl-debranched beet araban as substrate, and rate of hydrolysis is measured using the Nelson-Somogyi reducing-sugar procedure with arabinose as the standard. The substrate is physically and chemically stable in solution, and the assay procedure is simple, reliable and specific. Other assay procedures for the measurement of endo-arabinanase, which employ dyed debranched araban substrates, are also briefly described.

KEYWORDS

Debranched araban, endo-arabinanase, endo-galactanase, arabinofuranosidase, dyed debranched araban, fruit juice, arabinan haze.

INTRODUCTION

Commercial and consumer pressures on fruit processors to produce higher quality products at lower prices have necessitated the implementation of modern technology into traditional processes (Ducroo, 1987). In the processing of apples and pears, the yield of juice can be dramatically improved both by the use of enzymes to degraded pulp polysaccharides and by more exhaustive extraction of the pulp with diffusion equipment (Ducroo, 1987). Of course, these processes also significantly increase the amount of partially degraded polysaccharide which is solubilised. This polysaccharide material may be soluble as extracted, but subsequent changes in temperature and pH conditions can lead directly to precipitation (or crystallisation), or to chemical modification followed by precipitation. Such a problem has recently been experienced in the production of clear apple and pear juices, in which case, an arabinan haze material was identified. This material was shown to be microcrystalline linear 1,5-α-L-arabinan (Churms et al., 1983).

Arabinans, as present in cell-wall pectic-substances, have been shown to consist of a main chain of 1,5-α-linked L-arabinofuranosyl residues to which other L-arabinofuranosyl residues are linked (1-3)-α and/or (1-2)-α in either a comb-like (Voragen et al., 1987) or ramified (Aspinall, 1984) arrangement.

It has been suggested (Whitaker, 1984) that the problem of "arabinan hazes" could be resolved by the use of pectolytic enzyme mixtures devoid of α-L-arabinofuranosidase. This enzyme produces 1,5-α-L-arabinans from highly branched, very soluble arabinans, by cleaving the 1,3-α and 1,2-α-linked L-arabinosyl branch units. However, since branched arabinans are not stable in juice and juice concentrates at certain steps of the manufacturing process, due to the low pH of the juice and high temperatures (at the heat treatment step), debranching can occur in the absence of arabinofuranosidase. Consequently, it is now generally accepted that the best solution to this problem lies in the use of pectinase enzyme preparations containing high levels of both arabinofuranosidase and endo-1,5-α-L-arabinanase. The arabinofuranosidase removes 1,3-α- and 1,2-α-linked arabinofuranosyl residues allowing ready access of the 1,5-α-L-arabinan main-chain to attack and depolymerisation by endo-arabinanase. The implementation of this technology is, however, limited by the absence of a simple and specific procedure for the measurement of endo-arabinanase in technical pectinase preparations. The substrate generally employed is linear 1,5-α-L-arabinan (Voragen et al., 1987) which is recovered by filtration of "hazy" fruit-juice concentrates. This substrate is costly to produce and thus difficult to obtain. Furthermore, assays employing this substrate (e.g. Hplc assays) are not specific. The substrate and its degradation products are cleaved by certain arabinofuranosidases, making the assignment of the relative contributions by each enzyme difficult.

The aim of the current research was to investigate the possibility of making a specific endo-arabinanase substrate from enzymically and chemically modified sugar-beet araban. Sugar beet araban is composed mainly of a 1,5-α-L-arabinan which is highly branched with 1,3-α- and 1,2-α-linked L-arabinofuranosyl residues. However, typical beet araban preparations also contain about 5-10% uronic acid and 8-15% D-galactose (Rombouts et al., 1988; McCleary, unpublished).

MATERIALS AND METHODS

α-L-Arabinofuranosidase, endo-1,5-α-L-arabinanase, endo 1,4-β-D-galactanase and endo-polygalacturonanase were purified from a commercial pectinase preparation by ion-exchange and gel-filtration chromatographic procedures (McCleary, unpublished). Sugar-beet araban was purified from sugar-beet pulp according to the procedure of Jones and Tanaka (1965). This was further purified by application to a column of DEAE-Tris Acrylamide (pH 8.0) in 20 mM Tris/HCl buffer (pH 8.0) (cf. Tagawa and Kaji, 1969). The araban fraction eluted unbound, and was precipitated by the addition of 5 volumes of ethanol. Alternatively, the pH of this araban solution (2% w/v araban) was adjusted to pH 4.5, and the solution treated for 4 h at 40°C with endo-1,4-β-D-galactanase and arabinofuranosidase to effect complete removal of

all the galactan regions which are susceptible to endo-galactanase and of the 1,3-α- and 1,2-α-L-arabinofuranosyl branch units. Debranched araban was recovered by alcohol precipitation, washed with ethanol, acetone and hexane and dried in vacuo.
Samples of this debranched araban were modified further as follows:
1. Carboxymethyl Debranched Araban. The debranched arabinan was treated with chloroacetic acid as previously described (McCleary, 1980) to give carboxymethylation degree of substitution values of approximately 0.05, 0.10 and 0.20. Following carboxymethylation, the CM-arabans were dissolved and treated with sodium borohydride (0.1 g/g) to remove the background reducing colour.
2. Red Linear Araban. Debranched araban was dyed with Procion Brilliant Red MX5B using a procedure similar to that described by Babson et al. (1970) for the dyeing of amylopectin.
3. Remazolbrilliant Blue- and Remazolbrilliant Black-CM-Linear Araban. CM-Debranched Araban was dyed with Remazolbrilliant Blue R or Remazolbrilliant Black B by the procedure previously described for the dyeing of barley beta-glucan (McCleary, 1986).

RESULTS AND DISCUSSION

A. Carboxymethyl Debranched Araban

In developing a substrate for the measurement of endo-arabinanase, it is essential that the activity values obtained relate directly to the activity of the enzyme on arabinan as found in fruit-juice concentrates. It is also essential that the assay specifically and reliably measures endo-arabinanase, even in crude enzyme mixtures. Furthermore, it is desirable that the assay is simple to use and that the substrate is easy to prepare and is chemically stable and gives solutions which are physically stable i.e. the polymer remains in solution.

An almost pure linear 1,5-α-L-arabinan fraction (Linear Arabinan) can be obtained by filtration of certain hazy fruit-juice concentrates. However, this process is very costly and greatly restricts the supply of this arabinan material.

In the current studies we have produced a material, termed "Debranched Araban" (it is not pure arabinan), from sugar-beet araban by a combination of chromatographic, enzymic and chemical procedures. A solution of beet araban is passed through a column of DEAE-Tris Acrylamide to remove most of the uronic acid containing material plus coloured material. The araban is then debranched with arabinofuranosidase and treated with endo-galactanase to remove sections (of galactan) which are susceptible to hydrolysis by this enzyme. The polysaccharide is recovered, alcohol washed and dried. It is then subjected to controlled carboxymethylation to improve solubility properties. Excessive carboxymethylation must be avoided as this reduces the susceptibility of the substrate to cleavage by endo-arabinanase.

The action of endo-arabinanase in a crude technical preparation containing 27 mU endo-arabinanase/0.2 ml (i.e. per assay) and

20 mU arabinofuranosidase/0.2 ml on these substrates, is shown in Fig. 1. Carboxymethyl (CM) Debranched beet Araban with a degree of carboxymethylation of 0.1, is hydrolysed at approximately 90% the rate for Linear Arabinan recovered from fruit juice concentrates. Increasing the degree of substitution (DS) with carboxymethyl groups to 0.2 seriously affects the susceptibility of the substrate to hydrolysis by endo-arabinanase. In fact, the rate is reduced to 59% the rate for Linear Arabinan. Decreasing the degree of substitution to DS 0.05 renders the substrate slightly more susceptible to cleavage by endo-arabinanase, but this substrate is more difficult to dissolve (similar to Linear Arabinan), and like Linear Arabinan, it precipitates from solution on short term storage at 4°C. Thus,

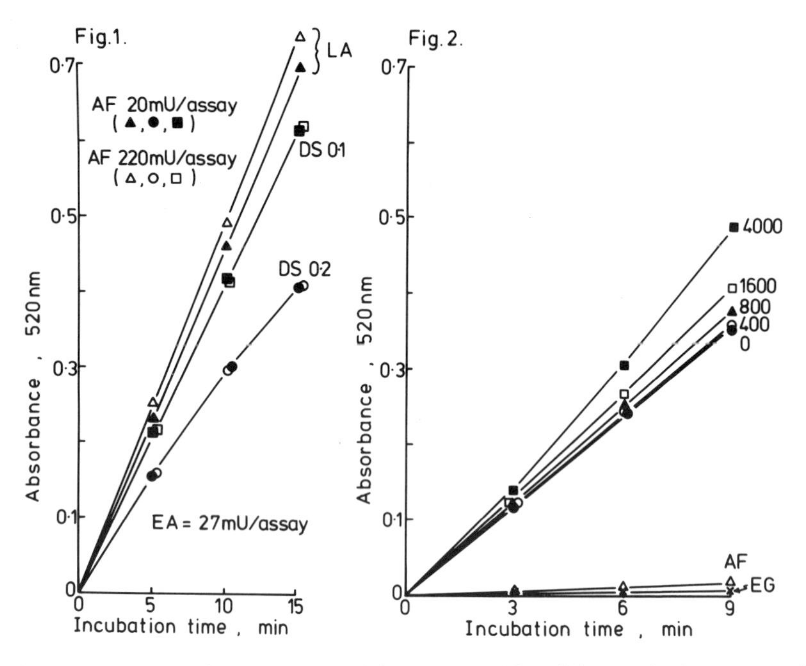

Fig. 1. Assay for endo-arabinanase (27 mU/assay) in a crude pectinase preparation using Linear Arabinan (LA) and Carboxymethyl Debranched beet Araban (DS 0.1 or 0.2) as substrate, in the presence of significantly different levels of arabinofuranosidase (20 or 220 mU/assay).

Fig. 2. Effect of arabinofuranosidase (0-4000 mU/assay) on the specificity of CM-Debranched Araban (DS 0.1) for the assay of endo-arabinanase (23 mU/assay). The effect of purified arabinofuranosidase (AF, 1000 mU/assay) and endo-galactanase (EG, 1400 mU/assay), in the absence of endo-arabinanase, on this substrate, is also shown.

the optimal degree of carboxymethylation is 0.1, as the substrate is easy to dissolve, remains in solution on dissolution, and is hydrolysed by endo-arabinanase at a rate very similar to that for the substrate of interest, namely Linear Arabinan from fruit-juice concentrates.

From Fig. 1 it is also evident that addition of purified arabinofuranosidase, at a level 10-times that in the crude preparation, has no effect on the absorbance values obtained for endo-arabinanase on the CM-Debranched Araban substrates. This clearly demonstrates the specificity of the assay procedure and substrate for the measurement of endo-arabinanase in the presence of arabinofuranosidase. [This arabinofuranosidase is type B described by Rombouts et al. (1988); it hydrolyses sugar-beet araban at about 50% the rate of para-nitrophenyl α-L-arabinofuranoside.] It is interesting to note that added arabinofuranosidase does have some effect when Linear Arabinan is the substrate.

The linearity of the assay procedure for highly purified endo-arabinanase (23 mU/assay) with time, on CM-Debranched Araban (DS 0.1), is shown in Figure 2. Also shown, is the action of arabinofuranosidase (AF, 1000 mU/assay) and endo-galactanase (EG, 1400 mU/assay) on this substrate, and the effect of added arabinofuranosidase (400-4,000 mU/assay) on the rate of hydrolysis by endo-arabinanase (23 mU/assay). It is evident that the substrate is resistant to attack by both endo-galactanase and arabinofuranosidase, and that contamination of endo-arabinanase with up to a 20-40-fold excess of arabinofuranosidase (400-800 mU/assay) has little effect on the specificity of the assay for endo-arabinanase. Consequently, the effect of arabinofuranosidase on the specificity of the endo-arabinanase assay procedure is of no practical significance. In a survey of a wide range of commercial pectinase preparations, the ratio of arabinofuraosidase to endo-arabinanase was, at maximum, 7.6:1.

The slight absorbance increase on treatment of CM-Debranched Araban (DS 0.1) with high levels of endo-galactanase are considered to be due to a very slight contamination of the endo-galactanase employed with endo-arabinanase (approximately (0.05%), rather than to a susceptibility of the substrate to endo-galactanase. A very similar absorbance increase was obtained on treatment of fruit-juice Linear Arabinan (which by g.l.c. is devoid of galactose) with this endo-galactanase preparation.

The major reaction products on cleavage of Linear Arabinan and Debranched Araban by endo-arabinanase are arabinobiose and arabinotriose, not arabinose. Since these oligosaccharides are not commercially available, it is convenient to use arabinose as the sugar standard in the reducing sugar assays. However, for this to be analytically valid, arabinose, arabinobiose and arabinotriose must give similar molar absorbance values with the reducing sugar method employed. The results shown in Figure 3 demonstrate that with the Nelson/Somogyi reducing sugar assay (Somogyi, 1952), this is the case.

The linearity of the assay procedure with concentration of _endo_-arabinanase is shown in Figure 4. The reaction curve is linear up to an absorbance of 0.8 i.e. with concentrations of purified _endo_-arabinanase up to 54 mU/assay under standard assay conditions (10 min, pH 4.5, 40°C).

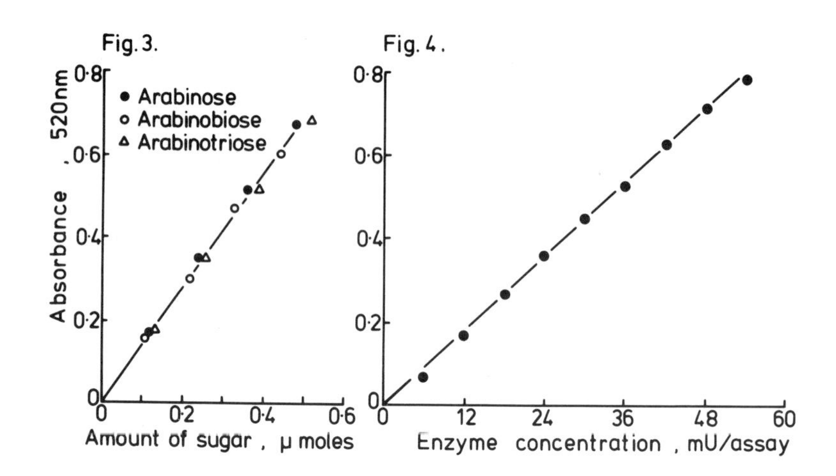

Fig. 3. Molar colour responses for arabinose, arabinobiose and arabinotriose with the Nelson-Somogyi reducing sugar assay.

Fig. 4. Standard curve relating _endo_-arabinanase concentration to Nelson-Somogyi colour response under standard assay conditions (pH 4.5, 10 min, 40°C) with CM-Debranched Araban (DS 0.1).

The effect of substrate concentration on the rate of reaction is shown in Figure 5. The concentration of CM-Debranched Araban (DS 0.1) for half maximal velocity is 0.37 mg/mL. Thus, a concentration of 2.0 mg/mL in the reaction mixture is adequate, and is routinely employed.

These studies demonstrate that the substrate termed CM-Debranched Araban (DS 0.1) is very suitable for the assay of _endo_-arabinanase in crude enzyme mixtures. The substrate is readily solubilised and remains in solution on extended storage at 4°C, and under these conditions it is chemically stable for several months. (Addition of 2 drops of toluene to the substrate prevents microbial infection.) The substrate is hydrolysed by _endo_-arabinanase at a rate very similar to that for Linear Arabinan (from fruit-juice concentrates), and it is resistant to cleavage by arabinofuranosidase and _endo_-galactanase. Also, _endo_-polygalacturonanase has no action on the substrate (the substrate contains a small percentage of galacturonic acid). One limitation of this assay procedure is that it can't be directly used to measure activity in enzyme preparations which contain a high level of reducing compounds (e.g. sugars; used as bulking agents in powder samples). With such samples, the reducing sugars must be removed before the assay is performed, or alternatively, a dyed-debranched araban substrate can be employed.

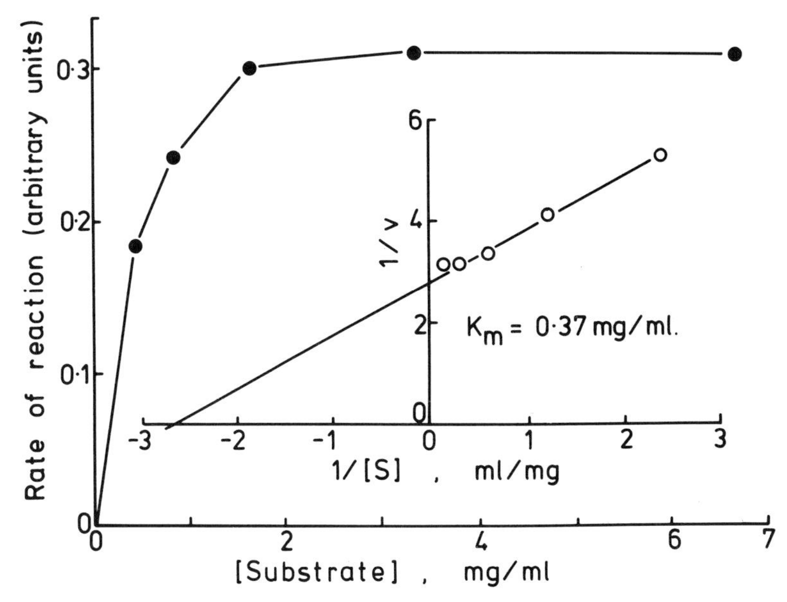

Fig. 5. Effect of substrate concentration on the rate of hydrolysis of CM-Debranched Araban (DS 0.1) by endo-arabinanase (purified from Gist Brocades C80 preparation).

B Dyed Debranched Araban

In the current studies, a range of dyed substrates have been prepared and these include Red Debranched Araban, (prepared with Procion Brilliant Red MX5B), Black CM-Debranched Araban (prepared with Remazolbrilliant Black B) and Blue CM-Debranched Araban (prepared with Remazolbrilliant Blue R). Each of these substrates is effective in the specific measurement of endo-arabinanase even in crude enzyme preparations containing high levels of reducing sugars. However, each suffers from certain limitations. With the Red Araban, the standard curve is sigmoidal, and it is very hard to produce substrates which give low background (blank) absorbances. The latter is not a problem with the Black and the Blue CM-Debranched Arabans. However, with these substrates it is difficult to get linearity over an acceptable absorbance range (i.e. about one Absorbance Unit). Work on these dyed substrates is continuing.

REFERENCES

Aspinall, G.O. and Fanous, H.K. (1984). Carbohydr. Polymers **4**, 193-214.

Babson, A.L., Tenney, S.A. and Megraw, R.E. (1970). Clin. Chem. **16**, 39-43.

Churms, S.C., Merrifield, E.H., Stephen, A.M., Wolwijn, D.R., Polson, A., van de Merwe, K., Spies, H.S.C. and Costa, N. (1983). Carbohydr. Res. 113, 339-344.

Ducroo, P. (1987). Liquid Fruit 5,265-269 and 280-281.

Jones, J.K.N. and Tanaka, Y. (1965) in "Methods in Carbohydrate Chemistry" (R.L. Whistler, Ed.) Academic Press, N.Y. Volume V, 74-75.

McCleary, B.V. (1980). Carbohydr. Res. **86**, 97-104.

McCleary, B.V. (1986). Carbohydr. Polym. **6**, 307-318.

McCleary, B.V., unpublished.

Rombouts, F.M., Voragen, A.G.J., Searle-van Leeuwen, M.F., Geraeds, C.C.J.M., Schols, H.A. and Pilnik, W. (1988). Carbohydr. Polym. **9**, 25-47.

Somogyi, M. (1952). J. Biol. Chem. **195**, 19-23.

Tagawa, K. and Kaji, A. (1969). Carbohydr. Res. 11, 293-301.

Voragen, A.G.J., Rombouts, F.M., Searle-van Leeuwen, M.F., Schols, H.A. and Pilnik, W. (1987). Food Hydrocolloids 1, 423-437.

Whitaker, J.R. (1984). Enzyme. Microb. Technol. 6, 341-349.

Part 5

MICROBIAL POLYSACCHARIDES

Fermentation technology of microbial polysaccharides

P. DELEST

Sanofi Bio Industries, 66 Avenue Marcedau, 75008, Paris, France

INTRODUCTION

Polysaccharides are a major constituent of the living world, mostly in plant species. Microbial polysaccharides have long been regarded as a nuisance, an infamous slimy layer associated with rotting and sickness, or, at least a problem in various fermented foods.

The recognition of their usefulness dates back to the 2nd World War with the use of dextran as a blood extender, and the real success came from the pioneering work achieved in Peoria by Miss A. Jeanes and her team, in the early 60's, which led to the Xanthan success story. Viscous strains can be found in all microbial classes, although yeasts are largely underrepresented in this respect.

Polysaccharides are generally divided into three categories : cell wall polysaccharides, intracellular or exocellular.

Cell wall components are generally difficult to extract and thus of little industrial interest except chitin and yeast glycan. Intracellular molecules are also difficult to recover, thus, all efforts have been devoted to exocellular polysaccharides.

The exact role played by polysaccharides in microbial metabolism is not fully understood, and it may depend upon the type of microbe and its physical situation.

They are supposed to create a more favourable environment to the living cell, preserving a high level of hydration, acting as a potential energy reserve and as a barrier against bacteriophages. It is also assumed that they may act as messengers in a host-recognition system. (1)-(2)

All these polysaccharides are water-soluble, generating very viscous solutions. They are potential candidates for a lot of commercial applications, ranging from food products to petroleum recovery and in competition with traditional thickeners such as natural gums (starch, arabic gums,) extract polymers (alginates, pectins) or synthetic products (CMC's, acrylates).

Despite numerous announcements by both scientists and industrialists it is estimated that only approximately fourteen polysaccharides have reached a significant industrial stage.

Xanthan is the best known and surpasses all other in terms of sales. Chitosan, which is presently extracted from shells could be produced by fermentation, and its unusual situation as the only cationic polysaccharide could induce a significant development.

I -SYNTHESIS MECHANISM

Polysaccharide production is a normal feature of the metabolism of the microorganisms studied. However, the mechanisms involved may be substantially different, since two cases are prevalent, exo and endo synthesis.

a) Exo synthesis

This is exemplified by dextran and levan polysaccharides. A simple enzymatic system (glucosyl transferase) is involved and can even work in absence of microorganism. The mechanism has been described by Robyt et al., and appears in fig 1 (3).

In such cases, the structure is rather simple, with obtention of an homopolysaccharide, generally non ionic.

b) Endo synthesis

This is not a perfectly appropriate description, since it is ascertained that, although monomers (or small blocks) are synthetized in the cell, they appear to be assembled at the outer surface involving extrusion sites.

The synthesis of the monomer is rather complicated, and, in the case of Xanthan (4) (fig 2) involves a dozen of different enzymatic steps. A total synthesis in vitro is thus difficult to predict.

Genetic engineering has been largely used to better understand these processes, and improve them. Results obtained are significant, according to studies published by Kelco, Syntro and Synergen (5) (6) (7) among the dozen of international teams who work on Xanthomonas genes. Gene mapping and plasmid construction have allowed design of new bacteria with improved performance, in terms of yield, rheology or better production. However, from a practical point of view, the long term stability of such strains in industrial conditions of production remains to be ascertained. Moreover, regulations are a major obstacle towards industrial use of such strains.

c) Excretion

The original symbiotic role of polysaccharides with their producing organism is a key factor to try to explain the physical structure of these polymers in solution. It is likely that the extrusion sites cited above perform an elongation function by "screwing" blocks generated in the cell to the chain already excreted. It is indeed unlikely that such high molecular weight polymers (a few million Daltons) could pass through the cell membrane. It looks also as if these sites play a predominant role in the fibrillar structure of polysaccharides in solution. They create an intertwisting of macromolecules leading to the well known double or triple helix configuration which is at the origin of the rod-like structure responsible for the high viscosity of most

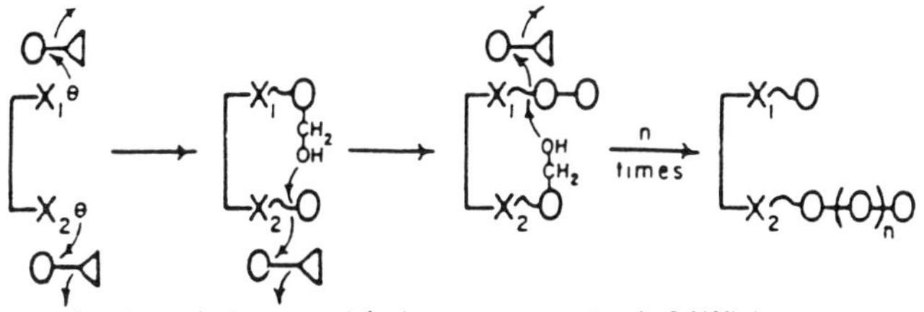

Reaction mechanism proposed for Leuconostoc mesenteroids *B-512F dextransucrase* X_1 *and* X_2 *represent nucleophiles at the active site.* ($\bigcirc$ - $\triangle$) *represents sucrose.* ($\bigcirc$ -) *is a glucosyl group;* (- $\triangle$) *is a fructosyl group; and* (—) *is an α-1,6-linkage*

FIG. 1
(from ROBYT & al - ref. (3)

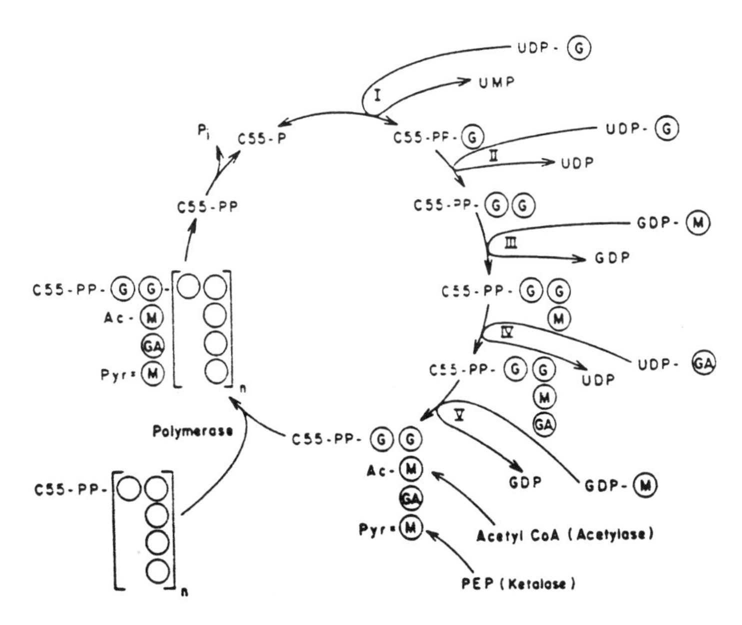

Pathway for xanthan gum biosynthesis. UDPG = uridine-5'-diphosphoglucose; GDPM = guanidine-5'-diphosphomannose; UDP-GA = uridine-5'-diphosphoglucuronic acid; GG = cellobiose; C55 = isoprenoid lipid carrier; PEP = phospho(enol)pyruvate; Acetyl CoA = acetyl coenzyme A; I-V = glycosyltransferases I-IV; Ac = acetate; and Pyr = pyruvate (refs. 1, 4-7).

FIG. 2
(from BETLACH & al - ref. (4)

polysaccharides in aqueous solution.

This excretion layer is characteristic of capsular polysaccharides such as those produced by human pathogenic strains, but appears also clearly from work by Brooker, Brown and Dargent (8) (9) for other industrial strains.

II - FERMENTATION PROCESSES

In this case, biomass is of no direct interest, and microorganisms have to be considered only as a factory. The efficiency of this factory depends upon three factors : the strain, its environment and the technology.

II - 1 - Strain

As explained above, the potential of genetic engineering will be very helpful. But, up to now, industrial strains have been obtained by selection. Industrial protection (by patenting) is extremely important when distinct features can be claimed, and modern techniques of DNA sequencing allow for precise determination of strain origin if some microorganisms, (even dead)are found in the final product.

Strain storage and preservation is a major problem. Long time stability which is absolutely required for a constant quality and performance is not always guaranteed, and wild strains seem to have a plus over manipulated cells. Yeasts and fungi are also normally less prone to degeneration than bacteria. Physical loss (by fire or contamination) is normally avoided by storing parent strains in various independent locations.

II - 2 - Environment

This includes medium and fermentation conditions.

All microorganisms used for polysaccharides are aerobic, and most use plain sugars (glucose, saccharose or starch) as carbon source. During fermentation, all cells are under constant stress to deviate as much as possible their metabolism from growth to metabolite production. This ratio of polysaccharide versus total biomass is one of the key factors for an efficient operation.

Metabolism control is mostly achieved by proper selection of medium components. Salts (especially trace metals) and nitrogen source are the best tools for this control. Nitrogen, which is necessary for the build-up of the proteins can be supplied either by mineral salts (ammonium and nitrates) or by organic sources (corn steep liquor for instance). Such complex mixtures generally bring also some growth factors (like vitamins) but complicate somewhat final separation although they are generally preferred (10). The balance between cell growth and polysaccharide production depends upon the natural cycle of the organism. In case of bacteria, it is sometimes possible to have a dual process ; in a first step, a maximum concentration of cells is reached with a preliminary medium and polysaccharide is then switched on by different conditions. With fungi (Scleroglucan) growth and production are co-current and cannot be separated (11) (12) fig 3.

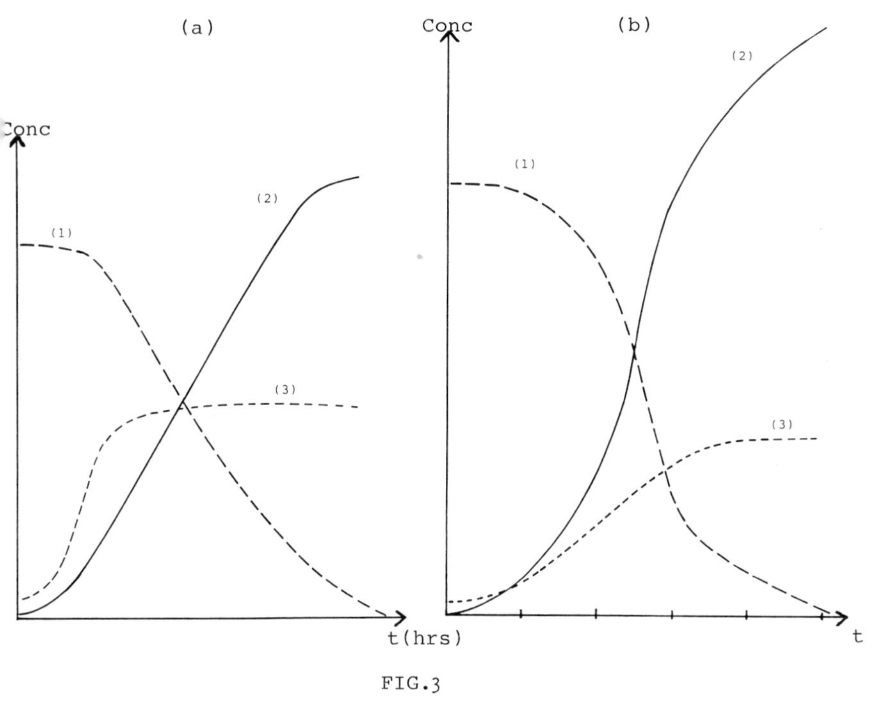

FIG.3

Comparison between Xanthan (a) and Scleroglucan (b)
(1) : Glucose - (2) : Polysaccharide - (3): Biomass

Medium cost is a significant part of total production cost, and sugar concentration is normally adjusted so that it falls as low as possible at the end of the cycle, but total depletion is avoided so that microorganisms do not start re-using polysaccharide as a substrate, which would decrease the overall yield.

Oxygen availability is obviously another key factor, which will be dealt with in the technology section.

Other parameters include temperature and pH control.

Temperature is normally kept constant, by coil exchangers or double-jacket fermentors. It is usually between 20° and 40°C, and a balance must be achieved between the highest production rate, the lowest viscosity and the microorganism survival.

For each strain, there is an optimum pH which is normally constant during the fermentation. Some strains (fungi) self-create the best pH which needs little adjustment, if at all.

II - 3 - Technology

Polysaccharide production requires a high level of sophisticated technology, since mass transfer problems are difficult to solve. The peculiar rheology of such systems (high viscosity, non-Newtonian fluids) have led to a considerable amount of research to improve final concentration which commands the best recovery costs.

Optimisation of sugar yields and fermentation capacity has been tested many times by continuous culture instead of traditional batch operation. However, such processes have been hampered by strain degeneration and contamination problems (13).

There is a direct link between agitation efficiency and oxygen transfer throughout the whole volume of the broth. Air supply at 1.0 VVM in 200 cubic meters fermentors require a lot of energy and significant cost reductions are looked for. Various systems have been tested :

- Thin film reactor (14)
- Broth recirculation with air injection in an outside pump (15) (16)
- Helicoïdal agitation ensuring a proper mixing at low speed, together with a high speed impeller for air solubilisation (17) (18).

In all cases, dissolved oxygen content, or direct oxygen consumption is monitored through O_2probes or effluent gas determination. A lot of improvements remain to be found in this important field of bio-chemical engineering.

Sophisticated programs allow combination of all physico-chemical parameters, including on line fluid viscosity, to optimise fermentation conditions and reduce total fermentor occupancy period.

III - RECOVERY AND PURIFICATION

After a thorough sterilisation, which can be achieved in the fermentor or by an outside continuous UHT apparatus, the highly viscous broth has a polymer concentration ranging from ten to sixty grams/litre. On top of dead cells, it contains all the salts and residues of the fermentation medium which may require a separation, depending upon various end-uses. It can also be necessary to submit the polymer to various physico-chemical treatments (such as deacetylation) to obtain the required performances.

III - 1 -Recovery

The standard commercial form of polysaccharides is a dry, free flowing powder. This is often obtained by precipitation, in a similar manner to the processes used for algal polymers (19) table 1.

For some special applications, or in order to decrease the precipitation cost, the polymer broth can be concentrated by ultra-filtration. Such a treatment eliminates most of the soluble salts, but leaves cell debris and proteinaceous materials in the polymer (20).

TABLE 1

Examples of techniques claimed in the postextraction processing of plant and algal gums

Operation	Example and reference number
Concentration	Evaporation[16] and ultrafiltration[17] of carrageenan
Solids removal	Centrifugation and filtration of carrageenan[16]
Colour removal	Resin treatment and sulphite or hypochlorite bleaching of agar[18]
Freeze thawing	Removal of salts and other soluble matter from furcellaran[19] and agar[18]
Salt precipitation	Pectin with aluminium salt[20] alginate with calcium salt[21]
Alcohol precipitation	Carrageenan[16] or pectin[20] with methanol or propan-2-ol
Electrolytic deposition of charged polymers	Alginate[22]
Drying of extract	Spray or drum drying of pectin[20] and carrageenan[16]

From Smith and Pace (19)

III - 1-a- The most usual technique is to add various solvents (acetone, ethanol and mostly isopropanol) to the broth until polymer separates as a fibrous mass. This mass is recovered, dried and ground to the required size.

All small molecules are dissolved and resulting polymer is substantially purified.

1-b- Other chemical or physical techniques can be used, depending upon polysaccharide nature :

- a substantial pH drop is used to precipitate alginates as alginic acid
- most polymers being anionic can react with polyvalent salts which will induce precipitation. It will be necessary to regenerate the polymer by an acidic solution or a complexing agent. Ca^{++} is largely used for alginates, but similar results can be obtained with xanthan gum, either with Ca or Al salts (21) (22) (23) (24) (25)
- They can also react with cationic organic compounds such as quaternary ammoniums, or with fatty amines (26) (27) (28) (29) (30) (31).
- Phase separation using polyethylene glycol can be used for both anionic (32) or non-ionic polysaccharides.
- A simple drying of the broth, by atomisation or drum drying gives a raw product, only for industrial use.

III - 2 - Purification

For some applications, i.e. food grade products, or enhanced oil recovery, it is desirable to eliminate most of the impurities and/or cell debris. Filtration with DE filter aids is theoretically possible, and sometimes industrially feasible. But it is often blocked by the high viscosity and the non-Newtonian character of the broth. The same reasons prohibit the use of centrifugation, except if a large dilution can be accepted.

In order to by-pass such problems, various solutions have been suggested :
- viscosity decrease by alcohol dilution (33) (34)
- solution heating above transition temperature to decrease viscosity (35)
- adsorption of X.Campestris cell debris on clay (36)

It is also possible to obtain solutions which look clear although dead cells remain present by treatment with various enzymes, essentially proteases (37) (38) (39) (40) (41) (42) (43).

III - 3 - Chemical modifications

It is sometimes desirable to modify the properties of the original polysaccharide. This is obtained by an alteration of the ratio between side chain functionnal groups.

3-a- A single heat treatment of Xanthan broth leads to improved viscosity, in relationship with pyruvate content.

Various conditions have been patented, indicating that original structure is somewhat dependent upon fermentation conditions (44) (45) (46) (47) (48) (49).

3-b- Deacetylation by alkali treatment modifies chain association, and this allows for better reactivity :
- with Xanthan gum, synergism with locust bean gum gives firm gels (50)
- with Gellan gum, similar treatment creates stronger gels and also a synergism with gelatin takes place (51) (52)

A global estimation of polysaccharide cost distribution, can be summarized as follows :

- Raw materials	25 %
- Energy	15 %
- Precipitation/Drying	15 %
- Manpower, G & A	45 %
	100 %

III - 4 - Safety

On top of traditional safety procedures, emphasis has to be put on studies about unwanted release of living organisms. It has a particular importance, when a microbe is a plant pathogen which is the case for many polysaccharides. Maximum care has to be exercised to sterilise all effluents, solids gases, and liquids.

TABLE 2

Name	Micro Organism	Ionic Character	Backbones & Substituents	Features
Curdlan	A. faecalis Agrobacterium spp	non ionic	Linear 1-3 glucose	Non thermo reversible gels
Baker's Yeast glucan (63)	Saccharomyces cerevisiae	non ionic	Glucose - Mannose	Low viscosity Food additive
Dextran	Leuconostoc spp	non ionic	1-3 glucose - 1-6 glucose	Low viscosity
Pullulan	Aureobasidium pullulans	non ionic	Linear 1-4/1-6 glucose	Flocculating agent
Schizophyllan	Schizophyllum commune	non ionic	Linear 1-3 glucose 1-6 branches	High viscosity
Scleroglucan	Sclerotium rolfsii	non ionic	Linear 1-3 glucose 1-6 branches	High suspending power
Alginate	Az. vinelandii/Ps. aeruginosa	strongly anionic	Linear 1-4 mannuronic acid	Identical to Algal product
Gellan (53) (54)	Aureomonas spp	anionic	Linear 1-4 glucose Rhamnose and glucuronic acid	Strong gels after deacetylation - Food additive
Rhamsan S 194 (55)	Alcaligenes spp	anionic	Very similar to Gellan	High suspending power, salt compatibility
Succinoglycan (58) (59) (60)	Pseudomonas Rhizobium Agrobacterium	slightly anionic	Galactose/glucose Glucose + acetyl - pyruvic and succinic acid	High viscosity Pseudo plasticity
Xanthan	X.campestris	anionic	Linear 1-4 glucose Mannose + glucuronic acid	High viscosity-Food additive Pseudo plasticity Acid stability
Welan S 130 (56) (57)	Alcaligenes spp	anionic	Similar to Gellan Rhamnose and mannose	Suspending power Thermal stability
Zanflo (61) (62)	Soil Bacteria	anionic	Glucose - Galactose - Fucose Glucuronic acid	Compatibility with cationic dyes
Chitosan (64) (65)	Various Fungi	cationic	D - Glucosamine	Complexation of metals

IV - MICROBIAL POLYSACCHARIDES : A REVIEW

Table 2 lists the fourteen polysaccharides which have been found significantly developed and ref (53) to (65) describe their structures and properties.

They represent a small tip of the wide range of molecules tested, and a smaller part of the potential. Hundreds of microorganisms have been studied for their capability to produce such polymers, and genetic engineering could create thousands more.

However, before a new polysaccharide participates significantly to the world market, it has to pass through a score of excruciating tests, in terms of cost, performance and acceptability.

The preliminary screening is obviously based on rheological performance. All major needs of industry are now fulfilled by natural, synthetic or microbial gums and a new entrant has to demonstrate either superior properties in a specific field, or a price advantage for the same properties.

Another primary possibility is a completely new field of application as for instance antitumor action of Schizophyllan and Scleroglucan (66) (67) (68)or a source of a new molecule, as for instance deoxy sugars (69)

The second step is based on preliminary cost estimates, based on yield, final polymer content, fermentation time and separation cost.

Somewhere along the decision process, strain stability and safety have to be assessed.

If rheological and physico chemical properties look suitable, then it comes the point to evaluate the potential as a food additive. This is a very difficult decision, since the cost and the time involved in an FDA (or any other major country) petitioning represent an exceptional burden, which amounts to many millions of dollars. Such a cost, linked with the requirement of "insufficiently filled need" limits very strictly the number of successful candidates.

CONCLUSION

Microbial polysaccharides have a definite number of advantages over their traditional competitors.

Being industrial, they are liberated from seasonal and climate constraints, they can be produced anywhere the industrial needs and capability exists, and their system of production guarantees an excellent product consistency.

Bio engineering represents a major opportunity to obtain new strains, more efficient, more specific. But the next challenge to the many dedicated researchers who are excited by polysaccharides, would be to be able to custom design a polymer according to technical needs. Up to now, polymer structure has been determined after polymer has been found adapted to miscellaneous applications. We can wonder whether, in a similar way to proteins, computer-aided design could lead to a de novo structure prediction which could be translated into a biochemical synthetic path way. Cloning of the required genes into a suitable host microorganism, selected for instance for non-pathogenicity and environmental safety would give birth to the perfect combination. Dream or reality of the 2000's ?

References

1. J.F. Wilkinson (1958) Bacteriol. Rev; 22 46-73
2. I.C.M. Dea, E.R. Morris, D.A. Rees & al (1977) Carbohydr. Res. 57 249-272
3. J.F. Robyt, B.K. Kimble, T.F. Walseth (1974) Arch. Biochem. Biophys. 165 634-640
4. M.R. Betlach, M.A. Capage, D.H. Doherty, R.A. Hassler, N.M. Henderson, R.W. Vanderslice, J.D. Marrelli , and M.B. Ward (1987), Industrial Polysaccharides : Genetic Engineering, Structure/Property Relations and Applications, Ed. by M. Yalpani, Elsevier NL
5. N.E. Harding, J.M. Cleary, D.K. Cabanas, I.G. Rosen, K.S. Kang (1987) J. Bacteriol 169 2854-2861
6. M.R. Betlach, D.H. Doherty, R.W. Vanderslice, Synergen, International Patent Publication Number WO 87/05937
7. L. Thorne, L. Tansey, T.J. Pollok (1987), J. Bacteriol 169 3593-3600
8. B.E. Brooker (1976), Surface coat transformation and capsule formation by Leuconostoc mesenteroides NCDO 523 in the presence of sucrose. Archives of Microbiology 111 99-104
9. R. Dargent (1977) Ph.D. Thesis, Univ. P. Sabatier-Toulouse, Fr.
10. S. Baig, M.A. Gadeer, S.R.A. Shamsi (1985) Pakistan J. Sci. Ind. Res 28 (2) 111-115
11. A.L. Compere, W.L. Griffith (1981) Adv. Biotechnol. Int. Ferment. Symp. 6 (1980) 441-446
12. A. Mulchandani, J.H.T. Luong, A. Leduy (1988) Biotechnology and Bio Engineering 32 639-646
13. D. Kidby, P. Sandford, A. Herman, M. Cadmus (1977) Appl. Environ. Microbiology 33 (4) 840-845
14. B.J. Lipps (1966), Esso Prod. Res. Cy. US 3 281-329
15. N.M.G. Oosterhuis, K. Koerts (1985), Coöp. Ver. Suiker Unie U.A. EP 185407
16. N.M.G. Oosterhuis (1987) Coöp. Ver. Suiker Unie U.A. EP 249288
17. A. Vincent (1985) Top. Enzyme Ferment. Biotechnol. 10 109-145
18. G.L. Solomons, G.A. Legrys (1978), Rank Hovis McDougal Ltd GB 1 584 103 (DE 2 823-923)
19. I.H. Smith, G.W. Pace (1982) J. Chemical Tech. Biotechnol. 32 119-129
20. L.Ho, R.J. Taylor (1983), Pfizer Inc. EP 69 523
21. C.L. Mehltretter (1965) , Biotechnol. Bioeng. 7 171-175
22. P.T. Cahalan, J.A. Peterson, D.A. Arudt (1977), General Mills Chemicals Inc. US 4 053 699
23. G.A. Towle (1977), Hercules Inc. US 4 051 317
24. H. Passedouet, F. Saint-Pierre (1972), Melle Bezons Fr. 2 106 731
25. G. Brigand, H. Kragen (1980), Ceca S.A. Fr 2 442 955
26. S.P. Rogovin, W.J. Albrecht (1964), The Secretary of Agriculture USA, US 3 119 812
27. G.P. Lindblom, J.T. Patton (1964), Jersey Production Research Cy., US 3 163 602
28. J.W. Gill, P.G. Lim (1969), Hercules Inc., US 3 422 085
29. C. Schroeck (1981), Lubrizol Corp., US 4 254 257

30. W.A. Jordan, W.H. Carter (1975), General Mills Chemicals Inc., US 3 928 316
31. E. Fischer, M. Schlingmann, W. Duersch, R. Rothert (1985) Hoechst A.G., DE 3 325 224
32. J.J. Cannon (1988), Pfizer Inc. EP 266 163
33. S.P. Rogovin, R.F. Anderson, M.C. Cadmus (1961), J. Biochem. Microbiol. Technol. Eng. 3 51-63
34. M.B. Inkson, C.K. Wilkinson (1980), Tate and Lyle, Talres EP 28 446
35. F.E. Corley, J.B. Richmon (1979), Merck and Co Inc., US 4 135 979
36. M.K. Abdo (1973), Mobil Oil Corp. US 3 711 462
37. Ministery of Agriculture and Forestry of Japan (1973) JP 48/33085
38. G.T. Colegrove (1976), Merck and Co. Inc. US 3 966 618
39. J.W. Drozd, A.J. Rye (1986), Shell Int. Res. Maatschappig B.V. EP 184 882
40. J.P. Gozoard, A. Jarry, A. Luccioni (1985), Rhône Poulenc Spécialites Chimiques Fr 2 551 087
41. T.J. Holding, G.W. Pace (1980) Tate and Lyle Ltd GB 2 065 689
42. M. Rinaudo, M. Milas, N. Kohler (1982) Institut Français du Pétrole Fr 2 491 494
43. I.H. Smith, K.C. Symes (1983) Kelco Biospecialites Ltd EP 78 621
44. P. Colin, V. Guibert (1969) Melle Bezons Fr 1 575 756
45. B. Eyssautier (1988) Sanofi Elf Bio Industries Fr 2 606 423
46. L. Ho, R.L. Miller (1980) Pfizer Inc. Fr 2 440 992
47. K.S. Kang, D.B. Burnett (1977) Merck and Co Inc. Fr 2 318 926
48. J.J. O'Connel (1967) Kelco Co. US 3 355 447
49. B. Seeber, P. Loewe, B. Richter, R. Zeitfuchs (1981) DD 147 948
50. G. Brigand (1979) Ceca S.A. Fr 2 401 951
51. K.S. Kang, G.T. Colegrove, G.T. Veeder (1982) Merck and Co Inc. US 4 326 052
52. J.L. Shim (1985) Merck and Co Inc. US 4 517 216
53. M.A. O'Neill, R.R. Selvendran, V.J. Morris (1983) Carbohydr. Res. 124 123-133
54. M.S. Kuo, AKJ. Mort, A. Dell (1986) Carbohydr. Res. 156 173-187
55. P.E. Jansson, B. Lindberg, J. Lindberg, E. Maekawa, P.A. Sandford (1986) Carbohydr. Res. 156 157-163
56. P.E. Jansson, B. Lindberg, G. Widmalm, P.A. Sandford (1985) Carbohydr. Res. 139 217-223
57. M.A. O'Neill, R.R. Selvendran, V.J. Morris, J. Eagles (1986) Carbohydr. Res. 147 295-313
58. L.P.T.M. Zevenhuizen (1987) in "Industrial Polysaccharides" S.S. Stivala, V. Crescenzi, I.C.M. DEA ed. 45-68
59. T. Harada, A. Amemura (1979) Carbohydr. Res. 77 285-288
60. A.J. Clarke-Sturman, D. Den Ottelander, P.L. Sturla (1988) Prepr. Am. Chem. Soc. Div. Pet. Chem. 33 11 25-29
61. K.S. Kang, G.T. Veeder, D.D. Richez (1977) in "Extracellular Microbial Polysaccharides" P.A. Sandford, A. Laskin ed. ACS Symp. Ser. 45 211-219
62. K.S. Kang, P. Kovacs (1974) 4th Int. Cong. Food Sci. Technol. Madrid

63.F. Paul, A. Morin, P. Monsan (1986) Biotech. Adv. 4 245-249
64.R.A.A. Muzzarelli (1985) in "New Developments in Industrial Polysaccharides", V. Crescenzi, I.C.M. Dea, S.S. Stivala Ed 207-231
65.W.J. McGahren, G.A. Perkinson, J.A. Growich, R.A. Leese, G.A. Ellestad (1984) Process Biochem. 19 88-90
66.J. Hamuro, G. Chihara (1973) Nature 245 (5419) 40-41
67.P.P. Singh, R.L. Whistler, R. Tokuzen, W. Nakahara (1974) Carbohydr. Res. 37 (1) 245-7
68.K. Nakajima, Y. Hirata, H. Uchida etc. (1982), Takara Shuzo Ltd Fr 2 501 232
69.H. Graber, A. Morin, F. Duchiron, P. Monsan (1988) Enzyme Microb. Technol. 10 (4) 198-206

Science, structure and applications of microbial polysaccharides

V.J.MORRIS
AFRC Institute of Food Research, Norwich Laboratory, Colney Lane, Norwich NR4 7UA, UK

ABSTRACT

Xanthan gum is the only microbial polysaccharide used widely by the food industry. Applications include use as a thickening and suspending agent and, in combination with certain galactomannans, use as a thermo-reversible gelling agent. Gellan gum has recently received food approval in Japan and is awaiting food approval in the U.K., Europe and USA. The major application is as a multi-purpose gelling agent although gellan-gelatin mixtures may have uses in encapsulation. The chemical structures and physical chemical properties of xanthan and gellan will be reviewed in an attempt to explain their functional properties. New microbial polysaccharides with potential future use in food will be described briefly.

INTRODUCTION

Bacterial polysaccharides provide a potential source of new biopolymers for the food industry. The major advantages over currently used plant and animal biopolymers are reproducible physical and physical-chemical properties, and a stable cost and supply. In addition such polymers may offer improved or new functional properties.

At present xanthan is the only bacterial polysaccharide used widely by the food industry. Gellan gum is likely to be the second bacterial polysaccharide with major food applications. The functional properties of these polymers, relevant to their food applications, will be described. Studies on the molecular properties of these polymers will be reviewed in an attempt to provide a molecular description of their functional properties. Finally, potential new food applications of bacterial polysaccharides will be discussed briefly.

XANTHAN

Xanthan is the extracellular polysaccharide produced by the aerobic fermentation of *Xanthomonas campestris* in batch culture(1). Commercial quantities were first produced by Kelco(2) in 1961 and USA food approval was obtained in 1969.

Xanthan is used widely as a thickener and stabiliser, and in certain formulations as a gelling agent.

Functional Properties

The requirements for a thickening and suspending agent are for a polymer which, at low concentrations, produces a high viscosity permitting suspension of particulates, and inhibition of emulsion droplet association. Upon shaking, stirring or pouring (shearing) the viscosity should decrease markedly (shear-thin) but recover completely upon removal of shear. There is often a need for the viscosity to be retained over extremes of pH, temperature and ionic strength.

Xanthan satisfies these conditions. In the absence of shear, and at sufficiently high polymer concentration and ionic strength, xanthan dispersions exhibit weak gel-like behaviour (3-5). This structure is disrupted upon shearing. At low shear rates the viscosity is high and the samples show reversible shear-thinning behaviour. Insensitivity to pH, temperature and ionic strength can be achieved by appropriate choice of experimental conditions.

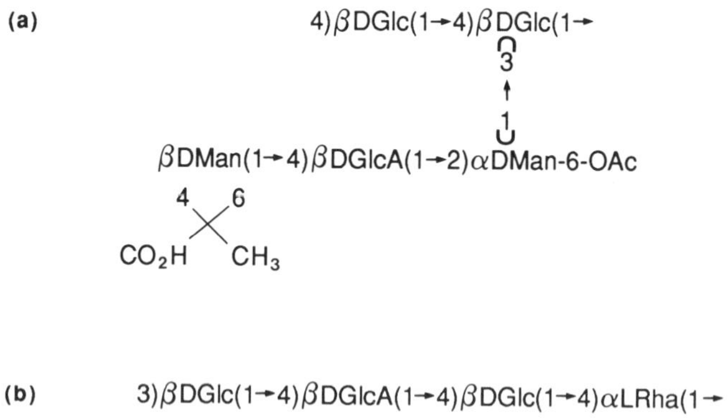

(a)

4)βDGlc(1→4)βDGlc(1→
3
↑
1
βDMan(1→4)βDGlcA(1→2)αDMan-6-OAc
4 6
CO_2H CH_3

(b)

3)βDGlc(1→4)βDGlcA(1→4)βDGlc(1→4)αLRha(1→

Figure 1. Chemical repeat units for (a) xanthan and (b) gellan.

Gelation of xanthan dispersions can be enhanced(6) by addition of divalent, trivalent or borate ions although only the former are likely to be of importance in food applications.

Xanthan will form strong transparent thermoreversible gels with certain glucomannans(7) (Konjac mannan) and certain galactomannans(8-12) (carob and tara). Enzymic treatment of guar with an α-galactosidase produces a modified polysaccharide which also gels with xanthan(13-15). Gelation of the mixtures will occur under conditions for which the individual components do not gel.

Molecular Properties

The chemical structure of xanthan consists of a (1→4) linked β-D-glucan (cellulosic) backbone substituted through position 3 on alternate glucose residues with a charged trisaccharide side-chain(16,17) (fig 1a). Various substrains of X.campestris have been identified which yield xanthans with differing levels of pyruvate and acetate substitution(18-20). The biosynthetic pathway for xanthan production is known(21) and the possibility exists for genetically controlling the carbohydrate structure and the substitution pattern of the polymer.

The effect of sidechain substitution on the geometry of the backbone has been studied(22-24) by X-ray fibre diffraction and molecular modelling. The normal 2-fold cellulosic ribbon-like structure is altered to a 5-fold helix (pitch 4.7nm). A single 5-fold helix has been suggested from modelling studies but the X-ray data is not good enough to preclude possible double-helical structures.

Is the xanthan helix retained in solution? If so, can this explain the unusual rheology of xanthan samples when compared with other simple cellulose derivatives? Evidence for a conformational transition of the molecule has come from qualitative studies using proton nmr(25,26), optical rotation(25,27-29), circular dichroism(25,26), differential scanning calorimetry (29), light scattering(28,29) and stopped-flow optical rotation (29). Optical rotation changes, induced by variations in ionic strength or temperature are primarily ascribed to changes in

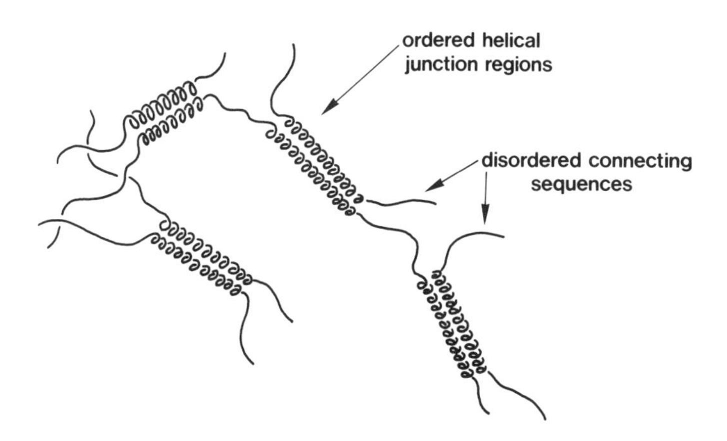

Figure 2. Model for xanthan self-association proposed by Norton et al.,(29) in order to account for "weak-gel" formation.

backbone geometry resulting from a helix-coil transition(25,26). Changes in the environment of the acetate residue, measured by nmr(25,26) and circular dichroism(25,26), are taken to favour binding of the sidechain to the helical backbone. The transition mid-point temperature (T_m) varies with the type and concentration of counterions in a manner consistent with the stabilisation of a charged helix.

Helix formation will obviously stiffen the molecule. The question is how stiff are xanthan molecules and how closely do they approach a rigid rod structure? Both elongational flow-(5) and electric-birefringence studies(5,30) of xanthan samples suggest an extended molecular structure. The rotatory relaxation for xanthan(5,30) are ~ 10^3 times those observed for typical cellulose derivatives(31,32). This suggests a persistence length q (a measure of stiffness) of ~ 100-300nm - about ten times as stiff as a cellulosic backbone. Light scattering studies by Burchard et al.,(33,34) (q ≈ 100nm) and Sato et al., (35-37) (q =120 ± 20nm) have confirmed that xanthan is one of the stiffest natural biopolymers. At low molecular weight (eg < 10^5) xanthan will behave as a rigid rod but at higher molecular weights it will resemble a stiff extended coil.

The stiffness of related molecules such as DNA, collagen or schizophyllan is often attributed to the formation of a multistranded helical structure. Thus is xanthan a single helix stabilised by sidechain - backbone binding, or a double helix stabilised by inter-chain binding and possibly sidechain-backbone binding? Experimental evidence favouring single and double helices has been reviewed elsewhere(38). The debate is current and active. At present the experimental evidence seems to favour a 'dimeric' or double helical model. Strong evidence for double helix formation comes from electron microscopy (39-41), estimates of mass-per-unit length from light scattering and combined sedimentation, viscosity and light scattering data. Stopped-flow optical rotation(42) exploits the sensitivity of the helix-coil transition to ionic strength in order to investigate the kinetics of the process. Such studies(29,43,44) have previously been taken to support a single helical model although more recent work(45) suggests that dimerisation may occur under certain experimental conditions.

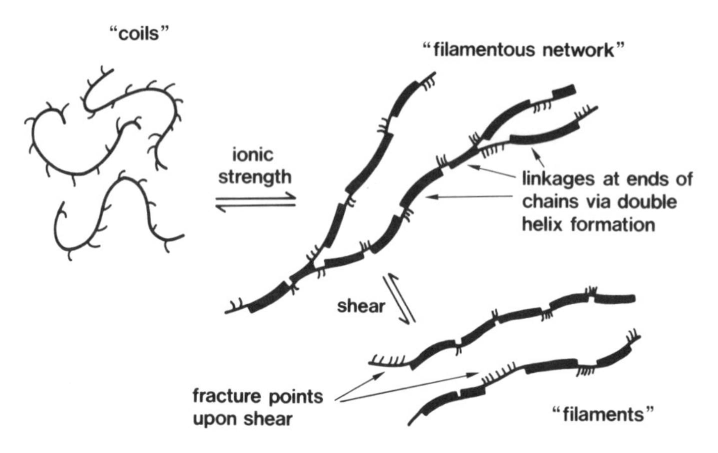

Figure 3. Filamentous model for "weak-gel" formation by xanthan self-association.

Weak Gels

There are two 'solution' regimes for xanthan rheology. In the disordered state(46,47), at low ionic strength, the polymeric solutions shear thin due to molecular distortion. Studies on solutions of the ordered (helical) polymer will show a higher degree of shear thinning due to molecular reorientation as evidenced by flow birefringence studies(5).

However, the weak gel-like properties of commercially thixotropic dispersions cannot be explained simply in terms of a dilute solution of stiff macromolecules. Southwick and co-workers(48-52) proposed that the thixotropy was due to entangled motion of stiff chains in a 'semi-dilute' solution. Theories of particle motion in 'semi-dilute' solutions suggest(53,54) that rotatory relaxation times should increase markedly with increasing concentration. Experimentally no such concentration dependence was observed(3,5,30).

Several studies(3,5,30,55) have stressed the need to distinguish between xanthan dispersions and true xanthan solutions. Heating thixotropic xanthan dispersions in the presence of 4Murea and then recooling to room temperature yields rheological properties appropriate to solutions of xanthan helices(4,5). This seems to suggest that xanthan thixotropy arises due to self-association of the molecules.

Two models for xanthan weak gels have been proposed. The model due to Norton et al.,(29) is based on the idea of single helices. Conformational ordering of the helix is considered to be incomplete. Molecules contain helical and disordered regions. The helical regions are considered to associate forming a network (fig 2). The mechanism of association is unspecified. The second model(12,56) is based on a double helical model for xanthan (fig 3). The primary mechanism is considered to be end-to-end association, via double helix formation, into fibrous aggregates. The fibres may thicken and bifurcate leading to weak network formation. In both examples network formation is favoured by conditions stabilising the xanthan helix accounting for the experimental insensitivity to pH, temperature and ionic strength.

Strong Gels

Synergistic mixtures of xanthan and carob are used by the food industry as thermoreversible gelling agents. Gelation also occurs with the galactomannans(8-12) tara and enzymically modified guar,(13-15) and with the glucomannan Konjac mannan (7). In all cases the mixtures gel under conditions for which the individual components will not gel. X-ray fibre diffraction studies(11-12) of the mixed gels and chromatographic studies (57,58) suggest that gelation arises due to specific binding between the two different polysaccharides.

A model by Dea et al.,(7,9,) proposed that the xanthan helix was bound to unsubstituted regions of the galactomannan backbone. The stoichiometry and size of the suggested junction zones is obscure and this model cannot easily predict new synergisms. Further, on the basis of this model, mixing of the galactomannan with xanthan in the helical form should lead to gelation. However, mixing experiments(11-12) suggest that gelation only occurs if the xanthan helix is denatured prior to cooling the mixture. This fact, together with an analysis of the X-ray data, suggest(11,12) that intermolecular binding involves a cocrystallisation of denatured xanthan chains within the galactomannan crystallite. Binding results from the stereochemical similarity of the cellulosic xanthan backbone and the galactomannan mannan backbone. $\beta(1\rightarrow4)$ linked glucans and mannans differ only in the orientation of the C(2) hydroxyl residues and form similar ribbon-like structures. The mixed crystallites probably act as strong junction zones consolidating the weak xanthan network (fig 3).

Gelation of xanthan-galactomannan mixtures is sensitive to the mannose-galactose (MG) ratio and xanthan-guar mixtures do not gel. Accommodation of xanthan sidechains into the galactomannan lattice will distort the "a"-dimension of the lattice. Sufficient defects, resulting from absent galactose residues, will also be necessary to accommodate xanthan sidechains. Thus enzymic modifications of guar to increase MG should and does promote gelation.

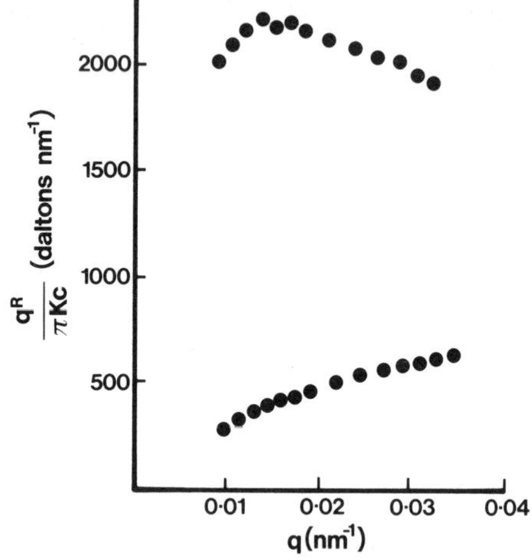

Figure 4. Holtzer plot for gellan samples (concentration 2mgcc^{-1}) after filtration through (a) 3μm filters and (b) 0.45μm filters. (TMA gellan in 0.075M TMA Cl).

Glucomannan backbones contain both glucose and mannose and can also bind to the xanthan backbone.

GELLAN

Gellan is the extracellular polysaccharide produced commercially by the aerobic fermentaion(59-61) of Auromonas elodea (previously called Pseudomonas elodea) in batch culture. The polysaccharide was first reported by Kelco in the late 1970's. Successful toxicity trials have been carried out and gellan received food approval in Japan in 1988. Food approval has been applied for in the USA, UK and Europe.

Functional Properties

Gellan can be employed as a broad spectrum gelling agent in a wide variety of food applications(62). Recipes using gellan normally require polymer concentrations substantially less than those required for algal or plant polysaccharides. Cold-setting gels can be prepared by adding monovalent or divalent cations to sufficiently concentrated aqueous gellan dispersions. The gels obtained are transparent and thermoreversible. The 'melting-point' of the gels increases with increasing ionic strength. Thus gellan can be used to prepare cold-setting or thermo-setting, thermoreversible or thermoirreversible gels, and is to be marketed as a multi-purpose gelling agent.

Molecular Properties

Gellan is a linear anionic heteropolysaccharide with a tetra-saccharide repeat unit(63,64) (fig 1b). The native poly-saccharide is esterified on the (1→3) linked glucosyl residue. Esterification amounts to incomplete C2 substitution with L-glycerate and ~50% C6 substitution with acetate(65). The commercial product "Gelrite" is subjected to alkaline conditions during preparation in order to produce a deesterified product.

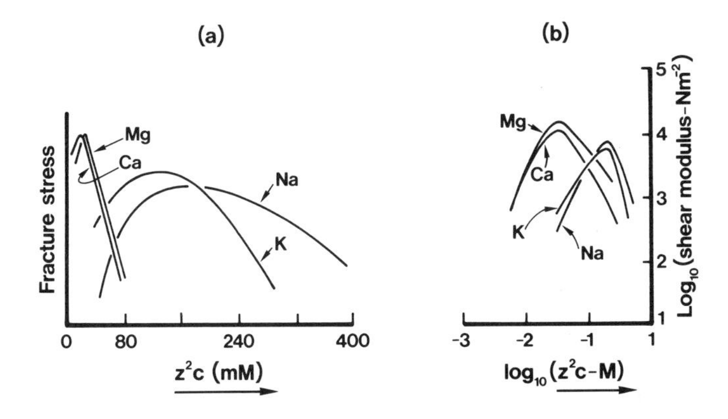

Figure 5. Effect of cations on (a) fracture stress and (b) low deformation storage modulus of gellan gels. Polymer concen-tration 0.6%. Z is cation valency, c cation concentration (a) mM and (b) M.

X-ray fibre diffraction studies of deesterified gellan(66-70) give highly crystalline patterns for the monovalent salt forms and poor patterns for the divalent salt forms. The highly crystalline patterns suggest a 3-fold helical structure with a pitch of 2.82nm: the axial rise per chemical repeat is approximately half the extended length of the repeat unit, suggestive of a double helix. Latest modelling studies suggest a left-handed 3-fold double helix(71).

Bulky tetramethyl ammonium (TMA) counterions inhibit gelation. Solutions of TMA gellan in TMA salts have been in-vestigated by a number of physical techniques including osmo-metry(72), differential scanning calorimetry(73,74), light scattering(72,75-77), viscosity(72-74), optical rotation(72-74) and circular dichroism(72-74) in order to probe the conformation, shape and size of the macromolecules.

Does the gellan double helix exist in solution? Viscosity, differential scanning calorimetry, optical rotation and circular dichroism measurements provide evidence for a conformational transition(72-74). Viscosity measurement(72) as a function of ionic strength revealed three types of behaviour: at low ionic strengths a decrease in viscosity consistent with shrinkage of a charged polymer was observed. At higher ionic strengths the

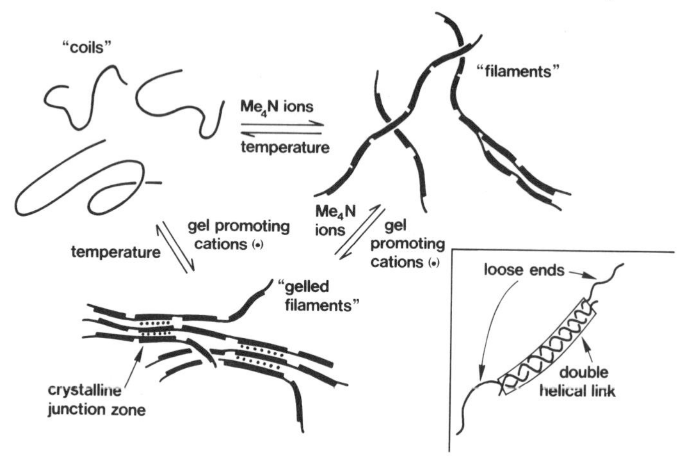

Figure 6. Model for the gelation of gellan gum. Insert shows an incomplete double helical segment obtained after breaking the filaments.

viscosity increased indicative of a conformational change to a stiffer molecular structure. At higher ionic strength, polymer aggregation was observed. Temperature dependent changes in viscosity at a fixed ionic strength provided further evidence (73,74) for a conformational transition. At higher polymer concentrations (>0.08%) evidence of thixotropy and polymer aggregation was obtained. Optical rotation, circular dichroism and differential scanning calorimetry provide further evidence (73,74) for a conformational transition of TMA gellan in TMA salt solutions. The ordered (helical) form can be obtained at room temperature at sufficiently high ionic strength. The disordered form can be obtained by heating such solutions or by a suitable reduction in ionic strength.

Light scattering studies(72,75-77) have been performed on TMA gellan under conditions favouring the ordered helical structure. Clarification of certain such solutions(72,76,77) involved filtration through 0.45μm filters. Static light scattering data is given in table 1. The calculated mass per unit length M_L = 710 daltons nm^{-1} is quite close to the expected value (766 daltons nm^{-1}) for the double helix. The estimated persistence length q $\approx$ 160nm is indicative of a stiff macromolecule. Filtration of gellan samples through 3μm filters yields clean solutions but evidence(75) of excess scattering at low angles is suggestive of polymer aggregates. Fig 4 shows Holtzer plots of TMA gellan in

0.075M TMACl obtained after filtration through 3μm and 0.45μm filters. These data suggest the presence of elongated aggregates (fibres) which are broken down into smaller units upon filtration through 0.45μm. Static light scattering on 0.45μm filtered solutions yields ($\langle M \rangle_w$ = 0.17x10^6 daltons, $\langle Rg^2 \rangle_z$ = 88nm) in reasonable agreement with the data in table I. The Holtzer plot (fig 8) suggests $M_L \approx$ 620 daltons nm^{-1} for the short segments. M_L values lower than the expected value of 766 daltons nm^{-1} could indicate dangling chain ends (fig 7).

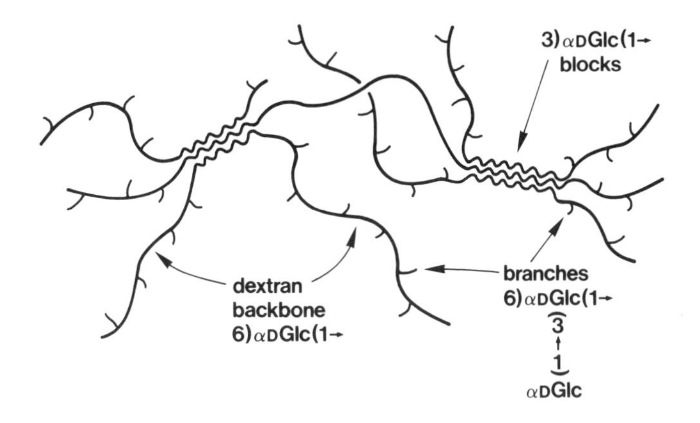

Figure 7. Effect of (1→3) α D Glc blocks on the gelation of dextran molecules.

Gels and Gelation

Gelation is sensitive to the type and concentration of cations in the medium(38). Fig 5 show the effects of common monovalent and divalent cations on the fracture strength and low deformation modulus of the gels. At equivalent ionic strengths (of added salts) divalent cations produce stronger gels with higher elastic moduli and higher melting-points. Detailed studies(70) of the effects of a range of monovalent and divalent cations suggests that gelation may be sensitive to cation type as well as valency. Measurements of selectivitiy coefficients for both monovalent and divalent cations have failed to reveal evidence for specific site binding of cations(72). With increasing ionic strength both the fracture strength and the modulus pass through a maximum and then decrease at higher ionic strength. At high ionic strengths the gels became noticeably turbid. Clarity can be restored by addition of sugars which match the refractive index of the aggregates.

Gellan will gel(72) upon addition of HCl. Gelation occurs at pH's below the pK_a of the carboxyl group and these acid set gels are the strongest gels produced at any polymer concentration.

A model for the gelation of gellan based on the above observations is shown in fig 6. In the TMA form inter-chain crystallisation is taken to be suppressed. Induction of the ordered helical form in solution is considered to involve end-to-end association of gellan molecules, via double helix formation, into fibrous structures. Filtration is considered to disrupt the fibres into smaller units (insert fig 6). Thickening or bifurcation of fibres would arise by the end of a gellan molecule linking to a middle region of another chain. Gel promoting cations are considered to induce inter-fibre crystallisation of short segments of the fibres.

The degree of esterification affects the rheology of gellan gels(78). Native gellan forms soft elastic gels. Progressive deesterification results in increasingly hard brittle gels. X-ray fibre diffraction patterns(66) of native gellan show poor crystallinity but good polymer alignment. Progressive deesterification(66) promotes crystallinity but doesn't alter the shape or size of the gellan helix. Esterification does not appear to inhibit helix formation but the larger glycerate substitution may inhibit helix-helix packing.

NEW BACTERIAL POLYSACCHARIDES

At present there are still market opportunities for introducing new polysaccharides. However, the cost is prohibitive both in time and money. In addition, licensing authorities are requiring that any such new products should be shown to be safe as well as necessary.

Encapsulation

The objective is to devise methods for entrapping volatiles, protecting delicate chemicals and for concentrating valuable components as slurries or powders.

Gellan-gelatin mixtures will form complex coacervates and the process can be developed to encapsulate oil soluble components (78) (eg vitamins).

Cyclodextrins(79), the cyclic (1→4) α D glucans, offer a more subtle method of encapsulation by entrapping molecules within a hollow central cavity. Cyclodextrins were first used in food in Hungary and Japan in the late 1970's, and food approval is being sought in certain European countries and the USA. They can be used(79,80) directly to stabilise flavours, fragrances, colours and pigments and/or to mask odours and taints. Polymerised cyclodextrins have been used as extraction media most noticably in the debittering of grape fruit juices.

Certain _Rhizobium_, _Agrobacterium_ and _Xanthomonas_ species produce cyclosophorans (cyclic β (1→2) D glucans) with degrees of polymerisation between 18 and 28. Modelling studies(83) suggest ring sizes overlapping and exceeding the cyclodextrins. Production was limited by the need to isolate these compounds from the viscous extracellular slimes (EPs) produced by the bacteria. EPs^- mutants facilitate cyclorsophoran production and methods are available for purifying and isolating individual ring sizes. Cyclosophorans are potential future encapsulating agents.

Fermented Foods and Ingredients

Another potential application for bacterial polysaccharides is in the development of naturally textured broths, produced by 'food grade microorganisms', and their use as food products, or as ingredients in processed foods. _Lactobacillus hilgardi_, isolated from starter cultures used to prepare sugary kefir, has been found to produce a gel-forming polysaccharide. Analysis shows it to be a dextran containing α (1→3) D glucose residues. Blocks of α (1→3) D glucose are known to form ribbonlike structures, insoluble at normal pH, the presence of which would serve to cross-link the dextrans (fig 7).

Optimisation of the production of other polysaccharides produced by lactic-acid bacteria used in dairy fermentations offers the possibility of developing new products such as naturally thickened yoghurts.

Vinegar is an important constituent of many acid-based foods which are often thickened, stabilised or gelled by addition of xanthan. Acetan produced(84-86) by certain Acetobacter duplicate xanthan rheology offering the prospect of developing naturally thickened or blended vinegars.

CONCLUSIONS

An attempt has been made to review the molecular basis for the functionality of xanthan and gellan. Possible new applications for bacterial polysaccharides in the areas of encapsulation and textured fermentation products have been discussed.

REFERENCES

1. Jeans, A., Rogovin, P., Cadmnus, M.C., Silman, R.W. and Knutson, C.A. (1974) ARS USDA Report ARS-NC-51.
2. Baird, J R., Sandford, P.A. and Cottrell, I.W. (1983) Bio Technology, 1, 778.
3. Morris, V.J. Franklin, D. and l'Anson, K. (1983) Carb. Res., 121, 13.
4. Frangou, S.A. Morris, E.R., Reed, D.A. Richardson, R.K. and Ross-Murphy, S.B. (1982) J. Polym. Sci., Polym. Lett., 20, 531.
5. Ross-Murphy, S.B., Morris, V.J. and Morris, E.R. (1983) Faraday Symp. Chem. Soc., 18, 115.
6. Sanderson, G.R. (1981) Brit. Polym. J., 13, 71.
7. Dea, I.C.M. and Morrison, A. (1975) Adv. Carb. Chem. Biochem., 31, 241.
8. Rocks, J.K. (1971) Food Technol., 27, 46.
9. Dea, I.C.M. and Morris, E.R. (1977) in Extracellular Microbial Polysaccharides, (eds. Sandford, P.A. and Laskin, A), p.174, ACS Symp. Ser. 45, ACS, Washington, USA.
10. Cairns, P., Miles, M.J., Morris, V.J. and Brownsey, G.J. (1986) Food Hydrocolloids, 1, 89.
11. Cairns, P., Miles, M.J. and Morris, V.J. (1986) Nature, 322, 89.
12. Cairns, P., Miles, M.J., Morris, V.J. and Brownsey, G.J. (1987) Carb. Res., 160, 411.
13. McCleary, B.V. and Neukom, H. (1982) Progr. Food Nutrition Sci., 6, 109.
14. McCleary, B.V., Amado, R., Waibel, R. and Neukom, H. (1981) Carb. Res., 92, 269.
15. McCleary, B.V., Dea, I.C.M., Windust, J. and Cooke, D. (1984) Carb. Polym. 4, 253.

16. Jansson, P-E, Kenne, L. and Linelberg, B. (1975) Carb. Res. 45, 275.
17. Melton, L.D., Mindt, L., Rees, D.A. and Sanderson, G.R. (1976) Carb. Res. 46, 245.
18. Sandford, P.A., Pittsley, J.E., Knutson, C.A., Watson, P.R., Cadmurs, M.C. and Jeanes, A. (1977) in Extracellular Microbial Polysaccharides, (eds. Sandford, P.A. and Laskin, A.) P.192, ACS Symp Ser. 45, ACS, Washington, USA.
19. Sandford, P.A., Watson, P.R. and Knutson, C.A. (1978) Carb. Res. 63, 253.
20. Smith, I.H., Symes, K.C., Lawson, C.J. and Morris, E.R. (1981) Int. J. Biol. Macromol., 3, 129.
21. Betlach, M.R., Capage, M.A., Doherty, D.H., Hassler, R.A., Henderson, N.M., Vanderslice, R.W., Marrelli, J.D. and Ward, M.B. (1987) in Industrial Polysaccharides, genetic engineering, structure/property relations and applications, (ed. Yalpini, M.) P.35, Elsevier, London, U.K.
22. Moorhouse, R., Walkinshaw, M.D. and Arnott, S. in Extracellular Microbial Polysaccharides, (eds. Sandford, P.A. and Laskin, A), p.90, ACS Symp. Ser. 45, ACS, Washington, USA.
23. Moorhouse, R., Walkinshaw, M.D., Winter, W.T. and Arnott, S. (1977) in Cellulose Chemistry and Technology (ed. Arthur, J.D.) p.133.
24. Okwyama, K., Arnott, S., Moorhouse, R., Walkinshaw, M.D., Atkins, E.D.T. and Wolf-Ullish, C.H. (1980) in Fiber Diffraction Methods (eds. French, A.D. and Gardener, K.D.) p.411, ACS Symp. Ser. 141, ACS, Washington, USA.
25. Morris, E.R., Rees, D.A., Young, G., Walkinshaw, M.D. and Darke, A. (1977) J. Mol. Biol., 110, 1.
26. Powell, D.A. (1979) in Microbial Polysaccharides and Polysaccharases (eds. Berkeley, R.E.W., Gooday, G.W. and Ellwood, D.C.) P.117, Academic Prerss, London, UK.
27. Holzworth, G. (1976) Biochem., 15, 4333.
28. Milas, M. and Rinaudo, M. (1979) Carb. Res. 76, 189.
29. Norton, I.T., Goodall, D.M., Frangou, S.A. Morris, E.R. and Rees, D.A. (1984) J. Mol. Biol., 175, 371.
30. Morris, V.J., l'Anson, K. and Turner, C. (1982) Int. J. Biol. Macromol., 4, 362.
31. Foweraker, A.R. and Jennings, B.R. (1975) Polymer 16, 720.
32. Foweraker, A.R. and Jennings, B.R. (1977) Makromol. Chem., 178, 505.
33. Burchard, W. (1983) Faraday Symp. Chem. Soc. 18, 233.
34. Coviello, T., Kajiwara, K., Burchard, W., Dentini, M. and Crescenzi, V. (1986) Macromolecules 19, 2826.
35. Sato, T., Norisuye, T. and Fujita, H. (1984) Polym. J., 16, 341.
36. Sato, T., Kojima, S., Norisuye, T. and Fujita, H. (1984) Polym. J. 16, 423.
37. Sato, T., Norisuye, T. and Fujita, H. (1984) Macromolecules, 17, 2696.
38. Morris, V.J. (1987) in Food Biotechnology (eds. King, R.D. and Cheetham, P.S.J.) p.193, Elsevier Appl. Sci., London, UK.
39. Holzworth, G. and Prestridge, E.B. (1977) Science, 197, 757.
40. Stokke, B.T., Elgsaaeter, A., Skjak-Braek, G. and Smidsrφd, O. (1987) Carbohydr. Res., 160, 13.

41. Foss, P., Stokke, B.T. and Smidsrφd, O. (1987) Carbohydr. Polym. 7, 421.
42. Goodall, D.M. and Cross, M.T. (1975) Rev. Sci. Instrum., 46, 391.
43. Rees, D.A. (1981) Pure Appl. Chem., 53, 1.
44. Norton, I.T., Goodall, D.M., Morris, E.R. and Rees, D.A. (1980) J. Chem. Soc. Chem. Commun. 545.
45. Goodall, D.M., Jones, S. and Norton, I.T. (this volume).
46. Thurston, G.B. (1981) J. Non-Newtonian Fluid Mechanics. 9, 57.
47. Thurston, G.B. and Pope, G.A. (1981) J. Non-Newtonian Fluid Mechanics, 9, 67.
48. Jamieson, A.M., Southwick, J.G. and Blackwell, J. (1983) Faraday Symp. Chem. Soc., 18, 131.
49. Southwick, J.G., McDonnell, M.E., Jamieson, A.M. and Blackwell, J. (1979) Macromolecules 12, 305.
50. Southwick, J.G., Lee, H., Jamieson, A.M. and Blackwell, J. (1980) Carbohydr. Res., 87, 287.
51. Southwick, J.G., Jamieson, A.M. and Blackwell, J. (1981) Macromolecules 14, 1728.
52. Jamieson, A.M., Southwick, J.G. and Blackwell, J. (1982) J. Polym. Sci., Polym. Phys. 20, 1513.
53. Doi, M. and Edwards, S.F. (1978) J. Chem. Soc., Faraday Trans. II, 74, 560.
54. Doi, M. and Edwards, S.F. (1978) J. Chem. Soc., Faraday Trans. II, 74, 918.
55. Dintzis, F.R. Babcook, G.E. and Tobin, R. (1970) Carbohydr. Res. 12, 257.
56. Morris, V.J. (in press) Food BioTechnology.
57. Cheetham, N.W.H., McClearly, B.V., Teng, G., Lum, F, and Maryanto (1986) Carbohydr. Polym. 6, 257.
58. Cheetham, N.W.H. and Mashimba, E.N.M. (1988) Carbohydr. Polym. 9, 195.
59. Kaneko, T. and Kang, K.S. (1979) Abstracts Ann. Meeting Amer. Soc. Microbiol., 101.
60. Kang, K.S., Veeder, G.T. Mirrasoul, P.J., Kanecko, T. and Coltrell, I-W. (1982) Appl. Environ. Microbiol., 43, 1086.
61. Kang, K.S. and Veeder, G.T. (1982) U.S. Patent 4,326,053.
62. Sanderson, G.R. and Clark, R.C. (1983) Food Technol. 37, 63.
63. O'Neill, M.A., Selvendren, R.R. and Morris, V.J. (1983) Carbohydr. Res. 124, 123.
64. Jannson, P-E, Lindberg, B. and Sandford, P.A. (1983) Carbohydr,. Res. 124, 135.
65. Kuo, M.S., Mort, A.J. and Dell, A. (1986) Carbophydr. Res. 156, 173.
66. Miles, M.J., Morris, V.J., and O'Neill, M.A. (1984) in Gums and Stabilisers for the Food Industry: 2, Application of Hydrocolloids, (eds. Phillips, G.O., Wedlock, D.J. and Williams, P.A.) p485, Pergamon Press, Oxford, U.K.
67. Carroll, V., Chilvers, G.R., Franklin, D., Miles, M.J., Morris, V.J. and Ring, W. (1983) Carbohydr. Res. 114, 181.
68. Carroll, V., Miles, M.J. and Morris, V.J. (1982) Int. J. Biol. Macromol. 4, 432.
69. Attwool, P.T., Atkins, E.D.T., Upstill, C., Miles, M.J. and Morris, V.J. (1986) in Gums and Stabilisers for the Food Industry: 3, (eds. Phillips, G.O., Wedlock, D.J. and Williams, P.A.), p135, Elsevier Applied Sci. London, UK.

70. Attwool, P.T. (1987) PhD thesis, Bristol University.
71. Chandrasekaran, R., Puigjaner, L.C., Joyce, K.L. and Arnott, S. (1988) Carbohydr. Res. 181, 23.
72. Grasdelen, H. and Smidsrφd, O. (1987) Carbohydr. Polym. 7, 371.
73. Crescenzi, V., Dentini, M., Coviello, T. and Rizzo, R. (1986) Carbohydr. Res. 149, 425.
74. Crescenzi, V., Dentini, M. and Dea, I.C.M. (1987) Carbohydr,. Res. 160, 283.
75. Chapman, H.D., Chilvers, G.R., Miles, M.J. and Morris, V.J. (1988) in Gums and Stabilisers for the Food Industry: 4 (eds. Phillips, G.O., Wedlock, D.J. and Williams, P.A.) p147, IRL Press, Oxford, Washington, U.S.A.
76. Dentini, M., Coviello, T., Burchard, W. and Crescenzi, V. (1988) Macromolecules, 21, 3312.
77. Denkinger, P., Burchard, W. and Kunz, M. (1989) J. Phys. Chem. 93, 1428.
78. Chilvers, G.R. and Morris, V.J. (1987) Carbohydr. Polym. 7, 111.
79. Szejtli, J. (1986) in Gums and Stabilisers for the Food Industry: 3 (eds. Phillips, G.O., Wedlock, D.J. and Williams, P.A.p351, Elsevier Appl. Sci, London, UK.
80. Pszczola, D.E. (1988) Food Technology (Jan.) 96.
81. Hisamatsu, M., Amemura, A., Koizumi, K, Utamura, T. and Okado, Y. (1983) Carbohydr. Res. 12, 31.
82. Amemura, A. and Cabrera-Crespo, J. (1986) J. Gen. Microbiol. 132, 2443.
83. Palleschi, A. and Crescenzi, V. (1985) Gaz. Chim. Ital. 115, 243.
84. Couso, R.O., Lelpi, L. and Danbert, M.A. (1987) J. Gen. Microbiol. 133, 2123.
85. Morris, V.J., Brownsey, G.J., Cairns, P., Chilvers, G.R. and Miles, M.J. (1989) Int. J. Biol. Macromolecules (in press).
86. Morris, V.J., Brownsey, G.J., Harris, J.E., and Stevens, B.J.H. (this volume).

A differential scanning calorimetric study of the conformational transition of xanthan in aqueous NaCl

SHINICHI KITAMURA, TAKASHI KUGE and BJØRN T.STOKKE[+]

Department of Agricultural Chemistry, Kyoto Prefectural University Shimogamo, Kyoto 606, Japan

[+]*Division of Biophysics, University of Trondheim, N-7034 Trondheim, Norway*

ABSTRACT

The thermally- induced conformational transition of xanthan polysaccharide in aqueous salt solutions has been studied by high-sensitivity differential scanning calorimetry(DSC). It was shown by electron microscopic analyses and light scattering measurements that our samples (degree of pyruvate substitution ca. 0.45) used for DSC, behaved as double strand in high ionic strength and at low temperature. The temperature at half completion of the conformational transition for these samples increased linearly with the logarithm of the cation(Na^+) concentration. The specific enthalpy for these samples, essentially independent of salt concentration above 20 mM, was 8.6 $\pm$ 0.7 J/g. Application of the Manning polyelectrolyte theory suggests that the double strand of xanthan is not separated into single coils even at the temperature where the endothermic peak is finished.

INTRODUCTION

Xanthan is an extracellular polysaccharide produced by the bacterium *Xanthomonas campestris* and composed of a cellulosic backbone and a trisaccharide side chain (β-D-mannopyranosyl-(1→4)-β-D-glucuronic acid-(1→2)-α-D-mannopyranoside-6-*O*-acetate) attached at C-3 to alternate glucose residues of the main chain. The terminal mannose residue of the side chain may have a pyruvate residue linked to the 4- and 6-positions.

It has been shown that xanthan in aqueous salt solutions undergoes a thermally-induced order-disorder transition, which has been detected by a number of physical methods, and that the transition temperature depends on the strength and nature of the counterion in the solution, and pyruvate and acetate contents of the side chain. However, the nature of the ordered structure of xanthan and of the thermally-induced conformational changes have been controversial despite extensive investigation.

The purpose of the present study is to investigate thermally-induced conformational transition of xanthan in aqueous salt solution over the wide range of salt concentration by high-sensitivity differential scanning calorimetry(DSC). We used double helix samples which were well characterized by electron microscopic observation and light scattering measurements. This enabled us to investigate the conformational changes of the double stranded helix with increasing temperature.

MATERIALS AND METHODS

Purified xanthan sample(NX) was prepared from a commercial "Xanthan gum powder" (Sigma Chemical Co., Ltd.) according to the method of Holzwarth(1). Low-molecular-weight samples(15-SX,80-SX) were prepared from NX by sonication of 300 mL of an aqueous xanthan solution (2.5 mg/mL) containing 1 volume % acetone with a sonicator(Ultrasonic,Ltd.,USD-350) for 15 and 80 min at 300 W and 20 kHz(2). Sonicated samples were repurified by the method of Paradossi et al.(3) to remove metal eroded from the sonicator probe. Their weight-average molecular weights(Mw) determined in 0.1 M aqueous NaCl by light scattering method(4) were 20.3, 5.7, and 3.7 X 10^5 for NX, 15-SX, and 80-SX, respectively. The degree of substitution with pyruvate groups per pentasaccharide unit, was determined by the lactate-dehydrogenase method (5) to be 0.44, 0.47, and 0.43 for NX, 15-SX, and 80-SX, respectively. These prepared solutions were stored under nitrogen gas as salt-free stock solutions.

The stock solutions of xanthan samples filtered through 0.45 μm membrane filters were concentrated to 5-8 mg/mL. The concentrated solutions were dialyzed against the aqueous salt solutions at 4 °C for 24-48 h before DSC measurements. DSC measurements were performed with a DASM-4 scanning microcalorimeter at a scan rate of 0.5 K/min. The calorimetric enthalpy was evaluated by planimeter integration of the DSC curves. Chemical baselines were drawn according to the method of Takahashi and Sturtevant(6).

Electron microscopic analyses were performed by the method described previously (7,8).

RESULTS AND DISCUSSION

Electron micrographs for NX, SX-15, SX-80 , which were obtained from xanthan solutions under ordering conditions, showed that every sample consisted of uniformly thick and convoluted linear chains and branched chains with very low frequency. The mass per unit length of the linear chains was determined from the data of weight-average contour length and Mw to be 2000, 2380, and 1880 nm^{-1} for NX, 15-SX, and 80-SX, respectively. These values happen to be very close to 2000 nm^{-1} obtained by Paradossi and Brant (3),and 1940 nm^{-1} by Sato et al(9) both from light scattering data. X-ray diffraction studies (10) show that crystalline xanthan could be described by a 5_1 double-stranded helix, in which each chain contains five pentasaccharide repeating units in a pitch of 4.7 nm, which corresponds to 1900 nm^{-1} for the double helix xanthan.Thus, it is concluded that our samples behave as double strands in high ionic strength and at low temperature.

Figure 1 shows DSC curves for the conformational transition of SX-15 in aqueous NaCl at various concentrations from 0.005 M to 0.1 M. Similar DSC curves were obtained for SX-15 in aqueous KCl and for NX and 80-SX in aqueous NaCl. It is seen that the transition temperature increases with increasing salt concentration. The plot of the reciprocal of the transition

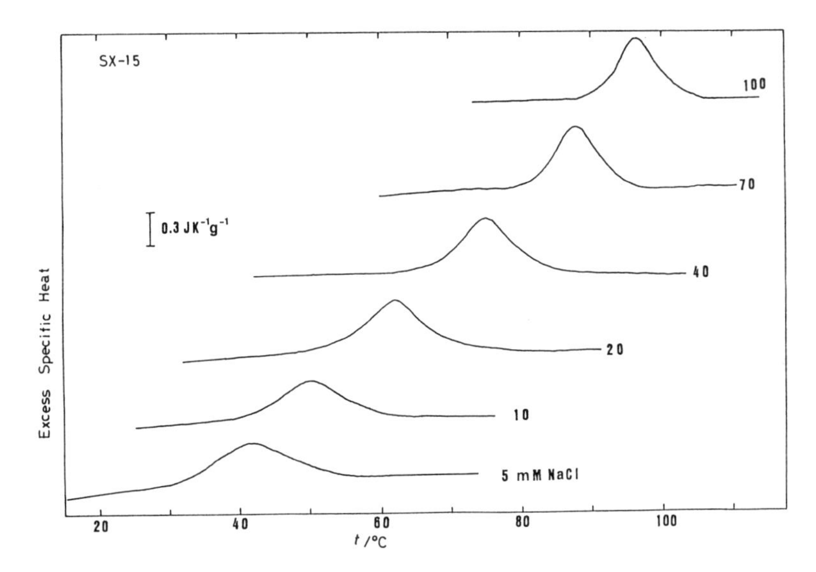

Figure 1. DSC curves for the order-disorder transition of xanthan (15-SX) in NaCl aqueous solutions.

temperature (Tm; defined as the temperature at half completion of the transition) versus logarithm of the sodium concentration in the solutions gave a linear relation for NX, SX-15, and SX-80 in aqueous NaCl.

$$10^3/T_m = 3.46 - 0.161 \ln[Na^+] \qquad (1)$$

The specific enthalpy (ΔH) for NX, SX-15, and 80-SX essentially independent of salt concentration above 20 mM, was 8.6 ± 0.7 J/g (8 measurements; mean ± standard deviation.) The value is much lower than those reported by Cesáro et al.(11) and greater than that by Norton et al.(12).

The observed enthalpy for the present system could be interpreted in terms of the Manning theory (13), since xanthan, which has charged carboxyl groups, can be regarded as a linear polyelectrolyte. We estimated the theoretical value of ΔH for the double helix - single coil transition based on the following assumptions. For the xanthan sample in the present study, about half of the side chains carry a pyruvate group in addition to the glucuronate residue, and thus the average charge per repeating unit is 1.50. The average separation of charges along the helix axis(b) for the double-strand model is 0.31 nm, since the chain has a helix pitch of 4.7 nm, with five pentasaccharide repeating units per turn. For the single coil, the fully-extended chain geometry of cellulosic backbone was assumed, and b is calculated to be 0.72 nm. Thus the double helix-coil transition is regarded as the transition between conformations with charge densities across the counterion condensation threshold. Using an equation (14) applied to this case and the slope of eq (1), the enthalpy change per carboxylic group for the transition is estimated to be 17.8 kJ/mol which is equal to 27.5 J/g. This value is much larger than that of observed ΔH

This discrepancy may be explained by assuming that the double strand of xanthan is not separated to single coils even at high temperature where the endothermic peak ends. Recently Liu *et al.* (15) investigated Mw, radius of gyration, and intrinsic viscosity of xanthan in 0.01 M aqueous solution at 25 and 80 °C. From the result obtained, they suggested that the ordered conformation of xanthan in 0.01 M aqueous NaCl at 25 °C is dimerized, double helices,and when the temperature is raised from 25 to 80 °C, the dimersdo not dissociate to single chains.

ACKNOWLEGEMENTS

The authors would like to thank Professor D.A.Brant and Dr. L.S.Hacche for light scattering measurements of our samples and for helpful discussion, and Professor J.M.Sturtevant of Yale University for the use of DASM-4 microcalorimeter. One of the authors(S.K) is indebted to Sugiyama Chemical & Industrial Laboratory for the research grant.

REFERENCES

1. Holzwarth,G.(1976) Biochemistry, 15, 4333-4339.
2. Igushi,T., Kitamura,S., Kuge,T. and Hiromi, K. (1986) Makromol.Chem.Rapid Commun.7, 477-503.
3. Paradossi,G. and Brant,D.A. (1982) Macromolecules, 15, 874-879.
4. Hacche,L.S., Washington,G.E., and Brant,D.A. (1987) Macromolecules, 20, 2179-2187.
5. Duckworth,M., and Yaphe,W. (1970) Chem. Ind. (London) 13, 747-745.
6. Takahashi,K. and Sturtevant,J.M. (1981)Biochemistry,200,6185-6190.
7. Stokke,B.T., Elgsaeter,A., and Smidsrød,O.(1986) Int.J. Biol. Macromol.8,217-225.
8. Stokke,B.T., Smidsrød,O. and Elgsaeter,A.(1989)Biopolymers 28, 617-637.
9. Sato,T., Norisuye,T. and Fujita,H.(1984) Polym.J. 16,341-350.
10. Okuyama,K.,Arnott,S.,Moorhouse,R.,Walkinshaw,M.D.Atkins,E.D.T. and Wolf-Ullish,C.(1980)ACS Sym.Ser.141,411-427.
11. Paoletti,S. Cesàro,A. and Delben,F. (1983)Carbohydr. Res. 123 173-178.
12. Norton,I.T., Goodall,D.M., Frangou,S.A. Morris,E.R. and Rees,D.A.(1984) J. Mol. Biol. 175,371-394.
13. Manning,G.S. (1978) Q.Rev.Biophys.11, 179-246.
14. Paoletti,S. Smidsrød,O, and Grasdalen,H. (1984)23 1771-1794.
15. Liu,W.,Sato,T.Norisuye,T. and H.Fujita,H.(1987)Carbohydr. Res. 160,267-281.

The functional properties and applications of microbial polysaccharides – a supplier's view

GEORGE R.SANDERSON

Kelco Division of Merck and Co. Inc., San Diego, California, USA

ABSTRACT

Hydrocolloids are used in a wide variety of food and industrial products. The selection of a particular hydrocolloid by the end user for a specific application depends on its ability to provide the required functional properties and cost. Cost considerations extend beyond price per pound, and include cost effectiveness, or price per product unit, quality, batch-to-batch consistency, and uninterrupted availability. For a supplier to obtain business, he must convince the potential user that his hydrocolloid meets these two criteria. For a supplier to expand his business, he must displace an established alternative hydrocolloid with his own by showing superiority in terms of these criteria. This is frequently done by demonstrating superiority in functional properties. The ability to pursue this approach implies that, in many applications, the hydrocolloid currently used does not provide all of the properties desired. This is indeed the case, and some functional requirements are not met by any of the established hydrocolloids. Microbial polysaccharides, present and future, offer the opportunity to obtain these desired properties.

Against this background, the commercial development of xanthan gum, gellan gum, welan gum, rhamsan gum, dextran and curdlan is described.

INTRODUCTION

A key employee in the food industry is the product developer. Unfortunately, to the uninitiated, the name conjures up the image of an individual who, almost at random, selects ingredients from his laboratory shelf in an unstructured attempt to create a new delight for the consumer. The reality is quite different, and it is the product developer's job to use his knowledge of the many ingredients available to the food industry to create new products which are not only acceptable to the consumer but also meet the many technical, financial and marketing constraints imposed by his company. This is a daunting proposition and the success rate, in terms of new products reaching the marketplace, is extremely low, in the order of one for every hundred concepts initially developed. The enormity of the product developer's task is somewhat alleviated, however, by the fact that his efforts usually focus on one type of food, for example soups, ice-cream, or beverages.

Suppliers of food ingredients also rely heavily on product developers because a standard method of illustrating the benefits and features of ingredients is through the development of food products. In the case of hydrocolloids, the products in which they are used are diverse, encompassing almost all processed foods. Thus, unlike his counterpart in the food industry, the product developer working for the hydrocolloid supplier must be conversant with the science and technology of many types of food systems. This can be best achieved through the use of specialists who deal with only a few selected application areas. The challenge of these specialists is ultimately to demonstrate to the end user that the hydrocolloid he is promoting meets the end user's product and process constraints.

In a sense, the product developer in the food industry and the hydrocolloid technologist have the same objective. They both, for different reasons, wish to develop a successful product. This can only be achieved through a strong relationship between the end user and the supplier. This allows the supplier to teach the 'tricks of the trade' in handling the hydrocolloid and the user to provide specific guidelines on process and product development. When such a relationship does not exist, the user frequently encounters problems handling the hydrocolloid while the supplier often finds himself trying to solve perceived rather than real problems. This is a fundamental issue. Successful applications of hydrocolloids are only developed by full cooperation between customer and supplier. Since there is invariably a strong personal element in this relationship, it is sometimes the case that hydrocolloids are not purchased for technical reasons alone. This sometimes explains why different hydrocolloids from different suppliers are used in the same application.

What exactly is the role of the supplier in his relationship with the food industry? Most importantly, he must be capable of supplying on demand and of consistently high quality the hydrocolloids upon which the industry depends. This is no easy task and requires the allocation of considerable resource to ensure that procurement of raw materials, isolation and purification of the hydrocolloid, quality control and shipment proceed at pace with industry demand. Secondly, it is the supplier's primary responsibility to develop an understanding of how his products function and where they may be used and to pass this information on to potential customers. It is sometimes said that the customer develops the applications for hydrocolloids. While this is true in some cases, the applications can only be developed through first having an understanding of the properties of the hydrocolloid. This understanding should come from the supplier. In the context of product understanding, it would be remiss not to acknowledge the important contributions made by the academic community. Such contributions promise to be even more important in the future in determining hydrocolloid functionality and utility.

Currently, there are a wide variety of hydrocolloids available to the food industry and also many hydrocolloid suppliers, some of whom provide similar products. This gives rise to a highly competitive market. As a result, it is difficult, but by no means impossible, for individual suppliers to achieve growth. This can be done by expanding the size of the overall market or by gaining market share at the expense of a competitor. Both approaches require the supplier to establish to the end user's satisfaction that his hydrocolloid best meets product and process requirements. While functional performance is the overriding criterion of utility, cost is also important. By cost is meant cost effectiveness or cost of hydrocolloid per batch of end-product as opposed to cost per pound. Table I reveals an interesting relationship between the relative volumes of hydrocolloids sold and relative prices per pound. Although the less expensive hydrocolloids are sold in greater volume, not insignificant amounts of the more expensive gums are also purchased, lending support to the importance of in-use performance characteristics and cost effectiveness. The ability to obtain business by replacing an established hydrocolloid implies that the latter is not entirely satisfactory in terms of either performance or cost, or both. Since, in many applications, the ideal hydrocolloid is expected to provide many beneficial features relating to dispersibility, hydration, in-process compatibility and end-product stability, it is not too difficult to appreciate that existing hydrocolloids often fall short of expectation. In other words, there are many situations in which there is a need for new hydrocolloids to provide requirements not fully met by those currently on the market. This has led to the development of a number of new polysaccharides from bacteria, sometimes referred to as biogums.

Table I. U.S. MARKET FOR FOOD GUMS

Gum	Volume (lbs x 10^{-6})	Price ($/lb)	Gum	Volume (lbs x 10^{-6})	Price ($/lb)
Guar	12	0.5	Pectin	3.7	3.0
Carboxymethyl-cellulose	12	1.5	Locust bean	3.0	6.0
Arabic	12	2.0	Agar	1.2	7.0
Xanthan	6.9	6.1	Methylcellulose	0.5	3.0
Carrageenan	6.0	3.0	Tragacanth	0.4	12
Algin	5.3	5.0			

Data compiled in 1983 from Chemical Marketing Reporter

DEXTRAN

Dextran, the first microbial polysaccharide to achieve commercial significance, is produced by a variety of microorganisms and also by enzymic synthesis. Dextran formation by bacteria is sometimes undesirable and can cause problems during wine processing and sugar production. Plaque formation on teeth, an occurrence familiar to all of us, involves the synthesis and deposition of dextran. It has been suggested that no polysaccharide has been studied more than dextran and produced fewer applications (1). While this may be true in terms of research on synthesis of the polymer and chemical modification, it is certainly not the case in rheological terms. Like other structurally complex polysaccharides, dextran has proved less than exciting as a substrate for such evaluations. Despite its lack of industrially important rheological features, dextran is, in some respects, a supplier's dream. For the applications in which it is used, it commands a high value and is sold by the gram rather than the pound. Although some applications of dextran in foods have been described, its important commercial outlets are not related to foods. These include its use as a blood plasma extender and a medium for chromatography. The latter application was developed by the Swedish company, Pharmacia, who discovered in the late fifties that useful molecular sieves could be obtained by cross-linking dextran under alkaline conditions with reagents such as epichlorohydrin (2). Following this initial work, the company has developed a highly successful venture in biotechnology, and the tradename, SEPHADEX*, is now well known. Their approach, which has been to extend and modify their basic invention to meet market identified needs, is one that is practiced by several of the hydrocolloid suppliers. However, it should be recognized that pursuit of such an approach is more difficult with the food industry as the target, since the opportunities to develop improved products through chemical modification are not available unless subsequent food approval is obtained for the new material, which is both time consuming and expensive.

XANTHAN GUM

Around the same time that the early work on the use of dextran as a base material for molecular sieves was being conducted, the USDA discovered another interesting microbial polysaccharide, B-1459 (3). This product came to be known as xanthan gum and was commercialized by Kelco beginning in the early sixties. Xanthan gum is now well known and used in the food industry on a worldwide basis.

Table II. KEY PROPERTIES OF XANTHAN GUM

- High viscosity at low shear rates
- Significant yield value
- High pseudoplasticity i.e. highly "shear thinning"
- Stable viscosity over a wide pH and temperature range
- Maintenance of viscosity in the presence of salts
- Resistance to enzymic degradation
- Synergistic viscosity increase with guar gum
- Gel formation with locust bean gum.

Table II shows the key properties of xanthan gum. Although some would consider that such a simple list greatly oversimplifies a much more complex situation, introduction of this basic information to the food industry has provided a clear understanding of xanthan gum's properties, an essential prerequisite for developing the applications. From these properties, many of the applications are obvious. The best example is salad dressings. These require a stabilizer which is stable in the presence of salt and acid and also able to maintain the stability of the oil-in-water emulsion for up to one year at room temperature. Xanthan gum not only meets these requirements admirably, it also provides, through its pseudoplastic rheology, the flow properties desired in this type of product. One of xanthan gum's most important properties, its ability to stabilize suspensions, emulsions and foams, originates from the solid-like structure it displays in the absence of shear, sometimes reflected in the nebulous term, yield value. This structure is readily detected by an instrument such as a rheometer. Figure 1 shows how the elastic modulus, a measure of structure in solution and hence ability to provide dispersion stability, compares for xanthan gum and carboxymethylcellulose. The reason why xanthan gum is such a good stabilizing agent is apparent.

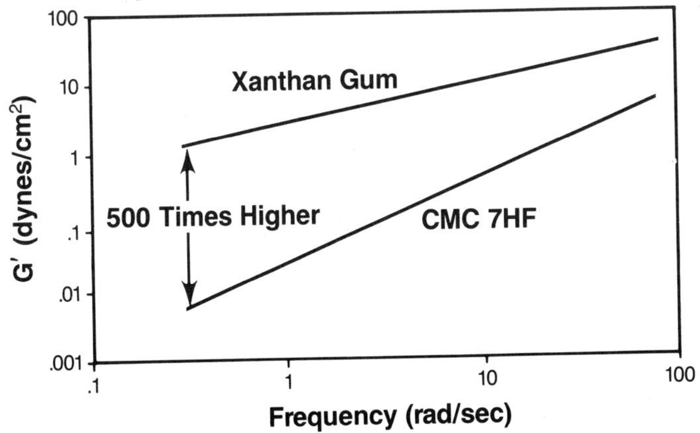

Figure 1. Effect of frequency on the elastic moduli of xanthan gum and carboxymethylcellulose (0.20% in 0.05M Na+).

Although an understanding of rheological properties is essential in developing the applications for a particular hydrocolloid, it is important to appreciate that rheological considerations derived from simple systems are sometimes not sufficient to establish utility. In other words, in some products, what might be expected from xanthan gum's basic rheology is not realized in practice. In these situations, establishment of xanthan gum as the hydrocolloid of choice is the responsibility of the product developer who must use his experience to manipulate the total system to achieve the desired result. Similarly, a great deal of insight would have been required to predict from xanthan gum's rheology its frequent use as a key component in, for example, dry mixes for cakes.

As hydrocolloids mature in the marketplace, and xanthan gum is no exception, new applications, obvious or obscure, become harder to find. A valid approach to developing new applications is to analyze where the product is currently deficient and come up with variants of the original product to overcome these deficiencies. This can provide genuinely new application opportunities or lead to replacement of established hydrocolloids in some existing applications. This approach has led to the development of differentiated xanthan gums such as clear xanthan gum, smoothflow xanthan gum (4), which has more Newtonian rheology than regular xanthan gum at low shear rates, and salt soluble xanthan gum, which can be hydrated directly in hostile environments such as soy sauce. Progress in this direction is thwarted by lack of a complete understanding of xanthan gum at the molecular level. The literature contains ample evidence of the controversy that still reigns over the exact conformation of xanthan gum in solution and the factors affecting conformation. Only when the answers to these questions are obtained will the full potential of xanthan gum be realized.

At this point, it is appropriate to emphasize that the supplier's view of hydrocolloid applications does not always coincide with the applications that are eventually developed. The real clues to where the new applications lie are in the food industry. Much of the skill in developing new applications is the supplier's ability to identify problems in the industry for which his particular hydrocolloids provide a solution.

WELAN GUM AND RHAMSAN GUM

The utility of dextran and, in particular, xanthan gum stimulated a great deal of interest in the use of bacteria to produce industrially useful hydrocolloids. Many microbial polysaccharides have been isolated since the early sixties (5) and, in the course of a ten year screening program, Kelco identified more than two thousand organisms capable of producing such materials. As might be expected, many of these biogums were too similar to xanthan gum to merit commercialization. The three exceptions are S-130 (welan gum), S-194 (rhamsan gum) and S-60, gellan gum. Interestingly, all three of these biogums have the same backbone, composed of a tetrasaccharide repeating unit consisting of glucose, glucuronic acid, glucose and rhamnose. Each differs in the nature of the substituent attached to the backbone. Rhamsan gum has a disaccharide side chain, welan gum a monosaccharide, and gellan gum smaller, acyl substituents, namely acetate and L-glycerate.

The viscosity/shear rate profiles for rhamsan gum and welan gum relative to xanthan gum are shown in Figure 2. The measurements were made with 0.25% gum in synthetic tap water, which is water containing 1000 ppm NaCl and 147 ppm $CaCl_2 \cdot 2H_2O$. As can be seen, both rhamsan gum and welan gum provide higher viscosities than xanthan in this shear rate range. Figure 3 shows the elastic moduli of the same solutions. The higher degree of structure provided by the rhamsam gum immediately suggests superior performance to xanthan gum as a stabilizer. This is sometimes but not always the case, in keeping with the fact that, as already mentioned, relative performance based on simple rheological measurements is not always mirrored in practice. However, rhamsan gum has generally better salt compatibility than xanthan gum. For example, it has better compatibility with polyphosphates, which has led to its use in suspension fertilizers. It is also useful in slurry explosives which contain a high concentration of ammonium nitrate. Rhamsan gum provides good suspension and recovery in textured coatings and it is possible to build a coating which is three times as thick as that obtained with hydroxyethylcellulose coatings. Perhaps less obvious from its rheology is rhamsan gum's ability to improve the suspension properties of paints based on synthetic, latex-based associative thickeners.

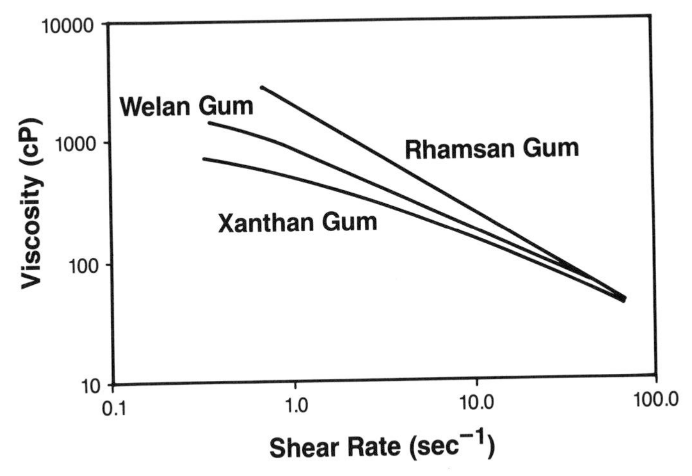

Figure 2. Viscosity vs. shear rate for rhamsan gum, welan gum and xanthan gum (0.25% in water containing 1000 ppm NaCl and 147 ppm $CaCl_2 \cdot 2H_2O$).

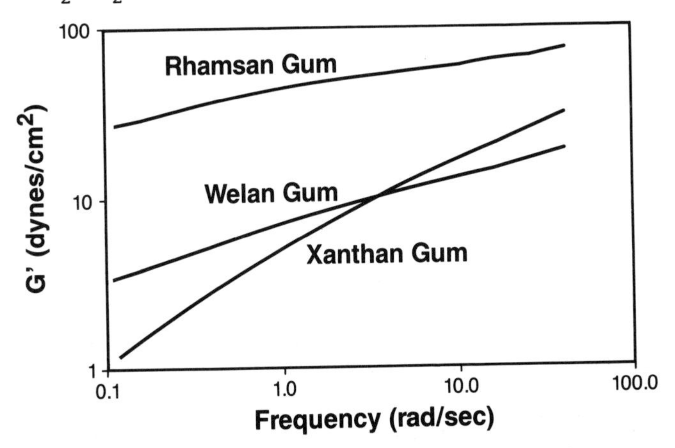

Figure 3. Effect of frequency on the elastic moduli of rhamsan gum, welan gum and xanthan gum (0.25% in water containing 1000 ppm NaCl and 147 ppm $CaCl_2 \cdot 2H_2O$).

Figure 4 shows the effect of temperature on the viscosities of xanthan gum, welan gum and rhamsan gum. The most stable viscosity on holding the solutions at 250°F for 1 hr. is provided by welan gum. Because of this property, welan gum is now finding use in oil well drilling fluids in which stable viscosity at high temperatures is required.

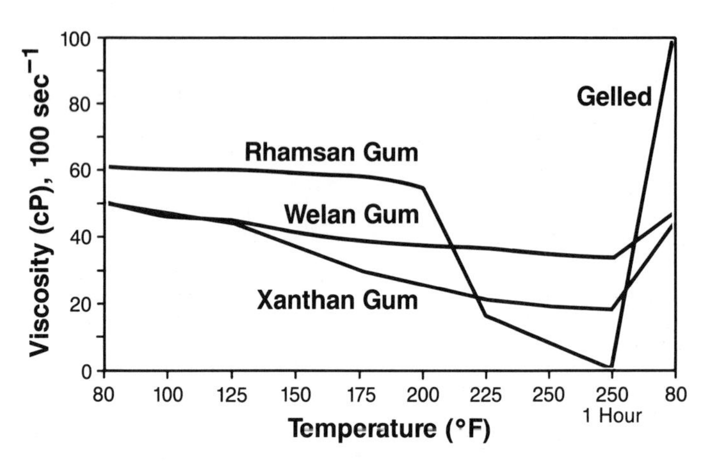

Figure 4. Effect of temperature on the viscosities of rhamsan gum, welan gum and xanthan gum (0.40% in seawater).

GELLAN GUM

It is well known that foods are generally a more lucrative market for hydrocolloids than industrial applications. However, approval to use a new hydrocolloid in foods takes a long time and is also very expensive. Kelco believes that, in the case of gellan gum, its potential in foods justifies this time and expense. All of the toxicological studies on gellan gum are essentially complete and the data are now being evaluated by the FDA. Although it was recently said that gellan gum is likely to be the last totally new food polysaccharide (6), it is important to emphasize that it is still not food approved at this stage. However, food approval could be realized in 1990.

There are two basic forms of gellan gum. The form containing the glyceryl and acetyl substituents, approximately one and a half respectively per repeating unit, is the end product of the fermentation. This has been referred to as high acyl or high acetyl gellan gum. Gellan gum in which the acyl groups have been removed is known as KELCOGEL*. KELCOGEL and its ultra pure version, GELRITE*, have been available to industry for testing for some time. Consequently, most of the research conducted to date relates to this form of the gum.

Table III. KEY PROPERTIES OF KELCOGEL GELLAN GUM

1. Is easy to use
2. Has good inherent stability; also good stability in products at acid pH
3. Forms gels at extremely low use levels
4. Provides gels with excellent flavor release
5. Gels are exceptionally clear
6. Gels can be made which either melt or do not re-melt on heating
7. Gels are texturally similar to those from agar and K-carrageenan
8. Use levels are normally one half to one third those for agar and K-carrageenan
9. A range of gel textures can be produced
10. Provides useful properties in combination with starches and gelatin.

The key properties of KELCOGEL, presented in similar fashion to the basic properties of xanthan gum, are shown in Table III. Inspection of these properties shows that KELCOGEL can both replace established gelling agents in existing applications and provide opportunities to develop new products. Replacement of existing gelling agents can be achieved through more cost effective performance or by eliminating deficiencies in performance. In regard to cost effectiveness, KELCOGEL is extremely efficient and can form free standing gels at a concentration as low as 0.05%. The efficiency of KELCOGEL relative to agar and K-carrageenan is shown in Table IV which shows gel firmness or modulus at several gum concentrations. A similar, although less pronounced, relationship is (7) seen when comparing gel hardness or rupture strength. K-carrageenan shows a substantial loss in rupture strength in going from neutral pH to the acid pH's encountered in many foods. KELCOGEL, in contrast, has good stability at acid pH and this is leading to its use in some acidic products in place of K-carrageenan. Another KELCOGEL property of emerging importance is the flexibility it provides in terms of gel melting points. Depending on the type and level of ions used to prepare the gels, they can be tailored to either re-melt on heating or remain heat stable (7). It is thus possible to make gelled products which are stable to retorting or microwave cooking. These represent new product opportunities for the food industry. Companies in the industry survive and differentiate themselves from their competitors through the development of new products.

Table IV. MODULI OF KELCOGEL, AGAR AND K-CARRAGEENAN GELS AT TWO pH VALUES

Concentration	KELCOGEL		K-Carrageenan		Agar	
(%)	4.0	neutral	4.0	neutral	4.0	neutral
0.25	5.3	6.3	0.35	0.56	0.18	0.23
0.50	15.3	16.4	1.2	1.3	0.54	0.56
1.00	41.0	39.8	4.7	4.6	1.8	1.9
1.50	64.0	61.6	9.9	9.9	4.3	4.3

Modulus values measured in Newtons/cm^2

One of KELCOGEL's most interesting features in terms of providing new product opportunities is the textural diversity it can produce. A range of textures can be obtained, for example, by combining KELCOGEL with xanthan gum and locust bean gum (8). It is perhaps not surprising that, since the acylated form of gellan gum provides gels similar in texture to those from xanthan gum and locust bean gum, combinations of this form of the gum and KELCOGEL also produce a similar spread of textures (9).

For gellan gum to approach the commercial success of xanthan gum, it is necessary for it to make some inroads into the markets currently enjoyed by gelatin and starch. Gelatin is not without defects. It requires refrigeration to induce setting and subsequent exposure to relatively low temperatures to prevent remelting. As with all anionic polysaccharides, the effects obtained in combining KELCOGEL with gelatin depend on a variety of factors such as pH, ionic strength, gum concentrations, and processing conditions. Under certain conditions, for example, it is possible to form coacervates (10). It is also possible to make products such as dessert gels in which the two gelling agents are compatible. Inclusion of KELCOGEL with the gelatin provides gels which set without refrigeration and do not re-melt on heating. One of gelatin's features is good flavor release. This is also one of KELCOGEL's strong points, not because the gels melt in the mouth but because of the way in which the gels break down in the mouth and release water readily.

KELCOGEL produces a number of interesting effects when combined with starch (8). These are shown in Table V. These benefits are obtained to a greater or lesser extent with all starches. However, the magnitude of the effects is more pronounced with native than modified starches. This is perhaps not too surprising since the reason starches are modified is to improve some of the properties mentioned. The addition of KELCOGEL to starch is thus a means of improving the performance of the starch without recourse to chemical modification.

Table V. FEATURES AND BENEFITS OF KELCOGEL GELLAN GUM WITH STARCH

1. Compatibility with existing starch processing equipment.
2. No increase in initial peak viscosity.
3. Additional structure, permitting lower starch usage levels and accompanying superior flavor release.
4. Improved paste stability, including reduced retrogradation and better water holding capacity.
5. Retention of the typical starch mouthfeel (the structure imparted by the KELCOGEL is not detected in the mouth).

Since ions promote chain association and gelation with KELCOGEL, reflecting an instability of the polymer molecules in the fully hydrated state, it follows that ions inhibit hydration. Fortunately, and in keeping with their relative efficiencies in promoting gelation, Ca^{2+} inhibit hydration much more than Na^{+}. It is therefore possible to hydrate KELCOGEL simply by using a sequestrant to remove Ca^{2+} (11). The levels of sequestrant required are such that the Na^{+} introduced from the sequestrant is insufficient to inhibit hydration. KELCOGEL is produced predominantly in the K^{+} form. However, the low level of divalent ions in the product is sufficient to prevent hydration in water below around 75°C. Using sequestrant, it is possible to hydrate KELCOGEL even at room temperature. The effects of sequestrant on hydration temperature and gel formation are shown in Table VI. As can be seen, to obtain the strongest gels, it is necessary to add back Ca^{2+} to compensate for the ions removed by sequestrant.

Table VI. GEL FORMATION WITH 0.25% KELCOGEL GELLAN GUM

Water Hardness (ppm $CaCO_3$)	[1]Added Citrate (%)	[2]Added Ca^{2+} (%)	Hydration Temperature (°C)	Gel Modulus (N/cm^2)
0	0	0.024	75	5.1
100	0.05	0.032	25	5.4
300	0.10	0.024	65	5.1
600	0.20	0.024	65	5.0

1. Sodium citrate was added to cold solution
2. $CaCl_2$ was added to hot solution

Control of the Ca^{2+} level is the basis of algin technology. The techniques used to manipulate KELCOGEL are therefore similar to those used with algin (11). There are, however, two key differences. KELCOGEL, at elevated temperature, will not gel in the presence of the level of Ca^{2+} required to produce the strongest gels. Thermal energy cannot prevent gelation under these circumstances with algin. When ions are introduced into KELCOGEL solutions in the cold, for example by diffusion, gelation will occur as with algin. However, one of the limitations of using KELCOGEL in the cold is its reactivity with monovalent ions. The levels of sequestrant sometimes needed to permit hydration in the presence of Ca^{2+} result in excessively high Na^+ levels, which themselves inhibit hydration. From a practical viewpoint, KELCOGEL and algin can be considered complementary. Use of the former normally involves heating and cooling; use of the latter, in many cases, is preferably in the cold.

CURDLAN

Curdlan, a β-1,3 glucan, is another microbial polysaccharide that has been known for many years (12). Like gellan gum, curdlan is a gelling agent. It is insoluble in cold water and undergoes hydration and subsequent gelation upon heating. The resulting gels do not melt below 100°C and are stable over the pH range encountered in foods. Gels may also be prepared by dissolving the gum in alkaline solution followed by neutralization or removal of the alkali by dialysis. Although there are indications that curdlan is beginning to find application in some food products in Japan, it is not approved for use in the U.S.A. or Europe. Aside from this major obstacle, a significant barrier to curdlan's use in foods is the fact that the conditions under which it forms gels are not those used in many of the processes in the food industry. In other words, the industry is more familiar with gels which are prepared by cooling rather than by heating or neutralizing alkaline solutions. Nevertheless, curdlan provides a good example of the benefits that can be derived from microbial polysaccharides. These are new properties which, with imagination and foresight, can lead to novel applications. The curdlan experience in Japan also illustrates another important aspect of the development of hydrocolloid applications. In the course of time, provided there is a dedicated commitment from the supplier and the potential end user, new applications can always be found.

ACKNOWLEDGEMENTS

The author wishes to acknowledge all of his colleagues at Kelco involved in the identification and development of new biogums. Special thanks are due to John Baird, Ross Clark and Walt Rakitsky for the provision of data.

REFERENCES

1. Whistler, R.L. (1973) Industrial Gums, 2nd edn., 807 pages, Academic Press, New York, San Francisco, London.
2. Flodin, G.M. and Ingelman, B.G.A. (1962) U.S. Patent 3,042,667.
3. Jeanes, A.R., Pittsley, J.E. and Senti, F.R. (1961) J. Appl. Polymer Sci., 5, 519-526.
4. Wintersdorff,P. (1981) U.S. Patent 4,269,974.
5. Sandford, P.A. and Laskin, A. (1977) Extracellular Microbial Polysaccharides, 326 pages, American Chemical Society, Washington, D.C.
6. BeMiller, J. (1989) Carbohydrate Polymers, 10, 64.
7. Sanderson, G.R. in Food Gels, (ed. Harris, P.) in press. Elsevier Applied Science, Barking, U.K.
8. Sanderson, G.R., Bell, V.L., Burgum, D.R., Clark, R.C. and Ortega, D. (1988) in Gums and Stabilizers for the Food Industry 4, (eds. Phillips, G.O., Wedlock, D.J. and Williams, P.A.) pp 301-308. IRL Press, Oxford, U.K.
9. Sanderson, G.R., Bell, V.L., Clark, R.C. and Ortega, D. (1988) in Gums and Stabilizers for the Food Industry 4, (eds. Phillips, G.O., Wedlock, D.J. and Williams, P.A.) pp 219-229. IRL Press, Oxford, U.K.
10. Chilvers, G.R. and Morris, V.J. (1987) Carbohydrate Polymers, 7, 111-120.
11. Sanderson, G.R., Bell, V.L. and Ortega, D. (1989) Cereal Foods World, to be published.
12. Harada, T. (1977) in Extracellular Microbial Polysaccharides, (eds. Sandford, P.A. and Laskin, A.) pp 265-283. American Chemical Society, Washington, D.C.

SEPHADEX*, Trademark of Pharmacia Fine Chemicals., Piscataway, NJ.
KELCOGEL* and GELRITE*, Trademarks of Merck & Co., Inc. (Rahway, NJ) Kelco Division, U.S.A.

Gellan gum – quick setting gelling systems for ‘jelly’ dessert products

G.OWEN

General Foods, Banbury, Oxon, UK

ABSTRACT

Being at the 'sharp' end of the hydrocolloids business requires a pragmatic approach to Product Development. With development times restricted to weeks rather than months as a direct result of trade and consumer time deadlines, a logical but effective route is necessary to achieve the desired result.

This paper shows a typical approach to evaluate the attributes of gellan, together with existing systems in a quick-set jelly, for the UK market. Gelatine is the undisputed textural reference standard to which we all aspire, and this programme uses it as the main point of reference.

Powder dispersibility, gel setting time, textural characteristics of the set gel, and flavour release, are some of the attributes studied in this development programme.

INTRODUCTION

The abstract hints at the restrictions imposed by the retail trade and the consumer on the product manufacturer. I will make frequent reference to the consumer during this paper since, after all, they are buying our products and expect the best in quality.

This paper aims to show you not only 'what' conclusions we draw, but also 'how' a commercial Product Development Group like ours approaches and executes development programmes.

This programme looks specifically at quick-set gelling systems.

Let's look at the overall objective for the project. This is a clear and concise statement of what we will achieve. It is important for this statement to be carefully constructed and agreed. To complete a commercial assessment of available gel systems suitable for a quick-setting jelly dessert.

This approach to project management uses management by objective techniques.

I need to briefly describe the product in which the quick-set jelly is used.

Trifle is peculiar to the UK market. It is a multi-component indulgent product and is used on special occasions e.g. dessert at Christmas and Easter, birthdays, children's parties and treats.

Traditionally, trifle consists of stale, dry sponge soaked in a gelatine-based jelly (canned fruit or sherry can be an optional extra). The second layer is a custard-blancmange, finally topped with a layer of cream (which might or might not be whipped) and some decorations. The time taken to make product is very long - the setting times of the jelly being particularly long (usually more than 2 hours). Bird's offer a more convenient, less time consuming component pack which can be made up within 1 hour. (The product, and the simple make-up technique used was demonstrated).

PROCEDURE
What gel systems are commercially available?

CARRAGEENAN
CARRAGEENAN/LOCUST BEAN GUM
GELLAN GUM
COLD-WATER SOLUBLE GELATINE
ALGINATE
XANTHAN/LOCUST BEAN GUM
PECTIN

This list represents what we believe is available. CARRAGEENAN - this is what we currently use. CARAG/LBG - this system was used before the sharp increase in LBG costs. It is a system which we believe is the most desirable, but cost restraints continue to prevent its usage. GELLAN - it could be argued that gellan is not commercially available, but this new ingredient is in an advanced stage of development and is now worthy of a closer investigation.

COLD-WATER SOLUBLE GELATINE	) All these systems produce
ALGINATE	) gels and this is an ideal
XANTHAN/LOCUST BEAN GUM	) opportunity to assess their
PECTIN	) potential

KEY TO ANY STUDY OF THIS NATURE IS TO DEFINE THE ATTRIBUTES AGAINST WHICH THE TEST SYSTEMS ARE TO BE ASSESSED.

Before I explain the key attributes it is important to emphasise that these attributes are defined by the UK consumer; not by us the manufacturer. They are:

1. One-stage make-up
2. Easily dispersible
3. Set within one hour
4. Texture similar to gelatine
5. Good flavour release

The remainder of the presentation will compare the success of each gel system within each category.

1. ONE STAGE MAKE-UP

A single-stage make-up procedure is vital in this product. The product is already multi-component, a further make-up stage is a detrimental step. Systems based on xanthan gum/LBG and pectin were rejected. They could be made to work if a two stage make-up was acceptable.

2. EASILY DISPERSIBLE

It is important that the minimum amount of energy should be used at make-up. There are two reasons for this:

- Excess energy can lead to aeration in the finished product. A jelly should be crystal clear

- Safety. Vigorous stirring of hot water is a potential hazard. For instance, we would not recommend a hand-held electric whisk to make up a jelly.

Formulas based on the recommended alginate type for a jelly required an excessive amount of energy for proper dispersion. Therefore alginate systems were rejected for poor dispersibility.

3. SET WITHIN ONE HOUR

A one hour set time is the absolute maximum. Since the trifle is being constructed, the two stages being the most time consuming are the custard (blancmange) make-up and cooling, and the setting of the jelly. Clearly, a reduction in the setting time would be a significant product improvement.

To make the test truly comparative it was necessary to adjust the concentration for each system to give the same gel strength.

The following parameters were used to define a satisfactory jelly:-

A minimum reading of 35g using the LFRA Stevens Gel Tester with 1 inch diameter plunger, penetrating a total depth of 4mm at a speed of 0.5mm/sec. within a 1 hour set time.

Carrageenan and carrageenan/LBG only just achieved the requirement, gellan was well in, the cold-water soluble gelatine was rejected (hot and cold water tried).

4. TEXTURE SIMILAR TO GELATINE

4.1 Taste panel evaluation. A taste panel comprised of a minimum 6 trained persons is used. No flavour or colour present in the samples. The sweetener is sucrose, and the citric acid is at a consistent level throughout the testing regime. A gelatine reference (unidentified) is present in all tests. Panelists are asked to express difference rating against each sample in the set. Various levels of gum systems are tasted against the consistent gelatine, in a series of taste sessions.

Main features from this work are:

- Carrageenan/LBG system is closest to gelatine.

- Small changes in concentration of gellan result in significant differences in 'closeness' to gelatine. Carrageenan/LBG is more 'tolerant' to changes in level.
- Gellan has the lowest concentration to achieve 'closeness' to gelatine.

From the results, the concentration of each system that achieves the closest rating to gelatine goes on to the next stage, the analysis by the 'Instron' texture analyser.

4.2 Instrumental measurement of texture using the 'Instron'. Briefly, the instrument continuously measures the load in grams during two reciprocal cycles of the penetration of a semi rigid sample like a jelly, with a shaped probe.

The resultant profile can be interpreted; now it is generally accepted that features from the profile have a correlation with some descriptive terms, for example hardness, elasticity, brittleness, etc. The following results were obtained:

		GELLAN	CARRAG/LBG	CARRAG	GELATINE
MODULUS	(Mean)	0.85	0.38	0.34	0.29
	(s.d.)	0.09	0.03	0.03	0.02
HARDNESS	(Mean)	1.43	1.74	0.44	2.03
	(s.d.)	0.37	0.16	0.15	0.39
BRITTLENESS	(Mean)	41.33	61.17	38.3	69.56
	(s.d.)	1.29	3.79	12.89	0.94
ELASTICITY	(Mean)	27.27	30.05	32.28	61.73
	(s.d)	10.93	3.52	8.37	10.23
COHESIVENESS	(Mean)	5.64	14.17	11.39	35.86
	(s.d.)	1.35	2.78	2.98	7.83

The closest system to gelatine is carrageenan/LBG. The next closest is carrageenan, the least close being gellan gum.

Overall, the facts show that none of these systems come very close to gelatine in both the eating characteristics and instrumental analysis. So pragmatism takes over and we rationalise that, after all, the jelly is part of a multi-component product and major differences in texture become less apparent! At the same time we throw down the gauntlet to the supplier by informing them of the drawbacks to their systems. Meanwhile we proceed to study the flavour release aspects of gellan, carageenan and carrageenan/LBG.

5. GOOD FLAVOUR RELEASE

These are the main criteria which influence flavour release:

SWEETNESS, FLAVOUR STRENGTH, ACIDITY

Throughout the remaining work programme the acidity is at a constant level. The optimised formulas for closeness to gelatine texture are used for each system, the sweetness and the flavour are changed (only one at a time) and measurement is by difference against gelatine in the panel test.

5.1 Varying sugar concentration

Panellists were specifically asked to rate difference for flavour only. Clearly in systems where the textures are inherently different it is very difficult to focus on one aspect alone and ignore the remaining obvious differences. This is a limitation of the test regime. Across all concentrations of sucrose, gellan gave the highest difference ratings against gelatine, followed by carrageenan and carrageenan/LBG. No statistical analysis was carried out but it was quite clear from the results that the differences were not statistically different. We would describe them as directional rather than significant. The key observation is the wide difference in flavour release between gelatine and the other three systems. This tends to bring the test systems closer together in the panel testing procedure. There is an opportunity to use less sucrose in gellan formulas than in other systems, to deliver the same flavour delivery.

5.2 Varying flavour concentration

A very similar profile is observed when flavour level is varied, with gellan giving the best flavour release followed by carrageenan and carrageenan/LBG.

Again, statistically I'm sure there is no significant difference between each test system and there is the opportunity to use less flavour in the gellan system.

In conclusion gellan is very interesting!

As a footnote to this programme a rough calculation shows that approx. 75% of our time was spent on the last two attributes. Clearly it is important to set aside some time for experimental design to identify failures, fast.

The use of xanthan gum in salad dressings and other products – the user's viewpoint

C.M.GORDON
H.J.Heinz Co. Ltd, Hayes Park, Hayes, Middlesex UB4 8AL, UK

Abstract

The use of xanthan gum and its properties in salad dressings and various emulsion products are discussed. The various functional properties are described along with product references where these properties are used to arrived at a desired end product. Possible factors which could affect the use of xanthan gum in the manufacturing environment are covered. Choice of hydrocolloid system and blending with other hydrocolloids for cost efficiency, synergy, textural characteristics and emulsion stabilization are discussed.

INTRODUCTION

Xanthan gum is widely used in formulations such as salad cream, low calorie dressings, mayonnaise, sandwich spreads. Examples of Heinz's products containing xanthan gum include, Salad Cream, Reduced Calorie Dressing, Vegetable and Potato Salad, Coleslaw, All Seasons Dressings (Herb and Garlic, Yoghurt and Chive, Thousand Island) and Spreads (Cucumber, Sandwich, Spicy, Tomato and Onion).

This paper will concentrate on the ways of using the functional properties of xanthan gum to produce at desired end product. Product references will be restricted to dressings and spreads.

General Properties of Xanthan Gum

Rheology

Solutions formed from xanthan gum are classified as non-Newtonian. Thus, at low shear rates these solutions would exhibit high apparent fluid viscosity and as the shear rate is increased the apparent viscosity decreases. This change in viscosity is reversible as soon as the shear is removed. This phenomenon is termed pseudoplastic and is differentiated from thixotropic, which refers to a change in viscosity of a solution with increasing shear and is time dependent. Both phenomena are temperature dependent.

Pseudoplasticity is a very useful property for developing emulsion products such as dressings and spreads where high apparent viscosity is required in the end product, but for ease of manufacture with particular reference to pumping and mixing operations, a low viscosity product is more desirable.

It is general practice for most manufacturing processes to be divided into unit operations. For example, a typical spread manufacture flow chart may consist of an aqueous stage, a vegetable mix stage, and an emulsion stage. To arrive at the emulsion phase the process could be described as follows:

a) Make egg mix and pump to pre-emulsion tank.

b) Form crude emulsion with oil (mix properly) and pump via size reduction equipment (eg colloid mill) to emulsion holding tank.

c) Pump the required amount for final batch from holding tank.

d) Mix final batch (emulsion+aqueous phase), pump to filling line.

There are three mixing and four pumping stages that the emulsion component passes through. If the emulsion is pseudoplastic, the mixture would shear thin during these operations and this would make it easier to perform the mixing and pumping.

A problem could occur where a final blend of emulsion and liquor could have an incorrect ratio if one or both of the components are viscous non-shear thinning phases. In this case the metering pumps may not be capable of delivering the standard amount of emulsion or liquor. This was encountered recently during a factory trial of a modified recipe which consisted of a thicker, liquor phase compared to the standard liquor used on that system. It became apparent that the piston pump was unable to deliver the quantity required and the final blend was affected. The ability of xanthan gum to shear thin would be of benefit in these circumstances.

With reference to establishing viscosity control limits on emulsions, the ability of xanthan gum solutions to rapidly recover from shear thinning allows the viscosity to be monitored before the final blend is done. The solution would recover from shear thinning in the stock tanks, from which samples can be drawn for quality control tests.

Solubility and Dispersibility

Xanthan gum is readily hydrated in hot or cold water to form viscous solutions which are slightly opaque.

Xanthan gum like most other hydrocolloids, would form lumps during dispersion in water if not correctly dispersed. Referring back to the emulsion manufacturing pathway, xanthan gum can easily be incorporated in the product via two avenues. The gum can be pre-blended with the sugar and other dry ingredients then added to the egg mix or it can be dispersed in the oil before forming the crude emulsion.

Various studies have reported the adverse effects of dispersants and salts on xanthan gum hydration. It has been reported that the best method of hydrating xanthan gum is in demineralised water resulting in a minimum hydration time of 2 minutes. In the presence of salts and/or vinegar, or using oil as a dispersant, hydration times can increase seven fold. However, it is the author's opinion that this increase in hydration times does not affect the choice of dispersion, since most manufacturing cycles would exceed the longest hydration time, in which case 90% of the gum would be hydrated. Another related point is that product viscosity is normally measured immediately after manufacture and 24 hours after filling when 100% hydration would have occurred. Nevertheless, it is very important that the gum is correctly dispersed since it is almost impossible to hydrate any formed lumps which will then affect final product viscosity. More importantly the product becomes unsalable due to the mouthfeel of the lumps formed and the poor visual presentation which results. Another point for consideration is that, if the gum was being used for stabilization of an emulsion and incomplete hydration occurred, then its functional role would be impaired.

Stability

Xanthan gum is known to be stable over a wide range of pH , salt and temperature conditions. From the end users point of view, the uniform viscosity of the gum over a pH range of 2-8 is adequate for his/her needs, (the pH range for most foods would be between 3-7). Xanthan gum is known to exhibit a near uniform viscosity between 2-60 °C, and above 60 °C there is decrease in viscosity which is reversible. This decrease can be of benefit to the Technologist since improved heat penetration can be achieved in those situations where a product containing the gum is being heated or sterilised.

Emulsion Stabilization.

An emulsion can be described as a system comprising of two immiscible liquids which are intimately mixed, with one liquid, the internal (discontinuous) phase consisting of small droplets or globules, being dispersed in the external (continuous) phase. Examples of oil (discontinuous phase) in water (continuous phase) emulsion are salad dressings, mayonnaise, sauces and soups.

The preparation of an emulsion involves reducing the interfacial tension between the two phases to aid mixing; mixing the two phases along with stabilisers, emulsifiers or surface active compounds and finally mechanically homogenise the crude mixture to reduce the internal phase.

Many hydrocolloids can be used as stabilisers within this system, since their main function is to increase the viscosity of the continuous phase and thus decrease the random movement of dispersed oil droplets. This minimises the chances of collision between molecules which could result in coalescence and ultimately phase separation. Xanthan gum is normally used but not in the strict capacity of a thickener, but as a thickener with other desirable properties such as mentioned above.

One such property not mentioned previously is the yield value. The yield value can be referred to as the shear stress that is applied to a viscous solution in order for flow to occur. It is this property which allows a salad dressing to cling to the salad and to appear to have 'body' and not a 'thin appearance '. The yield value can also be expressed as a measure of the suspending powers of a hydrocolloid, that is the greater the yield value the greater the suspending powers. Vinaigrette is a good example for exhibiting this property, where herbs, peppers and other pieces of vegetable are suspended in an aqueous phase. In terms of salad dressings, this category of product can be described as consisting of oil particles finely suspended in a continuous aqueous phase. Therefore a high yield value can be considered to have contributed to the overall stability of the system by fixation of the dispersed oil particles.

Xanthan gum is reported to have a higher yield value than guar gum, locust bean gum, carboxymethyl cellulose and sodium alginate, thus it would be the preferred hydrocolloid for any of the above mentioned functions.

Dressings Applications.

In dressings, it is normal to use a combination of gums to stabilise the emulsion. From experience a blend of xanthan gum and guar gum is a good combination for the following reasons: synergy, texture considerations, cost efficiency, and Guar Gum's ability to rapidly hydrate in cold-water systems, its pH stability and its complementary pseudoplasticity.

Xanthan gum and guar gum when used in combination exhibit a synergistic viscosity increase compared to solutions containing the individual gums. A blend of 80% guar to 20% xanthan has been reported as the optimum ratio for this effect at total gum levels of 0.5%. However, experience has shown that this optimum can and does shift depending on the ingredients within the food system. In some systems the ratio can move as far as 50/50.

From a textural point of view, using the optimum of 80/20, if that is the true optimum for the system, can introduce unwanted characteristics. Solutions of guar gum tend to have a shorter texture compared to solutions of xanthan gum and depending on the concentration of the gums used, guar solutions have a slimy perception on the pallet compared to xanthan gum solutions. Therefore using the optimum ratio of 80/20, if obtainable, may not be practical. From a cost point a view, it would be best to use the 80/20 blend which gives the greatest synergistic viscosity, since guar gum is significantly less expensive than xanthan gum and at the levels required to achieve this optimum it is worth considering. In low calorie dressings, the oil content is reduced and replaced by water and the combination of xanthan and guar is very useful in 'mopping up' the extra water. In these situations, it is cost effective to try and shift the blend ratio in favour of guar gum while striving to maintain the desirable characteristics of xanthan gum.

Having considered the various options in dressing applications, it has to be stated that the tried and tested approach is normally used to develop new emulsion products. That is, most new emulsion recipes are modelled on existing ones. There are several reasons for this approach.

Firstly, unless the new product range is to be manufactured on new lines, then the manufacturing cycle and equipment will be a fixed condition within which the new products will have to be manufactured. A variation on an existing theme obviates the need for major equipment modification.

Another consideration is the stability record of an existing product which would significantly influence the approach used for developing a new emulsion system. This influence would be greatest with respect to the gum blends and ratios. Testing a completely new emulsion system may take anything up to twelve months whereas a minor alteration of an existing stable system would not.

From a sensory perspective, the attributes of an existing gum blend ratio are already known and there will be inevitable reluctance to alter these unless the existing product/ blend is a poor performer in the market place. Most new product briefs would request that any recipe modifications be similar in character to an existing product.

Finally the development costs and time for a new recipe formulation can be prohibitive. Developing a new system normally requires several attempts before arriving at an approved product and this is costly compared to the other avenues available. This negating factor along with the tight schedules set by the Marketing Department, means that this route would have a low priority rating.

Other Applications

There are several other areas of food application where xanthan gum is used to aid processing, modify texture and maximise shelf life. It is used to improve freeze/thaw characteristics of frozen foods, in syrups to give cling and pourability to fruits and desserts. In fillings, it allows a cold make-up of the filling which is incorporated in the pastry prior to baking. In low calorie bakery products, xanthan gum can be used as a partial fat replacement giving good body, texture and mouthfeel to the end product.

The Food Technologist has found many applications for hydrocolloids and in particular xanthan gum. New applications are being investigated, for example low calorie cakes and microwaveable foods. The vast amount of published work detailing the functional properties of the gums and their interaction with each other and with starches makes this task easier. However, it must be stated that the complexity of the task becomes evident when the recipe formulation is tested in the experimental laboratory and the factory. This results from the fact that the published information does not and cannot account for the various recipe formulations which are normally confidential. For example, published ratios relating to the optimum synergy may not be applicable in the system being formulated. However, this is but a small price to pay for the wealth of information available on hydrocolloids.

Acetan – a new microbial thickening and suspending agent

V.J.MORRIS, G.J.BROWNSEY, J.E.HARRIS and B.J.H.STEVENS
AFRC Institute of Food Research, Norwich Laboratory, Colney Lane, Norwich NR4 7UA, UK

ABSTRACT

Acetan is an anionic heteropolysaccharide produced by Acetobacter xylinum strain NRRL B42. Aqueous dispersions of acetan are thixotropic suggesting applications as a thickening and suspending agent. Thixotropic vinegars for potential use in acid-based foods can be prepared by dispersing acetan in vinegar or blending clarified, cell-free A.xylinum broths with vinegar. The possibility of producing both vinegar and acetan in a single fermentation process, thus leading to the production of self-thickened vinegars, will be discussed briefly.

INTRODUCTION

A major use of U.K. produced vinegar is as a base for a number of acid-based foods such as mayonnaise, sauces, spreads, relishes, salad dressings. Xanthan gum, a microbial polysaccharide produced by Xanthomonas campestris, is used as a thickening and suspending agent and imparts unique thixotropic properties to these foods. The chemical structure of xanthan gum (1,2) consists of a cellulosic backbone substituted on alternate glucose residues with a trisaccharide sidechain (fig 1a). The trisaccharide sidechain alters the solid-state conformation of the molecule from a 2-fold ribbon-like structure to a helix with 5-fold symmetry (3,4). Retention of this helical structure in solution is crucial to explaining the rheology of xanthan dispersions. Acetan is a microbial polysaccharide produced by *Acetobacter xylinum* strain NRRL B42 (5). A partial chemical structure for acetan suggests (5) that it consists of a cellulosic backbone substituted on alternate glucose residues with a pentasaccharide sidechain (fig 1b). The

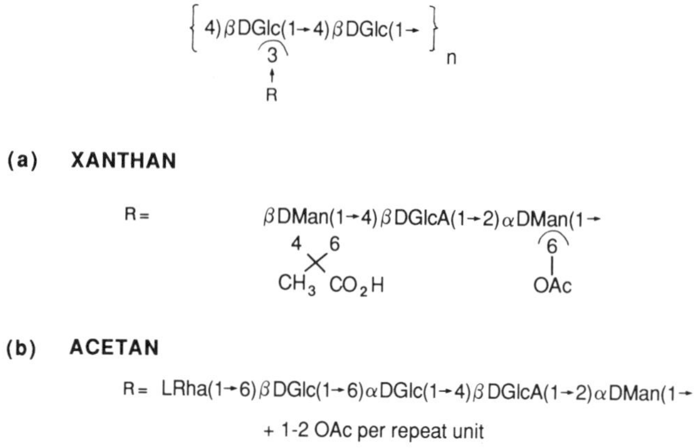

{ 4)βDGlc(1→4)βDGlc(1→ }n
(3 ← R)

(a) XANTHAN

R = βDMan(1→4)βDGlcA(1→2)αDMan(1→
(4,6: CH_3, CO_2H; 6: OAc)

(b) ACETAN

R = LRha(1→6)βDGlc(1→6)αDGlc(1→4)βDGlcA(1→2)αDMan(1→

+ 1-2 OAc per repeat unit

Figure 1. The chemical structure of the bacterial polysaccharides acetan and xanthan.

backbone, backbone-sidechain linkage, and the first two sugars in the sidechain are all identical to those of xanthan. Thus the possibility exists that acetan may form a similar ordered structure to xanthan and exhibit similar rheological properties. Preliminary physical and physico-chemical studies (6) do suggest that the solid-state conformation of acetan is a helix with 5-fold symmetry (pitch 4.8nm) and that aqueous dispersions exhibit optical rotation data consistent with a reversible helix-to-coil transition on heating. This article demonstrates the useful rheological properties of acetan in both water and vinegar preparations. Thixotropic vinegars can be prepared by dispersing purified (freeze-dried) acetan in vinegar or by blending cell-free, clarified *A.xylinum* broths with vinegar. The possibility of obtaining prethickened vinegars from commercial vinegar fermentations by using acetan producing strains of *Acetobacter* will be discussed briefly.

METHODS

Acetobacter xylinum strain NRRL B42 was grown in AJ medium containing (g/l) : yeast extract (10), glucose (50), K_2HPO_4 (0.1), KH_2PO_4 (0.1), $MgSO_4$ $7H_2O$ (0.25) and $FeCl_3$ (0.005). The final pH was adjusted to 6.0. Acetan was prepared from cell free culture broths by ethanolic precipitation as described by Couso *et al*. (5).

Wild type Acetobacter xylinum strain NRRL B42 produces both acetan and cellulose during growth. In unshaken cultures, a surface pellicle of cellulose forms containing most of the cells and some acetan. Disruption of the pellicle in a blender and subsequent removal of cellulose fibres and cells by filtration and centrifugation respectively, is necessary prior to ethanolic precipitation of acetan from the culture medium. However, it has been reported (7) that repeated transfer of Acetobacter xylinum in shake flask cultures selects for spontaneous cellulose⁻ mutants. Thus, to improve the procedure for the preparation of acetan, this method was used to select variants of strain NRRL B42 which do not form pellicles in shaken conditions, allowing immediate ethanolic precipitation from cell-free broths.

The colouration and smell usually present in Acetobacter xylinum culture broths can be removed by treatment with charcoal, either by passing cell-free broths through charcoal columns or adding charcoal to the broth and removing it later by filtration.

Rheological measurements of acetan dispersions in water or vinegar, and cell-free A.xylinum broths were made using an INSTRON 3250 mechanical spectrometer equipped with a cone and plate assembly (cone angle 2.4 degrees).

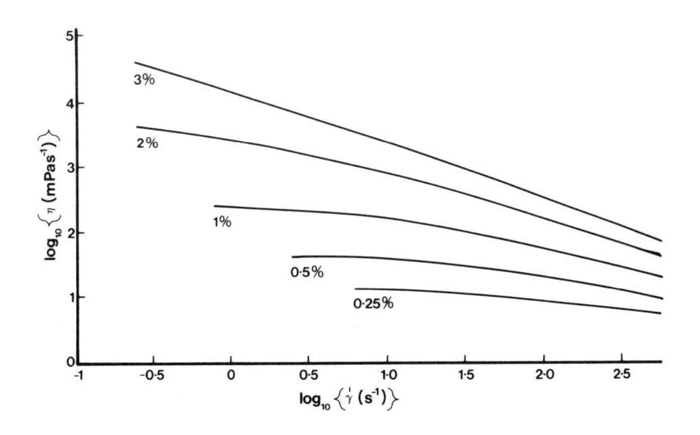

Figure 2. Viscosity (η)-shear rate ($\dot{\gamma}$) data for aqueous acetan samples.

RESULTS AND DISCUSSION

Initial rheological studies were carried out using freeze dried samples prepared from ethanol precipitates obtained from A.xylinum broths. Fig 2 shows viscosity-shear rate profiles for aqueous acetan dispersions. The dispersions exhibit high stable viscosity at low shear rate and show marked shear-thinning behaviour. The shear thinning is reversible as indicated by the coincidence of the experimental curves obtained upon increasing and then decreasing the shear rate.

Fig 3 shows the viscosity-shear rate dependence for a dispersion of 2% acetan in vinegar. The high viscosity and marked shear thinning behaviour is retained and slightly enhanced under acidic conditions, characteristic of a structured medium (gel-like).

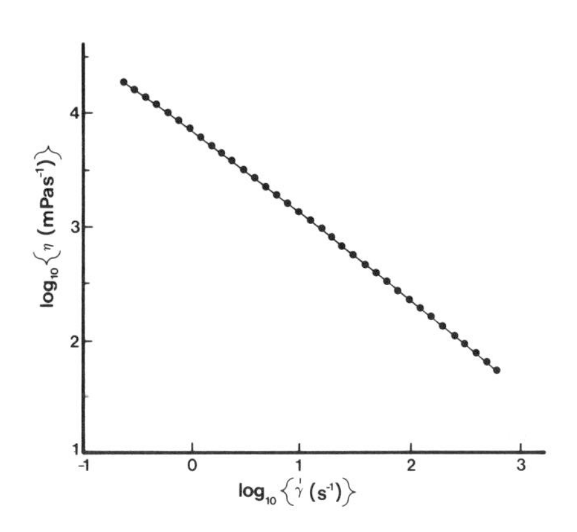

Figure 3. Viscosity (η)-shear rate ($\dot{\gamma}$) data for a 2% dispersion of acetan in vinegar.

Fig 4a shows viscosity-shear rate data on a typical A.xylinum (cell-free) broth. The viscosity profile corresponds closely to that observed for an aqueous acetan dispersion of ~0.2%. Fig 4b shows the effect of removing pellicle formation by growing selected strains under shaken conditions. This leads to a substantial increase in broth viscosity. The viscosity profile is most closely related to that obtained for a 0.5% aqueous acetan sample. Ethanol precipitation confirms that the increase in viscosity is due to increased total polysaccharide production. Fig 4c shows the effect of concentrating the broth shown in fig 4b by rotary evaporation under reduced pressure. Concentration by a factor of ~6 results in broth viscosity-shear rate profiles equivalent to 3% aqueous dispersions of acetan. Such dispersions may be subsequently diluted or blended with vinegar to produce thixotropic preparations with appropriate rheological properties.

Acetan producing strains of Acetobacter offer potential for the manufacture of 'novel fermented food ingredients', such as thixotropic acetan broths and self-thickened vinegars. However, Acetobacter xylinum strain NRRL B42 has two disadvantages for these applications. The first, cellulose production, can be overcome by the use of shake flask cultures to select cellulose$^-$

strains which yield broths with enhanced viscosities. Secondly, strain NRRL B42 shows poor tolerance to ethanol, which is a pre-requisite for the commercial vinegar making process. A genetic approach is being used in an attempt to solve these problems. Aims are to isolate stable cellulose $^-$ mutants, and to increase acetan yield and ethanol tolerance. An alternative

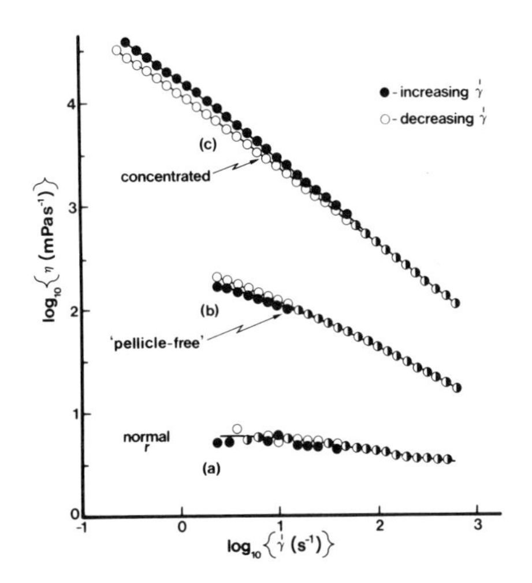

Figure 4. Viscosity (η)-shear rate (γ) data for A.xylinum broths (a)-cell-free broth of A.xylinum grown under unshaken conditions. (b)-cell-free broth of a pellicle free A.xylinum strain grown under shaken conditions. (c)-concentrated (x6) sample as defined under (b).

strategy would be to use genetic techniques to transfer acetan production to Acetobacter strains currently used in commercial vinegar fermentations.

CONCLUSIONS

It has been demonstrated that acetan, the microbial polysaccharide produced by A.xylinium, has rheological properties similar to xanthan gum. Thixotropic dispersions in water or vinegar can be prepared by dispersing the polymer in the appropriate solvent. Thixotropic cell-free clarified A.xylinum broths can be prepared and blended with vinegar to produce thickened vinegars. The possibility of obtaining genetically modified strains of Acetobacter suitable for production of 'self-thickened' vinegars and thixotropic acetan based broths as 'novel food ingredients has been discussed briefly.

REFERENCES

1. Jansson, P.E., Kenne, L. and Lindberg, B. (1975) Carbohydr. Res. 45, 275-282.
2. Melton, L.D., Mindt, L., Rees, D.A. and Sanderson, G.R. (1976) Carbohydr. Res. 46, 245-257.
3. Moorhouse, R., Walkinshaw, M.D. and Arnott, S. (1977) in 'Extracellular Microbial Polysaccharides' (Eds. Sandford, P.A. and Laskin, A.) ACS Symp. Sev., 45, 90-102.
4. Okwyama, K., Arnott, S., Moorhouse, R., Walkinshaw, M.D., Atkins, E.D.T. and Wolf-Ullish, C.H. (1980) in 'Fiber Diffraction Methods' (Eds. French, A.D. and Gardener, K.D.) ACS Symp. Sev. 141, 411-427.
5. Couso, R.O., Lelfi, L. and Dankert, M. (1987) J. Gen. Microbiol. 133, 2123-2135.
6. Morris, V.J., Brownsey, G.J., Cairns, P., Chilvers, G.R. and Miles, M.J. Int. J. Biol. Macromol. (submitted).
7. Schramm, M. and Hestin, S. (1954) J. Gen. Microbiol. 11 123-129.

Molecular structures of variants of xanthan gum with truncated sidechains

R.P.MILLANE, T.V.NARASAIAH and BOWEI WANG

The Whistler Center for Carbohydrate Research, Smith Hall, Purdue University, West Lafayette, Indiana 47907, USA

ABSTRACT

Variants of xanthan gum with truncated sidechains, produced by mutants of *Xanthomonas campestris*, have been trapped in ordered conformations in oriented fibers. X-ray diffraction patterns from these fibers show that the variant polymers have molecular repeats and helix symmetries identical to those of xanthan, but smaller molecular diameters. These results indicate that the conformations of the variant polymers are similar to that of xanthan, and show that the sidechain terminal sugar units are not critical determinants of the xanthan ordered molecular structure.

INTRODUCTION

Xanthan is an extracellular polysaccharide produced by the bacterium *Xanthomonas campestris*, that is widely used as a thickener and stabilizer in the food (and other) industries as a result of its unique rheological properties. The primary structure of xanthan is a $\beta(1\rightarrow4)$ linked D-glucose (cellulosic) mainchain with a trisaccharide sidechain attached to the 3-position of every second glucose unit, giving a pentasaccharide repeating unit (Fig. 1a). The sidechain α- and β-linked mannose units are specifically and variably acetylated and pyruvylated respectively. In aqueous solution, xanthan undergoes a thermally induced cooperative, conformational transition which has been monitored by a number of physical techniques (2). Although xanthan has been the subject of many physicochemical studies, the details of the ordered structure, particularly the number of chains involved, are still the subject of debate (3,4).

X-ray diffraction studies of xanthan show that it forms a helical structure with 5-fold screw symmetry and a molecular repeat distance of 47.0 Å in hydrated fibers (5,6). The diffraction data are insufficient however to define the molecular structure precisely, or the number of chains involved.

Recent studies of xanthan biosynthesis have identified mutant strains of *Xanthomonas* that produce variant gums that are polymers of truncated versions of the normal repeating unit of xanthan, some of which have rheological properties similar, but different, to xanthan (7). Two of these, referred to as

→4)-β-D-Glc-(1→4)-β-D-Glc-(1→

3

↑

1

β-D-Man-(1→4)-β-D-GlcA-(1→2)-α-D-Man-6-OAc

4 6

C

H_3C COOH

a

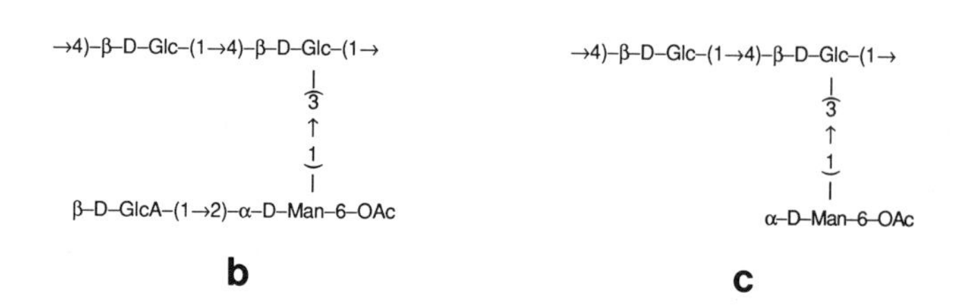

Figure 1. Primary structures of (a) xanthan, (b) polytetramer and (c) polytrimer.

"polytetramer" and "polytrimer", have repeating units identical to xanthan except that one and two of the sidechain terminal sugar units are absent, respectively (Fig. 1b,c).

We describe here x-ray fiber diffraction studies on these variants that allow comparisons to be made between their ordered structures and that of xanthan. Some results of x-ray diffraction studies on the polytetramer have already been reported (8,9). Structural studies of variant xanthan polymers may provide information on the molecular basis of their physical properties as well as shed light on the details of the ordered conformation of xanthan and side sidechain functionality, which are presently poorly understood.

EXPERIMENTAL

Samples of the variant polymers were supplied by the Synergen-Texaco Joint Venture (Boulder, Colorado). The polytetramer was converted to the lithium salt by dialysis against 0.4 M LiCl. Excess salt and any impurities were removed by additional dialysis against a large volume of distilled water, and the sample lyophilized. Samples were dry spun using standard methods, and x-ray diffraction patterns obtained at 75% relative humidity using a flat film camera and CuKα radiation. Spacings were calibrated by dusting the specimens with calcite. The diffraction patterns were digitized, the data processed and c-repeat determined using standard methods (10). Diffraction patterns were recorded from the sodium salt of wild-type xanthan (courtesy of M. Milas, CERMAV-CNRS, France) at 92% relative humidity, and spacings measured in the same way. Since the polytrimer is neutral, fibers were prepared directly from lyophilized material at 66% relative humidity and spacings determined as above.

RESULTS

Diffraction patterns obtained from wild-type xanthan and the variant polymers are shown in Fig. 2. The diffraction patterns show that the xanthan specimen is the most ordered and the polytrimer specimen the most disordered. Measurement of the layer line spacings show that the molecular repeat distances are 47.2 Å (xanthan), 47.5 Å (polytetramer) and 47.7 Å (polytrimer). The axial repeats are therefore identical within experimental error. The polytetramer diffraction pattern has meridional reflections on the 5th and 10th layer lines indicating that the molecule has 5-fold helix symmetry, as is the case with xanthan. The polytrimer pattern has a meridional reflection on the 5th layer line indicating 5-fold symmetry also. The first diffraction maxima on the equator of the patterns correspond to lateral spacings of 18.0 Å (xanthan), 14.4 Å (polytetramer) and 12.0 Å (polytrimer).

DISCUSSION

The diffraction patterns from the xanthan variants show that they adopt ordered conformations in fibers. Conservation of the c-repeat and helix symmetry with respect to xanthan, suggests that the molecular conformations are similar to those of xanthan. The identical c-repeats suggest rather strongly that the backbone conformations of the variants are identical to those of xanthan, as it is unlikely that significant changes in these would not be accompanied by changes in the c-repeats. The sidechains in these molecules probably interact with the mainchain, as is observed with other branched polysaccharides (11), so that any significant changes in the sidechain conformations (and therefore their interactions) would probably induce changes in the mainchain, thereby altering the c-repeats. The sidechain conformations of the variants are therefore likely (but not necessarily) to be similar to those of xanthan also.

The first diffraction maxima on the equator of the patterns are rather sharp, indicating some limited lateral packing of the molecules. There is a decrease in the lateral spacing implied by these reflections with increasing truncation of the sidechain. This is probably due to a reduction in the molecular diameters by about 4 Å and 6 Å, for the polytetramer and polytrimer

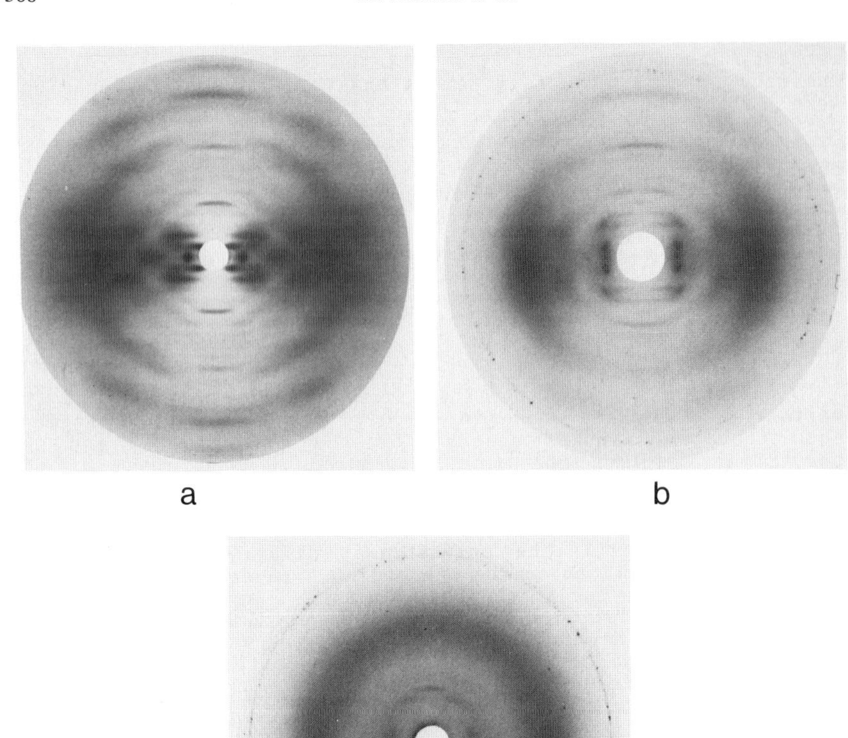

Figure 2. X-ray fiber diffraction patterns from (a) Na^{+} xanthan, (b) Li^{+} polytetramer (fiber tilted by 10° to the normal to the x-ray beam) and (c) polytrimer.

respectively, with respect to xanthan. This indicates that the outer 2 Å of the xanthan structure is occupied by the terminal mannose unit, and the outer 3 Å by the mannose and glucuronic acid units. This suggests that the sidechain may not be wrapped closely alongside the mainchain, but that the two sidechain terminal sugar units are displaced further from the mainchain than is the α-linked mannose unit.

The similarities between the ordered structures of the variants and xanthan indicate that this conformation is quite robust. Any interactions involving the terminal two sugar units appear therefore not to be essential for stability of the structure. The stability of this conformation is consistent with the xanthan structure (represented by the diffraction pattern in Fig. 2a) being the only one trapped (so far) in oriented fibers, even at different humidities, different types of salt, and different pyruvate contents.

It is interesting that although the polytrimer can be regarded as a rather lightly substituted cellulose, the backbone adopts the 5-fold structure characteristic of xanthan, rather than the 2-fold structure characteristic of cellulose. It appears therefore that branching at alternate 3-positions in the backbone is the critical factor controlling the xanthan conformation. This is probably a result of both steric compression at the (1→4) linkage due to the branching, and favourable interactions between the α-linked mannose unit and the mainchain. This is in spite of the fact that, purely on stereochemical grounds, branching at the 3-position of alternate glucose units does not preclude the backbone from adopting a 2-fold cellulose-like conformation (12).

ACKNOWLEDGMENTS

We are grateful to the Joint Venture between Synergen Inc. and Texaco Inc. for providing us with the xanthan variant samples, Dr. M. Milas for provision of the xanthan samples, the U.S. National Science Foundation for support (DMB-8606942 to RPM), Deb Zerth for word processing and Robert Werberig for photography.

REFERENCES

1. Janson, P.E., Kenne, L. and Lindberg, B. (1975) Carbohydr. Res., **45**, 275-282.
2. Morris, E.R., Rees, D.A., Young, G., Walkinshaw, M.D. and Darke, A. (1977) J. Mol. Biol., **110**, 1-16.
3. Norton, I.T., Goodall, O.M., Frangou, S.A., Morris, E.R. and Rees, D.A. (1984) J. Mol. Biol., **175**, 371-394.
4. Milas, M. and Rinaudo, M. (1986) Carbohydr. Res., **158**, 191-204.
5. Moorhouse, R., Walkinshaw, M.D. and Arnott, S. (1977) ACS Symp. Ser. **45**, pp. 90-102. Am. Chem. Soc., Washington, DC.
6. Okuyama, K., Arnott, S., Moorhouse, R., Walkinshaw, M.D., Atkins, E.D.T. and Wolf-Ullish, C. (1980) ACS Symp. Ser. **141**, pp. 411-427. Am. Chem. Soc., Washington, DC,
7. Betlach, M.R., Capage, M.A., Doherty, D.H., Hassler, R.A., Henderson, N.M., Vanderslice, R.W., Marrelli, J.D. and Ward, M.B. (1987) in Industrial Polysaccharides: Genetic Engineering, Structure/Property Relations and Applications, (ed. Yalpani, M.) pp. 35-50. Elsevier, Amsterdam.
8. Millane, R.P., Narasaiah, T.V. and Arnott, S. (1989) in Recent Developments in Industrial Polysaccharides: Biomedical and Biotechnological Advances, (ed. Stivala,

S.S., Crescenzi, V. and Dea, I.C.M.). Gordon and Breach, New York, 1989, in press.
9. Millane, R.P. and Narasaiah, T.V. (1989) Carbohydr. Polym., in press.
10. Millane, R.P. and Arnott, S. (1985) J. Macromol. Sci. - Phys., **B24**, 193-227.
11. Moorhouse, R., Winter, W.T., Arnott, S. and Bayer, M.E. (1977) J. Mol. Biol., **109**, 373-391.
12. Millane, R.P. and Wang, B. (1989) Carbohydr. Polym., in press.

Part 6

CELLULOSICS AND SEED GUMS

Structure/function relationships of galactomannans and food grade cellulosics

IAIN C.M.DEA

Leatherhead Food Research Association, Randalls Road, Leatherhead, Surrey KT22 7RY, UK

ABSTRACT

Galactomannans are plant food reserve polysaccharides based on a ß1,4-linked D-mannan backbone which is solubilised by substitution of α1,6-linked D-galactose stubs to certain of the mannose residues. There are a number of other plant food reserve polysaccharides which are based on water insoluble ß1,4-linked D-glycans. Of these, the acetylated ß1,4-linked D-glucomannans, and the amyloids or xyloglucans which are based on cellulose. The galactomannans, acetylated glucomannans and plant seed xyloglucans are industrially important for their thickening properties, their self-gelling properties, and their participation in gel formation with other polysaccharides. Differences in structure affect these important functionalities. In addition to molecular size, structural properties such as degree of substitution and distribution of substituents along the main chain are important. While most of the data discussed is from studies on galactomannans, comparisons are also made on the effect of O-acetylation in the glucomannan series, and the effect of D-xylose and D-galactose substitution along cellulose.

INTRODUCTION

A number of plant food reserve polysaccharides are based on structural polysaccharides from the plant cell wall. Probably the most well known of such reserve polysaccharides, from an industrial viewpoint, are the plant seed galactomannans, which are based on a ß1,4-linked D-mannan backbone which is solubilised by substitution of α1,6-linked D-galactose residues to certain of the mannose residues (1). However, other 1,4-linked structural polysaccharides also form the basis for groups of plant reserve polysaccharides. Thus the food storage polysaccharides from tubers of the *Amorphophallus* genus are based on a ß1,4-linked D-glucomannan backbone, which is solubilised by a limited level of O-acetylation along the molecule

(2). Similarly the pentosans from cereal seeds are storage polysaccharides based on a ß1,4-linked D-xylan backbone, which is solubilised by substitution of α-L-arabinofuranosyl residues along the main chain (3).

The plant seed amyloids and the galactomannans also have strong similarities in structure. In the amyloids, a cellulose main chain is solubilised by substitution with D-xylopyranosyl residues and 2-O-ß-D-galactopyranosyl-D-xylopyranosyl disaccharide units via α1,6-linkages to certain of the D-glucose residues (4). In this case the amyloids are a less heavily substituted form of the plant cell wall xyloglucans which carry fucose containing trisaccharide side chains attached to the cellulose backbone, in addition to the monosaccharide and disaccharide side chains found in the amyloids (5). This is in contrast with the galactomannans, O-acetylated glucomannans and pentosans, which are more heavily substituted versions of plant cell wall polysaccharides.

A large number of galactomannans from different plant sources have been investigated (1), and it is clear that within the galactomannan series there is a wide spectrum of chemical structure. This diversity of structure includes not only a wide variation in the degree of galactose substition (galactose: mannose ratios vary from 1:5 to 1:1), but also significant differences in the distribution of galactose substituents along the mannan chain. The amyloid series and the acetylated glucomannan series have been much less studied, but there are clear indications that similar structural variations, at least in the level of substitution, occur (6,7).

It is not immediately obvious what the underlying reasons are for these wide structural variations in the plant food reserve polysaccharides. One possible factor could be the control and manipulation of water absorption and water holding properties of these polysaccharides when the seeds are wetted (8). The importance of this environmental protection activity could vary from species to species, and might be controlled by variations in the substitution patterns along the essentially insoluble main chains. Whatever the reasons, the structural variety of these polysaccharides is a "windfall" for industry, since it provides a range of functional ingredients with defined structures and different functionalities.

The galactomannans, amyloids and acetylated glucomannans have a number of potential applications in the food industry, which rely on their thickening properties, self-gelling properties, and their varying ability to participate in mixed gels with other polysaccharides. The effect of structural variation on these functionalities will be discussed in this paper.

THICKENING PROPERTIES

In the food industry many polysaccharides are used as simple viscosifiers to give shear thinning solutions. Here the polysaccharide molecules exist as fluctuating random coil chains. The viscosity behaviour is non-specific, in that when molecular weight is normalised for, a general pattern describing the concentration dependence and shear dependence of all polysaccharides of this type is found (9). For such polysaccharides, double logarithmic plots of zero shear specific viscosity against concentration multiplied by intrinsic viscosity, show a pronounced increase in gradient above a specific critical value, indicative of a general behaviour. The break in this plot is at the transition from dilute solution behaviour (low degree of coil overlap) to concentrated solution behaviour (total interpenetration of random coil molecules). Polysaccharides which exhibit this type of behaviour include λ-carrageenan, sodium alginate and carboxymethyl cellulose.

Interestingly, galactomannans which have been characterised in this way (locust bean gum and guar gum) exhibit departures from this generalised relationship between viscosity and concentration. These galactomannans show an earlier onset of concentrated solution behaviour and substantially greater concentration dependence thereafter. This special behaviour of galactomannans is attributed to the occurence of a proportion of specific inter-chain associations, which are of longer timescale than the non-specific physical entanglements between fluctuating random coil molecules. This is undoubtedly related to the inherent tendency of the β1,4-D-mannan to self-association. The close similarity in backbone structure and substitution pattern of amyloids and acetylated glucomannans might suggest that similar special viscosity/concentration behaviour would occur in these cases.

This particular thickening functionality would result in galactomannans, all other factors being equal, possessing more effective viscosifying properties than other random coil polysaccharides. However, economic cost effectiveness is of over-riding consideration in the application of thickening polysaccharides, and molecular weight together with price are the parameters which are vital in determining the cost that needs to be paid for a desired thickening, and therefore the decision on which polysaccharide to select for a particular application.

In the case of the two commercially important galactomannans, guar gum is cheap and has a high molecular weight, while locust bean gum is expensive and is lower in molecular weight. This results in guar gum usually being selected for thickening applications. The situation with the amyloid from *Tamarindus indica* seeds (tamarind gum) is somewhat complex. Tamarind gum is very cheap indeed but has a very much lower molecular weight than the two galactomannans. The net result of these two factors is that, when thickening is the only objective, usage of tamarind gum is non-economic. There are however a number of special applications in the food industry (e.g. production of stable food emulsions; production of freeze-thaw stable sauces), where thickening using high levels of low molecular weight materials like tamarind gum or indeed deliberately degraded galactomannans, for example by the controlled use of irradiation (10), could be the preferred approach.

SELF GELLING PROPERTIES

Tamarind gum has about 70% of the glucose residues in the cellulose backbone substituted by sugar stubs. While in aqueous situations it behaves as a rather inefficient thickener, it has the novel property of gel formation in the presence of high levels of sugar. Because of this property, tamarind gum has been suggested as a substitute for high-methoxy pectin in the jam, jelly and marmalade industry. One of its advantages over pectin is its ability to form gels over a wide pH range, including pH 7 (11). The maximum gel strength for tamarind gum is obtained when the sucrose concentration is 65%-72%, and at these conditions of low water activity, polysaccharide concentrations of 0.7 - 1.0% give excellent gel properties. The mechanism of gelation in high sucrose concentrations is likely to involve chain-chain associations of regular two-fold cellulose ribbon-like conformations. These chain-chain associations are presumably favoured over chain-solvent interactions at these low water activities.

Chain-chain associations would also be expected to occur in the galactomannan series, and to depend on the degree of substitution by galactose along the mannan chain. Locust bean gum has a low level of galactose (mannose: galactose ratio of 3.5:1) compared to guar gum (mannose: galactose ratio of 1.5:1). Another potentially important galactomannan, tara gum, has an intermediate galactose content (mannose: galactose ratio of 3:1). There is indeed a correlation between galactomannan structure and the ability to form gels at low water activity (12). Thus in 60% (w/v) sucrose, locust bean gum forms weak cohesive gels down to 0.2% concentration, while tara gum requires a concentration of 0.5% for gelation. At higher concentrations both galactomannans give firm gels, locust bean gum being the stronger. Under such conditions guar gum precipitates.

To extend structure/function comparisons, we have recently studied the rheological properties of konjac mannan under these low water activity conditions. Konjac mannan is an acetylated glucomannan in which approximately one out of every six glycosyl residues is acetylated (2). The degree of substitution along the glucomannan backbone is thus lower than that along the mannan backbone of locust bean gum. Despite that, the gelation properties in 60% (w/v) sucrose are poorer for konjac mannan, in that the gels which are formed are weak and very prone to syneresis. It is likely that these gelation differences arise from the substitution by acetate moieties instead of monosaccharide stubs, since the acetates could promote hydrophobic interactions.

The influence of structure is also found in the case of freeze-thaw induced gelation. Thus on freezing and re-thawing guar gum solutions, no evidence of permanent association is obtained, since a solution with unchanged viscosity is obtained (12). In contrast tara gum forms weak gels at concentrations above 0.75%, and locust bean gum forms a weak but cohesive gel network at concentrations as low as 0.5%. At lower concentrations (0.1%), both locust bean gum and tara gum give flocculant precipitates on freeze-thaw treatment.

We have recently extended these freeze-thaw studies, and have found no evidence of gel formation for konjac mannan (up to 1.5% concentration) and tamarind gum (up to 5% concentration). However, at lower concentrations, freeze-thaw treatment results in a small degree of precipitation in both cases. It is therefore clear that freeze-thaw treatment induces a degree of self-association in both the acetylated glucomannan and the xyloglucan, but that it is not extensive enough to result in gel formation. In the case of konjac mannan, this could be related to the substitution by acetate groups rather than sugar residues, while for tamarind gum the much lower molecular weight could be the cause.

These findings raise an interesting comparison between the galactomannans and the amyloids. The cellulose backbone of tamarind gum is substituted by xylose residues, to virtually the same extent as the mannan backbone of guar gum is substituted by galactose. Despite that, its self-association properties, induced by both low water activity and freeze-thaw, resemble more closely those of the much less substituted locust bean gum. The key to this anomaly lies in the additional substitution by galactose in the tamarind gum molecule; roughly half the xylose residues are substituted at O-2 by ß-D-galactopyranosyl residues. It has previously been shown (13) that enzymatic depletion of galactose from tamarind gum at a 1% concentration, using ß-D-galactosidase free of endo-ß-glucanase activity, results in firm gel formation when about 50% of the galactose residues have been removed. This indicates that the terminal galactose residues of tamarind gum are the major structural features determining self-association and water-solubility of the molecule. The xylose substituted cellulose backbone of tamarind gum has properties comparable to the mannan backbone of galactomannans, in that they both require extra substitution to confer water solubility. It has been reported that the selective enzymatic removal of galactose side chains from galactomannans increases the self association of the mannan backbone until, at a degree of substitution of *ca*. 88%, the product becomes insoluble (14). Interestingly, gel formation induced by galactose depletion of tamarind gum occurs at a similar degree of substitution by galactose along the xylose substituted cellulose backbone. The greater similarity of tamarind gum to locust bean gum, rather than guar gum, in self-association properties is perhaps not so surprising, since the respective degrees of substitution by galactose along the main chains are 33%, 30% and 64%.

The freeze-thaw precipitation of galactomannans at low concentration can be used to examine further the relationship between structure and self-association. Table 1 indicates that in general the lower the galactose content of the galactomannan, the more extensive the interaction (15). Some anomalies are however apparent. Thus, the comparison between locust bean gum (25% galactose) and *Caesalpinia pulcherima* galactomannan (24% galactose) is interesting. Although the latter has a slightly lower galactose content than locust bean gum, it precipitates significantly less on freeze-thaw treatment. This significant difference in self-association is not a result of large molecular weight differences, since the two galactomannans have closely similar intrinsic viscosities. This indicates that in addition to galactose content, the fine structure of galactomannans may be important in controlling the extent of self-association. A measure of differences in the distribution of galactose substitution along the mannan backbone of galactomannans may be obtained from an examination of the characteristic array of oligosaccharides produced by digestion with *A.niger* ß-D-mannanase, together with the degree of hydrolysis obtained (16). When this approach is used to compare locust bean gum and *Caesapinia pulcherima* galactomannan, significantly different data are obtained, analysis of which indicates that the latter has a galactose distribution along the mannan chain closely similar to statistically random, while locust bean gum has a non-regular, non-statistically random distribution of galactose with a higher proportion of unsubstituted blocks of intermediate length (15). This higher proportion of unsubstituted blocks of intermediate length is consistent with the observation that locust bean gum self associates to a much greater extent than *Caesalpinia pulcherima* galactomannan.

TABLE 1

SELF-ASSOCIATION PROPERTIES OF A RANGE OF GALACTOMANNANS AFTER FREEZE-THAW TREATMENT

Galactomannan sample	Galactose content (%)	Intrinsic viscosity (dL/g)	Amount precipitated on freeze-thaw treatment (%)
Medicago sativa	48	13.0	0
Trigonella foenum-graecum	48	-	0
Leucaena leucocephala	40	11.0	0
Cyamoposis tetragonolobus	40	14.3	0
Caesalpinia vesicaria	29	13.3	7
Gleditsia triacanthos	27	13.8	9
Ceratonia siliqua	25	9.9	37
Caesalpinia pulcherima	24	11.1	6
Ceratonia siliqua (hot-water-soluble fraction)	19	11.7	88
Sophora japonica	17	15.7	82

MIXED POLYSACCHARIDE GELLING SYSTEMS

1β4-linked glycans play a central role in mixed gelling systems. The most commercially important interactions are with the agars, κ-carrageenan and xanthan gum. Galactomannans, acetylated glucomannans and amyloids not only have the ability to strengthen agar and κ-carrageenan gels, but their presence can cause non-gelling systems to gel. Thus addition of these 1β4-linked glycans can bring about gelation in non-gelling concentrations of agars and κ-carrageenan, non-gelling low molecular weight segments of agarose and κ-carrageenan, and the non-gelling bacterial polysaccharide xanthan gum (1).

The galactomannans have been most studied in this context, and in general it is found that their degree of interaction with agarose, κ-carrageenan and xanthan gum increases will decrease in galactose content. In the case of agarose, the gelling interaction with galactomannans can be monitored by a perturbation in the temperature dependence of optical rotation; the cooling curve of mixed agarose/galactomannan systems take the form of a composite of the usual negative contribution from the agarose coil to helix conversion, and a new positive contribution from a galactomannan transition which does not occur in the absence of agarose (17). The hysteresis now shows a complex form, and is interpreted as showing that, although the disorder-order transitions of the two polysaccharides occur together on cooling, on reheating the melting transition of the galactomannan occurs first (Figure 1). It is found that the ability of a galactomannan to gel a non-gelling concentration of agarose, and the positive contribution to the optical rotation change on cooling agarose/galactomannan mixtures, in general decreases with increasing galactose content (1,18).

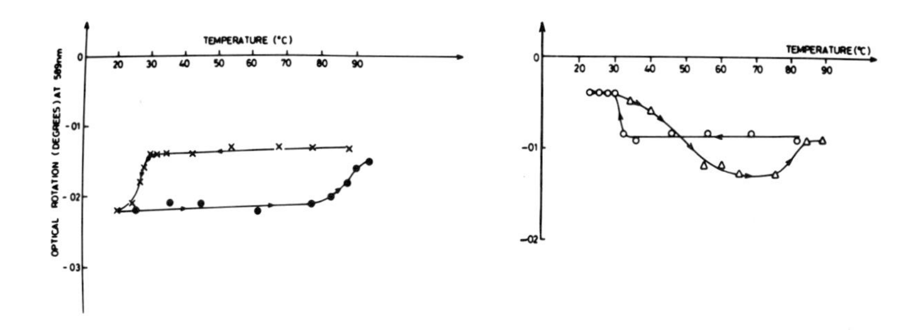

Figure 1. Optical rotation changes with temperature for agarose (0.05% w/v alone (left) and mixed with 0.1% (w/v) locust bean gum (right).

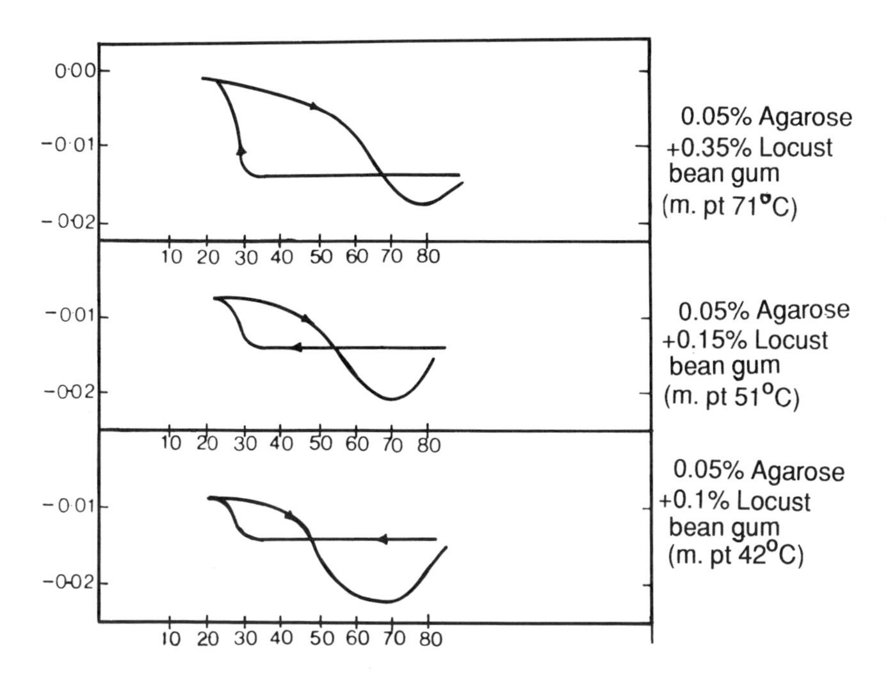

Figure 2. Comparison of the optical rotation variations with temperature for different agarose to locust bean gum ratios.

As in the case of galactomannan self-gelling properties, this is not the whole story; galactomannan fine structure has an important role to play. Thus the pair of galactomannans discussed earlier, locust bean gum and *Caesalpinia pulcherima*, interact to different extents with agarose. The lower degree of interaction of *Caesalpinia pulcherima* galactomannan is evidenced by the smaller positive change in optical rotation on cooling the mixed agarose/galactomannan system, and the significantly narrower hysteresis loop obtained on reheating (15). Again, the greater ability of locust bean gum to interact with agarose is consistent with the higher proportion of unsubstituted blocks of mannan backbone in the structure of this galactomannan compared with *Caesalpinia pulcherima* galactomannan.

The role of galactomannan/fine structure is also important in xanthan/galactomannan gels. Thus both these galactomannans form firm rubbery gels in admixture with xanthan. However, comparison of the storage modulus of mixed galactomannan (1%) xanthan (0.5%) gels, indicates that the gel formed with locust bean gum is significantly firmer (15). Furthermore, the melting temperatures of these gels can be determined by measuring the temperature dependence of the storage modulus, and the gel formed with locust bean gum is more temperature stable. (Melting point of mixed gel of 39°C, compared with 33°C for the *Caesalpinia pulcherima* galactomannan based gel).

The effect of galactomannan fine structure on mixed gelling properties is further illustrated by comparison of guar gum and the galactomannan from *Leucaena leucocephala*. These galactomannans have the same, high level of galactose substitution (39%) but have widely differing interaction properties. Addition of guar gum does not cause xanthan gum to gel; rather it enhances xanthan gums novel rheological properties. In contrast *Leucaena leucocephala* galactomannan interacts strongly with xanthan gum to give firm gels. Again this difference cannot be accounted for by molecular weight differences, since guar gum has by far the higher intrinsic viscosity. Examination of the oligosaccharide arrays produced by digestion with *A.niger* ß-mannanase indicates that the two galactomannans differ significantly in fine structure. The degree of hydrolysis is almost twice as great for the *Leucaena* galactomannan (10% compared with 5%) and the hydrolysate contains high proportions of trisaccharide 6'-α-D-galactosyl-ß-D-mannobiose (*ca*.16% of the total), indicating that there must be frequent regions of mannan chain in this galactomannan in which every second mannose residue is substituted by galactose. This would indicate that in the ordered two-fold conformation of the main chain, the molecule would have lengths of unsubstituted sides. Guar gum releases very small amounts of this trisaccharide on ß-D-mannanase hydrolysis, and consequently, will have few or any regions in the molecule which have smooth sides. Since these two galactomannans will have no substantial regions of mannan chain unsubstituted by galactose, the basis of their interaction with with xanthan gum arises from association with the xanthan gum ordered conformation via their unsubstituted sides. The difference in fine structure between the galactomannans, involving the extent of regions of unsubstituted sides in their structure, therefore accounts for their very different interaction properties.

Another important structural feature of these polysaccharides is the width of their molecular spectra. That locust bean gum has a wide molecular spectrum, is indicated by the fact that it can be fractionated on the basis of differential water solubility (17,19). The native locust bean gum (23%

galactose) can be fractionated into a cold water soluble fraction (25% galactose) and a hot water soluble fraction (20% galactose). This structural feature of locust bean gum affects its interaction with agarose. Thus, at high ratios of locust bean gum to agarose, the agarose interacts preferentially with the low galactose containing portion of locust bean gum to give stronger complexes. This is evidenced by wider hysterises loops in the temperature dependence of optical rotation plots, and higher gel melting temperatures, for the higher ratios of locust bean gum to agarose (Figure 2). Other galactomannans, such as tara gum, have much narrower molecular spectra, as indicated by fractionation using differential water solubility. As a result tara gum does not give mixed gels with agarose which differ significantly in melting temperature, when different galactomannan to agarose ratios are employed (17).

The interaction properties of the acetylated glucomannan, konjac mannan is worthy of special mention. Despite the fact that its degree of substitution by acetyl groups is less than the degree of substitution by galactose in locust bean gum, it interacts less well than locust bean gum with agarose. Thus a non-gelling concentration of agarose (0.05%) forms cohesive gels with 0.05% locust bean gum, while 0.25% konjac mannan is required for gelation. This is in line with the poorer self-gelling properties of konjac mannan mentioned earlier. In contrast, however, konjac mannan forms mixed gels with xanthan gum which are much stronger than those with locust bean gum. Thus a mixed gel of 0.25% xanthan gum and 0.25% konjac mannan melts at 63°C. By comparison, the highest melting temperatures observed for mixed xanthan gum/locust bean gum gels, under comparable conditions, is 40-43°C. Furthermore, recognisable xanthan gum/konjac mannan gels are formed at total polysaccharide levels as low as 0.02%.

Examination of the interaction properties of de-acetylated xanthan gum rationalises these observations. Thus, while de-acetylation of xanthan gum improves its interaction properties with galactomannans (20), it is detrimental to the interaction with konjac mannan. Mixed gels of de-acetylated xanthan gum and konjac mannan melt at around 40°C, compared with over 60°C for the native xanthan gum. This would indicate that hydrophobic interactions between the acetate groups in konjac mannan and xanthan gum play an important role in this interaction.

Finally, the molecular size of these polysaccharides has a major influence on their interaction properties. This is not surprising, since gelation or the enhancement of gel properties would be expected to be promoted by high molecular weight increasing the degree of cross-linking in the three dimensional network. Using chemically or enzymically degraded galactomannans it has been found that, with gelling concentrations of agarose and κ-carrageenan, either weaker gels or collapsed precipitates are formed (18). This can be rationalised by considering that the interaction of ß1,4-linked D-glycans with these gelling polysaccharides has two opposing effects. The molecular associations lead to more cross-linking of the double helix aggregates in the agarose and κ-carrageenan networks, and therefore tend to reinforce gel formation. However, the interaction also results in an increase in the degree of aggregation of these double helices and this has a tendency to collapse the gel. For high molecular weight ß-D-glycans there is enough cross-linking effect to counteract the tendency towards gel collapse. However, for low molecular weight versions the cross-linking effect is minimised and gel collapse dominates.

CONCLUSIONS

Wide structural variations occur within the galactomannan, acetylated glucomannan and amyloid plant storage polysaccharides. These variations in structure affect, in particular, the self-gelling and mixed gelling properties of these polysaccharides. A thorough understanding of these phenomena should permit the selection of polysaccharide structures best fitted for desired applications such as gel strength/texture and gel melting temperature.

REFERENCES

1. Dea, I.C.M. and Morrison, A. (1975) Adv. Carb. Chem. Biochem.,31, 241-312.
2. Kato, K and Matsuda, K. (1969) Agric. Biol. Chem., 33, 1446-1453
3. Aspinall, G.O. (1959) Adv. Carb. Chem. 14, 429-576.
4. Kooiman, P. (1960) Acta Botan. Neerl., 9, 208-219
5. Darvill, A., McNeil, M., Albersheim, P. and Delmer, D.P. (1980) In: The Biochemistry of Plants Vol.1, pp 91-160, ed. Tolbert, N.E., Academic Press, London.
6. Kooiman, P. (1961) Rec. Trav. Chim., 80, 849-865
7. Vernon, A.J., Cheney, P.A. and Stares, J. (1980) GB Patent 2,048,642.
8. Grant, G.T., McNab, C., Rees, D.A. and Skerrett, R.J. (1969) Chem. Commun., 805-806
9. Morris, E.R., Cutler, A.N., Ross-Murphy, S.B. and Rees, D.A. (1981) Carbohydrate Polymers, 1, 5-21
10. Marrs, W.M. (1988) In: "Gums and Stabilisers for the Food Industry 4", pp.399-408, IRL Press, Oxford, UK.
11. Savur, G.R. (1956) Chem. Ind., 212-214
12. Dea, I.C.M., Morris, E.R. Rees, D.A., Welsh, E.J., Barnes, H.A. and Price, J. (1977) Carbohydr. Res., 57, 249-272.
13. Reid, J.S.G., Edwards, M. and Dea, I.C.M. (1988) In: "Gums and Stabilisers for the Food Industry 4", pp. 391-398, IRL Press, Oxford, UK.
14. McCleary, B.V., Amado R., Waibel, R. and Neukom, H.(1981) Carbohydr. Res., 92, 269-285.
15. Dea, I.C.M., Clark, A.H. and McCleary, B.V. (1986) Carbohydr. Res., 142, 275 - 294.
16 McCleary, B.V., Clark, A.H., Dea, I.C.M. and Rees, D.A., (1985) Carbohyr. Res., 139, 237-260.
17. Dea, I.C.M., McKinnon, A.A. and Rees, D.A. (1972) J. Mol. Biol., 68, 153-172.
18. Dea, I.C.M. and Rees, D.A. (1987) Carbohydrate Polymers, 7, 183-224
19. Hui, P.A and Neukom, H. (1964) Tappi, 47, 39-42
20. Tako, M. and Nakamura, S. (1985) Carbohydr. Res., 138, 207-213

Production and applications of seed gums

WILLEM C.WIELINGA
Meyhall Chemical AG, Sonnenwiesenstrasse 18, CH-8280 Kreuzlingen, Switzerland

ABSTRACT

The availability of raw material of seed gums, locust bean-, tara- and guar gum and their production is described. Emphasis is laid on different qualities of these gums regarding stabilization, viscosity, purity and mesh analysis. The interaction of regular and coldswelling LBG with xanthan gum is dealt with. The applications of these gums in food systems are discussed briefly.

KEY WORDS

Lbg, tara- and guar gum, production, properties, applications, interaction.

INTRODUCTION

The main representatives of galactomannan gums are derived from the seeds of the carob tree (Ceratonia siliqua L.), of the guar plant (Cyamopsis tetragonoloba L.) and to a much lesser degree of the tara tree (Cesalpinia spinosa L.). The galactomannans are reserve carbohydrates, which are found in the endosperm of the seeds. These polysaccharides consist of a linear mannan backbone, the mannose units being linked together by β $(1 \rightarrow 4)$ glycosidic bonds. The galactose side stubs are attached as single side chains to the mannan backbone by $\alpha (1 \rightarrow 6)$ glycosidic linkages.

The galactomannans can be differentiated by their overall galactose content. The ratio of M/G, on an average basis, in lbg, tara- and guar gum is 4.5 : 1 3 : 1 and 2 : 1. These ratios are only a rough distinction since a fractionated precipitation of the in-water dissolved galactomannans of carob endosperms (from Portugese origin), by increasing amounts of ethanol, shows that the composition is heterogeneous (figure 1).

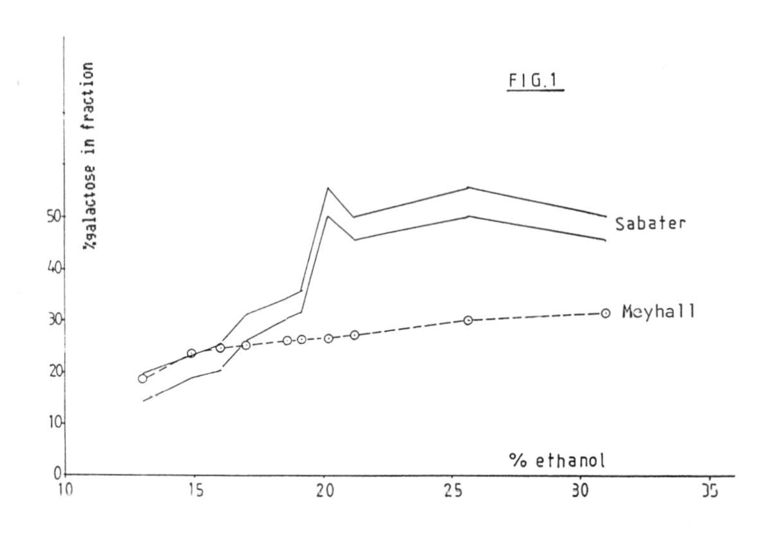

The surprising results of Sabater (1) could not be confirmed. In his work certain fractions showing galactose contents in excess of 50 % were obtained. Our findings of the M/G ratios remain above 2 : 1. Already in 1962 Hui (2) published M/G ratios of galactomannan fractions of a commercial lbg and guar product and these are given in Table 1.

Table 1.

Fraction	Amount wt%		M/G	
	guar	lbg	guar	lbg
Cold water soluble	69	35	1,3 : 1	1,2 : 1
Hot water soluble	10	52	1,7 : 1	3,3 : 1
Insoluble	17	10	7,0 : 1	5,2 : 1

A model of a galactomannan is given in figures 2 and 3.

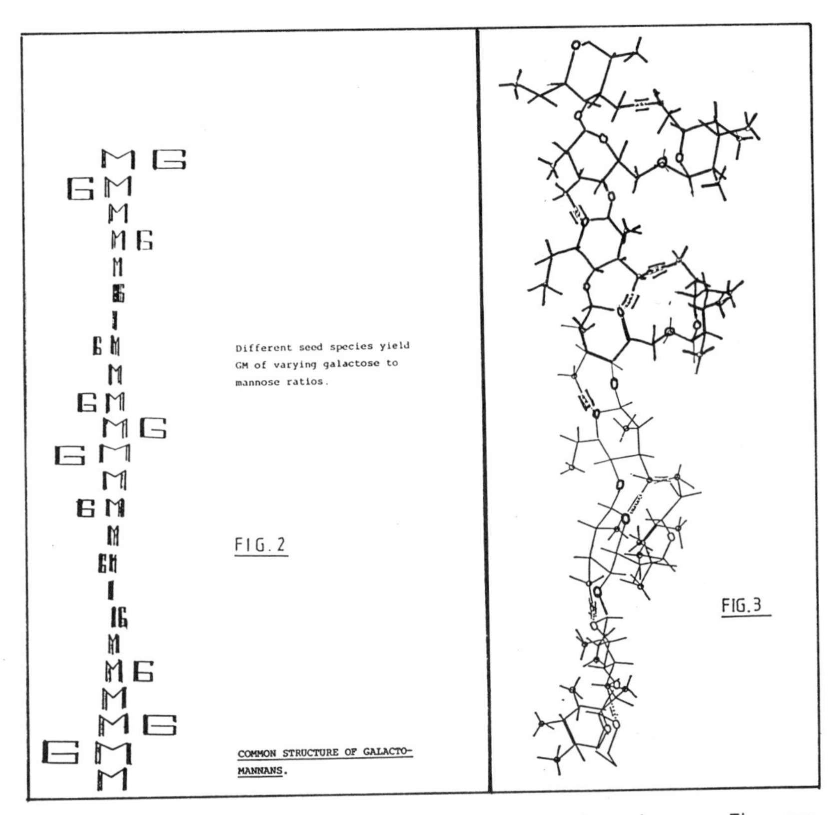

Commercial products do not always consist of the whole endosperm. They are specified mainly according to viscosity, protein content, acid insoluble residue and particle size distribution. Taking also into account the different growth and storage conditions (geographical locations, weather) and different seed varieties to mention only some factors, an exact definition of guar and locust bean gum becomes difficult.

This statement can be illustrated by the morphological cross section of a guar endosperm half (figure 4). It shows that after grinding and classification products of different cell compositions must be obtained.

The cells of the endosperm halves of the lb-kernel are not as different morphologically as those of the guar seed, but also here distinctions can easily be made.

These factors and others lead to the poor agreement about the compositions of the galactomannans between different workers, even more so when the analysis technique is not defined. In lbg a variety of M/G ratios have been determined, as shown in Table 2.

Table 2.

Worker	Year	M/G
Bourquelot and Herissey	1899	5,1 : 1
Williams	1928	2,6 : 1
Iglesias	1935	4,0 : 1
Lew and Gortner	1943	3,5 : 1
Wise and Appling	1944	3,7 : 1
Courtois and Dizet	1966	3,9 : 1
McCleary et al	1983	3,2 - 4,0 : 1

The overall M/G ratios of Guar Gum vary from 1.48 : 1 to 1.80 : 1.

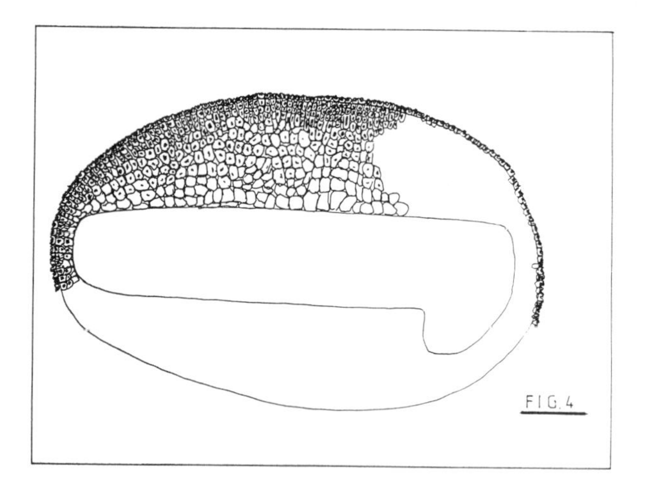

AVAILABILITY OF SEEDS

Annually about 10-12 thousand tons of lbg and up to 125 thousand tons of guar gum are produced worldwide. The amount of tara gum may be as high as about 1000 tons. The average carob fruit amounts to about 350'000 tons annually, of which 150'000 tons grow in Spain/Portugal (in the 50's 500'000 tons were collected there). Guar is mainly grown in the semi-arid zones of India and Pakistan. Since 1944 it is also grown in the USA in Texas. In order to have two crops a year, the guar plant has in recent years also been cultivated in the Southern Hemisphere, i.e. in Brasil, Africa and Australia. So far the harvest of guar seeds in India can be as high as 1,4 MM tons per year, or as low as 0,28 MM tons. This fluctuation is mainly due to weather conditions. The carob tree is growing around the Mediterranean Sea, in Morocco and Portugal. Recently much more attention is paid to the systematic cultivation of these trees. The tara tree grows in Peru. Efforts are made - in cooperation with international organizations, agricultural authorities and farmers - to set up programmes for the improvement of carob tree cultivation and a better organization of harvesting and utilization of carob pods.

The composition of the seeds can be seen from following table 3.

Table 3.

Component	wt%		
	Carob kernel	Guar seed	Tara kernel
Hull	30 - 33	20 - 22	ca. 38
Endosperm	40 - 50	32 - 36	ca. 22
Germ	20 - 25	44 - 46	ca. 40

PRODUCTION OF TARA AND LBG

Tara and carob kernels are difficult to process, since the hull is very tough and hard. Special processes had to be developed to peel the kernels without damaging the endosperm and the germ.

Acid process

The kernels are treated by sulfuric acid at elevated temperature to carbonize the hull. The remaining hull fragments are removed from the clean endosperm/germ "sandwich" in efficient washing and brushing machines. After a drying step these "sandwiches" are cracked and the crushed germ, being much more friable, can be sifted off from the unbroken endosperm halves. These halves then can be ground and the powder can be classified to obtain the commercial products.

Roasting process

The kernels are roasted in a rotating furnace, where the hulls more or less pop off from the rest of the kernels. The germ and endosperm halves are recovered as mentioned above. This process yields products of somewhat darker colour. The advantage, of course, is that no sulfuric acid is necessary and therefore no effluent. (3)

PRODUCTION OF GUAR GUM

The production of guar endosperms and guar meal from the seeds is depicted in flowsheet I (figure 5). This has been discussed previously (4) and is now only briefly described:

The whole seeds can be fed into an attrition mill, or any type of mill having two grinding surfaces travelling at different speed. The seed is split into two endosperm halves covered with hull and into fine germ material, which later can be sifted off. The crude crack (endosperm + hull) is heated to soften the hull and is fed into a mill, which can either abrade the hull away from the endosperm or into a hammermill to shatter the hull away. An idealized seed cracking operation is shown in figure 6. Any remaining germ parts are pulverized during this step and the resulting split after a sifting operation is essentially pure endosperm. The fine material is marketed as cattle feed and is called guar meal, which shows a minimum protein content of 35 % (%N x 6,25). The guar endosperm can be worked up by suitable milling and screening techniques to produce the commercial products.

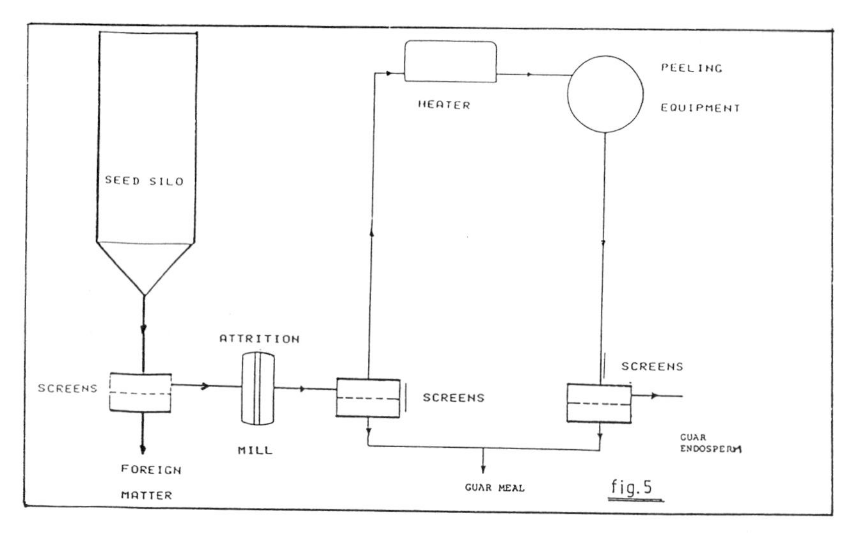
SEED SILO
SCREENS
FOREIGN
MATTER
ATTRITION
MILL
SCREENS
HEATER
PEELING
EQUIPMENT
SCREENS
GUAR
ENDOSPERM
GUAR MEAL

fig. 5

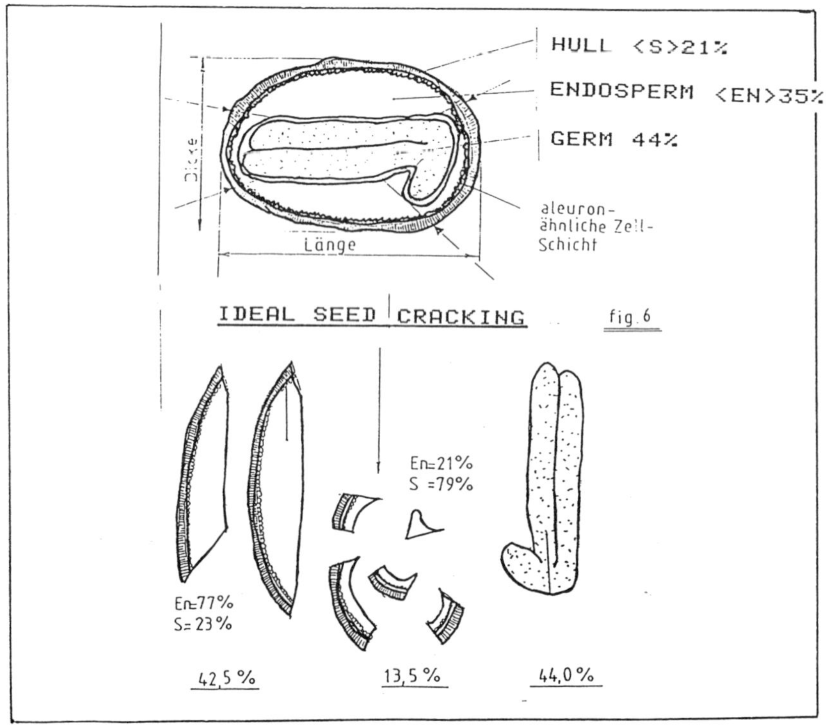
HULL <S>21%
ENDOSPERM <EN>35%
GERM 44%
aleuron-
ähnliche Zell-
Schicht
Dicke
Länge
IDEAL SEED CRACKING
En=21%
S =79%
En=77%
S= 23%
42,5 %
13,5 %
44,0 %

fig. 6

SPECIFICATIONS OF SEED GUMS

Typical analysis data of commercial lbg and guar gums, as well as EC specifications and FAO/WHO requirements for Tara-gum are shown below (table 4-6).

Table 4.

Specification	Typical analysis of lbg			
	High grade M 175 / M 100		Industrial	Technical
% H_2O	10,0 - 12,0		10,0 - 14,0	10,0 - 14,0
% Protein	6,5		5,7 - 7,0	6,0 - 13,0
% AIR	2,0		2,0 - 4,5	5,0 - 8,0
1 % Viscosity 10' at 86-89°C RVT Brookfield 20 rpm, 25°C mPa.s	min. 3000		1500 - 1800	500 - 1000
- M 80 (max.)	99 %	99 %	99 %	99 %
- M 200 (max.)	25 %	10 %	25 %	not specified
Metal				
As	0,2 ppm			
Pb	ca. 0,03 ppm			
Cu	2,5 ppm			
Zn	5,6 ppm			
Cd	ca. 0,05 ppm			

AIR = acid insoluble residue

Table 5.

Specification	Typical analysis of high grade guar gum				
	M 100	M 175	M 200	M 225	M 200/50
% H_2O (max.) % Protein (max.) % AIR (max.) % Ash (max.) pH 1 % solution	12,0 5,0 3,0 1,0 6,5				12,0 4,5 2,5 1,0 6,5
1 % Viscosity fully hydrated RVT Brookfield 20 rpm, 25°C) mPa.s (min.)	3000	3000	3000	3600	5000
+ M 100 (max.)	1 %	1 %	-	-	-
+ M 150 (max.)	-	-	1 %	1 %	1 %
+ M 200 (max.)	-	-	-	15 %	-
- M 150 (max.)	15 %	-	-	-	-
- M 200	-	max. 25 %	min. 80 %	-	min. 90 %
- M 250 (min.)	-	-	-	75 %	-

Metal content of guar gum is similar to that one of lbg (see table 4). EC specifications for E 410 (lbg) and E 412 (guar) are as follows (WHO/FAO requirements for tara gum are also shown-table 6):

Table 6.

		E 410	E 412	Tara gum
Galactomannan %	(min.)	75,0		75,0
Protein %	(max.)	7,0		3,5
AIR %	(max.)	4,0		2,0
H_2O %	(max.)	14,0		15,0
Ash %	(max.)	1,2	1,5	1,5
Starch		neg.	neg.	neg.
Heavy metals	(max.)	20	20	20
As	(max.)	3	3	3
Pb	(max.)	10	10	10
in ppm				

These specifications do not include the water soluble content of the gums. The galactomannan content does not specify which of these polysaccharides are water soluble and which are not. Tara gum is available in similar qualities regarding particle size distribution. Its viscosity (of 1 % aqueous solutions) is given in Table 7.

Table 7.

	M 175	M 200
Cold viscosity	3000 mPa.s	3000 mPa.s
Hot viscosity	4400 mPa.s	4000 mPa.s

The data so far allow the producer of seed gums to meet reproducibly the specifications. The viscosity illustrates only the thickening effect in water and of course not in a food system. The chain length of galactomannans and thus viscosity can be adjusted by enzymatic-, oxidative, hydrolytic- and thermal processes. Specific requirements, such as rheology and stabilization can be met by such depolymerized products. Hydrolytic depolymerizations of these gums occur, if acidic food systems are pasteurized for instance at pH values below 3.5. If hydrations can take place at 25°C at pH values of 2.5-3.5 no extensive acid hydrolysis will be obtained. However on extended shelf life at room temperature a gradual hydrolysis will take place.

SOME FUNCTIONAL PROPERTIES OF GUAR GUM

Hydration of guar gum

The hydration rate up to 120 minutes, as measured with a Brookfield RVT Viscometer at 20 rpm and 25°C for a 1 % aqueous solution of a fine mesh guar gum M 225 can be expressed by following equation:

$$\eta_a = 452 + 656 \ln t$$

where

η_a = apparent viscosity in mPa.s at 20 rpm
t = hydration time in minutes (max. 2 hours)

Viscosity/concentration relationship of aqueous guar gum solution

The apparent viscosity (Brookfield) after 2 hours and concentration in aqueous systems can be obtained using following equation:

Fine mesh guar M 225:

$$\eta_a = 3105 \times [C]^{3.11}$$

Coarse mesh guar M 100:

$$\eta_a = 2274 \times [C]^{3.27}$$

η_a = apparent viscosity in mPa.s (RVT, 20 rpm, 25°C)
C = concentration in wt%, as is.

These equations show that a twofold concentration yields a 8-10 fold viscosity.

Viscosity of aqueous guar gum M 200 solutions and temperature

A 1 % solution of guar has a lower viscosity at higher temperatures, as is also true for all other seed gums. This is illustrated in Table 8.

Table 8.

Temperature °C	Viscosity, 20 rpm, 25°C
20	4100 mPa.s
40	3150 mPa.s
60	2500 mPa.s
80	2050 mPa.s

SPECIAL GUAR GALACTOMANNANS

These commercial products are almost sterile, neutral in taste and are produced from selected raw material purified in special ways. The viscosity range of purified or depolymerized products is adjusted to 5 - 35'000 mPa.s for a 2 % aqueous solution, measured with a RVT Brookfield viscometer at 20 rpm and 25°C. The guar gum types A and B are purified and the types C, D and E depolymerized. The product with the lowest viscosity is called type F. This product is applied at spray embedding of sensitive vitamins, for instance. Experience has shown that during the production of the types C, D and E practically no functional groups, like carboxylic- or keto groups are introduced. The % mannose has not been changed, as illustrated in Table 9.

Table 9.

Guar gum type	% Mannose
C	65,8
D	65,5
E	65,8

No galactose was split off.

In 1976 the Swiss Health authorities, division Food department, have permitted to use guar gum F and E as thickening and stabilizing agents for food purposes (Art. 443 to paragraph 2 of the Swiss Food Law). As said before type F has the lowest viscosity i.e. 10 % = 700 - 1000 mPa.s (average molecular weight about 35'000). Table 10 illustrates the intrinsic viscosities of the different watersoluble guar galactomannans, determined in 1976, which can serve to estimate the average molecular weight, using the equation of Robinson et al (5);

$$[\eta] = 3.8 \cdot 10^{-4} \cdot M^{0.723}$$

Table 10.

Product	wt%	Viscosity Brookf. RVT,20rpm 25°C, after 1 h mPa.s	Insolub-les wt%	Conc. super natant wt%	Intrin-sic viscosity [η] dl/g	Mol. wt% x 10^6
Guar Gum A	0,96	4'550	26	0,70	14,58	2,19
B	1,10	5'050	29	0,79	13,31	1,93
C	1,60	3'850	21	1,27	8,23	0,99
D	3,00	4'600	19	2,40	4,69	0,46
E	3,63	2'250	16	3,09	3,41	0,29
F	9,22	600	19	7,96	0,74	0,035
LBG M-175	1,06	3'850	14	0,85	11,81	1,70
M-200	1,16	4'760	15	0,93	11,43	1,64
Guar Gum A	0,96	4'550	26	0,70	14,58	2,19
pH 4 10'	0,96	4'450	26	0,71	13,45	1,96
pH 4 30'	0,96	4'200	23	0,72	13,41	1,95
pH 9 30'	0,96	4'750	25	0,72	13,80	2,03
LBG M-200	1,16	4'760	15	0,93	11,43	1,64
pH 4 10'	1,16	4'450	16	0,92	11,07	1,59
pH 4 30'	1,16	4'400	15	0,93	11,00	1,58
pH 9 30'	1,16	4'800	15	0,93	11,72	1,68
Guar Gum 200/50	0,96	5'350	24	0,72	15,16	2,31
pH 4 30'	0,96	5'100	20	0,75	13,75	2,02
pH 9 30'	0,96	5'150	21	0,72	14,61	2,19

The products have been dissolved at 86-89°C for 10 minutes while stirring. Also the water insolubles of these products have been determined (spun out at 35'000 g for 30 minutes, without further washings). To estimate the average molecular weight of investigated lbg products, the equation of Doublier and Launay (6) has been used.

$$[\eta] = 9.28 \times 10^{-6} \cdot M^{0.98}$$

More work is needed to verify the use of these equations. McCleary et al (7), the following average intrinsic viscosity values, as well as average contents of galactose of hot water - (HWS) and cold water soluble fractions (CWS) and total soluble part of locust bean gum (table 11). 10 varieties of carob seeds, 4 commercial lbg's and 11 types of seeds from 11 countries were investigated.

Table 11.

Parameter	HWS	Total Carob	CWS
D-Galactose %	18 ± 1	22 ± 2	25 ± 1
[η], dl/g	13 ± 2	12 ± 2	10 ± 2

A remark was made that the amounts of HWS to CWS varied from as much as 1 : 2 to 4 : 1, which again pointed out the difficulty of defining lbg. Presently the guar gum types A, B, C, D and E are produced by an improved process as shown by the insoluble fractions, reported in Table 12.

Table 12.

Type guar gum	% Insolubles	
	1976	1988
A	26	20
B	29	18
C	21	12
D	19	10
E	16	7

Another way to adjust the viscosity of guar products is to depolymerize the gums by thermal hydrolysis. Products with a viscosity range of 500 - 20'000 mPa.s at 2 % concentration are made. Figure 7 depicts as an example the relationship between viscosity (Brookfield viscometer RVT) and concentration of the guar gum types A, B, C, D and E.

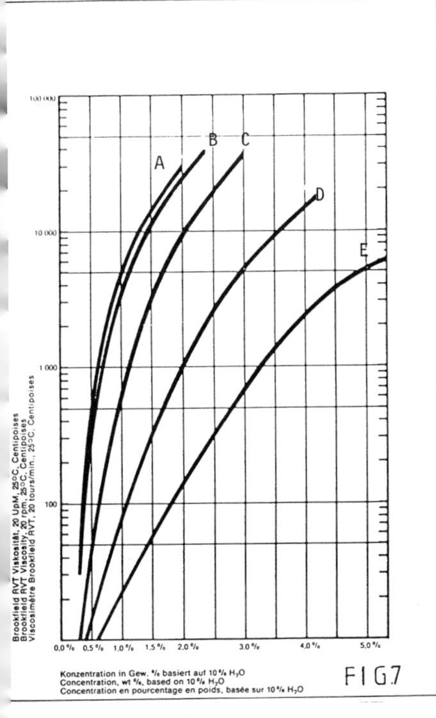

FIG7

Effect of concentration on Brookfield viscosity of Guar types A-E (own pH) prepared at 86-89 C for 10 min in demineralized H_2O viscosity measured at 25 C after 1 hour.

COLD SWELLING LBG

A selection of raw material and a multistage process enables the production of cold swelling lbg, at which 60 % of the viscosity of the fully hydrated gum (10 minutes at 86-89°C) is achieved at 25°C. Regular lbg with same overall composition and particle size distribution attains only 13-15 % of its hot viscosity at 25°C. Table 13 shows the amounts of spun-out material from 1 % solutions (30 minutes at 35'000 g) of regular- and cold swelling lbg.

Table 13.

Product	Insolubles wt%	Solubles, wt%	
		86-89°C	25°C
lbg M 200	17,0	44,0	39,0
cold swelling lbg	14,0	16,0	70,0

The data show that more than 80 % of the total water soluble galactomannan, present in the cold swelling version of lbg have become soluble, whereas in the regular lbg only 47 % dissolve at 25°C.

Lbg shows efficient interactions with carrageenans, especially with the Kappa type (table 14) and with biopolymers such as xanthan gum after heating of the aqueous system. With the cold swelling lbg such interactions occur already at lower temperatures.

Table 14.

Kappa Carrageenan / Lbg Synergism	
System	water gel strength (grams breaking force)
1. 0,5 % Kappa Carrageenan + 0,25 % KCl	60
2. 1 + 0,1 % of guar gum	57
1 + 0,1 % of lbg	214
3. 1 + 0,2 % of guar gum	56
1 + 0,2 % of lbg	314
4. 1 + 0,3 % of guar gum	57
1 + 0,3 % of lbg	390
5. 1 + 0,5 % of guar gum	62
1 + 0,5 % of lbg	481
6. 1 + 1,0 % of guar gum	86
1 + 1,0 % of lbg	555

INTERACTIONS OF LBG, ITS COLD- AND HOTWATER FRACTION AND COLDSWELLING LBG WITH XANTHAN GUM

1,5 % solutions containing 35 wt % of xanthan gum and 65 wt % of LBG, its fractions, or the coldswelling version were prepared at 86 - 89° C for 10 minutes, while stirring, cooled to 60° C, then filled into normal 250 ml. beakers. Next day the gelstrength was measured with the Boucher Jelly tester. Fractionation was obtained spinning out the insolubles after dissolution at 25° C or at 86 - 89° C for 10 minutes.

Galactomannan-component	Gelstrength Jelly tester Ø = 2,5 cm
1. Coldswelling LBG	150
2. Coldswelling LBG (+ 5 % sugar)	112
3. LBG M-200 (regular)	132
4. Coldswelling LBG	
4.1. Coldswelling fraction	156
4.2. Hotswelling + insoluble fraction	147
5. LBG M-200 (regular)	
5.1. Coldswelling fraction	110
5.2. Hotswelling + insoluble fraction	148

(Insoluble fractions of 4.2. and 5.2. are respectively 47 and 28 %)

Another commercial LBG, M-175 was used for interaction with xanthan gum varying the ratio, applying 1 % aqueous solutions at natural pH and at a pH of 3.0. Again a Boucher Jelly tester was used for gelstrength. The results are given in Table 15.

Table 15.

	Jelly Tester Gelstrength * (1 % solution)			
LBG M-175 Xanthan gum	30 70	50 50	75 25	90 10
Own pH	448	445	434	252
pH = 3.0	448	445	407	158

In presence of 0,1 % of KCl
* stamp diameter: 5,0 cm

RHEOLOGY OF SOME STABILIZERS IN WATER

The functionality of galactomannan products and others in aqueous systems can be described more precisely if reliable rheological data are available. Especially the behaviour at low and very low shear rates is important considering the long shelf life of some of the food products.

If the ascent of a test solution in a glass tube with an inner diameter of 2 - 5 mm is computed into shear rates and shearing stresses periodically, the rheology parameters can be obtained. The rheometer, which was used for this purpose, allows the investigation of such solutions.

LBG, Guar gum, depolymerized Guar gums, CMC, Na-Alginate and Xanthan gums were investigated in aqueous solutions showing a Brookfield RVT viscosity of 3000 m.Pa.S (20 rpm, 25° C). A simplified Cross-equation describes the flow of most of the investigated systems. Table 16 shows the calculated parameters of this equation from the best fits:

Table 16.

Product	%	Simplified CROSS EQUATION $\eta = \frac{\eta_0}{1 + \alpha . D^n}$						
		η_0 mPa.s	α (s)	n	r^2		a	b
					CROSS	$\eta = aD^b$		
Guar (straight)	0.81	11033	0.709	0.747	0.9996	-	-	-
Depol. Guar (1)	12.42	4627542	744.689	0.493	0.9822	0.9925	7257	- 0.5672
Depol. Guar (2)	4.38	4459	0.353	0.443	0.9931	-	-	-
Depol. Guar (3)	6.56	5563	0.625	0.372	0.9737	0.9785	3405	- 0.1906
LBG	0.99	3776	0.090	0.860	0.9976	-	-	-
Xanthan Gum	1.31	3292582	257.082	0.751	0.9978	?	14265	- 0.7665
CM-Starch	5.26	-	-	-	-	0.9940	5869	- 0.4283
CM-Guar	6.59	3490	0.091	0.762	0.9947	-	-	-
CM-Cellulose	1.65	3778	0.576	0.504	0.9963	-	-	-
Na-Alginate HV	1,91	9408	1.386	0.392	0.9855	-	-	-

Two shear thinning profiles can be seen from figures 8 and 9 (the former agrees well with literature information). These profiles show that the rheological behaviour can be adjusted more or less to desired need.

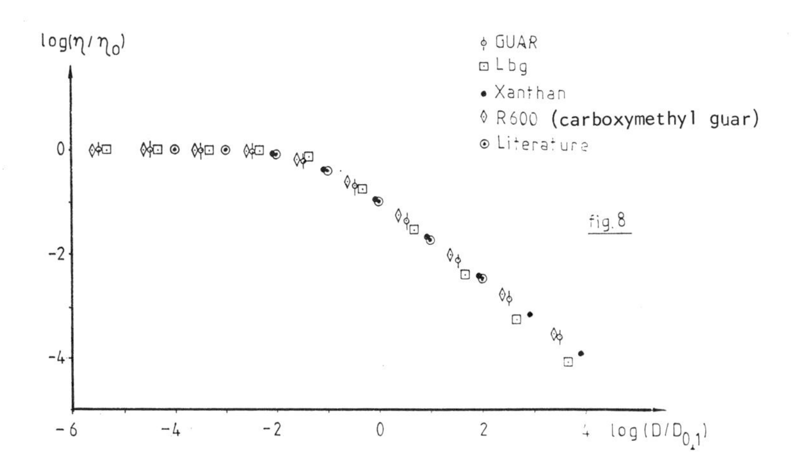

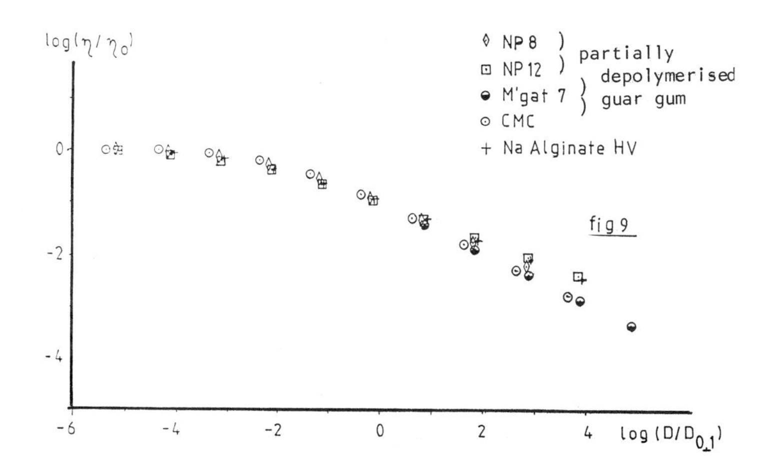

Figs.8 & 9

log relative viscosity against log $\frac{\text{shear rate}}{\text{shear rate required to reduce viscosity by a factor of 10.}}$

APPLICATION OF SEED GUMS IN FOOD SYSTEMS

All three gums have no restricted ADI-values and in common, they thicken aqueous systems very efficiently, thereby controlling the mobilization of water. This influences the consistency, body and shelf life of aqueous food systems as well as the stabilization of O/W and W/O emulsions. These seed gums in combination with minor amounts of carrageenan allow the production of ice cream with long shelf life, at least up to 18 months without changing the quality too much. These gums also influence the shape retention and meltdown of such frozen products and protect them in case of heat shocks.

The functionality of these gums can be adjusted enormously by their synergistic behaviour with carrageenans and biopolymers (pet food and dairy products).

The retrogradation of starches and modified starches in bakery products can be retarded or prevented.

Efficient stabilizers are obtained using blends with unmodified or modified starches in convenience food (salad dressings, prepared meals).

The passage of food through the stomach can be retarded, thickening its content. This can assist medical treatment of diabetics, since resorption of glucose for instance is slowed down.

Emulsification can be enhanced, if these gums are blended with starches and/or modified starches, as well as with protein like sodium caseinates, or with protein containing products like skim milk and whey powder (convenience food).

Depolymerized guar products are applied in pasteurised sour milk products for which no increase in viscosity is desired and the protection of protein in acidic media is required (quarq, processed cheeses).

Strongly depolymerized guar products achieve following effects at spray embedding:

- Considerable improvement in the shelf life of sensitive pharmaceuticals.
- Rate of dissolving and resorption of the active substance in the spray dried product is substantially increased compared to the behaviour of the same substance when ground in the conventional manner, due to its extremely fine distribution.
- The spray dried powders show good free-flowing properties, which facilitates further processing.
- Substances difficult to dose and handle, such as oily substances, become easy to process after spray embedding.
- Highly active substances can be homogenised in gamma quantities, if dissolved in aqueous solutions of such depolymerised guar products, before spray embedding.

Figures 10 and 11 show the course of Pallidin level in the blood in two in vivo tests on dogs. The resumption-rate of drugs sprayed in combination with such Guar products is substantially enhanced.

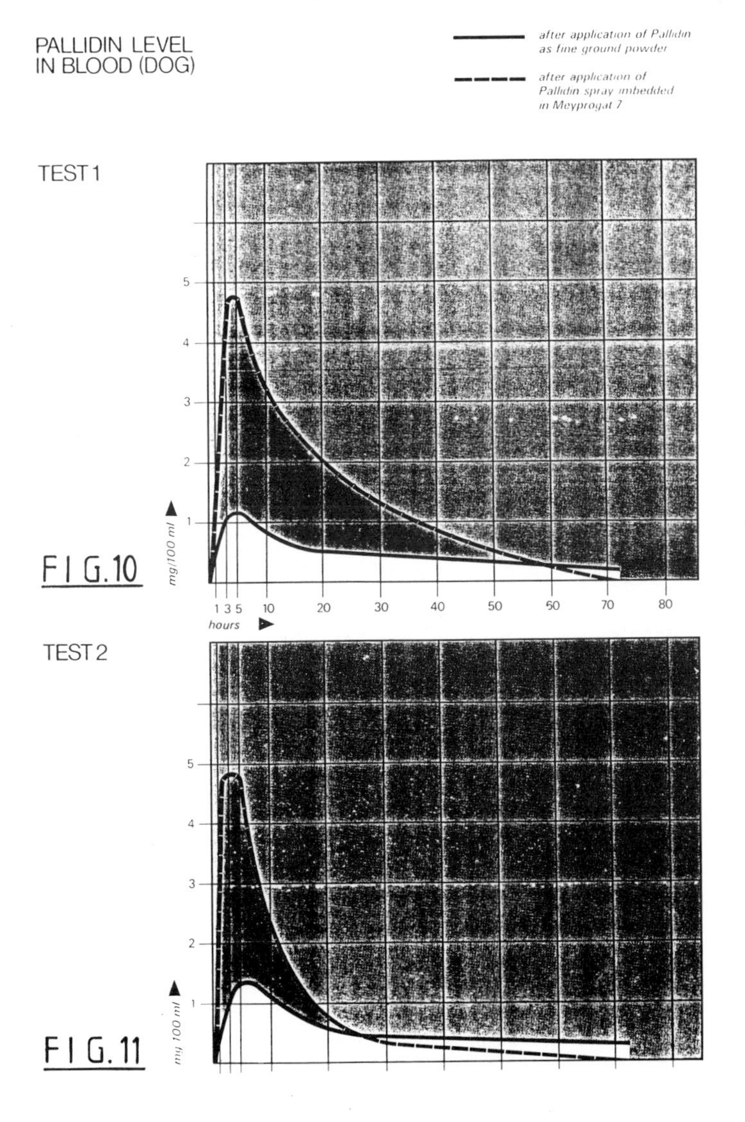

Further proof of the stabilising ability of these gums is demonstrated in figure 12, showing melt down behaviour of dairy ice cream. Furthermore the gradual hydrolysis of guar gum (MS-1) and of a blend of guar and lbg (MS-2) in the water phase of a salad dressing with the following composition is shown in figure 13.

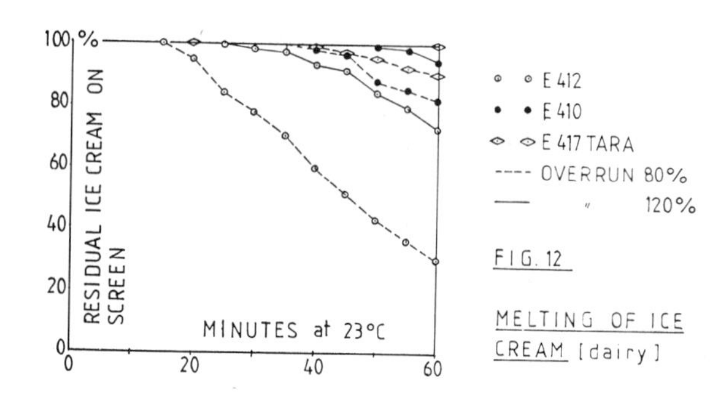

FIG. 12

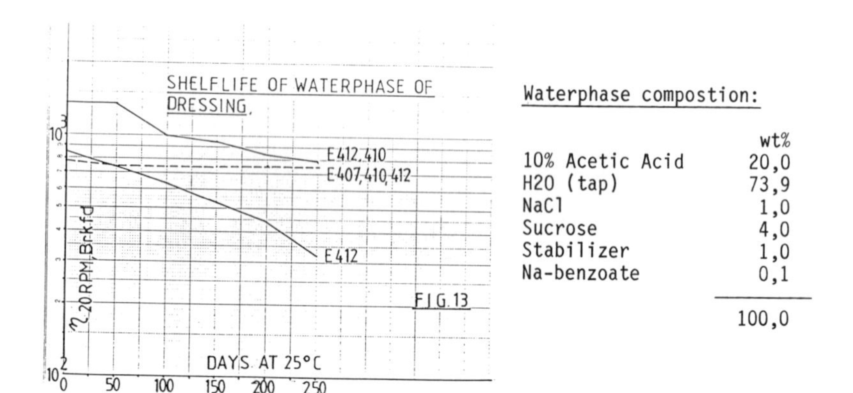

FIG.13

Waterphase compostion:

	wt%
10% Acetic Acid	20,0
H20 (tap)	73,9
NaCl	1,0
Sucrose	4,0
Stabilizer	1,0
Na-benzoate	0,1
	100,0

Also the contribution of carrageenan to such a blend of guar- and lbg is demonstrated thereby maintaining almost the initial viscosity. (MS-3)

References

1. Sabater de Sabates, A. Doctoral Thesis, University of Paris XI, Centre d'Orsay, 21-2-79

2. Hui P.A., (1962) Doctoral Thesis, ETH, Zürich Prom. No. 3297

3. Nittner E., Ullmanns Encyklopädie der technischen Chemie, Bd. 19, Chapter 12

4. Wielinga W.C. in "Gums and Stabilisers for the Food Industry 2" eds. (Phillips, G.O., Wedlock, D.J. and Williams, P.A.) publ.Pergamon Press (1984)p 251-276.

5. Robinson G et al. Carbohydrate Research (1982) 107, 17-32.

6. Doublier J.L. and Launay B., J.Texture Stud., (1981), 12, 151.

7. McCleary B.V., Dea, I.C.M. and Clark A.H. in "Gums and Stabilisers for the Food Industry" eds (Phillips, G.O., Wedlock, D.J. and Williams, P.A.) publ. Pergamon Press (1984), p33-44.

Application of microcrystalline cellulose

E.J.MCGINLEY and D.C.TUASON JR.
Food & Pharmaceutical Products Division, FMC Corporation, Princeton, New Jersey, USA

ABSTRACT

This paper describes the technology of colloidal microcrystalline cellulose (MCC) and the application of that technology to food industry product and processing problems with particular reference to the physical modification of the cellulosics to fit specific needs in diverse food systems.

1 INTRODUCTION

The application of MCC technology to food industry product and processing problems demonstrates properties uniquely different from most hydrocolloids and provides effective stabilization in a variety of food product systems. The practical importance of these properties has led to a highly developed application technology of colloidal microcrystalline cellulose and an extensive study of the nature of its functional properties. The following article describes several of these properties including ice crystal control, texture modification, emulsion stabilization, heat stability, foam stability, suspension of solids, and fat replacement relative to their utility in practical end uses.

2 THE TECHNOLOGY OF COLLOIDAL MICROCRYSTALLINE CELLULOSE

Cellulose, in the chemical sense, is a polysaccharide of sufficient chain length to be insoluble in water or dilute acids and alkalies at ordinary temperatures (1). It consists of anhydroglucose units linked together through the 1 and 4 carbon atoms with a beta-glucosidic linkage. Thus,it has the following repeating unit (2), with n having values ranging from about 50 to 5000 or more (Figure 1).

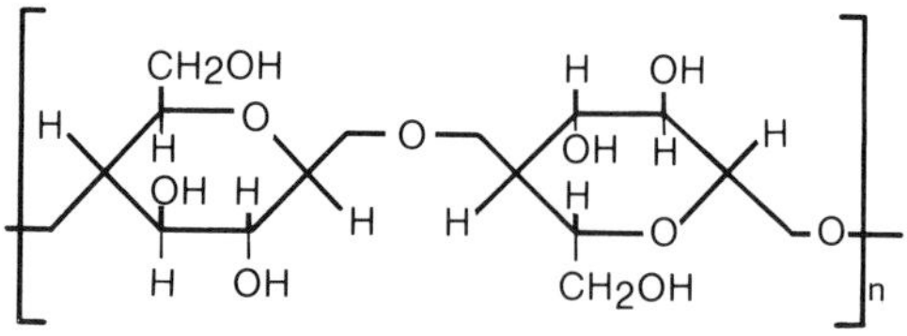

Figure 1

Fibrous cellulose is comprised of millions of microfibrils. The individual microfibril is composed of two regions, the paracrystalline region, which is an amorphous and flexible mass of cellulose chains, and the crystalline region, which is composed of tight bundles of cellulose chains in a rigid linear arrangement (Figure 2). Microcrystalline cellulose (MCC) is a purified, naturally occurring cellulose, produced by converting fibrous cellulose to crystalline cellulose or a redispersible gel. This is accomplished by a simple acid hydrolysis to a "level off degree of polymerization" (LODP), drying and/or coprocessing with various hydrophilic dispersants.

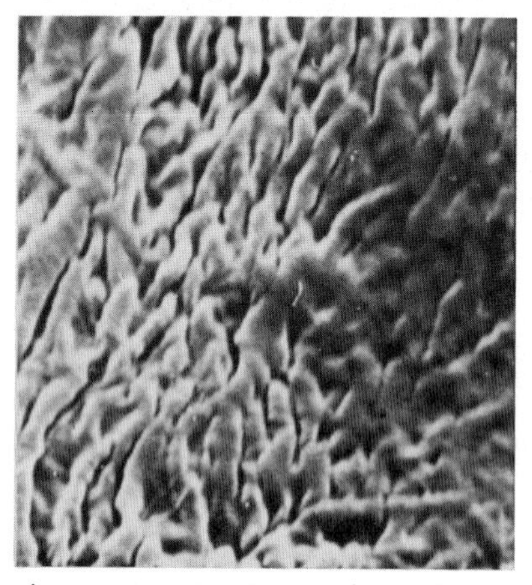

Figure 2 Cellulose Microfibril

3 MANUFACTURING PROCESS

3.1 Hydrolysis

To obtain crystalline cellulose, the fibrous plant material such as wood pulp, is hydrolyzed to remove the amorphous regions, leaving only the crystalline bundles (Figure 3). The first step requires cooking the purified pulp with a dilute concentration of mineral acid in water. This destroys the fibrous structure characteristic of the plant cellulose. The water-dispersible or crystalline cellulose, in fine particle form is produced. The process is carried to a point at which a levelling-off degree of polymerization (LODP) is attained. The LODP cellulose is then mechanically disintegrated to fragments of cell wall ranging from tenths of a micron to hundreds of microns.

3.2 Mechanical Disintegration

Upon completion of the hydrolysis, the cake obtained is neutralized and washed. The washed cake is then subjected to mechanical disintegration to break up the aggregates and release additional microcrystals. The attrition should be sufficient to produce a mass wherein about 1% by weight of the particles have an average length not greater than 1 micron as determined by electron microscopic examination (Figure 3). When this is accomplished, the material is then capable of forming a stable dispersion in aqueous and other media.

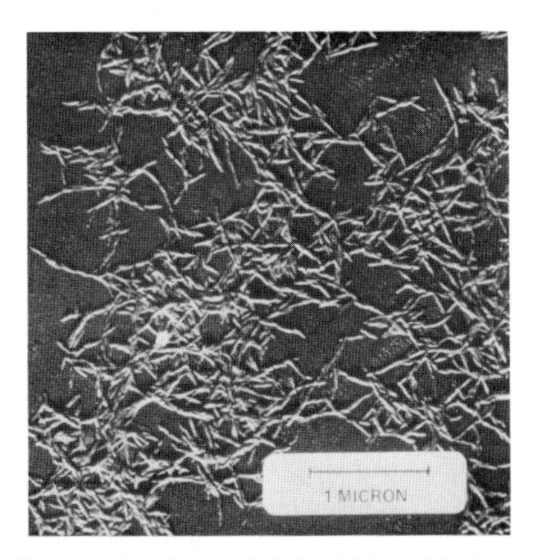

Figure 3 Colloidal MCC Cellulose

3.3 Coprocessing

After the mechanical disintegration step, the attrited microcrystals are coprocessed with a hydrophilic dispersant, sodium carboxymethylcellulose (CMC). The CMC prevents the microcrystals from reaggregating due to hydrogen bonding forces during drying (3). CMC also functions as a dispersant when the dry product is added to water.

After coprocessing with CMC, the material can then be dried in several ways. One method is bulk drying. Bulk dried colloidal products require homogenization to disperse the microcrystals and therefore are most suitable for use in whipped toppings and frozen desserts. A second drying method for the coprocessed attrited material is spray drying. These products require shear from a high-speed mixer for proper dispersion.

Another colloidal MCC product, referred to as stabilizer WC-595 is designed for use in food systems such as dry blended products where inadequate shear is available to achieve full dispersion. This easy-to-disperse stabilizer is manufactured by adding dried, sweet whey to the colloidal material before drying.

4 APPLICATION OF COLLOIDAL MCC TECHNOLOGY TO FOOD INDUSTRY PRODUCT AND PROCESSING PROBLEMS

The physical properties of colloidal dispersions are quite different from the properties of gum solutions, starch gels, and other water soluble materials commonly used in food technology. When colloidal grades of MCC are properly dispersed, the cellulose crystallites set up a network, with the majority of the particles less than 0.2 microns. It is the formation of this insoluble cellulose structural network that provides the colloidal MCC grades with their functional properties. The gel formed possesses certain solid-like properties of elasticity which exhibits relatively high yield stresses and a time-dependent type of flow behaviour (Figure 4). Thixotropic properties impart a variety of desirable characteristics suitable for products such as salad dressings and mayonnaise.

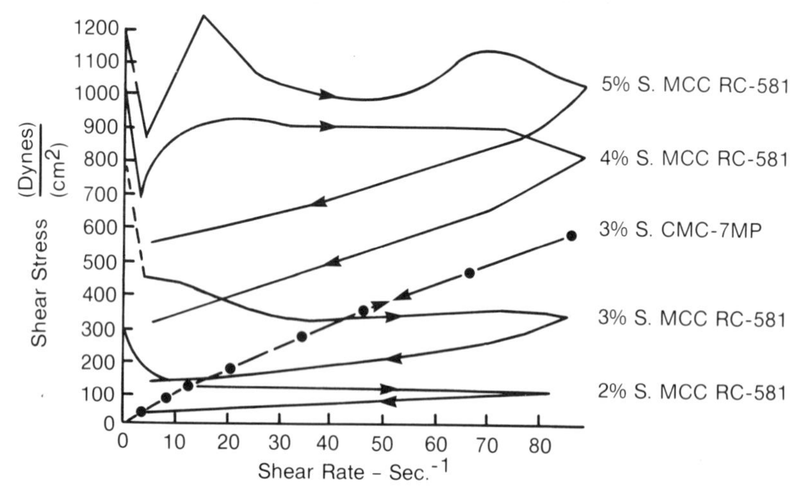

Figure 4 Shear Stress as a Function of Increasing and Decreasing Shear Rate for MCC RC-581 Gels and a CMC-7MC Solution (Measurements made on a RAO Viscometer)

4.1 Ice Crystal Control and Texture Modification

Body and texture are two of the most desirable characteristics that frozen dessert manufacturers try to impart to and maintain in their products. Effective levels of microcrystalline cellulose will produce frozen dessert dairy products with improved body and texture qualities, improved extrusion qualities, and good heat shock resistance. Heat shock, an industry term for the cyclical temperature conditions to which ice cream is exposed, causes partial thawing and refreezing of the product. This slow, quiescent process favors development of large ice crystals in the product. It is believed that free water generated by the melting crystals, is not reabsorbed immediately or completely by the other solids (MSNF, sugar, etc.) in the mix probably due to alteration of the solubility properties of these materials (4). Additionally, when the temperature decreases, the unabsorbed free water refreezes, preferentially on the larger crystals,contributing to the progressive textural deterioration of the product through an increase in the size of the ice crystals.

Microcrystalline cellulose with its unique water-adsorption capability will compensate for the inability or slowness of the other solids in the mix to reabsorb free water upon partial melting of ice crystals during heat shock. It is believed that colloidal MCC, with its tremendous surface area and wicking action, has the unique ability to manage the free water produced in the frozen product during heat shock (4). MCC also helps to minimize the agglomeration of these dairy solids ensuring product homogeneity. As little as 0.4% MCC in an ice cream mix can preserve the original texture of frozen desserts through numerous freeze/thaw cycles by maintaining the three phase system of water/fat/air. In addition to conventional butterfat-based ice cream, MCC utilized in conjunction with conventional stabilizers, i.e., CMC, locust bean gum, etc. supplements the body and controls the texture of frozen dessert novelties, vegetable fat-based products, as well as low solids, and/or artificially sweetened formulations.

4.2 Emulsion Stabilization

Effective stabilization against the coalescence of oil globules in an emulsion system can be obtained by using certain finely divided powders as an emulsion stabilizer. Microcrystalline cellulose functions as an emulsion stabilizer and thickener because it forms a colloidal dispersion of solid particles when sheared in water. In addition, the strong affinity of cellulose for both oil and water results in some orientation of the solid particulates at the oil-in-water interface. Colloidal MCC stabilized emulsions (o/w) were evaluated by means of brightfield and polarized light microscopy, freeze-etch electron microscopy and rheological measurements(5). These findings confirm that this colloidal network can provide a mechanical barrier at the interface of considerable strength and durability. Additionally, MCC thickens the water phase between the oil globules preventing their close approach and subsequent coalescence. This combined functional effect can be utilized to stabilize emulsion-based food products under adverse conditions of both processing and long term shelf stability.

4.3 Foam Stability and Syneresis Control

In aerated food systems, foam stability depends primarily on the types of additives present and their ability to produce the necessary structural strength in preventing the coalescence and subsequent collapse of the air bubbles.

While colloidal MCC is neither a whipping aid or a film

forming material, it provides effective foam stabilization in a variety of whipped and/or aerated food products. Colloidal dispersions of MCC not only thicken the water phase between air cells but provides added structural integrity to the protein film surrounding the air cells. Recent application studies have shown that the use of colloidal MCC improves body and texture, foam stiffness, and foam stability of both nondairy and dairy type whipped topping products. In addition, MCC is effectively used to stabilize marshmallow topping, confectionery products, and controls overrun in frozen desserts.

4.4 Heat Stabilization Properties

Temperature changes have little effect on the functionality and viscosity of colloidal MCC dispersions. Typical soluble gelling agents such as pectin and starch, when used in the preparation of conventional bakery jellies tend to lose their functionality under adverse baking conditions. Conventional pectin-based bakery gels will not usually withstand the effects of oven heating unless high levels of fruit (30 to 50%) are incorporated (6). This is generally a prohibitive practice because of the cost. Although there has been some success using modified starch in heat stable bakery jelly formulations, the starch level necessary to retain the gel structure usually contributes to a thick, pasty character after baking, which often is unacceptable to the baker. The addition of colloidal grades of MCC to both pectin-based and starch-based bakery fillings produces a modified gel structure with improved texture, spreadability, and heat stability characteristics. The modified gels produced, retain their texture and consistency when subjected to baking conditions (380-400°F for 15 minutes). The explanation for this outcome is somewhat speculative but it is believed that the formation of a second distinct gel system within the pectin water phase supports the fibrous pectin network as a heat stable structure. Application studies have also shown the importance of using colloidal MCC in retortable oil-based salad dressings. The consistency and emulsion stability of oil-based salad dressings products are often adversely affected by conventional methods of heat sterilization. Salad dressings stabilized with soluble hydrocolloids are susceptible to syneresis, and show emulsion instability such as creaming and/or "oiling" off after heat sterilization. Colloidal MCC has the ability to maintain viscosity, product consistency, and emulsion stability of oil-based dressing products prepared under sterilization processing conditions of 240°F for one-hour.

4.5 Fat Replacement

The inherent properties of colloidal MCC products dispersed in water have been used to simulate fat in various food applications including ice cream, salad dressings, sauces and gravies. The consistency of oil in water emulsions can vary from a thin fluid material at low oil levels, to a thick, viscous paste at very high oil levels. Studies have shown that there is a change in the rheological properties of the emulsions from simple Newtonian behaviour to non-Newtonian behaviour above an oil concentration of 50% v/v. By incorporating colloidal MCC into the system, the level of oil in emulsions can be effectively reduced, while still retaining their physical and rheological properties (7). It has been demonstrated that basic emulsions containing 60% soya bean oil have similar rheological properties and stability characteristics as 20% soya bean oil emulsions containing 1% colloidal RC-591 or 1.5% CL-611 (Figure 5). In addition, the properties of 30% soya bean oil emulsions can be simulated by incorporating either 1% RC-591 or 1.25% CL-611 into a 10% emulsion. The properties of the soya bean oil are in fact improved using lower levels of oil in conjunction with MCC. The emulsions acquire a yield value which makes the system stable against creaming. It is clear therefore, that microcrystalline cellulose has considerable potential for use as a fat substitute in the formulation of aqueous-based dietary food formulations.

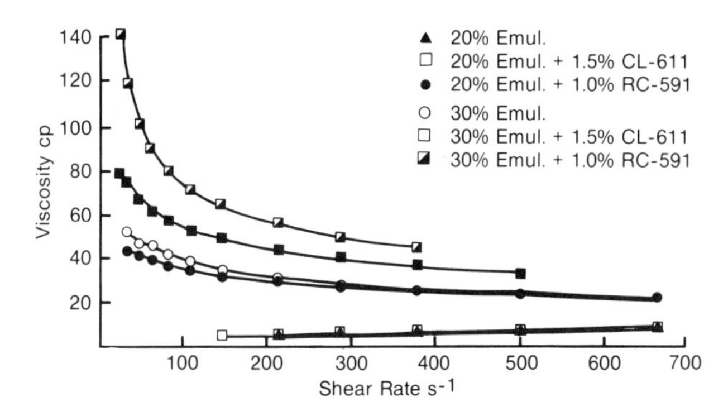

Figure 5 The Effect of MCC on the Viscosity of a Basic Emulsion

4.6 Suspension of Solids

In instant cocoa beverages, many soluble hydrocolloids have been blended and coprocessed with the cocoa powder but do not successfully provide suspension functionality under conditions of minimum agitation such as simple spoon stirring. These hydrocolloids swell, rather than dissolve, with minimum agitation. With proper dispersion, colloidal MCC forms a unique gel network. The network imparts the functional properties necessary to effectively suspend solids in food systems. Present commercial grades of colloidal MCC will redisperse under conventional shear conditions but will not redisperse rapidly enough to be utilized in dry mix food products that consumers would reconstitute by simple spoon stirring.

A codried MCC-based stabilizer system was developed primarily to provide instant colloidal MCC functionality in reconstituted cocoa drink beverages under simple spoon stirring conditions. The colloidal MCC-based stabilizing agent is a co-spray dried composition with three components: colloidal microcrystalline cellulose, starch, and maltodextrin. Starch is the key activator in the process in conjunction with the carbohydrate material and the sodium carboxymethylcellulose in proper proportions. The novelty of this development lies in the ability to achieve instant redispersibility of the cellulose microcrystals with simple spoon stirring. This feature is extremely useful for vending machine dispensed hot chocolate drinks.

5 PHYSICAL INTERACTIONS BETWEEN MCC AND OTHER FOOD APPROVED INGREDIENTS

The concept of intimate physical blending (alloying) of MCC and other food approved ingredients was recently investigated. To a gum technologist, the hydrocolloid alloying process may be defined as the special processing of two or more food approved components to provide properties or functionality not demonstrated by a simple mixture of the same materials. The study identified several potentially useful MCC-based blends, coprocessed with proteins and/or polysaccharides, to provide novel functionalities for specific uses. In all cases, the wet-end processing produced a physical interaction of the constituents and was reinforced by the subsequent drying step.

5.1 Colloidal MCC/Sodium Caseinate

Suspension of Solids/Unique Gel Texture. Certain physical interactions between MCC and proteins can be utilized to exhibit a given level of functionality at a lower cost-in-use. Heat sterilization experiments have shown that colloidal MCC and sodium caseinate physically interact at sterilization conditions to produce an effective cocoa suspending network useful in sterilized cocoa drink beverages. In terms of gel characteristics, heat sterilization treatment of sodium caseinate with MCC gives a turbid, caseinate gel with lengthy, stringy "mozzarella cheese-like" texture. The same treatment when applied to sodium caseinate alone yields a transparent gel and a short texture. Potential uses for this coprocessed product are in food applications where texture modification (imitation mozzarella cheese), suspension of solids (chocolate drinks), and enhanced protein film properties (batters and breadings) are important functional properties.

5.2 Colloidal MCC-Guar Gum Interaction

Fat Substitute Shear-stable Particle Configurations. Guar gum is a cold water swelling carbohydrate polymer containing galactose and mannose as structural building blocks. One of guar gum's major use is as a wet-end additive in the paper making process due to its ability to adsorb on hydrated cellulosic surfaces by hydrogen bonding (8).

In the wet state, guar gum physically interacts with colloidal crystalline cellulose by a hydrogen bonding type of mechanism, which usually results in the flocculation of the cellulose. In studying this physical interaction, codrying of the colloidal MCC/guar gum flocculates results in a new physical entity or hydrocolloid alloy. The resulting powder particles are essentially water-insoluble, shear resistant, and spherical. Spherical MCC/guar aggregates based on controlled aspects of this interaction are shown to have certain characteristics similar to those obtained when fats and/or oils are emulsified into an aqueous phase (Figure 6). In this shape and required particle size distribution, the particles behave like a continuous fluid creating the smoothness most closely approximating the physical and organoleptic properties associated with fat dispersed in water. It not only has a smooth, bland mouthfeel, but leaves no detectable aftertaste or residual effect.

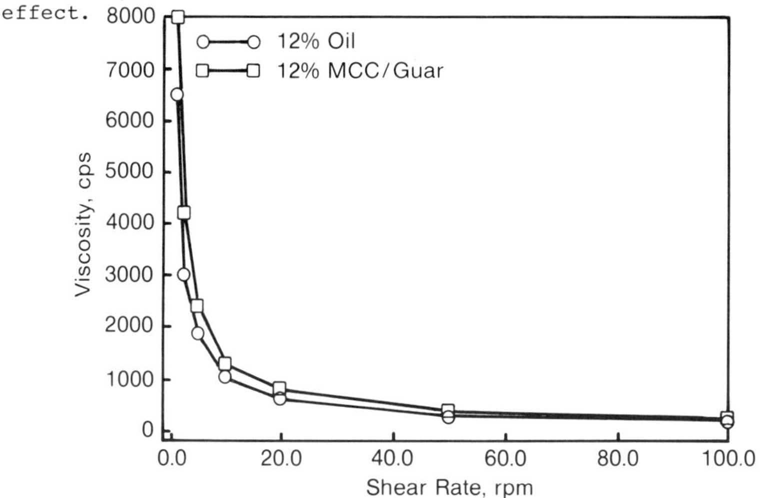

Figure 6 Rheological Properties-MCC/Guar vs Oil

The insoluble, spherical MCC/guar gum particulates can be agglomerated with a water-soluble hydrocolloid and subsequently dried to recover readily, water-dispersible spheroidal particles capable of forming a stable aqueous gel. A potential use for this dried, reconstitutible, spherical form of colloidal microcrystalline cellulose product is in the area of non-nutritive fat substitute in aqueous food systems. The non-nutritive aspects and water-insoluble form of colloidal MCC are key to the development of a successful product for this kind of use. A patent pertaining to a composition and method to produce a dried, spherical form of a colloidal microcrystalline cellulose for use as a non-nutritive and/or low calorie fat substitute in aqueous food systems is currently pending.

REFERENCES

1. Ward, K., (1954) Occurrence of Cellulose, Chapter II, Cellulose and Cellulose Derivatives, Part I, Vol. 5 pp. 9-27.
2. Ott, E., and Tennent, H.G., (1954) Introduction, Chapter I, Cellulose and Cellulose Derivatives, Part I, Vol. 5 pp. 1-8.
3. Durand, H.W., Fleck, E.G., and Raynor, G.E., (1970), Dispersing and Stabilizing Agent Comprising beta-1,4 Glucan

and CMC and Method for its Preparation, U.S. Patent No. 3,539,365.
4. Keeney, P.G. (1979), Confusion Over Heat Shock, Food Engineering Magazine reprint.
5. Oza, K.P., (1988) Colloidal Microcrystalline Cellulose-stabilized Emulsions, Ohio State University, Columbus, Ohio, USA From Diss. Abstr. Int. B 1988, 49 (2), 372-3.
6. McCormick, R.D. (1974), Controlling Heat Stability for Bakery Fillings, Toppings, Food Product Development Magazine reprint.
7. McGinley, E.J., Thomas, W.R., Champion, S ., Phillips , G.O., and Williams,P.A. (1984), The Use of Microcrystalline Cellulose in Oil in Water Emulsions," Gums and Stabilizers for the Food Industry, Vol. 2," Pergamon, Oxford, pp. 241-249.
8. Seaman, J.K. (1980), Guar Gum, Handbook of Water Soluble Gums and Resins, McGraw Hill, Inc., pp. 6-1 to 6-19.

Applications of carboxymethylcellulose and hydroxypropylcellulose in the food industry

ROGER VAN COILLIE
Aqualon France, Rueil Malmaison, France

SUMMARY

Cellulose derivatives are widely used for preparing food products. In this paper are summarized the properties and roles of two cellulose derivatives: sodium carboxymethylcellulose and hydroxypropylcellulose, and their various applications in food, as well.

INTRODUCTION

Sodium carboxymethylcellulose or CMC and hydroxypropylcellulose or HPC are two cold water soluble cellulose derivatives and therefore, develop common properties and functions. (1)(2)(3) However, their respective substitution in carboxymethyl or hydroxypropyl groups imparts to each of them, specific solubility properties and particular behaviours depending on the products or ingredients used.

This results in the use of one or the other derivative, as a function of the application involved.

CHEMISTRY AND PRODUCTION

Sodium Carboxymethylcellulose is a cellulose ether obtained by reacting cellulose with sodium monochloracetate while hydroxypropylcellulose results from the reaction of cellulose with propylene oxide.

Figure 1 shows the basic structure of cellulose which is a linear water-insoluble polymer made of cellobiose units, which consist in their turn, of two anhydroglucose molecules. In this structure, "n" is the number of anhydroglucose units or the degree of polymerisation (DP). The latter varies according to the origin of the cellulose. Cotton cellulose has the highest degree of polymerisation and is used to manufacture high viscosity type carboxymethylcelluloses and hydroxypropylcellulose while wood cellulose is used for medium and low viscosity types.

FIG. 1 : STRUCTURE OF CELLULOSE

Each anhydroglucose unit contains three hydroxyl groups. Sodium carboxymethylcellulose is obtained through substitution by carboxymethyl groups of certain hydrogens of these hydroxyl groups. The average number of substituted hydroxyl groups per anhydroglucose unit gives the degree of substitution or DS. Since there are three hydroxyl groups, the maximum theoretical degree of substitution ranges from 0.5 to 1.5 (Figure 2). This type of substitution imparts an anionic character to carboxymethylcellulose while hydroxypropylcellulose is non ionic. This accounts for the the fact that a number of their properties differs. In addition to this difference of ionic character, the substitution is expressed by the module of substitution (MS) since each fixed hydroxypropyl group contains an hydroxyl which in his turn can react with another molecule of propylene oxide. Here, the MS corresponds to the average number of fixed propylene oxide molecules per anhydroglucose unit, the latter ranging from 3.5 to 4.2 (Figure 3).

FIG. 2 : STRUCTURE OF CMC WITH DS OF 1.0

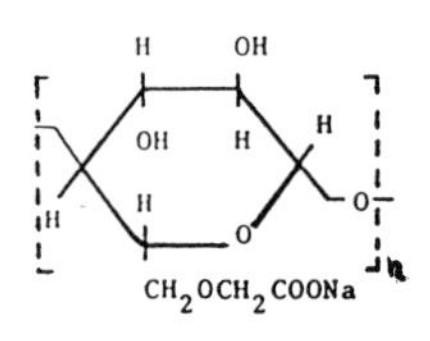

FIG. 3 : STRUCTURE OF HPC WITH MS OF 3

PROPERTIES

Carboxymethylcellulose and hydroxypropylcellulose are two cellulose derivatives which on dilution in cold water permit to generate more or less viscous solutions. The solution viscosity will depend on the DP of the cellulose used (high, medium or low DP), on concentration, on stirring conditions during dissolution. For instance, dissolving and putting in solution will be encouraged by a high shear if the carboxymethylcellulose has quite a low DS or if substitution is heterogeneous which may generate gelled aggregates. On the contrary, a rather high and homogeneous substitution produces a limpid and smooth solution.

Other factors can modify the solubility and the viscosity of CMC or HPC solutions, such as temperature, pH, the content in salts, sugars or other polymers.

A rise in the temperature of a CMC solution reduces the viscosity (Figure 4), but in the case of a short life heat treatment, the solution can recover its initial viscosity when back to its starting temperature. On the other hand, if the solution is kept at a high temperature over a long period (1 hour at 125° C), a severe viscosity drop will be noticed as a result of a depolymerization of the cellulose chain. This may occur when sterilizing certain products (meat).

Until roughly 40° C, HPC solutions behave like the CMC, but above this temperature, hydroxypropylcelluloses lose their solubility and precipitate. They recover their solubility after cooling (Figure 5).

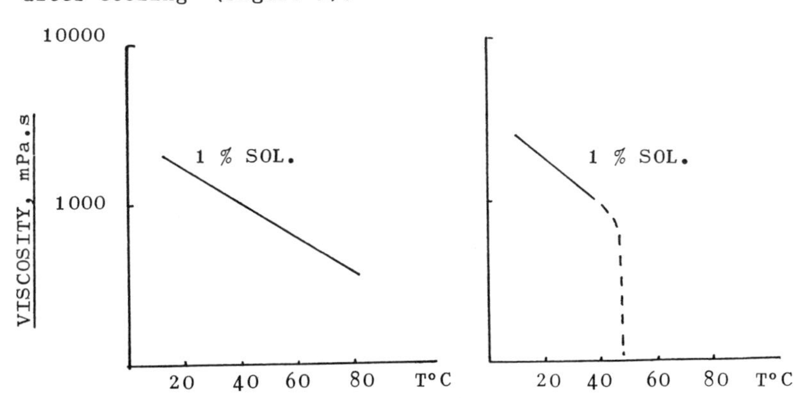

FIG. 4: EFFECT OF TEMPERATURE ON THE VISCOSITY OF AQUEOUS CMC SOLUTION

FIG. 5: EFFECT OF TEMPERATURE ON THE VISCOSITY OF AQUEOUS HPC SOLUTION

The pH deviation within the 2 to 11 range does not affect HPC solutions very much. Under similar conditions, CMC solutions are more sensitive to acid pH since salt functions can turn into acid ones and generate insoluble glycolic cellulose acid. In order to keep a good solubility in acid medium, it is often recommended to use CMC with a DS ranging to 0.8 / 0.9 and to dissolve CMC in water before adding acid.

The compatibility of hydroxypropylcelluloses with water-dissolved inorganic salts varies with salt. At a relatively high concentration, salt can reduce the solubility of HPC entailing a drop in viscosity, some haziness and a reduction of the precipitation temperature.

Thanks to its anionic properties, carboxymethylcellulose can react with monovalent salts to give a soluble CMC ; on the other hand, divalent and trivalent salts may involve a more or less important reticulation and entail either a drop in viscosity or a gelling or precipitation of the CMC. The effect of salts is generally speaking less important when CMC or HPC is first dissolved in water and salt is added afterwards.

The solubility of CMC and HPC may also be changed by the addition of solvents or of other products such as sugars, starches and other gums. Contrary to CMC, HPC,are very soluble in certain solvents like ethanol.

Aqueous solutions of hydroxypropylcellulose are as a general rule less thixotropic than the CMC ones, but both types may develop pseudoplastic behaviour under high shear rates. So, depending on the shear rate, a high viscosity type CMC solution may become less viscous than a medium viscosity one.

Compared to carboxymethylcellulose, hydroxypropylcellulose has two particular properties. Although both materials permit preparation of clear and smooth solutions, HPC solutions have a clearly lower surface tension than those obtained with CMC at an equivalent concentration. Therefore, HPCs are used as emulsifiers.

The second special property of HPC is thermoplasticity which enables extrusion of films or sheets which are used for food packaging. Films can also be obtained by evaporation of HPC or CMC aqueous solutions which are used, for example, as wrapping protecting for certain substances, such as fats, against oxidizing.

APPLICATIONS IN FOODS

In the preparation of foodstuffs, carboxymethylcellulose and hydroxypropylcellulose act as thickeners, stabilizers, binders, emulsifiers, water retention aids, protecting colloïds and assist in obtaining the desired texture and consistency, as well as the expected organoleptic properties.

So, thanks to these various functions, carboxymethylcellulose and hydroxypropylcellulose can be used in very different fields of the food industry as demonstrated here after.

FROZEN DESSERTS - ICE CREAM - WATER ICES:

Frozen desserts contain stabilizers which enable maintenance of the texture until the product has been consumed. Among these stabilizers, CMC is very often used in ice creams and in other frozen desserts. Several reasons explain this wide utilization of CMC. First, it dissolves rapidly when it is correctly dispersed, which results in the desired viscosity and a better control of overrun. On top of this, CMC is compatible with other stabilizers liable to be used and permits control of the formation of ice crystals and to keep a smooth texture, even if the product undergoes freezing/thaw cycles during storage. Though the utilization level of CMC is quite low, it brings excellent organoleptic properties (body and mouthfeel).

In low fat level ice creams and in milk ices, it is sometimes recommended to use CMC in blend with 10 - 15 % carrageenan in order to avoid whey separation in the mix before freezing. Generally speaking, the level of CMC is slightly increased when the content in fat decreases, in order to have an unctuous texture.

CMC is also used as a stabilizer in sorbets. In water ice, CMC enables a better flavour release and minimizes the trend to mask flavour and colour.

In soft frozen dairy foods, as in dry mixes, CMC is added as a stabilizer at a 0.2 % ab. concentration. This quantity is much higher in ripple syrups and can even reach 0.75 to 1 %.

Normally, the CMC concentration varies as a function of the ingredients contained in frozen desserts. It must be pointed out that in some countries, vegetable matters are authorized in place of milk fat. Besides, CMC can be used in ice creams where sugar has been replaced by an artificial sweetener and sorbitol, for example.

BAKERY PRODUCTS

Bakery products include many things ; for example: special breads, various cakes, pastry products (fillings, toppings, meringues), doughnuts.

In the preparation of special breads, cakes and of any product based on a dough, CMC, by its dissolution speed permits rapid binding of the different ingredients and gives, more rapidly, the desired dough consistency. In some cases, the use of CMC may require an adjustment of the ingredients' composition. Most of the time, more water must be added ; the rate can vary from 20 to 40 per gramm of CMC. The level of CMC depends on products, but generally it is in the range 0.1 to 0.4 % of the solids.

The CMC addition in these kind of products improves homogeneity of the dough and the distribution of ingredients such as raisins or crystallized fruits as well. While baking, these ingredients will remain evenly distributed througout the product.

In most cases, the additional quantity of water is kept during baking, generating a softer product even after a few days since CMC slows down staling. Often, it was observed that the volume yield increased involving a more aerated and softer crumb.

CMC improves the texture of fillings, toppings, icings, but at the same time, it avoids the syneresis of fillings and controls the sugar crystallisation of icings. In aerated products, toppings for instance, CMC and HPC, alone or in combination stabilize the texture. The use of HPC for the coating of chocolate chips, contained in a frosting avoids their softening.

SOFT DRINKS

CMC is used in a wide range of soft drinks to maintain the pulp in suspension, to bring or improve mouthfeel and body, to reduce the formation of an oily ring at the bottleneck, to hide the undesirable bitter aftertaste characteristic of artificial sweetener.

The efficiency of CMC in these products depends upon a number of parameters, such as CMC type : medium or high viscosity, use of CMC, type and composition of the soft drink.

In certain cases, the order of addition of ingredients or the homogenization treatment has only a slight effect on stability. In other cases, it is more advisable to add CMC at the end of the preparation which may improve stability. These differences of behaviour cannot be easily explained for, very often, the composition of bases or concentrates used is unknown.

Although the viscosity is not always the source of the suspension of pulp, it often proves easier to stabilize the pulp in a 25° Brix squash than in a 7 - 10 % soluble solids based ready-to-drink beverage.

DAIRY PRODUCTS:

Two categories of products must be considered: neutral products such as dessert creams and acid products like acidified or sour milk based drinks.

In neutral products, the addition of CMC allows preparation of milk desserts with different texture. Moreover, CMC suppresses the syneresis of starch or carrageenan containing gelled desserts and by CMC/carrageenan combinations, it is possible to prepare storage stable whipping creams.

The use of CMC in sour milk products is widespread due to its anionic character. It can react with milk proteins (casein) in the pH zone of the isoelectric point, i.e. 4.6 and can form soluble, heat treatment and storage stable complexes. Advantage is drawn from this complex formation for preparing and stabilizing a number of products such as yoghurt drinks, buttermilks, milk or whey based fruit drinks. Specific conditions must be respected to achieve a good stability. First, the quantity of CMC required to stabilize the drink has to be determined. It depends on the type of CMC (at an equivalent rate, a high viscosity type stabilizes better), on the rate of casein or of proteins contained in the medium, on the pH of the drink, on the fermenting or acidifying conditions which may generate more or less large casein aggregates. The consistency of stabilized products is dependent upon the CMC rate, fats, solids, but also upon mechanical treatments, such as the homogenization under pressure which allows reduction of the consistency, however without affecting the stability.

Acid cream, yoghurt and cream cheese sauces can also be stabilized by the addition of CMC.

Soluble complexes can be formed with other proteins such as soya proteins and gelatin.

SALAD DRESSINGS AND VARIOUS SAUCES:

The use of CMC in the preparation of salad dressings makes emulsion formation easier and improves stability, particularly others during an extended storage time under unfavourable conditions of temperature. Depending on the desired consistency and on the oil concentration a medium or high viscosity CMC grade might be recommended at a 0.5 - 1 % rate. Salad dressings can be prepared by diluting CMC in the aqueous phase with a gradual addition of oil with stirring. This mode of operation may also be used for instant preparations which require that all the dry ingredients are closely mixed and dispersed in water by means of a fork or of a whisk. After a few minutes mixing, the consistency of the blend allows addition of the oil progressively and leading to the formation of the emulsion. When the CMC dissolution in the aqueous phase is not technically possible, it is suggested to disperse it in oil, under the action of high shear, an emulsion will form during the addition of the aqueous phase containing the other ingredients (yolk of egg, vinegar, salt, etc...).

The use of CMC is also advised in various sauces intended for deep frozen dishes for it enables preparation of different textures (smooth, long or short), according to the CMC characteristics. Furthermore, its water absorptive power prohibits syneresis during thaw and when warming again in the oven.

In ketchups, CMC imparts the desired consistency and texture when used at a 0.5 to 1 % rate. This rate often depends on the amount of tomato concentrate used for the preparation and varies in the opposite direction.

WHIPPED CREAM PRODUCTS:

CMC and HPC are used as stabilizers in aerated structure products. In addition to its stabilizing effect, HPC is an excellent whipping aid in the foaming of whipped toppings prepared with vegetable fats. When the two cellulose derivatives are combined, one recommends a high viscosity type CMC in order to encourage the incorporation of air and a low viscosity HPC for the emulsifying effect.

The role of stabilizer is of outstanding importance in these types of toppings since it hinders clumping and coalescence of fat particles and layering during storage of the liquid base as well as stabilizing the foam against shrinkage and syneresis.

MISCELLANEOUS APPLICATIONS OF HPC:

The film forming properties of HPC allows coating of certain products such as peanuts with a view to protect them against oxidizing or to produce moulding compounds which can be used in the preparation of food, or to serve for the manufacture of low moisture permeability films permitting, for example, an increase in the shelf-live of some products. Film forming properties of HPC are also exploited for the encapsulation of flavour oil, encouraging a more uniform and thorough release in instant coffee and tea products (4).

MISCELLANEOUS APPLICATIONS OF CMC

Owing to its numerous functions (thickener, stabilizer, binder, water retention aid), CMC is still used in many other applications than those quoted above.

CMC has a low calorific value and therefore is used in formulations of low calorie products.

In instant products, CMC dissolves very rapidly, imparts the desired consistency and body and it also maintains certain ingredients in suspension, like for example, cocoa in chocolate drinks.

In meat based products, CMC is used as a thickener in gravies and avoids the separation of fats. It also permits to bind ingredients and improves the water retention and the texture while avoiding the shrinkage in sausage meat phenomenon.

CONCLUSIONS

The food applications of the carboxymethylcellulose and of the hydroxypropylcellulose are very numerous. However, due to the very attractive properties of these cellulose derivatives, studies are still carried out in order to discover new possibilities of utilisation.

REFERENCES

1. WHISTLER R.L. 1973 " Industrial Gums - Polysaccharides and their derivatives" 2nd Ed. Pages 649-672 and 695-729, Academic Press Inc.

2. PILNIK W. and NEUKOM H. (1980) "Gelling and thickening agents in foods" Page 163-174 - Forster Publishing Zürich.

3. GANZ AJ 1973 "Some effects of gums derived from cellulose on the texture of foods - Cereal Science today" page 398 - 416

4. MARMO D. ; ROCCO F.L. 1982 US Patent N° 4311720

The application of cellulose ether-starch interactions in food formulations

T.SCHWITZGUEBL

Dow Europe, Bachtobelstrasse 3, PO Box, CH-8810 Horgen, Switzerland

Methylcellulose and Hydroxypropylmethyl celluloses are well known food hydrocolloids manufactured and sold by The Dow Chemical Company under the tradename METHOCEL*. METHOCEL premium food gums show unique reversible thermal gelation in solution as well as other characteristics such as emulsification power, water binding, pH stability, lubricity, film formation and thickening power.

METHOCEL premium food gums also show specific interactions with starch and modified starches in liquid food products e.g. sauces, fillings, creams or soups. METHOCEL premium food gums type A give a synergistic build up of viscosity with starch and modified starches during the heating of liquid food systems. This viscosity synergism is useful, particularly in products like e.g. sauces for ready meals or fillings for baked goods as well as numerous microwavable foods to improve "hot cling".

Very often modified starches are used in food systems which need to be heat treated, frozen or processed at low pH. These modified starches can exhibit strong rheopectic characteristics. This could explain the stickiness and flavour masking property of these modified starches.
METHOCEL premium food gums show the ability to modify the rheological characteristics of these modified starches through specific interactions. This phenomenon would improve the flavour perception and texture of liquid food systems containing starch or modified starches.

* Trademark of The Dow Chemical Company

What are METHOCEL gums? What are the unique properties of these food hydrocolloids? How do they interact with starch? How could these interactions help in the formulation of food products?

That is a list of questions which will be discussed and answered in this paper.

What are METHOCEL gums? METHOCEL premium food gums are food grade modified cellulosics. These products were the first modified biopolymers produced to replace natural gums. They have been manufactured by The Dow Chemical Company for 50 years and find a wide range of applications in food products as well as in pharmaceutical, personal care and in industrial applications such as textile sizing, paint, building and polymerisation applications.

The starting material for the production of METHOCEL gums is the most abundant biopolymer on the face of the earth: cellulose. Natural cellulose is non-soluble in most common solvents and particularly in water, the first choice food solvent. The sugar building unit as well as the bonds between these sugar moieties explain the major cellulose properties, e.g. crystallinity, the rigidity of the polymer and the possibility to modify the chain by chemical reaction. The cellulose used to produce METHOCEL gums can be either of wood pulp or cotton linter origin.

Methyl cellulose (MC) is obtained by methylation of some of the hydroxyl functions of the sugar building units of cellulose. This type of modification was the first carried out on cellulose products for food applications. The number of methyl groups added to the sugar units is called the degree of substitution (D.S). The degree of substitution is specifically defined for food applications and is carefully controlled during the METHOCEL gum manufacturing process. The DS is a major factor in the final products' properties.

Hydroxypropyl methyl cellulose (HPMC) combines the methyl cellulose substitution with the hydroxypropyl substitution. The molecular structure of the hydroxypropyl methyl cellulose allows the possibility of adding modifying agents to the hydroxypropyl group itself as well as to the cellulose hydroxyl function. Therefore the number of hydroxypropyl groups is defined by a Molecular Substitution (MS) which represent the number of hydroxypropyl groups per cellulose sugar unit. Due to the increased number of possible substitution combinations, METHOCEL gum grades covering the HPMC family are more numerous than the MC family. This allows for a wide profile of physical characteristics to be available.

Legislative status

Modified cellulosics are commonly used in applications such as baked goods, sauces and creams. Due to varying food regulations from country to country, it is impossible to give a concise overview on this subject. Specific information can be obtained through local legislative offices.

METHOCEL A type gums have the E number E 461 following the EEC directive 74/3291 EEC and its amendments on methyl cellulose. METHOCEL type E, F and K gums have the number E 464 associated with the hydroxypropyl methyl cellulose which is also defined in the European directive 74/329/EEC and its amendment.

METHOCEL gums properties

The modifications made to the cellulose during the manufacturing of METHOCEL gums provide them with unique properties.

Thermal gelation

The thermal gelation of METHOCEL gums is the most interesting of these properties. The uniqueness of this property comes from the reversibility of the thermal gelation process of both MC and HPMC. When a METHOCEL gum aqueous solution is heated it gels. This gel remains as long as the temperature is kept high enough. On cooling the gel goes back to the initial gum solution state. Viscosity and other properties return to that of the starting solution. This process can be repeated as many times as needed without alteration of either the solution characteristics or the gel properties. This reversibility, as well as the overall process, is explained by the solubility equilibrium of METHOCEL gums with respect to temperature. The Incipient Gelation Temperature (IGT) is the temperature below which a METHOCEL gum in water is in the solution state; above this temperature METHOCEL gum solutions gel. What are the key characteristics of the hydrocolloid which explain this behaviour?

The degree of substitution is not uniform along the cellulose chain. This is partly due to the properties of the cellulose raw material which create zones along the chain with different hydrophilicity and hydrophobicity characteristics. Below the IGT, the sum of these two possible types of interactions with the solvent leads to an overall hydrophilic behaviour and the METHOCEL gum forms a transparent solution.

When the METHOCEL gum solution is heated more internal energy is given to the water molecules. Reaching the incipient gelation temperature the sum of the interactions favours hydrophobic interactions and the METHOCEL molecules form intra- and intermolecular bonds which build up a gel network. This purely physical phenomenon explains the total reversibility of the METHOCEL gum thermal gelation.

Gel temperatures and structures

METHOCEL gum gelation temperatures are defined for 2% aqueous solutions (this is a widely accepted definition). METHOCEL gums have a gelation temperature between 50°C and 90°C, depending on the type. As described previously, the gelation temperature is linked to the water affinity of METHOCEL gums which is itself related to the type of substitution. METHOCEL gum "A" has the lowest water affinity giving the lowest gelation temperature. This gel temperature increases in the order A-E-F-K. The temperature values are summarised in Table 1.
The gel structure is also related to the METHOCEL types in the same order as previously mentioned. Thus, METHOCEL A forms firm gels, METHOCEL K forms soft gels and METHOCEL gums E and F, semi-firm.

Factors modifying the gel temperatures

Gelation of METHOCEL gums is linked with water affinity. Food constituents which modify the water availability in a product can influence the gel temperature of METHOCEL gums. On the other hand food constituents which more specifically solvate the hydrophobic part of METHOCEL gums modify the gel temperature in the opposite direction. For example: sugars decrease the gel temperature of METHOCEL gum solutions by reducing the water activity in the solution. A polyol like sorbitol reduces the gel temperature more than sucrose at similar concentrations. On the other hand, ethanol increases the gel temperature by solvating the METHOCEL gum more efficiently.

Other typical additives decreasing the gel temperature are electrolytes like sodium chloride or similar products. The concentration of the gum also plays an important role on the gelation temperature. The gel temperature is decreased by increasing the METHOCEL gum concentration. In extreme conditions, for example when the METHOCEL concentration is below a critical value, no gel network can be formed. This is due to too great a distance between molecules, hence, the inter-molecular bonds cannot be formed. With a 2% METHOCEL gum solution a gel structure close to a gelatine strength for METHOCEL A is formed. The gels are semi-soft for METHOCEL E and F, and soft for METHOCEL K, as summarised in Table 1.

Factors modifying the gel structures

The gel structure is linked to the METHOCEL gums' viscosities for values below 400 mPa.s in a 2% aqueous solution (measured with a capillary viscosimeter). The gel hardness increases with the viscosity until a typical maximum value is reached. Above this, the viscosity value parameter has no influence. Other food constituents like hydrocolloids, starches or proteins can also be used to modify the gel texture in real food systems.

METHOCEL Premium Gum properties

Thermal gelation is the most unique of the METHOCEL gums properties but not the only one. The other key properties are all linked to the properties of cellulosics and with the type and degree of modifications of METHOCEL gums. This list of key properties includes the following: Thermal gelation, binding capacity, water retention, emulsification, suspension, surface activity, lubricity, pH stability, non-ionic, enzyme resistance, virtually tasteless and odourless and ability to thicken.

The METHOCEL Premium Gums/starch interactions

The specific function of METHOCEL gums within typical food ingredients is a key issue for food formulators and food producers. Starch is a major nutritive food constituent, thus its interactions with METHOCEL gums are of particular interest. The specific starch/METHOCEL gum interactions observed can be used by food formulators to tailor-make starch containing foods. This can be done under a number of different constraining conditions such as the end product texture or production cost optimisation.

Synergistic thickening power

METHOCEL gums show synergistic thickening power with starch under given conditions. These will be discussed and linked to some typical food products later.

When heat is applied to an aqueous solution of starch, viscosity reduction or thinning of this solution occurs (**see figure** 1). This loss of viscosity may vary from one starch to another and can be reduced with some modified starches, but this physical phenomenon can never really be overcome.

The METHOCEL gums' thermal gelation properties can be used in a liquid food system to alter the heat thinning characteristic of starch and to better control product viscosity over a wider range of temperatures. Shown in **Figures** 2 and 3 is the synergistic viscosity build up of METHOCEL gum/starch solutions in the range of hot served foods. METHOCEL A gum has a gelling temperature around 55°C in 2% solutions without other food constituents in the solution. When blended with starch or modified starch an apparent 2% or even higher concentration can be achieved with actual METHOCEL gum concentrations around 0.5%. The selection of the correct METHOCEL gum is also important, in order not to have too strong synergistic thickening through the formation of a hard METHOCEL gel. The relation described before between the gel hardness and METHOCEL gums' viscosities support the selection of low to medium viscosity products for this application. When METHOCEL gums are used in conjunction with the most heat stable starches, a strong increase in viscosity is observed during the heating phase. This

TABLE 1

GEL TEMPERATURES AND STRUCTURES

METHOCEL Premium product type	Approximate gel point (2% aqueous solutions)	Gel structure
Type A	50-55 C	Firm
Type E	58-64 C	Semi-firm
Type F	62-68 C	Semi-firm
Type K	70-90 C	Soft

FIGURE 1

Effect of Temperature on an aqueous solution of starch

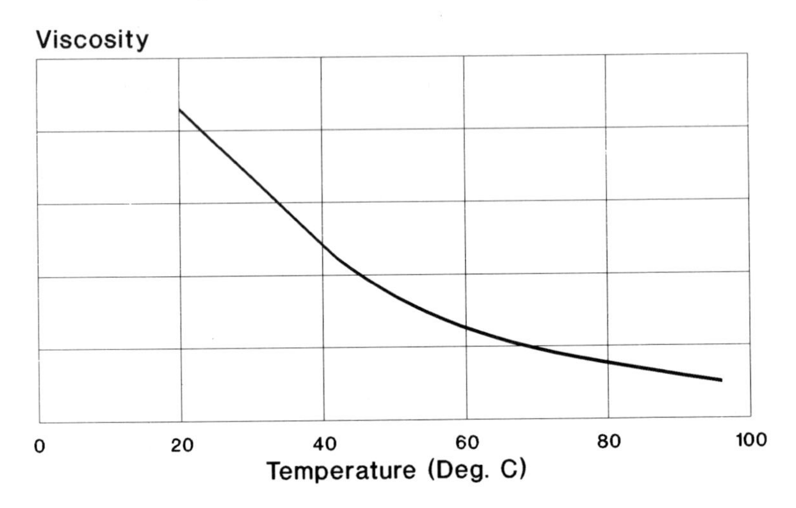

FIGURE 2

METHOCEL* A15-LV Premium and Kol Guard+ starch solutions

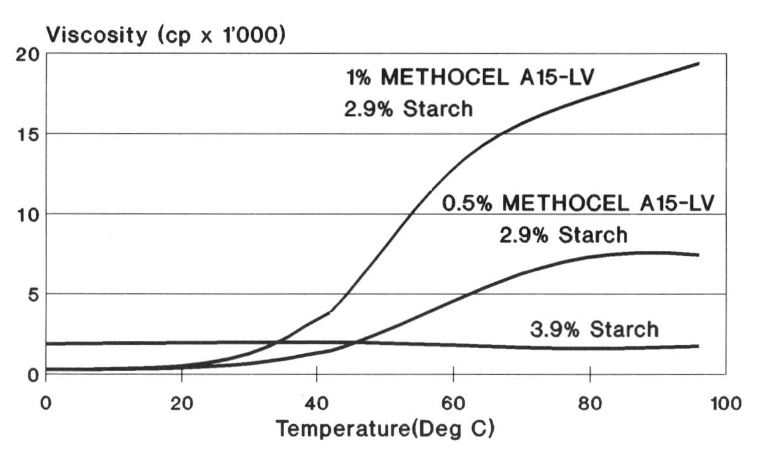

•Trademark of The Dow Chemical Company
+Trademark of The A.E.Staley Manufacturing Co.

FIGURE 3

METHOCEL* A15-LV Premium Thin-N-Thik 99+ starch solutions

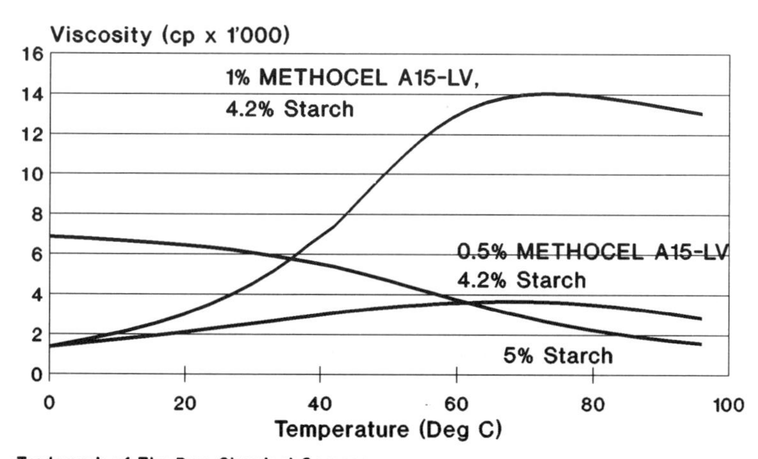

•Trademark of The Dow Chemical Company
+Trademark of The A.E.Staley Manufacturing Co.

provides a better hot cling with these types of liquid foods. When the main purpose is to control the viscosity of the product over a wider range of temperature than for starch alone, starch with a greater heat thinning behaviour should be selected to achieve the best control.

As already discussed, most food hydrocolloids/starch blends lose viscosity when heated (**Figure 4**). Such hydrocolloids are sometimes used to prevent syneresis of frozen starch containing products. The difficulty in such applications is when the product needs to be reheated for consumption. Often the viscosity loss is below the value needed for an optimum product. Using the METHOCEL gum thermal gelation phenomena, liquid food products containing starch can often be optimised with respect to their temperature-viscosity behaviour. Tailor-made solutions can be developed to control product quality from the frozen state to the hot serving one.

Rheology modification

METHOCEL gums show the ability to modify rheological behaviour of starch solutions. This property is important in improving food texture and flavour perception. Measuring the viscosity or shear stress of a starch solution by increasing and decreasing the shear rate applied to the solution gives curves similar to **Figure 5**. The strong rheopectic behaviour of certain modified starch could result in products which show texture and taste quality below those needed for optimum products.

Swallowing is a physical process with increasing-decreasing shear rate conditions, during which the rheopecticity of starch solutions can very often be expressed as "stickiness" in consumer terms or "starch length" by specialists. This undesirable property contrasts with the desired properties of modified starch e.g. heat stability or high viscosity.
When METHOCEL gums are added to starch solutions showing rheopectic behaviour, the lubricity of METHOCEL modifies the rheology. **Figure 6** shows the effect of a very low viscosity METHOCEL gum, (which does not contribute to the total viscosity of the solution) in efficiently reducing the solution rheopecticity. **Figure 7** shows the effect of a high viscosity METHOCEL gum which increases the total solution viscosity, but significantly decreases the starch rheopecticity. Both types of METHOCEL gums can be used in a food system, depending on the desired end result.

Designing foods with METHOCEL gums-starch interactions

This part will conclude the overview and act as the starting point for the formulation of optimised liquid food systems containing starch (i.e. soups, sauces, creams and fillings). These food types will be discussed following classification by the related METHOCEL gums-starch interactions. Further subdivision will be made for both the eating characteristics and the processing requirements of the final product.

FIGURE 4

EFFECT OF TEMPERATURE ON SOLUTION VISCOSITIES OF REZISTA STARCH-GUM BLENDS

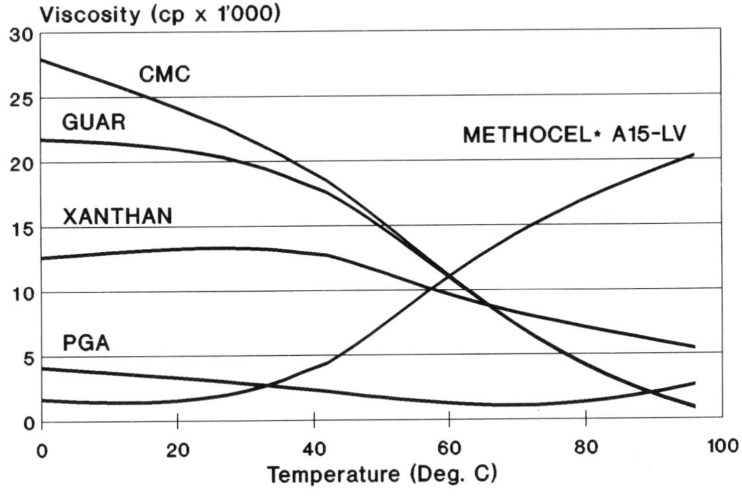

*Trademark of The Dow Chemical Company

FIGURE 5

RHEOLOGY OF AN AQUEOUS SOLUTION OF MODIFIED STARCH

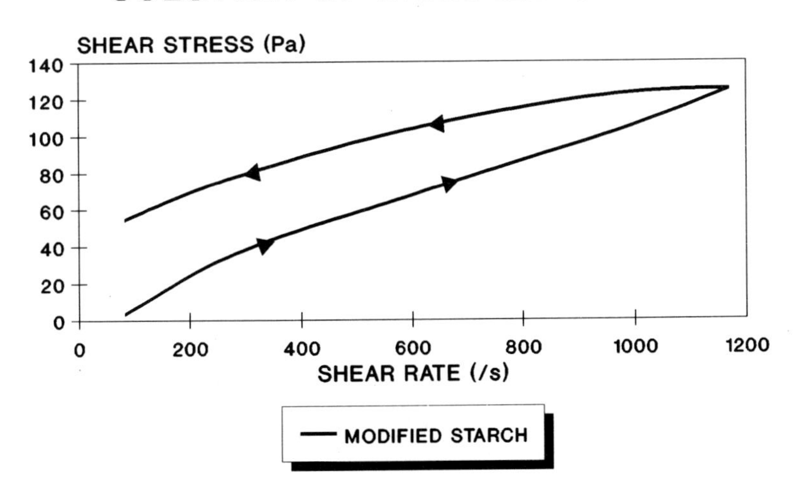

FIGURE 6

FLOW CURVES
OF METHOCEL* GUM - STARCH SOLUTIONS

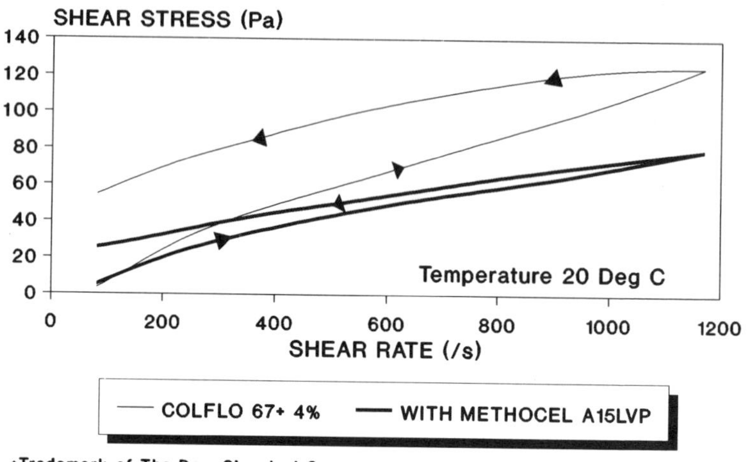

*Trademark of The Dow Chemical Company
+Trademark of National Starch Co.

FIGURE 7

FLOW CURVES
OF METHOCEL* GUM - STARCH SOLUTIONS

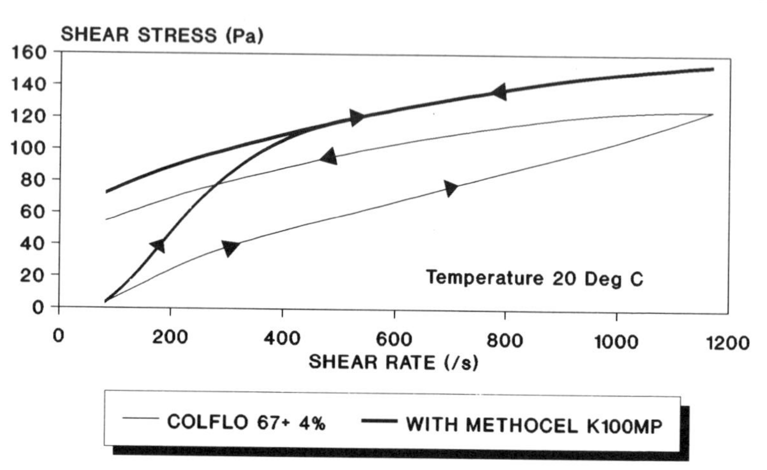

*Trademark of The Dow Chemical Company
+Trademark of National Starch Co.

Synergistic thickening power

The METHOCEL gum-starch synergistic viscosity build up takes place at elevated temperatures. This implies that to take full benefit of the possible reduction of the starch level, the food must be consumed hot e.g. sauces or soups. This synergism is observed in METHOCEL A type products which have the lowest gel temperature as well as the firmer gel texture. Therefore, too highashear rate mixing conditions during the food heating step will reduce the viscosity build-up. A non-exhaustive list of food products which could be optimised by using low viscosity METHOCEL A types are: sauces for ready-made meals, creams, fillings for bakery and liquid starch containing foods which are heated in a microwave or conventional oven.

Rheology modification

The rheology modification of starch with METHOCEL gums is particularly effective in the case of modified starches, which generally show the highest rheopecticity. Products like frozen sauces, in which modified starches are used to reduce syneresis, or sterilised sauces, which need to support high temperature treatment, have a tendency to show stickiness or too long texture in the mouth. These food products can be better optimised by using METHOCEL gums in conjunction with starch as a rheological modifier. This leads to a product with a stronger taste perception and flavour impact as well as giving a cleaner texture.

METHOCEL gums, pH stability

This property is particularly important for certain food products and is related to the two previous properties. Typically a low pH food system which must be heat treated needs high levels of modified starch to prevent the drop in thickening after being heated.

As METHOCEL gums are stable under these conditions their interactions with starch are hardly affected. These interactions still give either an effective reduction of the starch content or an improvement in the modified starch texture. It is also evident that these effects can be used together to optimise food design and processing conditions. A further benefit is a potentially better flavour impact. The combination of these three properties is particularly evident in the low pH sterilised foods, e.g. tomato sauces containing heat sensitive ingredients and a wide range of fruit products in the form of sauces, creams or fillings (for frozen foods, sterilised products, or fine bakery).

Properties of modified polysaccharides from soybean hulls

G.DONGOWSKI, G.STOOF and W.BOCK

Akademie der wissenschaften der DDR, Zentralinstitut fur Ernahrung, Potsdam-Rehbrucke, Arthur Scheunert Allee 114-116, Bergholz-Rehbrucke 1505, German Democratic Republic

ABSTRACT

Soybean hulls form rigid, elastic filled gels with xanthan. The gel strength is increased after pretreatment of the hulls with mineral acids and by intensive grinding and/or after partial thermal hydrolysis. The gel strength is influenced by the morphologically grown structure and the water binding capacity of the hull material. Sucrose, salts and other admixtures are compatible in a wide range. The producible gels belong to a new type of particle gels with a high content of dietary fiber and with interesting properties for the modern food technologist. A reason for the gelation effect is the galactomannan component of the hulls. It was isolated, purified and comparatively characterized. Both types of gels with xanthan were rheologically characterized.

INTROCUCTION

Galactomannans (GM) were found in the endosperm of the seeds of Leguminoseae as reserve polysaccharides in high concentrations (1, 2). Locust bean gum, guar gum or tara gum are such GMs with great importance to food technology.

The occurrence of GM in soybean hulls is a botanical exception (1). The yield of GM isolated was about 2 %. The GM had a Man/Gal ratio of 1.4 to 2.35 and a relatively low molecular weight (2, 3).

The present paper shows that soybean hulls or the GM from the hulls, also form gels with xanthan.

MATERIALS AND METHODS

Soybean hulls

Commercial samples of the hulls were defatted and ground (< 0.4 mm). Composition of the hulls: Dry matter = 90.43 %, raw protein = 17.25 %, soluble protein = 7.56 %, total pectin = 3.98 % (degree of esterification = 43.8 %), hemicellulose-pentoses 10.18 %, hemicellulose-hexoses 9.14 %, total hexose polysaccharides 39.98 %, dietary fiber = 80.2 %, ash = 3.57 %, K = 1.35 %, Ca = 0.40 %, Mg = 0.28 %, Na = 0.014 % and Fe = 0.011 %. The water binding capacity was 3.94 g H_2O/g.

Gel formation and characterization

A suspension of 12 g soybean hulls (or extracts from the hulls) and 1 g xanthan (Jungbunzlauer, Vienna/Austria) has been boiled down at pH 4.5 from 220 to 200 g for 3 min. The hot mixture was poured into a metal ring (height = 3 cm, diameter = 7.5 cm) placed on a glass plate. After 24 h the gel firmness was measured with a penetrometer (measuring device: hemisphere shape; diameter = 4 cm). One penetrometer unit (PU) is the penetration in 0.1 mm. Additionally water releasing (WR) values were measured by a filter paper suction test with a gel cylinder (30 g, basis = 9 cm^2) under standard conditions.

Analytical methods

Galacturonan was analysed with m-hydroxydiphenyl, methoxyl with chromotropic acid, hexoses with anthrone, pentoses with orcine, soluble protein (pH 8.0) with folin reagent, raw protein with micro-Kjeldahl (N x 6.25), dietary fiber according to Southgate (4), water binding capacity by capillary suction method (5) and minerals by atomic absorption spectroscopy.

Rheology

The flow behaviour was measured with a Haake-Rotovisco RV 100. Rheological measurements on gels were carried out with a rheometer with Searle geometry and compression cell or with a parallel plate shear device.

RESULTS AND DISCUSSION

Gel formation between soybean hulls and xanthan

Soybean hulls form rigid, elastic filled gels with xanthan. Previous extraction of the hulls with hot water reduces or prevents the gelation effect. The producible gels belong to a new type of particle gel with a high content of dietary fiber and with interesting properties for the modern food technology.

Gel strength with xanthan is dependent on the concentration and the morphologically grown structure of the hull material. Standard conditions for the following investigations are 12 g soybean hulls and 1 g xanthan/200 g gel.

Gelation occurs within a wide pH range ($<$ pH 3 to $>$ pH 6) and is optimal at pH 4.5. Under these conditions we found a gel strength of about 54 PU and a WR value of about 20 cm^2.

The gels become more rigid by addition of sucrose and starch syrup (more than 50 % dry matter of the gel). Adhesiveness increases in presence of high sugar concentrations (Table I).

Table I. EFFECTS OF SUCROSE AND STARCH SYRUP ON THE STRENGTH OF SOYBEAN HULL-XANTHAN-GELS (200 g)

Sucrose	(g)	0	40	60	80	100	100	100
Starch syrup	(g)	0	0	0	0	0	20	40
Strength	(PU)	55	54	49	40	39$^+$	38$^+$	37$^+$

$^+$adhesive

The addition of sodium chloride (up to 5 g) is compatible and causes a small increase of gel strength. Most other salts or Ca^{2+} alone or in combination with LM-pectins (2 g) have no significant effect on gel strength.

The reactivity of soybean hulls can be improved by different pretreatments. Short vibration grinding ($\leq$ 4 h) or intensive milling of the hulls do not change the strength of the gels with xanthan or lead to an increase. In this case it is possible to prepare gels with higher sensory quality. Simultaneously the extractability of soluble fractions and the water binding capacity of the hulls will be improved.

The gel forming mechanism of soybean hulls is quite thermostable. Gels with high strength (47,6 PU) were prepared with xanthan after 20 min autoclaving (121 °C) of hull suspension. Besides maceration of the plant tissue, release of reactive components from the hulls takes place. Short-time thermal pretreatment up to 220 °C had no negative effect on the gel formation with xanthan (Table II).

Table II. EFFECTS OF TEMPERATURE AND TIME DURING PRETREATMENT OF SOYBEAN HULLS ON THE STRENGTH OF GELS WITH XANTHAN

Temperature	(°C)	0	135	135	135	220	220	220
Time	(h)	0	1	1.5	3	0.5	1	1.5
Strength	(PU)	53	50	48	47	52	85	90

Pretreatment of soybean hulls in low moisture conditions with mineral acids, for instance HCl or H_3PO_4, at pH 1 to 2.5 acts positively on the gel formation with xanthan. Soybean hull-xanthan-gels were rheologically characterized for dependence on the shear stress using a parallel plate shear device. The shear modulus G_o is relatively low. The loss modulus represents only a minor part of the storage modulus. With increasing shear stress there is a decrease of loss angle tan δ (loss modulus/storage modulus), that means an increase of the elastic part. It was impossible to reach rupture strength with the overall strain of 150 % of the device. Gelatine gels show such viscoelastic properties, too. An increase of G_o and storage moduluswas found in gels of mechanolytic pretreated soybean hulls with xanthan. These gels are firmer and more elastic.

Isolation and properties of extracts from soybean hulls containing galactomannans

A reason for the gelation effect of soybean hulls with xanthan is a galactomannan component. GM can be extracted together with other neutral polysaccharides, pectin and protein. Yield and composition of these components depends to a high degree on the extraction conditions. A lot of extraction variants including the influence of pH and temperature-time-effects or enzyme action on the extractability of GM and their gelation with xanthan was tested. Extraction with cold or hot water is ineffective. Application of special salt or buffer mixtures gives higher yield of GM. Combinations of acid pretreatment in a low moisture state, partial mechanolysis or intensive milling and/or thermal treatment (autoclaving at pH 4.5) were found to be the most effective extraction procedures. For instance, 250 g soybean hulls was first treated with 720 ml diluted H_3PO_4 (2 h, 20 ^oC, pH < 2), then autoclaved (20 min, pH 4.5) and finally extracted with hot water. The extract was concentrated in vacuum (dry matter = 14.2 %). It shows pseudoplastic flow behaviour (flow behaviour index n = 0.885, consistency index k = 0.294 Pa/s).

Elastic, homogeneous gels were obtained with an extract concentration of 12 g native hulls with 1 g xanthan (PU = 83.6, WR = 27.6 cm^2). With permanently increasing shear stress the resulting strain and the shear modulus of the gels were estimated. The strain is proportional to the shear stress up to 50 Pa. Further increase of shear stress led to an increase of G_o, that means the gel offers a greater resistance to deformation. The strain becomes relatively lower. Gel rupture occurs at a shear stress of 200 Pa (maximum shear modulus 268 Pa, elastic limit strain about 75 %).

The strength of gels between GM-containing extracts and xanthan can be improved by addition of extracted soybean hulls to the gel system.

Purification and properties of GM from soybean hulls

Purification of the GM extracts can be made by precipitation with alcohol, separation of pectin with Cu^{2+} and dialysis. The content of hexose-polysaccharides is > 50 % after this procedure. Furthermore, protein, pectin and pentosepolysaccharides are present in the samples.

Relatively pure GM were isolated by chromatography on DEAE-Sephacel. The saccharide composition of the purified GMs were analysed by GLC/MS. A typical GM preparation contained 26.5 % Gal, 66.5 % Man and 0.4 % Glc (Composition of the original, ethanol-precipitated extract: 20.3 % Gal, 51.0 % Man, 6.8 % Ara, 1.4 % Rha, 0.6 % Xyl, 0.8 % Glc and 12.0 % GalA). In dependence on the extraction conditions, Gal/Man ratios between 1:2.2 and 1:3.0 were estimated in purified GM. In most cases minor quantities of Glc were detected in the preparations. The purified GM have relatively low intrinsic viscosities (between 85 and 148 ml/g) and a broad molecular weight distribution. After heating, GM solutions show practically Newtonian flow behaviour.

The purified GM form comparatively weaker, homogeneous gels than soybean hulls with xanthan. The GM-xanthan-gels possess a gelation temperature between 39 and 41 ^oC and a melting temperature of 41 ^oC. Their elastic deformation was measured up to more than 150 % without gel rupture.

REFERENCES

1. Dey, P.M. (1978) Adv. Carbohyd. Chem. Biochem. 35, 341-367.
2. Aspinall, G.O., Whyte, J.N.C. (1964) J. Chem. Soc. 5058-5063.
3. Sharman, W.R., Richards, E.L., Malcolm, G.M. (1978) Biopolymers 17, 2817-2833.
4. Southgate, D.A.T., Hudson, G.J., Englyst, H. (1978) J.Sci. Food Agric, 29, 979-988
5. Baumann, H. (1966) Seifen, Fette, Anstrichmittel 68, 741-743.

Interaction of galactomannans with ethoxylated sorbitan esters – surface tension and viscosity effects

DOV REICHMAN[+*] and NISSIM GARTI[*]

[+]*Adumim Chemicals Ltd, Mishor Adumim, Israel 90610*

[*]*Casali Institute of Applied Chemistry, The Hebrew University of Jerusalem, Israel*

ABSTRACT

Interactions between proteins or synthetic polymers with monomeric emulsifiers at interfaces can result in synergism or antagonism. However, there is little information in the literature on such interactions between natural hydrocolloids and food emulsifiers. Our long-term interest in hydrocolloid-emulsifier interactions is related to the mechanism and interfacial properties of nonionic constituents, and their implications regarding food systems. In the present work, surface tensions of LBG and guar gum have been studied. It has been demonstrated that galactomannans, although highly hydrophilic in nature, can adsorb at interfaces, reducing surface tension to levels whereby they can be categorised as surfactants. Surface tensions of LBG solutions with ethoxylated sorbitan trioleate indicate the formation of a "new species" (complexes) with improved surface properties. The viscosities of aqueous solutions of guar gum, LBG and their mixtures with ethoxylated mono- and trioleates (Tween 80 and 85) were measured at different shear rates. At high emulsifier concentrations (5.0 gr/dl) a significant increase in viscosity is shown with Tween 85 in comparison to a sharp decrease with Tween 80. Both LBG and guar gum exhibited similar performance. The results are discussed in relation to the molecular structure of the galactomannan-emulsifier complex and the hydrophobicity of the constituents.

INTRODUCTION

Formation of complexes between surfactants and polymers, and their rheological and interfacial properties, is of great importance for the stability of dispersions, emulsions and other formulations in many industrial fields (1-4).

Greener et al. (5) studied the interaction between anionic surfactants and charged gelatin (above its isoelectric point) over a wide range of conditions. The interactions, which are governed by electrostatic forces, lead to a large increase in viscosity in some systems, as a function of surfactant type, surfactant/gelatin composition, and ionic strength.

The effect of charged gelatin on the surface tension of anionic and nonionic surfactant was reported by Knox et al. (6) and the nonionic surfactant was found to not interact with gelatin. To the contrary, SDS is shown to react with gelatin to form complexes with lower surface tension and dislocated CMC.

Schwuger's measurements of surface tension (7) show the difference between the interactions of cationic and anionic surfactants with nonionic polymers (polyglycol ethers). In water, only weak reversible hydrophobic bonding between cationic surfactants and nonionic polymers is formed while complexing of nonionic polymers with anionic surfactants is very strong.

In this study we report for the first time the interactions obtained between nonionic hydrocolloids (LBG and guar) and nonionic food emulsifiers for applications in food systems.

MATERIALS AND METHODS

Materials

Gums (LBG and guar) and emulsifiers (Tween 80 and 85) were obtained from Sigma Chemicals and were used without further treatment.

Guar gum was dissolved while stirring in distilled water at room temperature. LBG solution was prepared by mixing at room temperature for 30 minutes, heating to 80°C for 15 minutes, and cooling to room temperature. Both gum solutions were stirred for 3 hours to ensure complete dissolution. All the experiments were repeated using freshly prepared gum solutions (kept no longer than 24 hrs).

Methods

Surface Tension Measurements. The surface tension was monitored with a Lauda Tensiometer (Lauda, West Germany) using a Wilhelmy platinum plate. Measurements were made at 20±2°C. In measurements of mixtures of gum and emulsifier the Tween solutions were injected gently to the bottom of the gum solution after 5 minutes of resting. The resting time enabled the gum to reach the water/air interface and eliminated the possibility of competitive adsorption between the emulsifier and the gum. The surface tension was measured at constant gum concentrations (0.4gr/dl) as a function of surfactant concentration. It was essential to read equilibrium measurements since the surface tension showed time dependence at various preparations (7). Pure gum solutions, as well as mixtures, with emulsifiers were aged for periods of up to 2 hours to achieve steady state values. Since LBG and guar gum solutions contained undissolved matter, samples were centrifuged to obtain transparent solutions. Surface tension measurements were unaffected by this action.

Viscosity Measurements. Viscosity was measured using a Brookfield cone and plate viscometer (Model DV-II) at shear rates from 38.4 to 384 sec^{-1}. Measurements were made at 30±0.1^{o}C using a Brookfield EX-100 constant temperature bath. Samples of 1 ml were aged 3 minutes on the plate prior to taking measurements to ensure temperature equilibrium. Mixtures of gums and emulsifiers were prepared by mixing at room temperature for 1 hour. The viscosity was measured as a function of surfactant concentration at constant gum concentration (0.7 gr/dl).

RESULTS

Surface tensions of LBG and guar gum are given in Figures 1 and 2, respectively. Surface tensions were reduced to levels of 50 dyne/cm without indication of existence of CMC. Solutions with gum concentrations of not more than 0.7 gr/dl were measured to avoid viscosity effects on surface tension readings.

The possible synergism between the two gums was tested by measuring surface tension of mixtures of LBG and guar at a 50:50 weight ratio. No additional reduction of the surface tension relative to the two individual components of the mixture was detected.

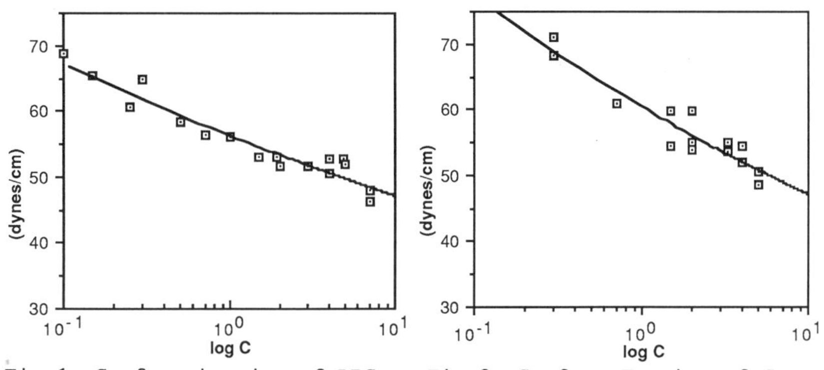

Fig 1. Surface tension of LBG   Fig 2. Surface Tension of Guar

The resulting γ-log C curve of LBG and Tween 80 mixtures is shown in Figure 3. The measurements were made at LBG concentration of 0.4 gr/dl. At this level the pure gum solution gave a surface tension of 50.3 dyne/cm. Thus, synergism is found at Tween concentrations between 2×10^{-4} to 1.5×10^{-3} gr/dl. In this range of concentrations, the surface tension values are lower than pure Tween and pure LBG.

In Figure 4, the viscosity of mixtures of gums and emulsifiers is given at a shear rate of 76.8 sec^{-1} at 30°C. At high levels of emulsifier (5 gr/dl), viscosity dropped sharply with Tween 80 followed by the precipitation (gel seperation) of the complex, while a higher increase in the viscosity was observed with Tween 85. The same trend was detected both for LBG and guar gum.

The viscosity values did not change by increasing or decreasing the shear rate. The viscosities obtained after a few hours from the preparation of the gum and emulsifier mixture did not change with time over the course of several days. (Sodium azide, 0.05%, was added as a preservative.)

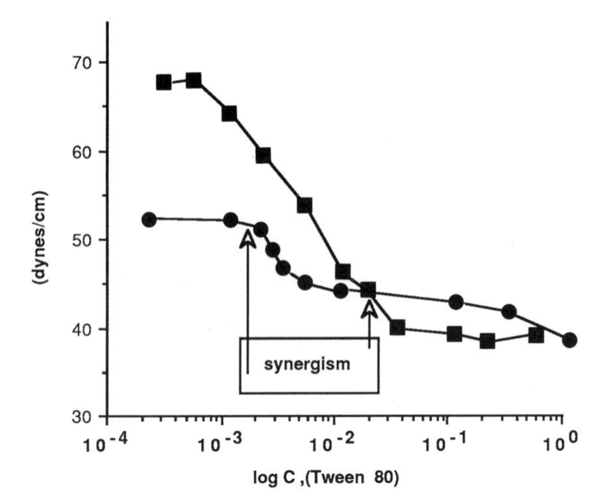

Figure 3. Surface tension of LBG (0.4 gr/dl) and Ethoxylated Sorbitan Monooleate (Tween 80)
-■- Tween 80
-●- Tween 80 with 0.4 gr/dl LBG

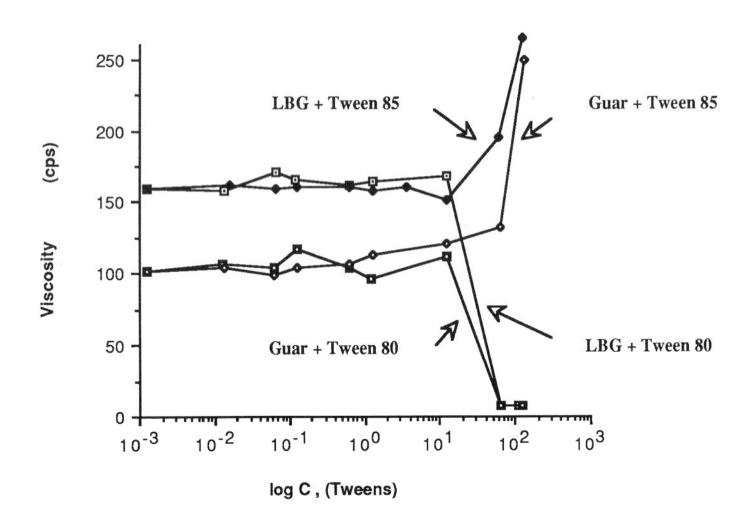

Figure 4. Viscosity measurements of galactomannans (0.7gr/dl) with increasing concentrations of ethoxylated sorbitan mono- and trioleate (Tween 80 and 85) solutions. Data taken at a shear rate of 75.8/sec and 30°C.

DISCUSSION

The γ-log C curves of the galactomannans presented above show that these hydrocolloids, besides modifying the flow properties of aqueous systems, can adsorb at water/air interfaces. The concentrations needed to reduce surface tension to 50 dynes/cm are low enough without strongly affecting the viscosity. This property can be of great advantage for practical industrial applications since no extreme viscosities are expected at levels for which the surface activity of the gum is sufficient.

No significant differences in the surface properties between guar and LBG were detected although the molecular structure of the two gums is different. The LBG has some nonanchored areas of galactose on the mannose backbone (smooth regions), while the guar is totally substituted.

Formation of a possible complex between LBG and Tween 80 is visualised since a moderate γ-log C curve is obtained, giving improved surface characteristics to the system in comparison with the pure substances.

Because both emulsifiers and gums are nonionic it seems that the interactions that govern the formation of such an active complex are hydrophilic (hydrogen bonds) or hydrophobic, unlike in other studied systems which are electrostatic in nature. The results, obtained by viscometry, emphasise the effect of this type of interaction. The trisubstituted Tween 85 is more hydrophobic (consisting of fewer hydrophilic sites than Tween 80) and does not seem to compete with the gums for the hydrogen bonding with the water molecules. Rather, it interacts with the gums to "stretch" the polymeric structure of the gum molecule resulting in an increasing number of sites for hydrogen bonding of gum with water, hence increasing the viscosity. Tween 80, with a more hydrophilic nature, competes with the gum for hydrogen bonding with the water molecules. Its strong interaction with the hydrophilic sites of the gum displaces the hydrogen-bonded water molecules along the polymer chain. This interaction significantly decreases the viscosity of the solution as shown in Figure 4, and the resulting complex eventually separates from solution.

The scheme in Figure 5 is an attempt to illustrate the possible organisation of the gum-emulsifier complex at the interface and in the bulk.

The interactive forces between the gum and the emulsifier seem to be stable under the hydrodynamic forces generated by the applied shear rate. The interaction is also stable with time as no change of viscosity was observed during several days of aging.

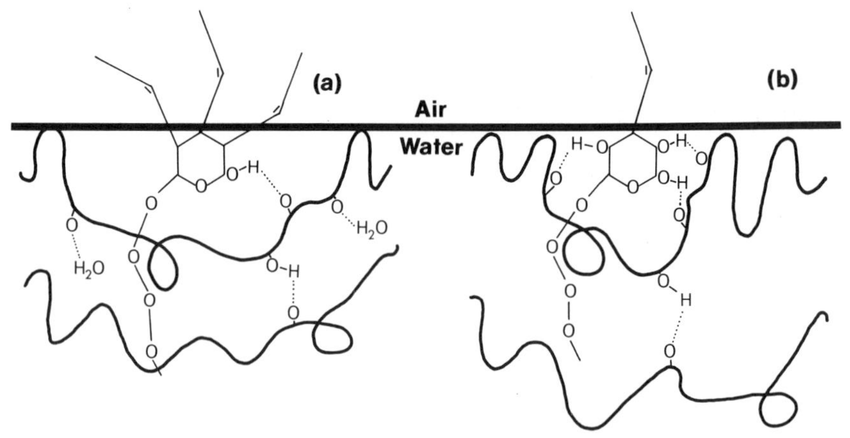

Figure 5. Schematic representation of possible organisation of the gum-emulsifier complex at the water/air interface and in the bulk
(a) In the presence of Tween 85
(b) In the presence of Tween 80

Although further work is needed in order to clarify the mechanism of the synergistic effects, even at this stage, the industrial formulator can significantly benefit from the existence of such interactions between the gums and the emulsifiers to control rheological properties.

REFERENCES

1. Tadros, Th.F. (1976) in Theory and Practice of Emulsion Technology, (ed. Smith, A. L.) pp 281-299, Academic Press, London, UK.
2. Bergenstahl, B. (1988) in Gums and Stabilisers for the Food Industry (eds. Phillip, G. O., Wedlock, D. J., and Williams, P. A.) pp 363-369, IRL Press, Oxford, UK.
3. Lipatov, Y. S. (1988) in Polymer Science Library 7: Colloid Chemistry of Polymers, (ed. Jenkins, A. D.) pp 85-107, Elsevier, Amsterdam, the Netherlands.
4. Reeve, M. J. and Sherman, P. (1988) J. Dispersion Sci. Technol., 9, pp 343-354.
5. Greener, J., Contestable, B. A., and Bale, M. D. (1987) Macromolecules, 20, pp 2490-2498.
6. Knox, Jr., W. J. and Parshall, T. O. (1969), J. Colloid Interface Sci., 33, pp 16-23.
7. Schwuger, M. J. (1972) J. Colloid Interface Sci., 43, pp 491498.

On the molecular weight distribution of dextran T-500

ABIGAIL BALL, STEPHEN E.HARDING and NEIL J.SIMPKIN
University of Nottingham, Department of Applied Biochemistry & Food Science, Sutton Bonington LE12 5RD, UK

ABSTRACT

The combined size exclusion chromatography/ low speed sedimentation equilibrium method is used to obtain an absolute molecular weight distribution for a commercially used dextran (T-500, Pharmacia). The form of the distribution and the weight average molecular weight for unfractionated T-500 [$(0.50\pm.02) \times 10^6$] are in good agreement with observations from, for example, on-line size exclusion chromatography/ multi-angle laser light scattering.

INTRODUCTION

One of the most important uses of dextrans is in medicine, for thickening blood plasma. For the food industry dextrans are mainly used as standards for analytical purposes for monitoring the performance (and in some cases calibrating) chromatographic separation processes. One of the most commonly used "standards" is dextran T-500 (so called because it has a molecular weight of ~500kD). In this short study we determine the molecular weight distribution of dextran T-500 by using a combined approach involving size exclusion chromatography (SEC) and low speed sedimentation equilibrium (LSSE) in the analytical ultracentrifuge.

MATERIALS

Dextran T-500 was supplied commercially from Pharmacia (Milton Keynes, U.K.) The solvent used for all analyses was a standard phosphate chloride buffer (pH 6.5, I 0.30). The relevant proportions of Na_2HPO_4 and KH_2PO_4 were made up to a combined ionic strength of 0.05. A total ionic strength of 0.30 was achieved by adding relevant proportions of NaCl in accordance with Green (1).

METHODS

Size exclusion chromatography

A column of Sephacryl S-400 (Pharmacia) was used, of 1.6cm internal diameter and 73cm height. Loading concentrations were ~ 1mg/ml, and sample volumes were 5ml. The flow rate was 10ml/hour.

Fractions of 2ml were collected and assayed for total sugar content using a phenol sulphuric acid procedure similar to that described by Dubois et al (2). Elution volumes were determined by weight. The void volume was determined using blue dextran 2000 and the total volume using sucrose. Column recoveries were between 90 and 100%.

Sedimentation Equilibrium

The low speed method (see ref. 3) was used in a Beckman Model E analytical ultracentrifuge equipped with an RTIC temperature measuring system, Rayleigh interference optics and a 5 mW He-Ne laser light source using procedures described by Creeth & Harding (4). All determinations were made in 30mm optical path length cells at the lowest possible loading concentrations (0.1-0.5 mg/ml) to minimise possible effects of thermodynamic non-ideality and associative phenomena. At the concentrations used, non-ideality effects are likely to contribute less than ~ 5% error in the measurement (see Table 2.2 of ref 5) which is of the same order as the precision of the measurement. A value of 0.613 ml/g was used for the partial specific volume (6).

Molecular weight determinations (apparent weight averages, M_w) of the fractions were mainly performed in groups of three using an Yphantis-style multi-channel 30mm path length cell. All samples had been made up in the solvent as described above and dialysed against this solvent. Whole cell weight average molecular weights were extracted using the M* function as described by Creeth & Harding (4).

Calibration of the SEC column using low speed sedimentation equilibrium

Fractions of 2ml volume were isolated from the eluate, and fractions of equal elution volume from a series of three or four runs were combined and concentrated using centrisart tubes (Sartorious Ltd.), to a concentration of ~ 0.2mg/ml. The molecular weights of these fractions were then determined by LSSE, as described, and a plot of the logarithm of the (apparent) weight average molecular weight, M_w versus elution volume obtained. The calibration plot and the elution profile were then used to determine the molecular weight distribution. This is the method described by Ball et al (7).

RESULTS

Molecular weight distribution

The plots of elution volume, calibration and molecular weight distribution are given in Figure 1 (a,b,c respectively).

A linear calibration appeared reasonable (Figure 1b) within

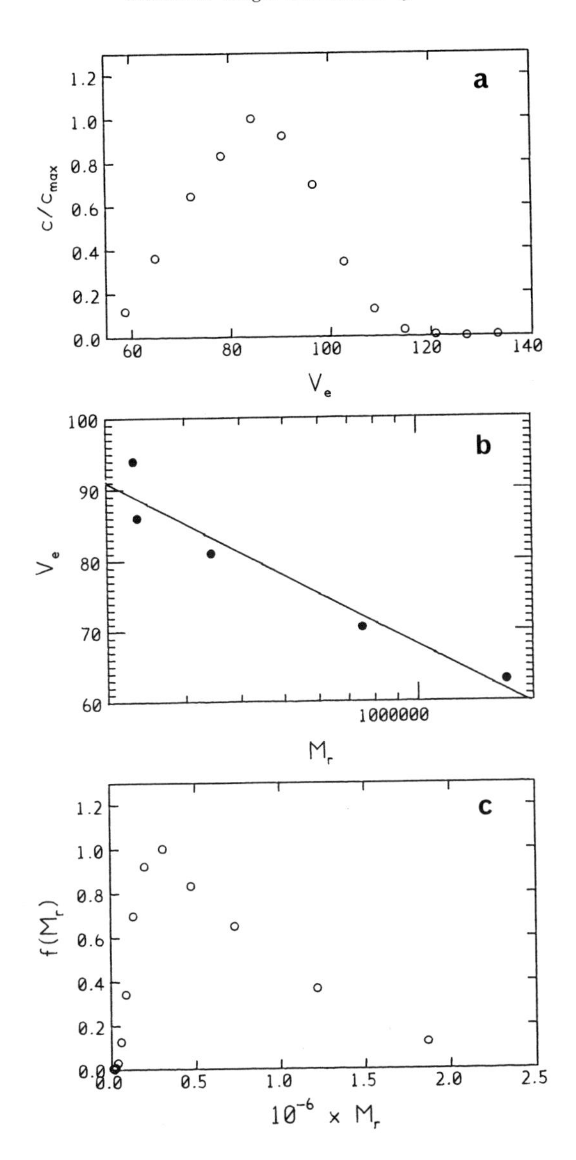

Figure 1. Calibrated Gel Chromatography for Dextran T-500. (a) Elution profile; (b) Calibration plot from low speed sedimentation equilibrium; (c) Molecular weight distribution

the range of molecular weights examined. The distribution is of a log-normal type with a mode molecular weight of ~.3 x 10^6. Molecular species with an M_w as high as 2.0 x 10^6 were present. From the form of this distribution, an approximate value for the weight average over the whole distribution of ~0.4 x 10^6 was obtained.

Weight average molecular weight of unfractionated material

LnJ vs ξ plots, [where J is the absolute concentration in fringe number units and ξ is a radial displacement squared parameter] were appreciably more upward curved for unfractionated than fractionated material - corresponding to the greater heterogeneity. A weight average molecular weight over the whole distribution for the unfractionated material of (.50±.02) x 10^6 was obtained.

CONCLUSIONS

The molecular weight of 500kD quoted by the commercial manufacturer for dextran T-500 was confirmed. Polydispersity was however rather high with species of M_w as high as ~2.0 x 10^6 present. The data for the average molecular weight and the distribution are in reasonable agreement with independent measurements using on-line SEC/ multi-angle laser light scattering (8), and also earlier sedimentation equilibrium studies (9).

REFERENCES

1. Green, A.A. (1933) J. Am. Chem. Soc. 55, 2331-2336
2. Dubois, M., Gilles, K.A., Hamilton, J.K., Rebers, P.A. and Smith, F. (1956) Anal. Chem. 28, 350-356
3. Creeth, J.M. and Pain, R.H. (1967) Prog. Biophys. Mol. Biol. 17, 217-287
4. Creeth J.M. and Harding, S.E. (1982) J. Biochem. Biophys. Meth. 7, 25-34
5. Harding, S.E., Varum, K.M., Stokke, B.T. and Smidsrød, O. (1989) Adv. in Carbohyd. Anal. (C.A. White ed.) 1, JAI Press, London - in press
6. Gibbons, R.A. (1972) in Glycoproteins: Their Composition, Structure and Function (Gottschalk, A. ed.), 5A, pp31-157, Elsevier, Amsterdam
7. Ball, A., Harding, S.E. and Mitchell, J.R. (1988) Int. J. Biol. Macromol. 10, 259-264
8. Dawn F Application Note (1988), 7/8/88 No. 3, Wyatt Technology Corp. (Santa Barbara, USA)
9. Edmond, E., Farquhar, S., Dunstone, J.R. and Ogston, A.G. (1968) Biochem. J. 108, 755-763

A rheological investigation of some polysaccharide interactions

R.O.MANNION, J.R.MITCHELL+, S.E.HARDING+, N.E.LEE and C.D.MELIA

Department of Pharmaceutical Sciences, and +Department of Applied Biochemistry and Food Science, University of Nottingham, UK

ABSTRACT

This publication contains work on several aspects related to polysaccharide synergy. Firstly a model for predicting the viscosity of non-interacting mixed polysaccharide solutions based upon coil occupancy was derived and experimentally tested for a series of non-ionic cellulose ethers. Good agreement was found between the predicted and experimental values. The model was adapted to accommodate the coil expansion of a polyelectrolyte occurring when it is mixed, and hence diluted, with a non-ionic polysaccharide in aqueous solution. The validity of this approach was examined using solutions of sodium carboxymethylcellulose and hydroxypropyl methylcellulose. The results showed that the enhanced viscosities observed on mixing non-ionic and anionic cellulose ethers can be explained in terms of coil expansion without invoking a mechanism based upon hydrogen bonding, and this was confirmed by sedimentation coefficient measurements. Rheological and ultracentrifugational techniques were used to examine the interaction between xanthan and the fraction of locust bean gum soluble at 35°C observed when solutions of the polysaccharides are mixed cold. Both techniques indicate that significant synergism occurs.

INTRODUCTION

The aim of this work was to examine polysaccharide synergism from a variety of angles. In order to determine clearly that two polysaccharides have interacted from viscosity measurements, it is necessary to possess some way of predicting or estimating the viscosity which a mixed polysaccharide solution would have in the absence of interactions. Various models have been proposed to predict the viscosity of mixed solutions of non-interacting polysaccharides (1,2). Many of these are entirely empirical in nature and all have relatively limited adaptability. This work describes the derivation and experimental validation of a model for the prediction of the viscosity of mixed polysaccharide solutions based upon coil occupancy, an approach suggested by Morris (3).

It has been shown(4) that at concentrations sufficient for coil overlap to occur, the viscosity of a solution of a single polysaccharide is given by equation 1.

$$\log \eta^{o}_{sp} = X\log(C[\eta])+\log K \qquad \text{(Eq.1)}$$

where η^{o}_{sp} is the zero shear specific viscosity, C is the total polysaccharide concentration, $[\eta]$ is the intrinsic viscosity, and X and K are constants specific for each polysaccharide.

By adapting this equation to a two polysaccharide system (A and B) the following relationship is obtained:

$$\log \eta^{o}_{sp(A+B)} = (f_A X_A + f_B X_B)\log(C(f_A[\eta_A]+f_B[\eta_B]))+f_A\log K_A+f_B\log K_B \qquad \text{(Eq.2)}$$

where f_A and f_B are the weight fractions of polysaccharides A and B respectively and subscripts denote the polysaccharides to which the other terms refer.

Eq.2 has been adapted to account for the changes in intrinsic viscosity of an ionic polysaccharide,"B", which occur due to alterations in the total ionic strength, I, of the solution, which is given by the following relationship:

$$I = I_s + Cf_B\alpha \qquad \text{(Eq.3)}$$

Where I_s is the ionic strength of the solvent and α = the weight fraction counterion content of the polyelectrolyte divided by the counterion molecular weight.

$$\log \eta^{o}_{sp(A+B)} = \qquad \text{(Eq.4)}$$

$$(f_A X_A + f_B X_B)\log(C(f_A[\eta_A]+f_B([\eta^{*}_B]+S.I^{-0.5})))+f_A\log K_A+f_B\log K_B$$

Where $[\eta^{*}_B]$ is the intrinsic viscosity of the polyelectrolyte at infinite ionic strength, and S is a measure of polyelectrolyte chain stiffness, equal to the gradient of a graph of intrinsic viscosity against square root of the ionic strength (5).

For weak gel structures the use of steady shear viscosity is less appropriate and the study of synergism by dynamic rheological methods is well established (6,7). A system of particular interest is xanthan gum with locust bean gum where it has been suggested that xanthan has to be heated for the disordered form to be attained before the interaction can occur (8). Preliminary dynamic rheological results are presented for unheated mixtures of these polysaccharides.

Finally, the use of ultracentrifugation to study the interaction between cellulose ethers and that between xanthan and locust bean gum is demonstrated.

MATERIALS AND METHODS

Hydroxypropylmethylcellulose, HPMC, (Methocel K4M) was donated by Colorcon. Hydroxyethylcellulose, HEC, (Natrosol 250HHX), and Hydroxypropylcellulose, HPC, (Klucel H) were donated by Hercules. Sodium carboxymethylcellulose, NaCMC, (Blanose 12M31FD) was donated by Aqualon. Xanthan (Keltrol F) and locust bean gum were purchased from Kelco and Sigma respectively.
The fraction of the galactomannan soluble at 35°C was isolated using a method based upon that of Gaisford *et al* (9). This temperature fraction was used in all subsequent rheological and ultracentrifugational studies.

Concentrated solutions (>0.1%) were prepared by dissolving the polysaccharide in the solvent using a high shear Silverson mixer. Dilute solutions were prepared by adding the correct weight of polysaccharide to 80% of the total solvent, shaking until dissolved, then adding the remaining solvent.

All mixtures of polysaccharide solutions were prepared using single polysaccharide solutions which had been hydrated for 24 hours. The solutions were mixed with an Ultra Turax mixer and allowed to stand for a further 24 hours prior to rheological measurements being performed.

For each of the cellulose ethers the viscosities of a series of dilute solutions (<0.1%) were determined using an Ostwald viscometer with a flow time for water of 70 seconds. The intrinsic viscosity for each was then obtained using a Huggins plot extrapolation to zero concentration.

Zero shear viscosities for the concentrated cellulose ether solutions were obtained using a Deer rheometer fitted with concentric cylinder geometry (MG7101).

The storage and loss moduli for 1.0%w/w solutions of xanthan, locust bean gum, and mixtures of the two were obtained using a Weissenberg rheogoniometer, model R19, fitted with 2° cone and plate geometry. A frequency range of 0.05 to 0.5 Hz was used.

All rheological measurements were performed at 25°C.

Sedimentation studies were performed with a MSE Centriscan analytical ultracentrifuge.

RESULTS AND DISCUSSION

Figures 1 and 2 show the values predicted by equation 2 relative to those obtained experimentally for two sets of mixed

Figure 1. Predicted and Determined Values of Specific Viscosity for Methocel K4M / Natrosol 250 HHX mixtures (Total polysaccharide concentration 1% w/w)

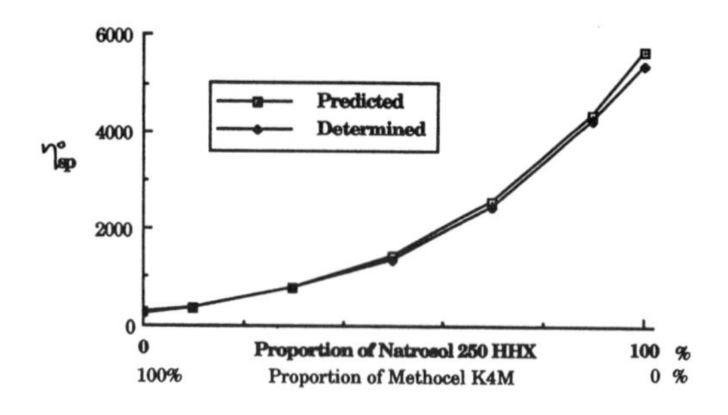

Figure 2. Predicted and Determined Specific Viscosity values for Methocel K4M / Klucel H mixtures (Total polysaccharide concentration 1% w/w)

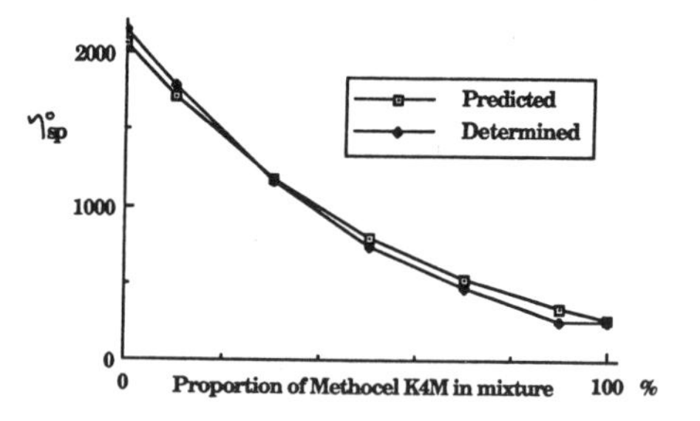

Figure 3. Predicted and Determined Values of Specific Viscosity For Methocel K4M / Blanose 12M31FD mixtures. (Total polysaccharide concentration 1%)

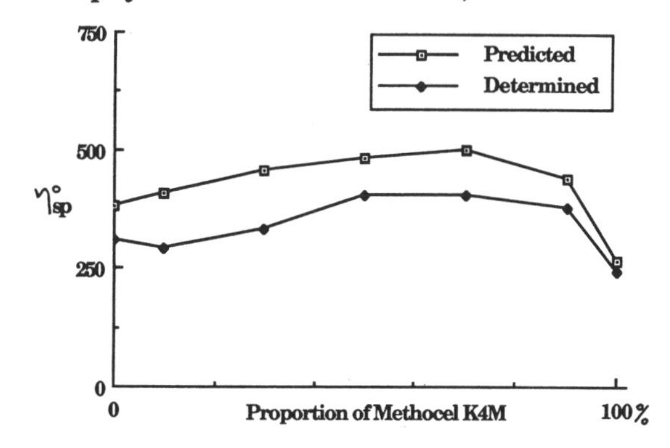

Figure 4. Storage and loss moduli for cold mixed solutions of xanthan and the fraction of locust bean gum soluble in water at 35°C.

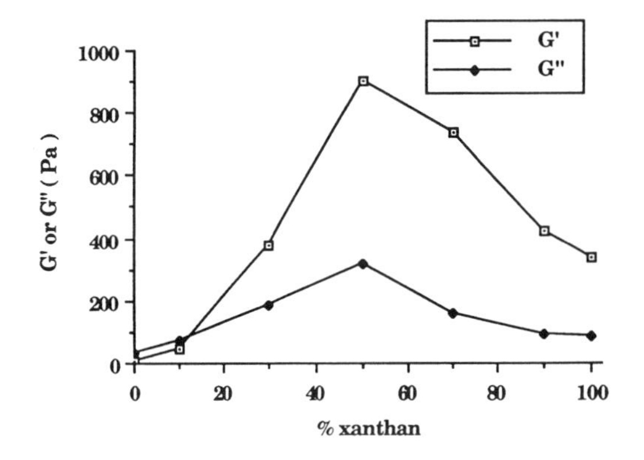

TABLE 1. Sedimentation coefficients of single polysaccharides and mixed systems.

POLYSACCHARIDE	SEDIMENTATION COEFFICIENT (Svedbergs)
Methocel K4M	2.0
Blanose 12M31FD	2.4
Methocel K4M (in mixture)	1.6
Blanose 12M31FD (in mixture)	2.2
Xanthan	7.4
Locust Bean Gum (LBG)	3.2
Xanthan/LBG mixture	Approx 60

solutions of non-ionic cellulose ethers. In each case the solvent was 0.1M sodium chloride and the total polysaccharide concentration was 1.0%w/w. There is excellent agreement between predicted and determined viscosities in each case indicating the applicability of the model to these systems.

Figure 3 shows the values predicted by equation 4 relative to those obtained experimentally for mixtures of HPMC and NaCMC. The solvent was distilled water, and the total polysaccharide concentration was 1.0%w/w. Agreement between the two curves is limited especially where the polysaccharide content is predominantly the polyelectrolyte. The values of the constants X and K for the NaCMC were obtained from a graph of Log zero shear specific viscosity against Log coil overlap parameter ($C[\eta]$) at an added ionic strength of 0.2M. A high ionic strength was chosen to minimise the effects of errors introduced by the estimation of the sodium content of the polyelectrolyte. However the discrepancy between the zero shear specific viscosity of the 1.0%w/w NaCMC solution predicted using these values and that obtained experimentally is significant. Therefore it appears that the behavior of the polyelectrolyte coils is not independent of ionic strength; with increased coil-coil repulsion leading to a decrease in the number and lifetime of entanglements and hence a reduction in measured viscosity as the total ionic strength is reduced.

Importantly, the results show that the enhanced viscosities produced when an anionic and a non-ionic polysaccharide are mixed in aqueous solution can be accounted for in terms of polyelectrolyte coil expansion due to counter-ion dilution on mixing and it is unnecessary to invoke a mechanism involving molecular association, as other workers have done (1).

The results of the ultracentrifugation studies, shown in Table 1, indicate that there is no specific association between HPMC and NaCMC. The sedimentation coefficient of each in the presence of the other varies little from that obtained in isolation and these small differences can easily be accounted for by Johnston-Ogston and viscosity effects. In contrast, the sedimentation coefficient of the xanthan/locust bean gum mixture is an order of magnitude greater than either of the individual polysaccharides. This, along with the relative turbidity of the mixed solutions signifies that an interaction involving molecular association has occurred.

Figure 4 shows the storage and loss moduli for various mixtures of locust bean gum and xanthan solutions in 0.1M sodium chloride at a concentration of 1.0% w/w. There is an optimum storage modulus for the mixture containing equal parts of each polysaccharide; the value of this is approximately 500% greater than that of the more elastic component (xanthan), at the same total concentration.

The dynamic rheological and sedimentation data both indicate that a synergistic interaction of significant magnitude has occurred between xanthan and locust bean gum. At this ionic strength and in the absence of heat, the xanthan will be fixed in its native form. According to the model previously proposed for this interaction (8) it should not occur in such circumstances. Hence it appears that these two polysaccharides can interact by an additional mechanism. A further report into the nature of this interaction is being prepared.

REFERENCES

1. Walker, C.V. And Wells, J.I. (1982) Int. J. Pharm., 11, p309-322.

2. Kaletunc-Gencer, G. and Peleg, M. (1986) J. Text. Studies, 17, p61-70.

3. Morris, E.R. (1984) in "Gums and stabilisers for the food industry volume 2" edited by Phillips, G.O., Wedlock, D.J., and Williams, P.A. p57-78. Pergamon Press, Oxford.

4. Morris, E.R., Cutler, A.N., Ross-Murphy, S.B., and Rees, D.A., (1981) Carbohydrate Polymers, 1, p5-21.

5. Smidsrod, O., and Haug, A., (1971) Biopolymers, 10, p1213-1227

6. Clark, R.C., (1988), in "Gums and stabilisers for the food industry volume 4" edited by Phillips, G.O., Wedlock, D.J., and Williams, P.A. p165-172. IRL Press Ltd, Oxford.

7. Dea, I.C.M., Morris, E.R., Rees, D.A., Welsh, J.E., Barnes, H.A., and Price, J., (1977) Carbohydrate Research, 57, p249-272.

8. Cairns, P., Miles, M.J., and Morris, V.J. (1986) Nature, 322, p89-90.

9. Gaisford, S.E., Harding, S.E., Mitchell, J.R., and Bradley, T.D., (1986) Carbohydrate Polymers, 6, p423-442.

Dependence of the specific volume of konjac glucomannan on pH

KAORU KOHYAMA and KATSUYOSHI NISHINARI
National Food Research Institute, Tsukuba, Ibaraki 305, Japan

ABSTRACT

Solution properties of konjac glucomannan were examined by measurement of specific volume at various temperatures, concentrations and pH. Apparent partial specific volume as a function of temperature increased with increasing temperature from 5 °C to 40 °C and then levelled off. Apparent partial specific volume as a function of pH was constant between pH=3 and 11 and increased steeply at about pH=11 with increasing pH. It is suggested that the change of molecular structure is necessary for gelation of konjac glucomannan.

INTRODUCTION

Konjac glucomannan is a major component of the tuber of Amorphophallus konjac C. Koch. Its main chain is composed of β-1,4-linked D-mannoses and D-glucoses. The ratio of mannose to glucose is known to be about 1.6 (1). It has some branching points at C3 (2). After dissolving in water, it gives highly viscous solutions.

The sol of about 2% is changed into Japanese traditional food "konnyak" under alkaline conditions (pH=11.3-12.6) by heating. Maekaji proposed that the gelation was initiated by elimination of carbonyl groups from the glucomannan molecules in the presence of alkali reagents (3,4). However, since refined konjac glucomannan has little or no acetyl groups, the mechanism of the gelation remains still enigmatic in some points. Recently many researchers have taken an interest in this material, since the texture of the gel is unique and it is known as a water soluble dietary fibre.

The physicochemical properties have not been elucidated mainly because of the difficulty to obtain well-fractionated konjac glucomannan samples (5,6). In the present work, solution properties of konjac glucomannan were examined by measurement of specific volume at various temperatures, concentrations and pH.

MATERIALS AND METHODS

The konjac flour was isolated from the tuber of

Amorphophallus konjac C. Koch cv Zairai harvested at Agatsuma, Gunma, Japan, in 1987. The flour was washed with methanol until the supernatant became clear soon after stirring. Then it was washed with ether and dried at room temperature.

Alkali hydroxides, hydrochloric acids and all other chemicals were special grade and were used without further purification.

The apparent specific volumes were evaluated by densimetry of solutions (7). Densities of solutions were measured using an Anton-Paar Precision Density Meter (Type DMA20D). Temperature was controlled by a circulated water system using a Lauda Low-temperature Thermostat (Type, RMS6) within ±0.005°C.

To prevent the bacterial degradation, all measurements were done in 0.02% sodium azide solution as a solvent. The partial specific volumes in the sodium azide solution agreed with that in water within±0.01$g^{-1}cm^3$ when the measurement was carried out just after the preparation of the solution. Solution viscosities were measured with an Ubbelohde-type viscometer at 25.0°C±0.02°C. The flow time for the solvent was 240sec.

RESULTS AND DISCUSSION

We could not obtainamore concentrated solution than 0.6% because of the low solubility of konjac glucomannan in the neutral region. In this region, the density of the solution was a linear function of the concentration of konjac glucomannan. Therefore, the partial specific volume was independent of concentration of glucomannan, and could be determined without extrapolation to zero concentration. We examined only one solution concentration, about 0.2w/w%, varying other conditions hereafter.

Figure 1 shows the dependence of the apparent partial specific volume on temperature. The apparent partial specific

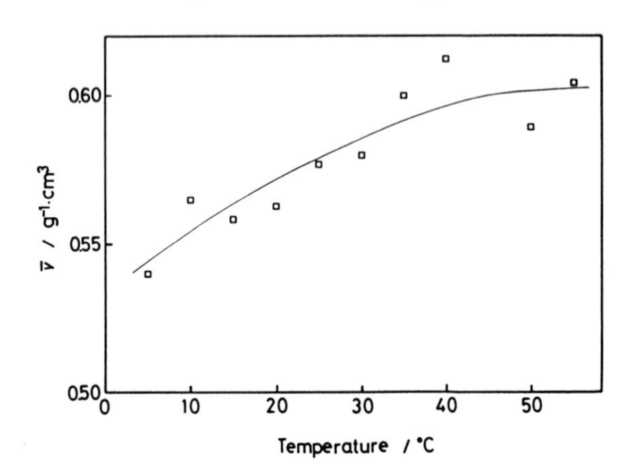

Figure 1. Temperature dependence of apparent partial specific volume. Concentration of konjac glucomannan: 0.205w/w%.

volume increased with increasing temperature from 5 °C to 40 °C, and then levelled off. Since the specific volume $\bar{v}$ decreased with cooling from 55 °C, large values of $\bar{v}$ at higher temperatures do not mean the degradation of konjac glucomannan molecules by heating. Both dextran (8) and pullulan (9,10) showed almost the same value $0.61 g^{-1} cm^3$ for the partial specific volume at 25 °C. The apparent partial specific volume of konjac glucomannan at 25 °C was $0.576 g^{-1} cm^3$ and smaller than that of those neutral polysaccharides.

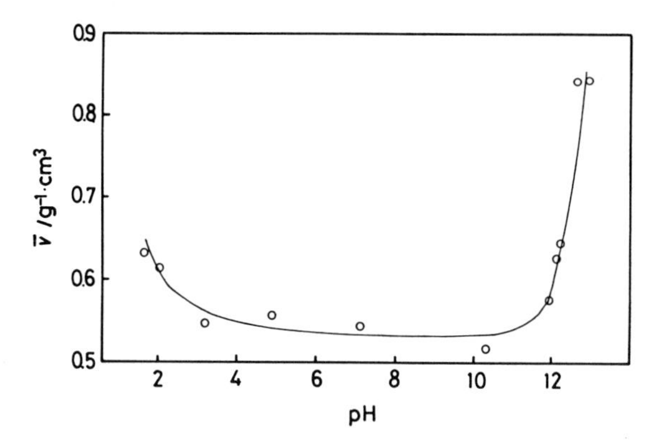

Figure 2. Apparent partial specific volume of konjac glucomannan as a function of pH at 25.0 °C. pH was adjusted by addition of HCl or NaOH. Concentration of konjac glucomannan: 0.221w/w%.

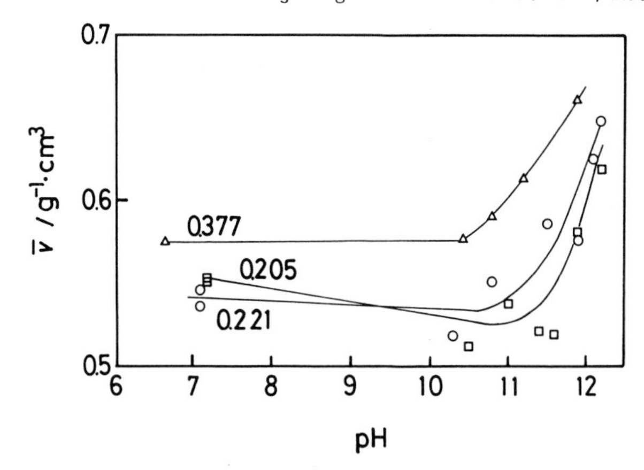

Figure 3. Dependence of apparent partial specific volume on pH at 25.0 °C. pH was adjusted by several alkali reagents: ○, NaOH; □, KOH; △, $Ca(OH)_2$. The numbers beside each curve represent the concentration of konjac glucomannan in w/w%.

The dependence of the apparent partial specific volume on pH is shown in Figure 2. The specific volume was almost constant between pH=3 and 11 and then increased steeply at about pH=11 with increasing pH. This phenomenon depends upon only pH. The results shown in Figure 3 suggests that any alkali hydroxide increases the apparent partial specific volume in this pH region. The change of the specific volume was independent of the size or valency of alkali ion. Since it is reported that the gelation of konjac glucomannan occurs from pH 11.3 to 12.6, the change of the specific volume at this pH range is very important. It was suggested that the change of molecular structure is necessary for gelation of konjac glucomannan.

The apparent partial specific volume decreased reversibly when pH of the solution was neutralized from pH=12. The molecules of konjac glucomannan were not decomposed by alkali hydroxides but they undergo some conformational change.

The partial specific volume in solution consists of three parts: (1) the constitutive atomic volume, (2) the void volume due to imperfect atomic packing and (3) the volume change for solvation. Since the constitutive atomic volume should be constant at normal pressure, the increase of the specific volume should be influenced by the change of the void volume which follows the conformational change of konjac glucomannan molecules or/and the solvation at different pH values.

At present, it is not clear why the specific volume increases in alkaline condition. Further investigations are required in order to clarify this point.

ACKNOWLEDGMENTS

The authors would like to thank Konnyaku Station for Production and Breeding, Gunma Agricultural Research Centre for kind supply of konjac flour.

REFERENCES

1) Kato, K. and Matsuda, K. (1969) Agr. Biol. Chem., 33, 1446-1453.
2) Maeda, M., Shimahara, H. and Sugiyama, N. (1980) Agric. Biol. Chem., 44, 245-252.
3) Maekaji, K. (1974) Agr. Biol. Chem., 38, 315-328.
4) Maekaji, K. (1978) Nippon Nogeikagaku Kaishi, 52, 513-517.
5) Nishinari, K., Kim, K. Y. and Kohyama, K. (1987) Proceedings of 2nd International Workshop on Plant Polysaccarides, Grenoble, 2
6) Mitsuyuki, H., Ohta, K. and Kawahara, K. (1987) Proceedings of 10th Japanese Carbohydrate Symposium, Tokyo, 45-46.
7) Kratky, O., Leopold, H. and Stabinger, H. (1973) Methods Enzymol., 27D, 98-110.
8) Gekko, K. and Noguchi, H. (1971) Biopolymers, 10, 1513-1524.
9) Kawahara, K., Ohta, K., Miyamoto, H. and Nakamura, S. (1984) Carbohydr. Polym., 4, 335-356.
10) Nishinari, K., Kohyama, K., Williams, P. A. and Phillips, G. O. (1989) presented at 2nd Euro-American Conference on Macromolecules, Oxford.

Determination of the relative molecular mass of konjac mannan

S.M.CLEGG, G.O.PHILLIPS and P.A.WILLIAMS
The North East Wales Institute, Deeside, Clwyd, CH5 4BR, UK

ABSTRACT

The molecular weights of two different konjac flours, a regular flour and a hydro-processed flour, have been investigated and compared using a variety of techniques, namely light scattering, intrinsic viscosity and gel permeation chromatography (gpc). The results show only minor differences in the molecular weight averages of the two flours and furthermore, the molecular weight distributions, as determined by gpc, indicate the same broad distribution of molecular weights in both samples.

Sonication of solutions of both konjac samples was seen to significantly reduce their viscosities indicating the presence of aggregates of only partially dissolved material in unsonicated samples. The existence of such aggregates could explain the large variations in molecular weights obtained by other investigators working in the field.

INTRODUCTION

Konjac mannan (KM) is the major polysaccharide constituent of konjac flour, a flour produced from the tubers of Amorphophallus konjac C. koch which is a member of the family Araceae found mainly in Japan and China. The major use of KM at present is as a food product in Japan where it is eaten in the form of a gel called konnyaku but its viscosity and gelling ability make the study of its properties of great interest to the food industry generally.

The polysaccharide is a gluco-mannan with a glucose: mannose ratio of approximately 1:1.6 [1-3] but the exact nature of the backbone linkages has been debated for some time. Okimasu and Kishida[4] drawing on all the published data consider the structure illustrated in Fig.1 to be the most likely repeating unit.

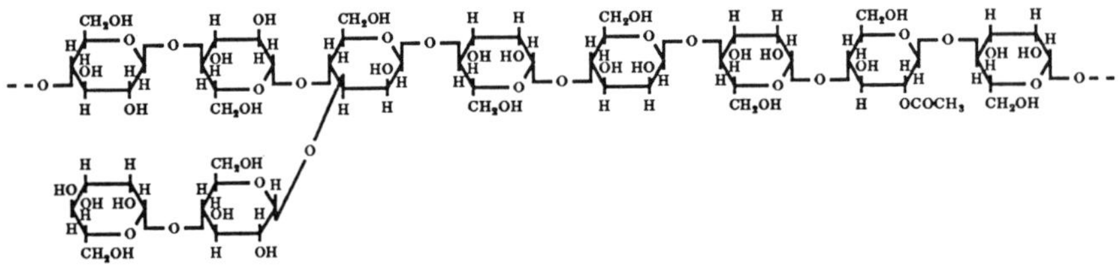

Figure 1. The most likely repeating unit of konjac mannan according to Okimasu and Kishida[4].

The actual molecular mass of KM molecules has been the subject of a number of studies [5-7], but again various conflicting values have been reported. Torigata et al[5] carried out light scattering and viscosity measurements on an isoamyl acetate solution of a nitro-KM and found a value of 2.71×10^5 for the molecular weight and a value of 56.3nm for the radius of gyration, R_G. These values are in close agreement with those of Sugiyama et al[6] who found a value of 2.7×10^5 for the molecular weight and 57nm for the R_G of a konjac mannan prepared by the deacetylation of acetylated-KM with alkali. However, Sugiyama et al[6] also determined molecular weights and R_G's of konjac mannans extracted from tubers of various strains produced in various regions of Japan, by carrying out measurements on reportedly soluble samples which were prepared by a freeze-dried method. These results indicated that both the molecular weight and R_G were dependent on the strain and place of production, the actual values for the molecular weights being two to four times larger than those obtained for their own deacetylated samples and for those of Torigata et al[5]; they explained this difference as being a result of degradation during the nitration and acetylation procedures.

Kishida et al[7] also carried out light scattering and viscosity measurements on konjac mannans. They used partially methylated derivatives and obtained molecular weight values of a similar magnitude to those obtained by the freeze drying method of Sugiyama et al[6], but found little dependence on type of strain. By partial acid hydrolysis of their KM samples Kishida et al[7] also determined the Mark-Houwink parameters, K and a for konjac mannan dissolved in water at 30°C ($K = 6.37 \times 10^{-4}$ and $a = 0.74$).

We now report our findings on the molecular weights, obtained using various techniques, of two konjac flours.

MATERIALS AND METHODS

Materials

The two samples of konjac flour were obtained from FMC Marine Colloids Division and were classified as a hydroprocessed flour, coded HP and a regular flour, coded R.

Methods

Solutions were prepared by dissolving the flour in the solvent at 80°C with stirring for 2 hours, followed by continued stirring overnight at room temperature. The insoluble material was then centrifuged away leaving the soluble konjac mannan. The molecular weights of the two samples were then investigated by gpc, intrinsic viscosity measurements and total intensity light scattering.

1 Gel Permeation Chromatography

This was carried out on a column of length 90cm and diameter 2.6cm packed with Pharmacia Sephacryl S-500. A flow rate of approximately 30mlh^{-1} was utilised with eluent detection being carried out using a differential refractometer and by monitoring the UV absorption at 218nm. Fractions were also collected and analysed for sugar content using the anthrone method[8].

2 Intrinsic Viscosity Measurements

These were performed on samples dissolved in water using an Ubbelohde type capillary viscometer (Cannon 75) immersed in a water bath at 30°C. Solution concentrations giving relative viscosities between 1.24 and 2.30 were used to obtained the required extrapolations.

3 Total Intensity Light Scattering

Total intensity light scattering was carried out on the samples dissolved in 0.02% sodium azide using a Malvern photon correlation spectrometer PCS 100 SM equipped with a He-Ne laser (λ = 632.8nm) in conjunction with a Malvern 'Loglin' dual function digital correlator. The refractive index increment of KM was determined as 0.155 using an Abbé differential refractometer. Prior to measurements the samples were filtered several times through a 0.2μm filter to remove dust particles.

RESULTS AND DISCUSSION

Preliminary investigations of the solution properties of both the hydroprocessed sample, HP and the regular sample, R showed that sonication of the solutions produced a marked decreased in their viscosities, which after approximately 3 minutes sonication remained constant with length of sonication time. Furthermore, on standing the solutions did not regain their original viscosities, even when allowed to stand for a time period of greater than 24 hours. This loss of viscosity on sonication was attributed to the break up of aggregates and consequently, in all molecular mass determinations, the sample was sonicated for 5 minutes prior to measurement.

The molecular weight distribution profiles of both samples were determined by gpc, results of which are presented in Figs. 2a and 2b. The two detection techniques give closely superimposable molecular weight distributions in the case of each sample and show both samples to possess approximately the same, broad molecular weight profile. Dextran and pullulan samples of known well defined molecular weights, ranging from 40,000 to 900,000, were used as calibrants for the gpc column, giving a good linear calibration curve. The distribution profiles were analysed so as to obtain the weight average, $\bar{M}w$, and number average, $\bar{M}n$, molecular weights presented in table 1, by splitting the curve into numerous strips, each assumed homogeneous in terms of molecular weight. The values obtained for $\bar{M}w$ are similar for both samples but the $\bar{M}n$ of sample HP is significantly lower than that of sample R which is probably a result of the hydro-processing treatment to which sample HP was subjected.

The intrinsic viscosities, [η], of samples HP and R are also listed in table 1 along with the viscosity average molecular weight obtained using the K and a values of Kishida[7]. Figs. 3a and 3b show the extrapolations used to obtain the intrinsic viscosities. Again, as with the gpc profiles, the values of intrinsic viscosity for the two samples are approximately the same indicating the same molecular size for both samples. The actual values obtained for the viscosity average molecular weight are in good agreement with $\bar{M}w$ and $\bar{M}n$ calculated from the gpc data.

Finally, total intensity light scattering measurements were carried out on the two samples to obtain their weight average molecular weights. These values are listed in table 1 and again show the similarity in molecular weight of the two samples, the actual values supporting those obtained by gpc and intrinsic viscosity measurements. R_G's from light scattering are also listed in table 1 and agree well with the values obtained by Torigata et al[5].

TABLE 1.

SAMPLE	HP	R
$\bar{M}_W$	477000[a] 454000[b]	490000[a] 438000[b]
$\bar{M}_n$	226000[a]	285000[a]
$[\eta]$ dlg^{-1}	7.6	7.3
$\bar{M}_{visc}$	322000[c]	305000[c]
R_G nm	72 [b]	58 [b]

a Results from gel permeation chromatography.
b Results from light scattering measurements.
c Results obtained using values of 6.37×10^{-4} and 0.74 for the Mark-Houwink constants, K and a, obtained by Kishida et al[7]. using a partially methylated konjac sample.

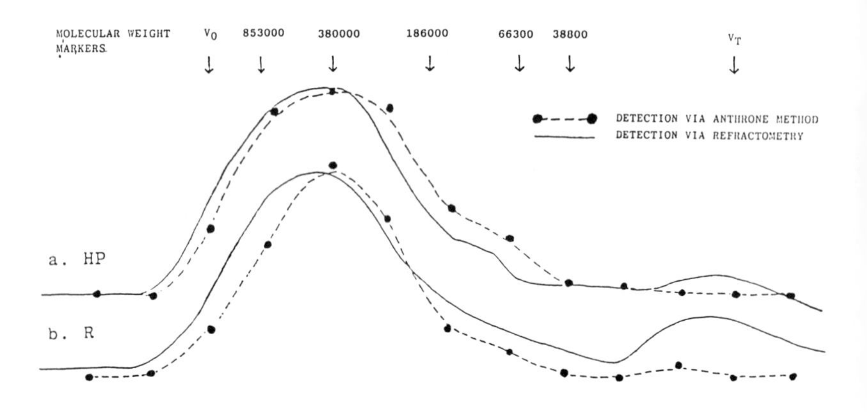

FIG.2 Gel permeation chromatograms of sample HP and sample R.

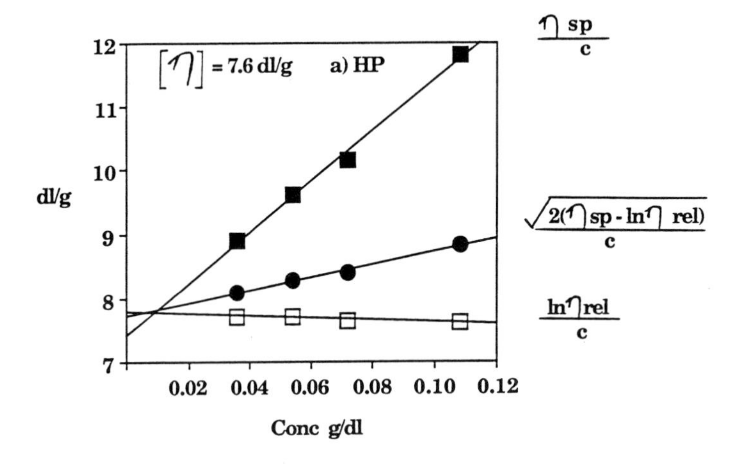

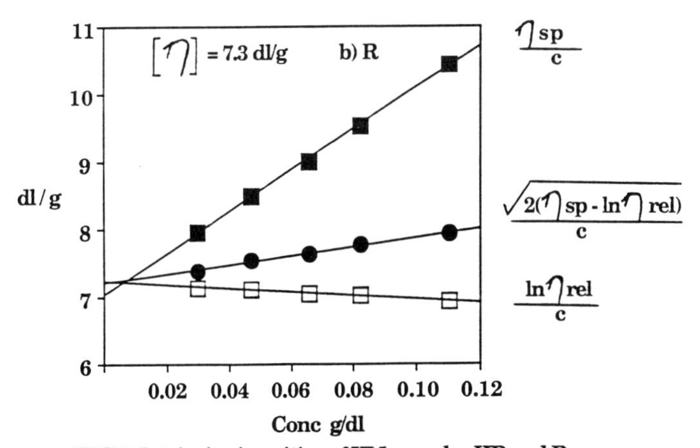

FIG 3. Intrinsic viscosities of KM samples HP and R

CONCLUSIONS

Despite the different treatments to which the two samples were subjected, there seems to be only minor differences in molecular weight and molecular weight distribution.

Our results, while being in good agreement with the values obtained by Torigata[5], are somewhat lower than those of Kishida[7] and Sugiyama[6]. This difference was originally explained as degradation of the polysaccharide chain during the nitration process to approximately a quarter of its original size[6]. An alternative explanation would be that the higher molecular weights obtained are a result of incompletely dissolved konjac mannan existing in some sort of partially aggregated form, which is particularly supported by the observations of the effect of sonication on the samples discussed above.

ACKNOWLEDGEMENTS

We acknowledge the financial support and supply of samples from FMC Marine Colloids Division.

REFERENCES

1. F. Smith and C. Srivasta, J Am.Chem.Soc. (1959) 81 1715.

2. M. Maeda, M. Shimahara and N. Sugiyama, Agric.Biol.Chem. (1980) 44 2 245.

3. K. Kato and K. Matsuda, Agric.Biol.Chem. (1969) 33 (10) 1446.

4. S. Okimasu and N. Kishida, Hiroshima Joshi Daigaku Kaseigakubukiyo (1983) 19 1.

5. H. Torigata, H. Inagaki and N. Kitamo, Nippon Kagaku Zasshi (1951) 7 30.

6. H. Sugiyama, H. Shimahara, T. Andoh, M. Takamoto and T. Kamata, Agric.Biol.Chem. (1972) 36 8 1381.

7. N. Kishida, S. Okimasu and T. Kamata, Agric.Biol.Chem. (1978) 42 9 1605.

8. M. D. Graham and G. Mitchell J. Food Sci. (1963) 28 546.

Part 7

MARINE POLYSACCHARIDES

Seaweed extracts – sources and production methods

JENS K.PEDERSEN
The Copenhagen Pectin Factory Limited, DK-4623 Lille Skensved, Denmark

ABSTRACT

By definition, alginates are extracts from certain families of brown seaweeds, whereas agar and carrageenan are defined as being extracts of (different) red seaweed families. Seaweeds may be collected from the beach, and seaweed beds in the sea may be harvested by hand or mechanically. Lately, most of the carrageenan raw material is ''farmed'' on coral reefs in Pacific tropical waters. Production of marine polysaccharides comprises three main steps: 1) extraction of the seaweed, 2) purification/clarification of the liquid extract, 3) isolation of the polysaccharide from the liquid extract. The isolation procedures for the three extracts show the biggest difference, the procedures chosen being direct consequences of the functional (and in turn - structural) characteristics of the polymers. Agar is thus isolated by gelling its solution and utilizing the tendency of the gel to exude water after a freeze/thaw cycle - or by mechanical compression. Alginate is precipitated from solution with acid or calcium salt, utilizing the insolubility of alginic acid - or calcium alginate. Alginate is subsequently converted into its soluble Na-salt form. All carrageenan types may be isolated by alcohol precipitation. Strong kappa-type carrageenan may be isolated by the agar processes, provided sufficient potassium salt is present (added) to produce the gel. The future growth of the seaweed extractives will depend on the possibility of procuring sufficient raw material at competitive costs.

INTRODUCTION

This paper endeavours to review those aspects of raw materials and production methods for marine polysaccharides which I consider of interest to the users of these polysaccharides in foods.

Within my subject area, I shall point out some characteristics of marine polysaccharides in general - where they differ from land plant polysaccharides. I shall point out similarities and differences within the group of alginates, agar and carrageenan.

You may not learn much new from this review paper - but I hope that I may present my information from another angle than you have experienced before - and that I may contribute to giving you an understanding that although the seaweed extractives are old and established products, they are also products with a future in food products.

DIMENSIONS

The 3 marine polysaccharides are produced at annual rates given in the table below. Typical sales prices are given as well.

It is characteristic that agar and carrageenan, the more expensive polysaccharides, are used almost exclusively in foods and related areas, whereas the less expensive alginates find use in non-food applications as well.

Table I. SEAWEED POLYSACCHARIDES
DIMENSION OF THE MANUFACTURING INDUSTRY

Polysaccharide	World production t/year in 1987	Typical sales price GBP/kg	Percentage of prod. used in food (cosmetics,diagnostics)
Alginate	20 - 24,000	6	30
Agar	7 - 8,000	11	> 90
Carrageenan	13 - 16,000	8	> 90

RAW MATERIAL

Botanical classification

The description of alginates requires that they are produced from brown seaweeds, whereas agar and carrageenans are extracts of certain (different) families of red seaweeds.

This does not mean that all extracts of the mentioned seaweed classes are carrageenans/agars/alginates. Also some structure/chemical composition requirements must be met.

What makes a seaweed qualify as raw material

A long list of conditions must be met, if a seaweed species

is to find use as a raw material in a <u>commercial</u> extraction process

- The seaweed must be available in sufficient quantity, quality, and at acceptable cost.
- The seaweed must be found in areas where it can be collected/harvested, and the adjacent land area must be inhabited by people who are willing to do so.
- The seaweed must grow in sufficiently pure beds/at sufficient density to allow harvesting/collection. It must be possible to preserve the seaweed from harvesting/collection to extraction - e.g. by drying.

<u>Brown seaweeds/alginate raw material</u>

Alginate raw materials are found in cool waters - around the coasts of Norway and Iceland, and in warmer waters, along the coast of California. The seaweed species are different, though - and so are the extracts. Today, these seaweeds are mostly harvested mechanically by specially equipped boats operated by the extracting companies, who own licenses to harvest certain areas.

Having the cost advantages of being able to harvest a wild crop mechanically, the alginate extractors face other restrictions: They depend on consistant regrowth of the seaweed beds. With intervals, shift of ocean currents with subsequent changes in water temperature has affected the Macrocystis bed along the Californian coast. Hard winter storms have hindered harvesting along the Norwegian coast for weeks. Extraction plants must preferably be placed in the seaweed area to take advantage of the fresh seaweed/not having to dry the raw material to preserve it.

<u>Red seaweeds/agar raw material</u>

Two families of red seaweed dominate as agar raw materials: Gracilaria and Gelidium. They are characterized by

- Being harvested by hand - to some extent by divers
- Being available in limited quantities in geographically widespread areas.

This raw material situation is probably the background for the structure of the agar industry that is characterized by many small companies, each with limited capability of technical or commercial development.

<u>Red seaweeds/carrageenan raw material</u>

The original carrageenan raw material was Chondrus crispus, which grows in temperate climates, along the coasts of Maine in the USA and the Maritime Provinces of Canada - and here in Europe along the coasts of Brittany in France, the Iberian Peninsula and Ireland.

As Chondrus is a low, parsley-like plant that grows attached to rocks/stones, it does not lend itself to mechanical harvesting, and it is either

- raked with hand rakes from boats, or
- collected from the beach when carried to the shore after storms.

As this seaweed can only be harvested in a relatively short

season from July to September/October, it must be preserved to allow an extraction process to run all year. Chondrus is, therefore, most often dried, either in the sun or in mechanical dryers. Once dried, Chondrus can be transported to extraction plants located far away from the seaweed harvesting area.

Other red seaweeds containing carrageenan are found in tropical waters. Examples are Eucheuma types, which grow along the coasts of the Philippines, Indonesia, Zanzibar etc. It is - like Chondrus - hand-picked and dried. In the beginning of the 1970'es, these Eucheuma seaweeds were losing importance as carrageenan raw materials, because overharvesting reduced the crop to an insignificant level.

That was the background for the development of a seaweed cultivation technique. It was found that Eucheuma species could be made to grow attached to nylon ropes suspended between bamboo poles that were driven into the coral reef along the coast of the Philippines. Over a period of ten years, cultivated Eucheuma seaweeds became the dominating carrageenan raw material providing a living for thousands of people in the Philippines. During the very latest years, this seaweed cultivation has spread to Indonesia.

The situation today can, therefore, be characterized as follows.

Eucheuma-type (kappa- and iota-type) carrageenan raw material can be made available according to demand. The supply is independent of a natural crop. Very much labour is involved in cultivating these seaweeds and competitiveness of the seaweed cultivation business - and in turn the carrageenan industry - depends on the labour being available at a low price.

With most of the carrageenan-bearing seaweeds being collected in areas without an industrial infrastructure, it is not surprising to find the carrageenan extraction plants being located in industrial countries without any own raw material sources. Also, to be able to offer a full range of carrageenan products, a substantial part of the raw material must be imported regardless where the extraction plant is built.

PROCESS

The manufacturing process can be split into three distinctive process steps shared by all three extracts

- Extraction from the seaweed
- Purification of the liquid extract
- Isolation of the hydrocolloid from the liquid extract.

Extraction

All seaweed hydrocolloids are structural materials, as cellulose and pectin in land plants. Pre-treatment/extraction processes are generally alkaline, reflecting the higher stability of the polymers under alkaline conditions. An alkaline pH may also serve to loosen the hydrocolloid better from the plant material. Agar and carrageenan may undergo a desulfation, as galactose-6-sulphate groups are alkali-labile and converted into 3,6-anhydro-galactose units. This desulfation leads to a molecular structure which is more favourable for double helix formation, i.e. material with higher gel strength.

The pre-treatment/extraction conditions may differ with regard to pH in the alkaline area, temperature and time - with the most extreme conditions (strong base - temperature above 100 deg. C) being used for agar and carrageenan, milder conditions (soda ash - room temperature) used for alginate extraction.

The hydrocolloid to water ratio resulting from the extraction process varies with a) the viscosity of the extract, b) the method subsequently used for isolation of the hydrocolloid from the liquid extract.

<u>Purification of the liquid extract</u>

Purification is mainly a separation of all insoluble material from the extract. Filtration is normally used - and is always the last "polishing" step. The quality of the polishing filtration determines the solution transparency of the hydrocolloid. Centrifugation may be used as a first step to remove coarser insolubles, if density differences and viscosity allow.

Viscosity and to a lesser extent quantity and character of the insoluble material determines the hydrocolloid concentration at the purification stage. If required, colour and odour may be reduced here.

<u>Isolation</u>

It is at the isolation stage in the process that the three hydrocolloids show the biggest differences. Here the specific functional properties of the individual hydrocolloids are taken advantage of.

<u>Alginate</u> is isolated by a chemical precipitation process, utilizing that alginic acid - or calcium alginate - is practically insoluble.

Acid precipitation - using sulfuric acid - is normally used for isolation of so called G-unit rich alginates (the strongest gelling - least soluble), eliminating the acid washing step, which is required if the calcium precipitation process is used. Acid precipitation produces products with lower Ca-content/ better solubility.

Calcium precipitation is used to isolate alginates high in M-units (more soluble). It requires a subsequent acid wash of the Ca-alginate to convert it into alginic acid and always results in alginates with a higher Ca-content than those made by the acid precipitation process.

The calcium or acid precipitation of alginate is principally stoichiometric, whereby the consumption of chemicals is proportional with the quantity of alginate being produced - <u>not</u> with the extract volume.

Further purification - removal of colour and odour - may take place at the calcium alginate/alginic acid stage prior to the conversion to soluble alginate salts.

The conversion of alginic acid into Na-alginate is done by a stoichiometric neutralization with solid sodium carbonate. The

Na-alginate is then dried.

Agar is traditionally isolated from its solution by a freeze/thaw-process, where the lack of freeze/thaw-stability of the agar gel is utilized. The gel is frozen; by subsequent thawing, the gel releases about 90% of its water. The concentrated gel is then dried in hot air to bring the water content down to 10-20%.

A newer technique is now used, which also uses the relatively poor water binding of the agar gel. The gel is put under pressure in a hydraulic press for an extended period of time, resulting in water exuding slowly from the gel, leaving an agar gel skeleton, which is dried to desired residual water content in hot air. This gel pressing technique dominates in agar processing today.

Furcellaran - or Danish/British agar can be (and was) isolated by the agar techniques, the freeze/thaw or the gel pressing technique. Gelation of furcellaran was obtained by pouring the liquid extract into a cold potassium chloride solution, utilizing the ability of K+-salts to gel furcellaran.

Furcellaran no longer exists as a separate seaweed extract; it is included in carrageenan, reflecting a very close similarity between composition of carrageenan and furcellaran, but also that the raw material for furcellaran (Furcellaria Fastigiata) is no longer available in commercial quantities.

Carrageenan is the designation for a family of sulfated galactans with a wide variation in the degree of sulfation and thereby also in functional properties.

Kappa-carrageenan, the strongest gelling carrageenan, produces strong gels with potassium salts, gels that give off water under pressure and after a freeze/thaw process. Kappa-carrageenan, therefore, may be isolated by the agar process.

Iota-carrageenan gels with K- and Ca-salts; the resulting gels are syneresis-free and freeze/thaw-stable. Iota-carrageenan thus cannot be isolated by agar techniques.

Lambda-carrageenan is non-gelling, regardless of the presence of K- and/or Ca-salts.

All carrageenans can be isolated by drying solutions or precipitating the polymer in alcohol.

Drum drying was used much in the past but is hardly used any more. Drum drying leaves all soluble material from the liquid extract (colours, odours etc.) in the carrageenan. Besides, so called "release agents" must be added to allow the carrageenan film to be scraped off the drum dryer. Release agents are di/monoglycerides, which are responsible for the carrageenan solution being hazy.

Alcohol precipitation is a universal technique for isolating polymers from solution; pectins and microbial gums are also isolated with alcohol. The alcohol concentration required to obtain precipitation varies with the carrageenan type - specifically the hydrophilicity of the carrageenan. The cost of alcohol precipitation is determined by the amount of alcohol required to isolate a kilogram of carrageenan. Concentration (by evaporation) before alcohol precipitation is, therefore, normally done to reduce the cost; it is less costly to remove one litre of water by evaporation than through alcohol precipitation.

Alcohol precipitates the polymer, exclusively, leaving any soluble low-Mw material dissolved in the alcohol/water mixture. Counter current washing of the precipitated polymer in pure alcohol further reduces the level of accompanying substances.

The alcohol used is normally isopropyl alcohol, ethanol may be used where the cost picture is advantageous; methanol is also permitted.

Alcohol precipitated material must subsequently be dried to a residual alcohol content below 0.1%. For safety reasons (explosion risk) drying of material containing alcohol must take place in vacuum or in an inert atmosphere.

The alcohol process offers the advantage of high microbiological quality as the alcohol prevents any growth at the critical last part of the process, the isolation step.

Purity and purity specifications

Being extracts, agar, alginate and carrageenan are very pure polymer products containing no cellulosic fiber material from the seaweed source (separated at the filtration step) or low molecular soluble material (separated at the isolation stage). Purity is ensured by specifications. Thus, an Acid Insoluble Material (AIM)-specification of max. 2 per cent ensures that less refined products (insufficiently purified or not being extracts at all) cannot be sold as agar/ alginate/carrageenan.

THE FUTURE OF SEAWEED EXTRACTS

The increasing need for functional properties displayed by the seaweed extracts is hardly questioned. The unique properties of these materials are the results of the activity of an array of enzymes in the seaweeds, producing polymers of rather complicated molecular structures. There is little chance that identical material will be produced by microbial synthesis.

A possible, unfavourable cost development for seaweed extracts - compared to alternatives - may some day exclude them from new developments in the food industry, and thereby from participating in the market growth.

The dominating factor in the cost development is the raw material, the seaweed. The challenge of the seaweed extract industry is, therefore, to develop better raw materials and more efficient ways of cultivating them.

Comparison of the properties and function of alginates and carrageenans

EDWIN R.MORRIS

Department of Food Research & Technology, Cranfield Institute of Technology, Silsoe College, Silsoe, Bedford MK45 4DT, UK

ABSTRACT

The basic structural unit in alginate and carrageenan gels is an ordered dimer: a co-axial double helix in carrageenan and a side-by-side dimer in alginate. Both polymers are biosynthesised as a soluble precursor and converted to the gelling form by enzymic modification: C(5) epimerisation of D-mannuronate to L-guluronate in alginate and formation of an anhydride bridge in carrageenan. In both cases the sugar ring is inverted from the $^4C_1$ to the $^1C_4$ chair form, with consequent drastic change in chain geometry. The dimeric polyguluronate junction zones in alginate gels are stabilised by incorporation of an array of site-bound divalent counterions. Counterions also enhance the thermal stability of the carrageenan double helix and promote network formation by helix-helix aggregation. Binding to iota carrageenan is probably limited to non-specific 'atmospheric' attraction: due to the high charge density of the polymer, divalent ions are particularly effective in promoting gelation. Kappa carrageenan, which has lower charge, shows less preference for divalent ions, but interacts with large Group I ions (K^+, Rb^+, Cs^+) by direct site-binding and, unlike iota, is very sensitive to lyotropic modification of solvent quality by anions.

INTRODUCTION

Alginates and carrageenans are the major polysaccharide components of certain species of marine brown and red algae, respectively (1). Both are linear polymers, but quite different in primary structure. The carrageenans are based on a disaccharide repeating sequence of alternating 1,3-linked β-D-galactose and 1,4-linked α-D-galactose (Fig. 1), with varying extents and patterns of sulphation (2,3). Alginate is a 1,4-linked block co-polymer of β-D-mannuronate and α-L-guluronate, with residues grouped in long, homopolymeric sequences of both types (Fig. 2) and in heteropolymeric sequences where the distribution of the two sugars can vary from near-random to near-alternating (3-5). Alginate gives thermally-stable gels with calcium (or larger Group II cations); carrageenan gels melt on heating and reform on cooling. Despite these apparent dissimilarities, however, there are also striking parallels in the way in which the two families of polysaccharides form gel networks.

ORDERED DIMERS

The first point of similarity is that the basic structural unit in both types of gel is an ordered dimer; a co-axial double helix in carrageenan and a side-by-side dimer in alginate. In both cases, detailed description of the ordered structures came from x-ray diffraction in the solid state, with other physical techniques then showing that the same structures can form under hydrated conditions (in solutions and gels).

The carrageenan double helix

The commercial gelling carrageenans, iota and kappa, have primary structures approximating to the idealised disaccharide repeating units shown in Fig. 1. In both cases the 1,4-linked galactose residues occur (predominantly) as the 3,6 anhydride. The principal difference between them is charge, with two sulphate groups per disaccharide in iota and one in kappa.

Analysis (6,7) of x-ray diffraction from stretched fibres of iota carrageenan gave clear evidence of a three-fold (3_1) double helix structure, with two parallel strands exactly staggered, so that the overall helix pitch (1.33 nm) is half that of the individual strands. In one salt form (Ca^{2+}) the diffraction data are of sufficient quality to allow the counterions to be located within the ordered structure; the ions lie in long channels, sandwiched between three double helices, with each calcium co-ordinated to two sulphate groups from different helices (7).

Diffraction patterns for kappa carrageenan are of poorer quality (6), but the best-fitting model (8) is again a right-handed, three-fold double helix similar to that of iota, but with a slightly shorter pitch (2.50 nm for a full turn of each strand, in comparison with 2.66 nm in iota) and with the two strands displaced by 28° from the fully-staggered arrangement.

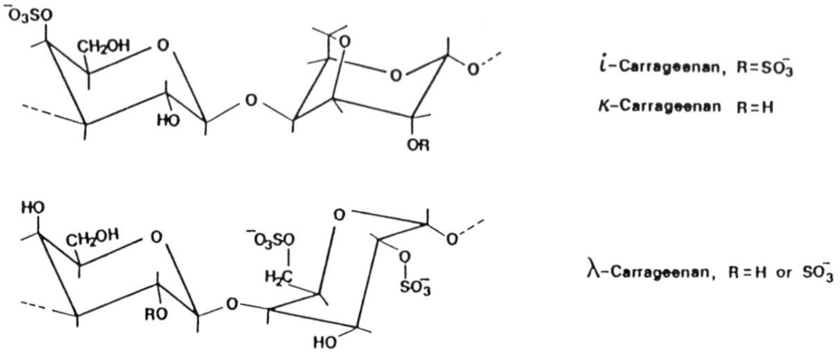

Figure 1. Idealised disaccharide repeating sequences for carrageenans. Note the difference in geometry of the 1,4-linked ring in gelling and non-gelling carrageenans (top and bottom structures, respectively)

'Kinking' residues

In naturally-occurring samples of gelling carrageenans a small proportion of the 1,4-linked residues (typically ∿10%) lack the anhydride bridge (3). As discussed in greater detail later, this radically alters the shape of the sugar ring, to a form that is incompatible with incorporation in the double helix structure (2). Most of these anomalous 'kinking' residues can be converted to the helix-compatible anhydride form by treatment with alkali (9,10), with the residual proportion of unbridged rings depending on the pattern of sulphation (10). 'Alkali modification' of carrageenan substantially enhances gel properties (9), indicating that helix formation and gelation are intimately linked.

Helices in solutions and gels

The first direct experimental evidence that formation of carrageenan gels is accompanied by adoption of the double helix structure was the observation (11) of a sharp change in optical rotation on cooling through the sol-gel transition, since it was well established for other biopolymer systems (notably proteins) that optical activity gives a sensitive index of chain geometry (12). Structurally-regular segments of carrageenan prepared by chemical cleavage (periodate oxidation) at 'kinking' residues (10) showed (13) similar changes in optical rotation on heating and cooling, but without the complication of gel formation. The magnitude of optical rotation change was in good agreement with semi-empirical calculations (10) of the change expected for conversion from a fluctuating, disordered coil at high temperature to the double-helix conformation at low temperature. Adoption of a rigid, ordered structure on cooling was also indicated (14,15) by a collapse in high resolution NMR spectra (^{1}H and ^{13}C) and by direct measurement (15) of the underlying changes in spin-spin relaxation time (T_2). Evidence that the ordered structure is dimeric, and therefore almost certainly the double helix identified in the solid state, has come from several independent lines of investigation.

1) The rate of conformational ordering of both iota (16,17) and kappa (18,19) carrageenan (monitored by polarimetric stopped-flow) follows second-order kinetics, showing the involvement of two chains.

2) The ionic-strength dependence of transition temperature for both iota and kappa carrageenan is in good agreement with values predicted (20) by Poisson-Boltzmann theory for adoption of the double helix structure, but not for other models such as the single-stranded structure that has also been proposed (21).

3) Small angle x-ray scattering from solutions of segmented iota carrageenan at high and low temperature gives values for the average cross-sectional radius of gyration in excellent agreement with those calculated for the disordered coil and double helix, respectively (results by A.H. Clark, C.D. Lee-Tuffnell and I.T. Norton, reported in ref. 22).

4) Iota carrageenan segments show an almost exact doubling of molecular weight on cooling through the disorder-order transition (17,23,24).

Evidence has also been presented (25-27) of an alternative single-stranded structure in the presence of iodide ions at concentrations above 50-60 mM. Under more normal salt conditions, however, there seems little doubt that the conformational transition accompanying carrageenan gelation is conversion from a disordered coil form to a double helix, with 'kinking' residues promoting formation of a three-dimensional network by terminating helix propagation and allowing each chain to form further helices with other partners (2).

<u>The alginate 'egg-box'</u>

There is strong evidence that formation of calcium alginate gels occurs predominantly (or solely) by association of polyguluronate sequences.

1) Gel strength increases systematically as the guluronate content of the alginate increases (4).

2) Alginate gelation is accompanied (28) by large changes in circular dichroism (CD). Closely similar changes are observed for polyguluronate; the corresponding changes for heteropolymeric sequences are an order of magnitude smaller, and no change is detectable for polymannuronate (28).

3) Addition of polyguluronate blocks to solutions of alginate prior to gelation causes a drastic reduction in final gel strength, indicating 'competitive inhibition' of polyguluronate binding-sites on intact chains (29). Equivalent concentrations of polymannuronate or heteropolymeric sequences have a far smaller effect.

X-ray fibre diffraction studies (30) of alginate with a high content of polyguluronate show a two-fold repeating structure with a pitch of ∿0.87 nm per disaccharide, close to the fully-extended chain conformation of polyguluronate. Closely similar results have been obtained for the acid form, and for various salt forms (Li^+, K^+, NH_4^+, Mg^{2+}, Ca^{2+}). When calcium alginate gels are dried to (transparent) solid films, there is virtually no change in circular dichroism (14), indicating that the two-fold ordered structure is also present in the gel (since CD, like optical rotation, is highly sensitive to changes in molecular geometry).

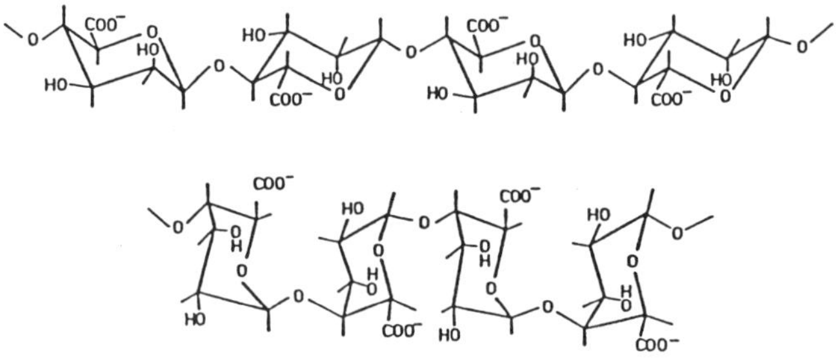

Figure 2. Homopolymeric sequences for alginate.
Top: Poly-D-mannuronate Bottom: poly-L-guluronate

Because the inter-residue linkages at C(1) and C(4) are both axial, the two-fold conformation of polyguluronate is highly buckled (as shown in Fig. 2), with large cavities between adjacent residues. Physical and computer model building (31) shows that these cavities can accommodate a calcium ion (or other ion of similar size), and have oxygen atoms well placed for cation chelation. It has therefore been proposed (31) that the inter-chain junctions in alginate gels involve polyguluronate sequences locked in the two-fold conformation identified by x-ray, with arrays of site-bound cations sandwiched between them like eggs in an egg-box. Both parallel and antiparallel arrangements of polyguluronate give suitable co-ordination sites (30), but the actual geometry of packing is not yet known. Studies of calcium ion activity and CD for solutions of polyguluronates of varying chainlength show (32) that the 'egg-box' structure is fully stable above a DP of 28 residues (corresponding to an array of 14 site-bound calcium ions). Polymannuronate, by contrast, shows only 'atmospheric' binding of Ca^{2+} (32).

<u>Calcium polyguluronate dimers</u>

The proportion of the calcium ions that are very tightly bound within the gel network (resistant to displacement by much higher concentrations of monovalent ions) is equivalent to half the total stoichiometric requirement of polyguluronate, suggesting preferential formation of a dimeric structure in which only the inner faces of each of the participating two-fold chains are involved in calcium-binding (28). Electron microscopy of calcium alginate gels (4) supports this interpretation, showing structures with the width expected for dimeric junctions, but no evidence of any much larger assemblies. The ability of polyguluronate segments to inhibit alginate gelation (29) is also consistent with interchain association occurring predominantly through dimeric junction zones: binding of a short segment to a polyguluronate region of an intact chain would prevent that region forming a dimeric crosslink with another chain, but would not inhibit formation of a larger aggregate. The preferential stability of the dimer structure can be explained by the site-bound counterions lowering the overall charge density and therefore reducing the relative stability of cation binding to the outer faces of the dimer.

BIOLOGICAL CONTROL OF GEL PROPERTIES

As already discussed, polymannuronate shows no specific affinity for divalent ions, in contrast to the strong intermolecular site-binding by polyguluronate. This difference can be traced to the difference in chain geometry indicated in Fig. 2. In polymannuronate the inter-residue linkages are diequatorial, giving rise to a flat, ribbon-like structure with no pockets for cation co-ordination, in contrast to the highly buckled, axially-linked structure of polyguluronate.

β-D-Mannuronate and α-L-guluronate differ only in the geometry of attachment of the carboxyl group at C(5). The requirement to place the bulky carboxyl group in an equatorial location, to avoid steric clashes with other axial substituents, however, forces inversion of ring geometry from the $^4C_1$ chair form in D-mannuronate to the alternative $^1C_4$ chair in L-guluronate. As indicated in Fig. 3, this converts all other axial substituents to equatorial and *vice versa*. In particular, the inter-residue linkages at C(1) and C(4) are converted from diequatorial to diaxial, with the consequent radical changes in chain geometry and cation-binding properties outlined above.

Alginate is biosynthesised initially in the (soluble) poly-D-mannuronate form, with subsequent conversion of a proportion of the residues to L-guluronate by enzymic action at the polymer level (3,33). The extent of conversion varies throughout the plant, tailoring physical properties to biological function. Thus in the holdfast, where mechanical rigidity is required for anchorage to crevices in the rock, there is a very high content of polyguluronate, whereas fronds, which need flexibility to resist wave damage, have a higher residual mannuronate content, with stipes having alginate of intermediate composition.

The conversion process is also sensitive to calcium. High levels of Ca^{2+} promote formation of heteropolymeric sequences, whereas polyguluronate is formed preferentially at lower concentrations of Ca^{2+}. Thus when the calcium content of the seawater is low, the alginate is converted to the form that can utilise it most effectively; conversely, at high Ca^{2+} concentrations where a high content of polyguluronate would make the tissue too brittle, enzymic activity is switched to production of sequences that are less effective in gel formation (3).

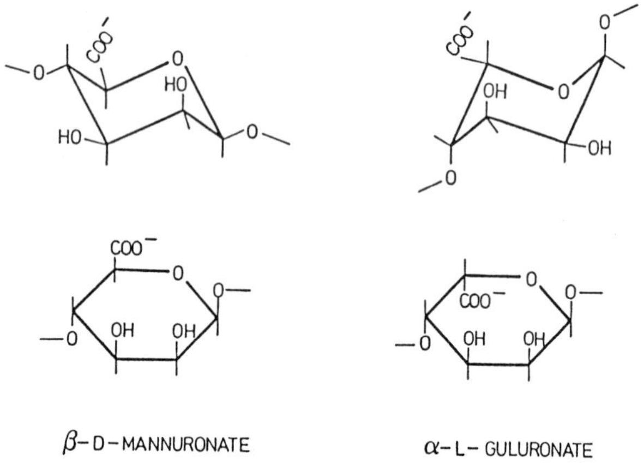

Figure 3. Conversion of alginate β-D-mannuronate residues to α-L-guluronate. The bottom structures show that change in configuration is confined to epimerisation at C(5). The top structures show the change in ring geometry required to retain the carboxyl group in an equatorial location.

A closely similar mechanism of biological control exists for carrageenan. As discussed above, adoption of the double helix structure and, consequently, the ability to form gels, requires the 1,4-linked residues to be in the bridged anhydride form. Thus lambda carrageenan, in which the anhydride bridge is absent (Fig. 1), exists in solution as a disordered coil under all conditions of temperature and ionic strength. The gelling carrageenans are also biosynthesised in a 'fully kinked', soluble form. As for alginate, they are then converted to the 'active' form by enzymic modification (34), in this case by closure of the anhydride bridge, which is possible only in the $^{1}C_4$ conformation where C(6) and O(3) are brought together in axial locations, rather than in the normal $^{4}C_1$ form where both are equatorial (Fig. 1). Adoption of the alternative chair conformation then changes the inter-residue linkages at C(1) and C(4) from diaxial in the soluble precursor to diequatorial in the helix-compatible form.

EFFECT OF SALTS

There are three main ways in which ions can affect the behaviour of charged polysaccharides.

1) Specific site-binding, as in the alginate 'egg-box'.
2) Non-specific suppression of electrostatic repulsion (35).
3) Modification of solvent quality. This is a general effect often referred to as 'salting-in' or 'salting-out' behaviour. It applies to both charged and uncharged polymers and, indeed, to non-polymeric systems. Anions and cations influence solvent quality independently, with each following a well established order of effectiveness: the lyotropic series (36,37).

Ionic environment has a dominant effect on the solution and gel properties of both alginates and carrageenans.

Alginates

Magnesium ions do not promote formation of alginate gels (presumably because they are too small for efficient chelation within the 'egg-box' structure) and have therefore been used (4) as a standard for comparison of the strength of binding of other divalent cations. Selectivity coefficients for binding to polyguluronate have maximum values of ∿40 for Ca^{2+}, ∿250 for Sr^{2+} and ∿3000 for Ba^{2+} (relative to a reference value of 1 for Mg^{2+}), indicating a progressively better fit of 'eggs' in the 'egg-box' with increasing ionic radius. Corresponding values for binding to polymannuronate or heteropolymeric sequences are normally less than 5, with no major differences between Group II cations, indicating far less specific binding. The turbidity of alginate gels prepared by dialysis against a high concentration of Ca^{2+} increases systematically with increasing content of mannuronate (4), suggesting that at high ionic strength polymannuronate sequences may form large aggregates by non-specific suppression of electrostatic repulsion. High concentrations (e.g. 0.5M) of Na^{+} induce a weak thixotropic network (38). This is accompanied by CD changes consistent with limited formation of weak 'egg-box' junctions. Other monovalent ions (Li^{+}, K^{+}, Rb^{+}, Cs^{+}, NH_4^{+} show no such effects.

Iota carrageenan

As in other polyelectrolyte systems, the disorder-order transition of iota carrageenan, and the associated sol-gel transition, move to higher temperature with increasing salt concentration (24), as adoption of the ordered structure is facilitated by suppression of electrostatic repulsions between the participating chains. The conformational transition is fully reversible, with no thermal hysteresis between heating and cooling (39). The slight hysteresis reported previously for the K^+ salt form by the present author and colleagues (24) arose from the presence of some kappa carrageenan in the sample used (17).

Divalent cations are far more effective than monovalent (20,40). For example, at fixed temperature (18°C) the concentrations of Na^+ and Ca^{2+} required to induce the onset of conformational ordering are ∿400 mM and ∿3 mM, respectively (20). The differences between different monovalent cations and between different divalent cations are, however, small (giving transition temperatures varying by a maximum of ∿5°C), and in both cases follow the lyotropic series (40), arguing against specific binding effects. Indeed, it has recently been suggested (20) that the large difference in effectiveness of monovalent and divalent cations arises from the very strong, but non-specific, electrostatic attraction of divalent ions to the highly-charged iota carrageenan double helix.

For each salt form, the reciprocal of the transition-midpoint temperature ($1/T_m$) varies linearly with the logarithm of cation activity (counterions to the polymer plus added salt). As a rough rule-of-thumb, doubling the cation concentration raises the transition temperature by ∿10°C (24). Changing the co-anion of the added salt has no effect on either the transition temperature or the rate of conformational ordering (17), in marked contrast to the behaviour described below for kappa carrageenan.

Kappa carrageenan

As in the case of iota, the ordered structure of kappa carrageenan is stabilised by salt, with a doubling of cation concentration again raising the transition temperature by ∿10°C (24). Certain monovalent ions (Li^+, Na^+, Me_4N^+) show virtually no difference in their stabilising effect (41), consistent with non-specific suppression of electrostatic repulsions within the double helix (20). The concentration of divalent cations required to give equivalent stabilisation is lower, by a factor of ∿6-8 (20,41), but different divalent ions (Mg^{2+}, Ca^{2+}, Sr^{2+}, Ba^{2+}, Co^{2+}, Zn^{2+}) increase the transition temperature by almost the same amount (41). It has again been proposed (20) that the greater effectiveness of divalent cations is due simply to their higher charge and consequent stronger electrostatic attraction to the double helix. In terms of this proposal, the much greater preference of iota carrageenan for divalent cations is explained by its higher charge density.

In contrast to these non-specific 'ion atmosphere' effects, larger Group I cations (K^+, Rb^+, Cs^+) show clear evidence of specific site-binding to kappa carrageenan. Firstly, they are much more effective in stabilising the ordered structure.

For example, the transition midpoint temperature for melting of the kappa carrageenan double helix in the presence of 0.1M LiCl, NaCl or Me_4NCl is ∿10-12°C; 0.1M KCl, RbCl and CsCl, by contrast, give T_m values of ∿51°C, ∿59°C and ∿51°C respectively (41). These ions also give gels with much higher rigidity (42 and yield stress (24).

The most direct evidence of specific binding of large Group I ions to the double helix has come from heteronuclear NMR (42-46). Extreme broadening of ^{39}K (44,45), ^{87}Rb (44,45) and ^{133}Cs (43-44) linewidth is observed when solutions of kappa carrageenan in these salt forms are cooled through the sol-gel transition. This broadening is accompanied by large changes in chemical shift, indicating site-binding rather than non-specific 'atomosphere' binding (45). No such effects are observed for lambda carrageenan (43,44) nor (44) for kappa in the sodium salt form (by ^{23}Na NMR). Experiments with added salt show exchange of site-bound ^{133}Cs with unbound K^+ and Cs^+, but not with Li^+ or Na^+, confirming that these smaller Group I ions cannot occupy binding sites on the helix. Iota carrageenan shows some broadening of ^{39}K and ^{87}Rb linewidth (42,44), but it has been demonstrated recently that this is probably due to small amounts of kappa-like structure in even the purest natural samples of iota (46).

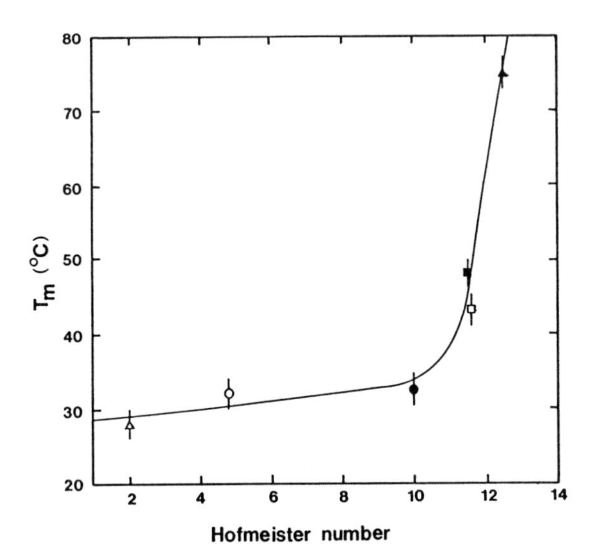

Figure 4. Dependence of transition mid-point temperature (T_m) for kappa carrageenan (Me_4N^+ salt form) on the Hofmeister number of the co-anion present. The results shown were obtained on cooling in the presence of 0.5M tetramethylammonium sulphate (Δ), fluoride (O), chloride (●), nitrate (□), bromide (■) and iodide (▲). From Norton, Morris and Rees, (47), with permission.

In contrast to iota, the behaviour of kappa carrageenan is very sensitive to co-anions. The rate of salt-induced conformational ordering is strongly dependent on the anion used (19). Equilibrium properties are also anion-dependent. Figure 4 shows midpoint temperatures for the disorder-order transition of the Me_4N^+ salt form in the presence of a fixed concentration (0.5M) of various tetramethylammonium salts (47). The T_m values vary by more than 45°C, and follow the anion lyotropic series: $SO_4^{2-} < F^- < Cl^- < NO_3^- < Br^- < I^-$.

As reported previously (27), adoption of the ordered conformation is accompanied by extreme line-broadening in ^{127}I NMR. Analogous broadening is also observed for ^{81}Br and ^{35}Cl. Although the magnitude of the effect decreases systematically (47) with decreasing "Hofmeister number" (36,37), the observations form a smooth progression, arguing against any unique behaviour for iodide.

For ions of high Hofmeister number (I^-, Br^-, NO_3^- and Cl^-) optical rotation values obtained on heating and cooling are closely superimposable at all accessible salt concentrations, and the samples do not gel. With increasing concentration of sulphate (and, to a lesser extent, fluoride), however, a weak, but cohesive, gel structure is formed, accompanied by a progressive increase in turbidity and in thermal hysteresis between formation and melting of the ordered conformation. These effects can all be attributed to helix-helix aggregation, promoted by non-specific screening of electrostatic repulsions and reduction in solvent quality, with aggregation stabilising the double helix to temperatures above those at which it forms. Thus the anions that are most effective in stabilising the ordered structure are least effective in promoting aggregation and gel formation, and *vice versa*.

Similar effects are observed for kappa carrageenan in the K^+ salt form (47). Anions of low Hofmeister number (e.g. F^-) promote rapid formation of extremely turbid gels: those of high Hofmeister number (I^- or SCN^-) give very slow development of clear gels, with a smooth progression of intermediate behaviour through the lyotropic series. The transition temperature again increases systematically with increasing Hofmeister number, but the salt concentrations required to give equivalent T_m values are much lower than in the Me_4N^+ salt form and the gels are much stronger, indicating that site-bound cations are far more effective than 'atmospheric' counterions both in stabilising the double helix and in promoting helix-helix aggregation.

Development of an aggregated network structure during the gelation of kappa carrageenan (K^+ salt form) has recently been demonstrated directly in an elegant series of electron-microscopy studies (48). Samples prepared from the disordered form at high temperature show no discernable structures: isolated chains are below the limit of resolution. On cooling through the temperature-range of the transition, however, a progressive sequence of structures is observed.

1) fine, wispy strands with a width consistent with unaggregated double helices
2) linear rods with lengths of the order of 100 nm and widths consistent with helices associated into dimers.
3) a 'course' network of larger aggregates, shown in Fig.5.

In summary, therefore, the initial process in gelation of kappa carrageenan is double helix formation, with exchange of partners at 'kinking' residues promoting formation of small, branched 'domains' (24). Under suitable ionic conditions, these may associate into a continuous network by helix-helix aggregation. Aggregation may be promoted by non-specific suppression of electrostatic repulsion or, more efficiently, by site-binding of large Group I cations, and by reduction of solvent quality.

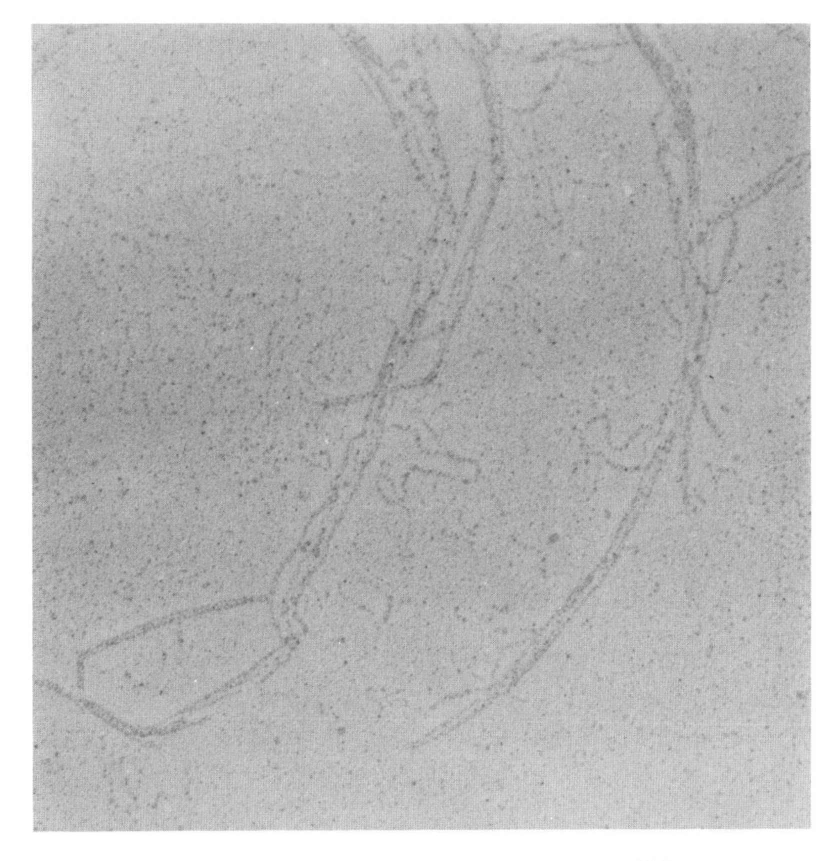

100 nm

Figure 5. Electron micrograph of the final, highly aggregated, network formed by kappa carrageenan in 0.1M KCl. From Hermansson (48), with permission

In iota carrageenan, there seems to be no direct site-binding (possibly because of the somewhat different geometry of the double helix), but atmospheric binding (particularly of divalent cations) is greatly enhanced by its very high charge density. Small amounts of kappa-like structure in iota carrageenan may have a significant effect in cross-linking domains and inducing gel formation. There is, however, no evidence of formation of double helices involving both iota and kappa: when the two are present together, discrete transitions for both can be observed under appropriate salt conditions (39).

SOLUTION PROPERTIES

The disordered forms of alginate and carrageenan (i.e. monovalent salt forms of alginate, except at very high concentrations of Na^+; lambda carrageenan under all conditions; iota and kappa carrageenan under conditions, described above, that do not promote helix formation) share the general solution properties of other 'random coil' polysaccharides (49).

In dilute solution individual coils are free to move independently, but with increasing concentration they are forced to interpenetrate one another and can then move only by wriggling ('*reptating*') through the entangled network of neighbouring chains. The onset of entanglement is accompanied by a sharp change in the concentration-dependence of solution viscosity from $\sim c^{1.4}$ to $\sim c^{3.3}$ (i.e. doubling concentration increases viscosity by a factor of ~ 2.5 in dilute solution, but gives a ten-fold increase for entangled networks).

At low shear rates ($\dot{\gamma}$), solution viscosity (η) has a fixed, maximum value (η_o), but decreases at higher shear rates. Shear thinning of entangled 'random coil' polysaccharide solutions has a general form, described quantitatively by:

$$\eta = \eta_o / [1 + (\dot{\gamma}/\dot{\gamma}_{\frac{1}{2}})^{0.76}]$$

where $\dot{\gamma}_{\frac{1}{2}}$ is the shear rate at which η is reduced to $\eta_o/2$. Thus viscosity at all shear rates can be characterised by just two parameters: η_o and $\dot{\gamma}_{\frac{1}{2}}$. Both can be obtained (50) from a simple linear plot of η *vs* $\eta\dot{\gamma}_{\frac{1}{2}}{}^{0.76}$.

The concentration (c) of polymer required to give an entangled network depends on the size (hydrodynamic volume) of the individual coils, which can be conveniently characterised by intrinsic viscosity, $[\eta]$. For most polysaccharides, including alginate and carrageenan, the onset of entanglement occurs (49) when $c[\eta] \cong 4$. The viscosity at this point is usually very close to 10 mPa s, so that double logarithmic plots of η_o *vs* $c[\eta]$ are closely superimposable for different 'random coil' polysaccharides.

The intrinsic viscosity of alginates and carrageenans, in common with other polyelectrolytes, decreases with increasing ionic strength (I) due to suppression of electrostatic repulsions within the coil. Quantitatively, $[\eta]$ varies linearly with $I^{-\frac{1}{2}}$, with the slope giving an indication of chain flexibility (35).

ACKNOWLEDGEMENTS

I am grateful to Drs. A.H. Clark, I.T. Norton, A.-M. Hermansson, L. Piculell and V.J. Morris for helpful discussions and generous access to experimental results prior to publication.

REFERENCES

1. Pedersen, J.K. (1989) This Volume, preceding paper.
2. Rees, D.A. (1972) *Biochem. J.*, 126, 257-273.
3. Painter, T.J. (1983) in *The Polysaccharides* (ed. Aspinall, G.O.) Vol. 2, pp 195-285. Academic Press, Orlando, USA.
4. Smidsrød, O. (1974) *Faraday Discuss. Chem. Soc.*, 57, 263-274.
5. Boyd, J. and Turvey, J.R. (1978) *Carbohydrate Res.*, 66, 187-194.
6. Anderson, N.S. Campbell, J.W., Harding, M.M., Rees, D.A. and Samuel, J.W.B. (1968) *J. Mol. Biol.*, 45, 85-89.
7. Arnott, S., Scott, W.E., Rees, D.A. and McNab, C.G.A. (1974) *J. Mol. Biol.*, 90, 253-267.
8. Millane, R.P., Chandrasekaran, R., Arnott, S. and Dea, I.C.M. (1988) *Carbohydr. Res.*, 182, 1-17.
9. Stanley, N.F. (1963) U.S. Patent No. 3094517.
10. Rees, D.A., Williamson, F.B., Frangou, S.A. and Morris, E.R. (1982) *Eur. J. Biochem.*, 122, 71-79.
11. Rees, D.A., Steele, I.W. and Williamson, F.B. (1969) *J. Polym. Sci., Part C*, 28, 261-276.
12. Djerassi, C. (1960) *Optical Rotatory Dispersion: Applications to Organic Chemistry*, McGraw-Hill, New York.
13. McKinnon, A.A.. Rees, D.A. and Williamson, F.B. (1968) *Chem. Commun.*, pp 701-702.
14. Bryce, T.A., McKinnon, A.A., Morris, E.R., Rees, D.A. and Thom, D. (1974) *Faraday Discuss. Chem. Soc.*, 57, 221-229.
15. Ablett, S., Clark, A.H. and Rees, D.A. (1982) *Macromolecules*, 15, 597-602.
16. Norton, I.T., Goodall, D.M., Morris, E.R. and Rees, D.A. (1983) *J. Chem. Soc., Faraday Trans. 1*, 79, 2501-2515.
17. Austen, K.R.J., Goodall, D.M. and Norton, I.T. (1985) *Carbohydr. Res.*, 140, 251-262.
18. Norton, I.T., Goodall, D.M., Morris, E.R. and Rees, D.A. (1983) *J. Chem. Soc., Faraday Trans. 1*, 79, 2489-2500.
19. Austen, K.R.J., Goodall, D.M. and Norton, I.T. (1988) *Biopolymers*, 27, 139-155.
20. Nilsson, S., Piculell, L. and Jönsson, B. (1989) *Macromolecules*, in press.
21. Paoletti, S., Smidsrød, O. and Grasdalen, H. (1984) *Biopolymers*, 23, 1771-1794.
22. Morris, E.R. (1986) in *Gums and Stabilisers for the Food Industry 3*, (eds. Phillips, G.O., Wedlock, D.J. and Williams, P.A.) pp 3-16. Elsevier, London.
23. Jones, R.A., Staples, E.J. and Penman, A. (1973) *J. Chem. Soc., Perkin Trans. 2*, pp 1608-1612.
24. Morris, E.R., Rees, D.A. and Robinson, G. (1980) *J. Mol. Biol.*, 138, 394-362.

25. Smidsrød, O., Andresen, I.-L., Grasdalen, H., Larsen, B. and Painter, T. (1980) *Carbohydr. Res.*, 80, C11-C16.
26. Slootmaekers, D., De Jonghe, C., Reynaers, H., Varkevisser, F.A. and Bloys van Treslong, C.J. (1988) *Int. J. Biol. Macromol.*, 10, 160-168.
27. Grasdalen, H. and Smidsrød, O. (1981) *Macromolecules*, 14, 1845-1847.
28. Morris, E.R., Rees, D.A., Thom, D. and Boyd, J. (1978) *Carbohydr. Res.*, 66, 145-154.
29. Morris, E.R., Rees, D.A., Robinson, G. and Young, G.A. (1980) *J. Mol. Biol.*, 138, 363-374.
30. Mackie, W., Perez, S., Rizzo, R., Taravel, F. and Vignon, M. (1983) *Int. J. Biol. Macromol.*, 5, 329-341.
31. Grant, G.T., Morris, E.R., Rees, D.A., Smith, P.J.C. and Thom, D. (1973) *FEBS Lett.*, 32, 195-198.
32. Kohn, R. (1975) *Pure Appl. Chem.*, 42, 371-397.
33. Madgwick, J., Haug, A. and Larsen, B. (1973) *Acta Chem. Scand.*, 27, 3592-3594.
34. Lawson, C.J. and Rees, D.A. (1970) *Nature*, 227, 392-393.
35. Smidsrød, O. and Haug, A. (1971) *Biopolymers*, 10, 1213-1227.
36. McBain, J.W. (1950) *Colloid Science*, Chapter 9. Heath, Boston, USA.
37. Von Hippel, P.H. and Schleich, T. (1969) in *Structure and Stability of Biological Macromolecules*, (eds. Timashef, S.N. and Fasman, G.D.) pp 417-574. Dekker, New York.
38. Seale, R., Morris, E.R. and Rees, D.A. (1982) *Carbohydr. Res.*, 110, 101-112.
39. Piculell, L., Håkansson, C. and Nilsson, S. (1987) *Int. J. Biol. Macromol.*, 9, 297-301.
40. Norton, I.T. (1989) This Volume.
41. Rochas, C. and Rinaudo, M. (1980) *Biopolymers*, 19, 1675-1687.
42. Belton, P.S., Chilvers, G.R., Morris, V.J. and Tanner, S.F. (1984) *Int. J. Biol. Macromol.*, 6, 303-308.
43. Grasdalen, H. and Smidsrød, O. (1981) *Macromolecules*, 14, 229-231.
44. Belton, P.S., Morris, V.J. and Tanner, S.F. (1985) *Int. J. Biol. Macromol.*, 7, 53-56.
45. Belton, P.S., Morris, V.J. and Tanner, S.F. (1986) *Macromolecules*, 19, 1618-1621.
46. Piculell, L., Nilsson, S. and Ström, P. (1989) *Carbohydr. Res.*, 188, 121-135.
47. Norton, I.T., Morris, E.R. and Rees, D.A. (1984) *Carbohydr. Res.*, 134, 89-101.
48. Hermansson, A.-M. (1989) *Carbohydr. Polym.*, 10, 163-181.
49. Morris, E.R., Cutler, A.N., Ross-Murphy, S.B., Rees, D.A. and Price, J. (1981) *Carbohydr. Polym.*, 1, 5-21.
50. Morris, E.R. (1989) *Carbohydr. Polym.*, submitted.

Structural investigation of carrageenans and oligomers using Raman spectroscopy

T.MALFAIT, H.VAN DAEL, K.VANNESTE[+], H.REYNAERS[+] and F.VAN CAUWELAERT

Interdisciplinary Research Centre, Katholieke Universiteit Leuven, Campus Kortrijk, Kortrijk B 8500, Belgium

[+]Katholieke Universiteit Leuven, Laboratorium voor Macromoleculaire structuurchemie, Celestijnenlaan 200F, Heverlee B 3030, Belgium

ABSTRACT

The use of model compounds for the Raman structural analysis of kappa carrageenan segments is shown. On the level of the primary structure,differences between the spectra of the monomer and polymer are due to the involvement of the hydroxyls at the C1 and C4 position in glycosidic linkages. Conformational differences between the oligomers and the polymer are related with the skeleton vibrational freedom. Polymer transitions, as monitored with optical rotation and conductivity measurements, result in a decrease of vibrational states for the skeleton accompanied with an increase of order.

INTRODUCTION

Although a variety of techniques and experimental conditions have already been used in conformational studies of carrageenans, the molecular mechanism of the gelation of carrageenans is not yet completely understood. A few years ago we introduced Raman spectroscopy (1) as a supplementary tool for the structural determination of galactans. Raman spectra of these molecules can contain information about the primary structure, the molecular conformation and about the associated ionic interactions. The complexity of the interpretation of the spectra of polysaccharides led us to this study of different oligomers from kappa carrageenan and other model compounds.

EXPERIMENTAL

The Raman equipment was already described elsewhere (1). Optical rotation measurements were performed with a Perkin-Elmer Spectropolarimeter 141 with a 10 cm pathlength thermostated cell. For conductivity measurements an Ingold electrode (type 109803010) was used, coupled with a Philips PW 9505 conductivity meter. The oligomers of kappa carrageenan, indicated with their degree of polymerisation (dp), were obtained from Grampian Enzymes. Fragments of kappa carrageenan (Sigma, C 1263) were prepared using the periodate oxidation (2).

RESULTS AND DISCUSSION

In a previous paper the Raman spectra of kappa carrageenan and its monomer (dp1) were compared (1). Both spectra showed large similarities which indicated that almost all observed vibrations in the region 700 → 1500 cm^{-1} can be attributed to modes within the repeating unit. Differences in bandwidth, intensity and frequency in the polymer and monomer spectra are due to long-range skeleton vibrations, intra-molecular interactions between chain segments, conformational changes and accompanied ion-specific interactions.

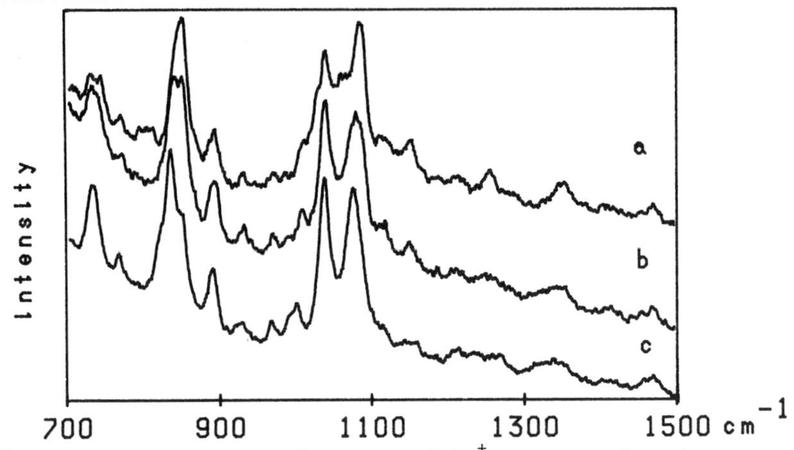

Figure 1. Raman spectra in water of Na^+-segments from kappa carrageenan (a), Na^+-dp2 (b) and Na^+-dp1 (c) at room temperature.

The spectra of segmented Na^+-kappa carrageenan and Na^+-dp1, dp2 are given in figure 1. A survey of the most important bands with their assignments is given elsewhere (3). A first difference between the polymer and the monomer spectra appears at the 734 cm^{-1} band. In the latter spectrum it is sharp and symmetrical while in the former splitting and/or shoulders appear. Since the effect observed for the polymer is already present for dp2 it cannot be due to conformational changes of the polymer solution. Previously we ascribed the band to a complex ring vibration of the basic repeating unit. Since the present observations now suggest that the band-shape depends on the chain-length, the band in the polymer spectrum must be partially due to a skeleton vibration which ranges at a longer distance than 1.04 nm, the projection length of one disaccharide (4). Comparing different ionic forms of the polymer it was also observed that the band became sharper for K^+-kappa carrageenan segments. Optical rotation and conductivity measurements (figure 2) revealed that the latter sample had undergone a conformational transition at ambient temperature. The spectrum indicated that the conformational change resulted in a decrease of possible vibrational states of that particular skeleton vibration. This is in accord with the idea that during the transition the chain becomes 'frozen' and stiffens (4).

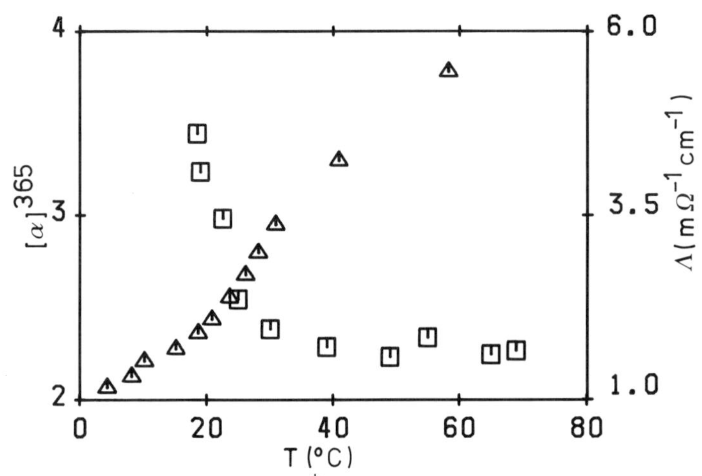

Figure 2. Cooling curves of K^+-kappa carrageenan segments in water (3.2 %) obtained with optical rotation (◘) and conductivity measurements (△).

Large differences in peak intensity are observed at 850 cm^{-1} for dp1 on one hand and dp > 1 on the other hand. In addition the latter all showed a band at 1010 cm^{-1} which was absent in the spectrum of dp1 (figure 1). Since the β 1→4 linkage was only present for dp > 1 it seems reasonable to assume that the band corresponds with this glycosidic vibration. The different intensity between the peaks at 840 and 850 cm^{-1} therefore reflects the involvement in glycosidic linkages of the hydroxyl at the C4 and C1 positions. The presence of OH bending vibrations in the band at 840 cm^{-1} is confirmed by the observation that its intensity strongly decreased in D_2O. The β 1→4 glycosidic linkage appears at a higher frequency than the α 1→3 at 965 cm^{-1}. Moreover, the latter is only weakly active. Therefore the motion of the α 1→3 binding is more restricted. Molecular models show that this is probably due to the vicinity of the large axial sulphate.

Large spectral differences between the segmented polymer and the oligomer were also observed in the region 1000-1100 cm^{-1}. The spectrum of dp1 shows two, well-separated, peaks at 1039 and 1076 cm^{-1}. With increasing dp, the latter peak broadens and shifts towards higher wavenumbers. For dp4 it was observed as a broad band near 1084 cm^{-1}. Band broadening and loss of resolution is due to the existence of a broad distribution of intramolecular hydrogen bonding energies as a consequence of the increased chain-length. However, the band at 1084 cm^{-1} for the Na^+-polymer was relatively small. Since all Raman measurements were performed above the critical overlap concentration all coils were interpenetrated which disturbs the intramolecular interactions. Strong ordered structures were probably not yet found because the conductivity measurements of the Na^+-polymer segments showed no conformational transition in the region 10°-70°C. In contrast a transition was observed for the K^+-segments (figure 2). In the region 1000-1100

cm^{-1} at 20°C (figure 3), very sharp bands were observed for the K^+-segments, which indicated the existence of ordered structures.

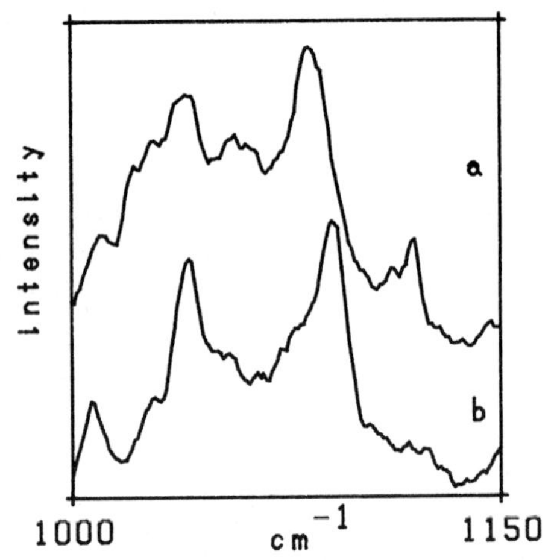

Figure 3. Raman spectra between 1000 and 1150 cm^{-1} of K^+-kappa carrageenan segments in water (6.5 %) at 60°C (a) and at room temperature (b).

Narrowing of the bands is accompanied with bandshifts from e.g. 1084 to 1091 cm^{-1}. At 60°C the spectrum of the K^+-segments equalled that of the Na^+-form which remained unaffected as a function of the temperature. The changes observed in Raman correspond with the transitions observed with other techniques. Work is now going on with Raman spectroscopy to gain insight in the molecular changes occurring during the conformational transition of carrageenans.

ACKNOWLEDGEMENTS

This work has been supported by a grant of the Belgian N.F.W.O.

REFERENCES

1. Malfait, T., Van Dael, H. and Van Cauwelaert, F. (1987) Carbohydr. Res., 163, 9-14.
2. Dea, I.C.M., McKinnon, A.A. and Rees, D.A. (1972) J. Mol. Biol., 68, 153-172.
3. Malfait, T., Van Dael, H. and Van Cauwelaert, F. Int. J. Biol. Macromol. : in press.
4. Vreeman, H.J., Snoeren, T.H.M. and Payens, T.A.J. (1980) Biopolymers, 19, 1357-1374.
5. Smidsrød, O., Andresen, I., Grasdalen, H., Larsen, B. and Painter, T. (1980) Carbohydr. Res., 80, C11-C16.

Kinetics of ion binding and structural change in alginates

E.BERGSTROM, D.M.GOODALL and I.T.NORTON[+]

Chemistry Department, University of York, Heslington, York YO1 5DD, UK

[+]Unilever Research, Colworth Laboratory, Sharnbrook, Bedford MK44 1LQ, UK

ABSTRACT

This paper reports kinetic investigations of the primary processes which occur when sodium alginate and calcium ions are mixed. Conductimetric stopped-flow studies reveal the presence of three kinetic processes. From reaction progress curves and orders of reaction with respect to calcium and alginate, a mechanism for the reaction is deduced. The slower event is ascribed to reaction between dimeric calcium(alginate)$_2$ strands, and occurs only in the presence of excess calcium ions.

INTRODUCTION

Sodium alginate is a natural biopolymer produced by marine algae. Chemically it is a binary copolymer of β-D-mannuronate (M) and α-L-guluronate (G) arranged in homopolymeric blocks and with regions approximating to an alternating sequence along a linear chain (1), Fig.1a.

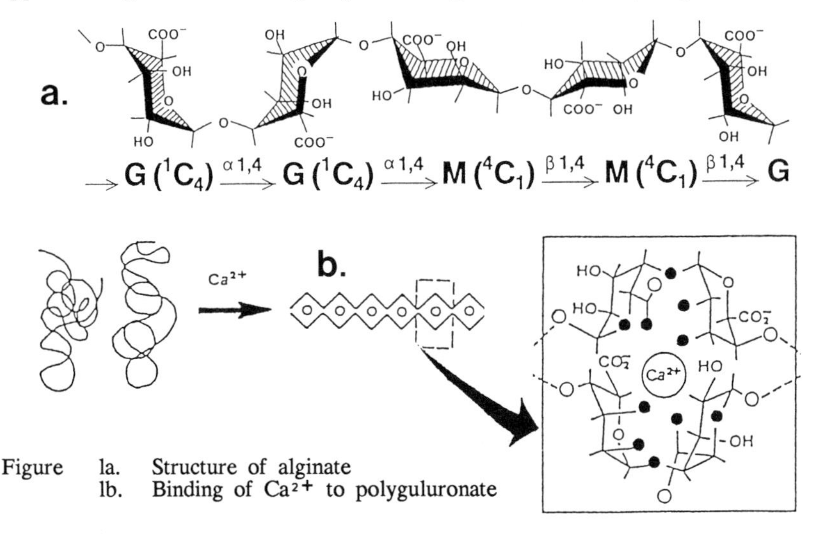

Figure 1a. Structure of alginate
1b. Binding of Ca^{2+} to polyguluronate

Alginates are highly selective ion exchangers which bind calcium ions particularly strongly (2). The binding of calcium ions to sodium alginate causes structural ordering and gel formation. Gel formation is the main biological function of alginates and is also of particular interest in food science, for example in reformed fruit (3).

The main contribution to gel strength comes from the homopolymeric G blocks. Binding of calcium ions to the G blocks is thought to be a cooperative process where an "egg box" like structure of polyanion chain and calcium ions is formed (4). Fig.lb.

MATERIALS, METHODS AND RESULTS

Materials

Alginates are characterized by chain length, composition in terms of M:G ratio, and distribution of the homopolymeric and mixed blocks. The material used in the work was SS/LD/2 supplied by Kelco, and M and G blocks obtained by partial hydrolysis of this alginate (5). ^{1}H and ^{13}C NMR (6), circular dichroism (CD) (7) and elemental analysis were used to characterize these materials. The intact alginate is composed of 56% G, 22% M and 22% MG with an M:G ratio of 0·5. The G block prepared is 94% poly G, and the M block is 64% poly M.

Equilibrium Studies

Methods. The extent of Ca^{2+} binding onto Na alginate has been studied in titrations between $CaCl_2$ and various alginate solutions with added salt. The two methods of monitoring the titrations were (i) conductivity and (ii) Ca^{2+} activity measurement by ion selective electrode (ISE). The added salt in the experiment, tetramethylammonium chloride (TMACl), stabilises the ionic strength of the solutions, which is necessary for the operation of the ISE. The solutions were thermostatted to 303K and titrated with small volumes of high concentration $CaCl_2$ solution while being stirred.

Conductivity measurements were made using a Philips PW 9527 digital conductivity meter and a PW 95 12/60 electrode in a quartz cell. The solutions were degassed with helium and kept under helium throughout the titration.

Calcium activity measurements were made with an Orion calcium ISE with a polymer membrane, of the liquid ion exchange type, and a Corning calomel double-junction reference electrode with the outer chamber filled with 0·1M KNO_3. This was used with a Corning 150 ion analyzer.

Results. The concentration of bound Ca^{2+} in Fig. 2a was calculated assuming the Ca^{2+} activity in the titration solution measured by the ISE is equal to the concentration of unbound Ca^{2+}. Both conductivity and ISE titration curves in Fig.2 have a similar shape; this indicates that conductivity measurement is a suitable method of monitoring the kinetics and equilibrium of Ca^{2+} binding to the alginate polyanion.

The dotted line in Fig.2a is the concentration of bound Ca^{2+} expected if the reaction obeys the simple equilibrium equation shown. The best fit values of the equilibrium constant, K, are 1300, 300 and 65 for the intact alginate, G block and M block respectively. As expected, K for the G block is higher than for the M block. The intact alginate binds Ca^{2+} more strongly than do the blocks, and the curve cannot be fitted well to the simple theoretical model. This is probably due to additional binding in the intact polymer.

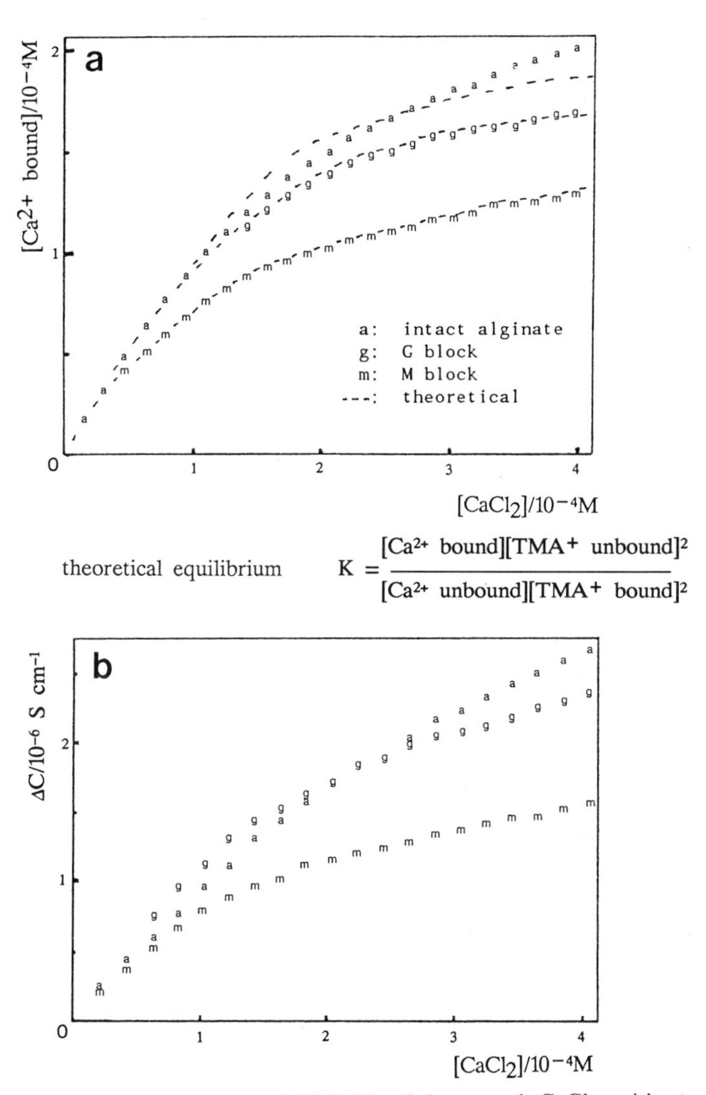

Figure 2. Titrations between 0.01% Na alginate and $CaCl_2$ with 1.6 mM TMACl to maintain constant ionic strength.
a. Ca^{2+} activity measurement by ISE.
b. conductimetric titrations

Kinetic Studies

Methods. The kinetic work has been carried out using conductivity stopped flow (CSF). The CSF apparatus (Unisoku Inc) uses a pneumatic drive which forces the solutions through a seven-stream mixer into a conductivity cell. This produces rapid mixing; the dead time has been measured to be 5·1 ms. The solutions are degassed with helium prior to mixing to reduce cavitation in the cell and the problems of dissolved CO_2. The apparatus is thermostatted and the temperature change during the reaction is < 0·01 K. The cell forms part of a resistance detector for signal retrieval. This allows a conductivity change (ΔC) of less than 10 ppm to be measured.

Results. The binding of Ca^{2+} to Na alginate induces structural ordering and leads immediately to gel formation at polyanion concentrations of > 0·1% w/v. On mixing $CaCl_2$ solutions with sub-gelling concentrations of Na alginate in molar ratios > 4:1, approximately 90% of the ΔC expected from equilibrium studies occurs within the dead time of 5 ms. This is probably due to non-specific counterion condensation. On the CSF timescale two processes are observed. The first occurs within 100 ms and the second with a time scale of seconds.

The reaction progress curve for the faster observable process is first order in polymer. Due to the timescale of this process, a probable explanation is the specific binding of ions which is coupled to a conformational change within a single alginate chain. We aim to clarify this by using stopped flow with CD detection.

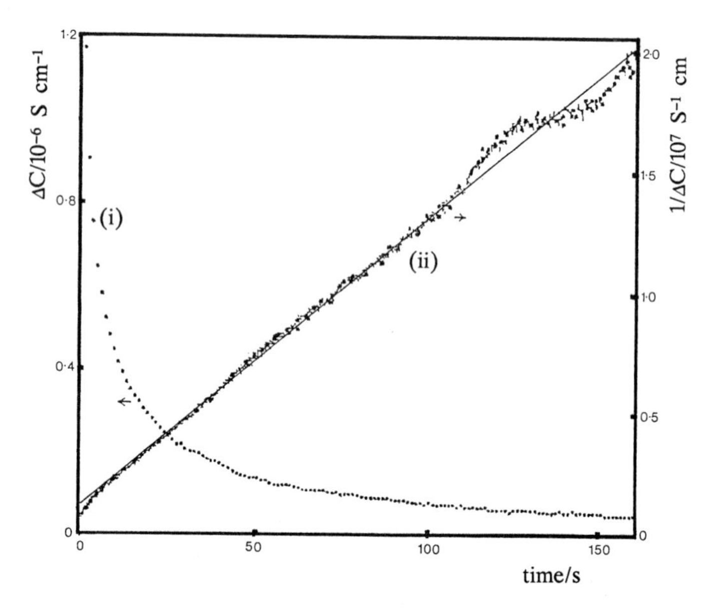

Figure 3. (i) Conductivity relaxation in 0·005% Na alginate with 0·5 mM $CaCl_2$ at 303K.
(ii) Fit to second order process.

The slower process shows a dependence of ΔC with time characteristic of second order kinetics with respect to polymer, Fig.3. The half life is inversely proportional to Ca^{2+} concentration, Table 1. As yet the polymer concentration dependence of the relaxation is unresolved and will form part of further investigations.

Table 1. DETAILS OF THE SLOWER OBSERVABLE RELAXATION

[Polymer]/% w/v	$[CaCl_2]$/mM	$t_{½}$/s
0·01	0·5	22
0·01	1·0	11
0·01	2·0	5·4
0·01	3·0	3·1

CONCLUSION

Conductivity is a useful technique for studying the equilibrium and dynamics of the reaction between calcium and alginates. A scheme which accounts for our findings is:-

	Process	Timescale
1	Non specific counterion condensation	< 5 ms
2	Specific Ca^{2+} binding coupled with a conformational transition	100 ms
3	Further Ca^{2+} induced structural ordering ("egg box" formation? (7))	seconds

ACKNOWLEDGEMENT

We thank the SERC for a CASE award to ETB.

REFERENCES

1 Haug, A., Larsen, B., and Smidsrod, O., (1974) Carbohydr. Res., 32, 217 - 225.
2 Smidsrod, O., and Haug, A., (1968) Acta Chem.,Scand., 22, 3098- 3102.
3 Imeson, A.P., (1989) Gums and Stabilisers for the Food Industry 5.
4 Smidsrod, O., Haug., A., and Whittington, S., (1972) Acta. Chem. Scand., 26, 2563 - 2566.
5 Penman, A., and Sanderson, G.R., (1972) Carbohydr. Res., 25, 273 - 282.
6 Grasdalen, H., and Larsen, B., (1981) Carbohydr. Res., 92, 163 - 167.
7 Morris, E., Rees, D., and Thom, D., (1982) Carbohydr. Res., 100, 29 - 42.

The role of the cation in the gelation of kappa-carrageenan

DAVID OAKENFULL and ALAN SCOTT
Food Research Laboratory, CSIRO Division of Food Processing, PO Box 52, North Ryde, NSW 2113, Australia

ABSTRACT

Single ion forms of κ-carrageenan (the caesium, rubidium and potassium salts) were prepared by ion exchange. The helix-coil transition was studied by optical rotation. The formation of the gel network was studied by measurement of the concentration dependence of shear modulus. All measurements were made in the presence of the appropriate alkali metal chloride at a concentration of 0.02 M. The helix-coil transition was only weakly influenced by the choice of cation. In contrast, the shear modulus data indicated a very strong influence of cation on the gelation step. Calculations from these data suggest that the difference between the potassium and rubidium salts is primarily that, in the case of the rubidium salt, the junction zones involve fewer polysaccharide chains with a correspondingly less negative free energy of association. K^+ and Rb^+ both bind specifically on gelation of κ-carrageenan. Presumably K^+ fits the interstices between the molecules better than Rb^+, better promoting the formation of stable junction zones. The results for the caesium salt are quite different. In this case more extensive aggregation occurs in the step preceeding gelation.

INTRODUCTION

In the formation of carrageenan gels, the polysaccharide chains are believed to associate in two steps: (i) Pairs of chains associate by forming intermolecular double helices. (ii) Interaction occurs between these helical structures to produce a gel network (1,2). Because of the ionic nature of the polymer, gelation is strongly influenced by the presence of electrolytes. The presence of a suitable cation, typically potassium or calcium, is an absolute requirement for gelation to proceed but the precise role of the cation is still not fully understood. For both ι-and κ-carrageenan, the alkali metal ions (Li^+, Na^+, K^+, Rb^+ and Cs^+) are all capable of inducing gelation but K^+ and Rb^+ are considerably more effective than the other ions, inducing gelation at much lower concentrations of both the cation and the carrageenan (3,4). The dependence of the elasticity modulus of the gel on cation type follows a Hofmeister series, $Cs^+ > Rb^+ > K^+ >> Na^+ > Li^+$, suggesting that the ions, by acting as solvent structure-makers or structure-breakers, may influence gelation through their effects on the solvent properties of water (5,6). Such effects could promote or inhibit the the formation of hydrogen bonds that stabilise the helices. Specific ion effects are also important, with evidence from NMR studies (3,7,8) that K^+, Rb^+ and Cs^+ bind to the carrageenan on gelation.

The nature of the cation appears to influence both steps in the gelation process. The helix-coil transition temperature varies with both cation type and concentration (9). In this first step, though, the effect seems simply to be a non-specific electrostatic screening since there is no great difference between the osmotic coefficients of K^+ and Na^+ as counterions of κ-carrageenan in dilute solution (10). The specific binding of Cs^+ Rb^+ and K^+ revealed by NMR (3,7,8) appears only to occur in the course of aggregation of the helices in the second step.

We have extended these studies by investigating, for κ-carrageenan, the effect of cation type on the helix-coil transition by polarimetry (11) at concentrations below the gel threshold. Then, for the overall gelation process, we have derived information about the gel network from measurements of the shear modulus of dilute gels just above the gel threshold (12,13).

MATERIALS AND METHODS

Single-ion forms of κ-carrageenan

Pure single cation forms of κ-carrageenan were prepared by ion exchange (14). The purity was checked by atomic absorption spectroscopy.

Polarimetry

Measurements were made at 589 nm using a Perkin-Elmer 141 polarimeter.

Gel Preparation and Shear Modulus Measurement

The required concentration of the κ-carrageenan was dissolved in the appropriate salt solution (0.02 M) at 90°, cooled and held at 25° for 18 h. Shear modulus was measured using the method described by Oakenfull, Parker and Tanner (15).

RESULTS AND DISCUSSION

The effect of cation on the helix-coil transition

In Fig. 1 we show the temperature dependence, both heating and cooling, of the optical rotation of the κ-carrageenan (2 g/kg) in the presence of either KCl, RbCl or CsCl (0.02 M). The shapes of the curves are characteristic of cooperative thermally induced order-disorder transitions. We have analysed these data as described by Norton *et al.* (16), calculating, from each cooling curve, the change in specific rotation at the transition ($\Delta[\alpha]_{589}$), the transition midpoint temperature (T_m) and the apparent enthalpy change from the Van't Hoff isochore (ΔH_{app}). These results are given in Table 1.

The results in Table I firstly show that the extent of helix formation is independent of cation type, there being no significant differences between the observed changes in specific rotation at the transition. Secondly, the results confirm the observation of Rochas and Rinaudo (8) that the transition temperature (t_m) is higher in the presence of Rb^+ than K^+ or Cs^+. In other words, Rb^+ stabilises the helical conformation more than do the other two ions. Thirdly, the Van't Hoff enthalpy change (ΔH^{o}_{app}) was smaller in the case of Cs^+ than for K^+ or Rb^+ (there was no significant difference between the values for K^+ and Rb^+). This suggests a smaller cooperative unit in the transition (16). The Cs^+ form also differs from the other cation forms in that the curves in Fig. 1 show significantly greater hysteresis. In detailed comparative studies of the K^+ form, ι-carrageenan showed little hysteresis compared with κ-carrageenan and this has been interpreted as indicating more extensive aggregation on cooling than simply the formation of double helices (15). Thus the greater hysteresis of the Cs^+ form of κ-carageenan, seen in Fig. 1, suggests that even more extensive aggregation might be occurring in this case.

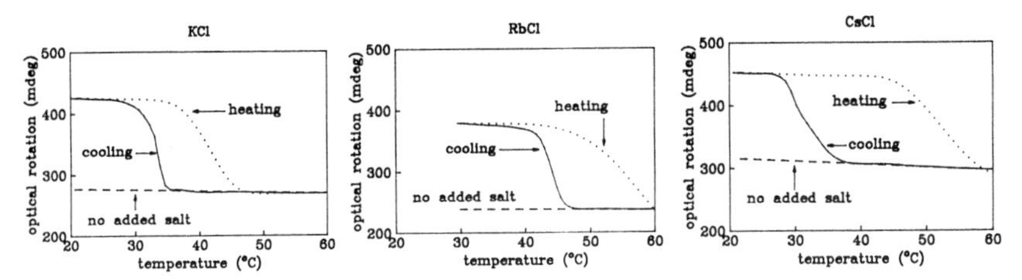

Figure 1. Optical rotation in a 10 cm pathlength cell for κ-carrageenan *(2 g/l) in the presence of 0.02 molar alkali metal chloride.*

Table I. *Parameters for the helix-coil transition of* κ-carrageenan *(2 g/kg) in the presence of different alkali metal chlorides (0.02 M)>*

Cation	t_m (°C)	$\Delta[\alpha]_{589}$ (deg cm^{-1})	ΔH_{app} (kJ mol^{-1})
K^+	33.4	7.5	1,140 ± 120
Rb^+	44.5	6.8	1,460 ± 190
Cs^+	33.9	7.0	232 ± 40

The effect of cation on the formation of the gel network

In Fig. 2 we show shear modulus *vs* concentration for the three cation types of κ-carrageenan. As has previously been reported for ι-carrageenan (3), the shear modulus for a given concentration depends strongly on the cation type. K^+ produces stronger gels than Rb^+ which are stronger than those produced by Cs^+. From these data we have estimated (11) the size and thermodynamic stability of the junction zones, with the results given in Table II.

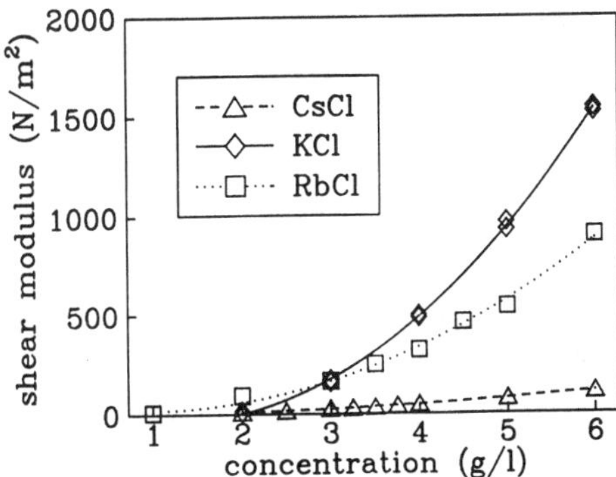

Figure 2. Shear modulus vs concentration for κ-carrageenan *in the presence of 0.02 molar alkali metal chloride.*

The results suggest that the major difference between the potassium and rubidium salts is that, for the rubidium salt, the junction zones involve fewer polysaccharide chains (smaller value for n) with a correspondingly less negative free energy of association (ΔG^o_j). Presumably K^+ fits the interstices between the molecules better than Rb^+, better promoting the formation of stable junction zones. The results for the caesium salt are quite different. The theoretical curve can be made to fit the experimental data points only if the molecular weight of the network forming polymer is about three times its value for the K^+ or Rb^+ forms. This is concordant with the polarimetric data which suggest that some association of double helices occurs in the pre-gelation step.

Analysis of the same data using the model of Clark and Ross-Murphy (13) indicates the same overall picture. In carrying out the calculations from this model, we treated the molecular weight of the primary chain (M_1) as an adjustable parameter. The value for M_1 which best fitted the experimental data was greater for the Cs^+ salt than the K^+ salt by a factor of 12, again suggesting more extensive aggregation of the Cs^+ form in the pre-gelation step.

Table II. *Junction zone parameters calculated from the model of Oakenfull (12) from shear modulus vs concentration for the gelation of* κ-carrageenan *in the presence of different alkali metal chlorides (0.02 M).*

Cation	M^a	M_j^b	n^c	K_j^d	$\Delta G^o_j{}^e$
K^+	105,000	8,900	6.42	1.66×10^{23}	-132
Rb^+	104,000	8,240	2.31	3.60×10^5	-31.7
Cs^+	318,000	175,000	1.93	3.62×10	-37.4

[a]*Number average molecular weight of the polysaccharide.* [b]*Number average molecular weight of the junction zones.* [c]*Number of associating units per junction zone.* [d]*Association constant for junction zone formation (in units of mole fraction).* [e]*Free energy of formation of junction zones (kJ/mole).*

REFERENCES

1. Clark, A.H. and Ross-Murphy, S.B. (1987) Adv. Polymer Sci. 83, 57-192.
2. Oakenfull, D. (1987) CRC Crit. Rev. Food Sci. Nutr., 26, 1-25.
3. Belton, P.S., Chilvers, G.R., Morris, V.J. and Tanner, S.F. (1984) Int. J. Biol. Macromol., 6, 303-308.
4. Watase, M. and Nishinari, K. (1981) J. Texture Stud., 12, 427-445.
5. Eliot, J.H. and Ganz, A.J. (1975) J. Food Sci., 40, 394-398.
6. Morris, V.J. and Chilvers, G.R. (1983) Carbohydr. Polym., 3, 129-141.
7. Morris, V.J. and Belton, P.S. (1980) J. Chem. Soc. Chem. Commun., 983-984.
8. Grasdalen, H. and Smidsrød, O. (1981) Macromol., 14, 229-231.
9. Rochas, C. and Rinaudo, M. (1980) Biopolymers, 19, 1675-1687.
10. Rinaudo, M., Karimian, A. and Milas, M. (1979) Biopolymers, 18, 1673-1683.
11. Rees, D.A. (1969) Adv. Carbohydr. Chem., 24, 267-332.
12. Oakenfull, D. (1984) J. Food Sci., 49, 1103-1104 & 1110.
13. Clark, A.H. and Ross-Murphy, S.B. (1985) Br. Polymer J., 17, 164-168.
14. Morris, V.J. and Chilvers, G.R. (1981) J. Sci Food Agric., 32, 1235-1241.
15. Oakenfull, D.G., Parker, N.S. and Tanner, R.I. (1989) J. Texture Stud., 19, 407-417.
16. Norton, I.T., Goodall, D.M., Morris, E.R. and Rees, D.A. (1983) J. Chem. Soc. Faraday Trans. 1, 79, 2489-2500.
17. Flory , P.J. (1953) Principles of Polymer Chemistry, p.31, Cornell University Press, Ithaca, NY.

The influence of ionic environment and polymeric mixing on the physical properties of iota and kappa carrageenan systems

I.T.NORTON

Unilever Research, Colworth House, Sharnbrook, Bedfordshire MK44 1LQ, UK

ABSTRACT

The effects of ionic concentration and salt type on the formation and stability of the ordered structures in pure iota carrageenan, pure kappa carrageenan and mixtures of iota and kappa carrageenan have been investigated. The data obtained on the pure forms are similar to previous studies. The larger Group I ions show the greatest stabilisation of the ordered forms of Kappa carrageenan. For iota carrageenan only small effects are observed with Group I ions which are consistent with the position of the cation in the lyotropic series and Group II ions inducing very stable ordered forms. Anions are observed to influence the stability of kappa carrageenan with the trend fitting the lyotropic series, but having no effect on iota carrageenan. The results obtained for the pure forms of carrageenan have been used to predict the effect of ion type and concentration when iota/kappa mixtures are studied.

INTRODUCTION

The effects of different cations and anions on the stability of the ordered states (the helix and gel) of pure iota and kappa carrageenan have been studied extensively.[1-12] These measurements have shown that at a fixed salt concentration orders of selectivity for the different cations occur,[2] with the proposal that for kappa carrageenan, ions can be termed gelling or non-gelling,[1] and that although the response of iota carrageenan to monovalent ions is smaller than that for kappa carrageenan[12], the same series is followed. In iota carrageenan the divalent ions appear to be more effective at stabilising the ordered form than any of the monovalent ions.[2]

Studies on the effects of anions on these polyanion materials have led to proposals of special effects in iodide salts,[7,8] or no response to anion type.[2] For kappa carrageenan these studies have been rationalised by an extended study[10] which shows that the response followed the well established lyotropic (Hofmeister) series.[13]

In these studies on the role of ions there has been little consideration of the effect of mixing of iota and kappa carrageenan and how this might effect the response to ionic enviroment. This is unfortunate as many commercial samples are blended to produce a required rheological parameter ('gel strength'). In order to use these mixed samples for other purposes, knowledge of the way that the mixing changes the effect of other added ingredients is therefore important. As the response of iota and kappa carrageenan to both cations and anions is quite different, mixtures of the two polymeric materials could respond differently to either of the pure materials in isolation or be predicted once the pure materials have been studied.

In the present work the effect of ions on very pure samples of iota and kappa carrageenan, and a commercial sample containing both, has been studied. The data obtained for the pure kappa carrageenan is largely in agreement with previously reported information.[2,10] The results obtained on the mixed system are similar to those reported previously for 'iota-carrageenan'[1,3,4,5] while for the pure iota carrageenan studied only a very slight dependence on cation form is observed, and has been shown to follow the lyotropic series. The results obtained for the mixture in the presence of different salts are used to predict the midpoint temperatures for the mixture in the presence of different salts and at various ionic strengths.

SAMPLES

The pure carrageenans (iota batch no SE2280/G and kappa batch No. B5466) were obtained from CECA SA France. A commercial mixture of iota and kappa carrageenan (Auby gel X52) was supplied by Perrefitte Auby. Low molecular weight sugars and excess salt were removed by extensive dialysis and specific salt forms produced by repeated ion exchange on Amberlite IR 120 resin.

Absolute purities were determined from elemental analysis (Butterworth Microanalytical Consultancy Ltd.), ^{13}C-nmr spectroscopy and Infrared spectroscopy. The kappa carrageenan was found to have less than 5% 'iota character' and the iota carrageenan had less than 5% 'kappa character'. The Aubygel X52 was shown to contain 60% iota carrageenan and 40% kappa carrageenan.

EFFECT OF CATIONS

Kappa Carrageenan

The disorder-order transition of kappa-carrageenan has been studied by optical rotation, for various salt concentrations and a number of different cations. Defining the midpoint temperature (T_m) for the transition as the point at which half the maximum optical rotation change has occurred, the change in stability of the ordered state with salt concentration and also with salt type may be conveniently compared. Figure 1 shows that for kappa carrageenan a number of cations can be used to produce the ordered state, and that at constant ionic strength; the stability of the helix is dependent on the cation used. In addition it can be seen that the increase in stability caused by increasing the salt concentration is different for the various ions, demonstrating the different involvement of the ions in the ordering process.

Figure 2 shows how the melting temperature varies, at a constant salt concentration (0.25M), with cation size (expressed as the ionic radius of either the solvated or non-solvated species). It appears from this analysis that there could be a preferred ionic size for interaction (i.e. Rb^+ is slightly more effective than K^+ or Cs^+ which are more effective than Na^+ or Li^+). From this representation of the data it can be seen that the divalent ions, are more effective at stabilising the helix than are monovalent ions of the same size. As this ordering event in kappa carrageenan entails an increase in charge density on formation of the double helix and aggregated helices, the

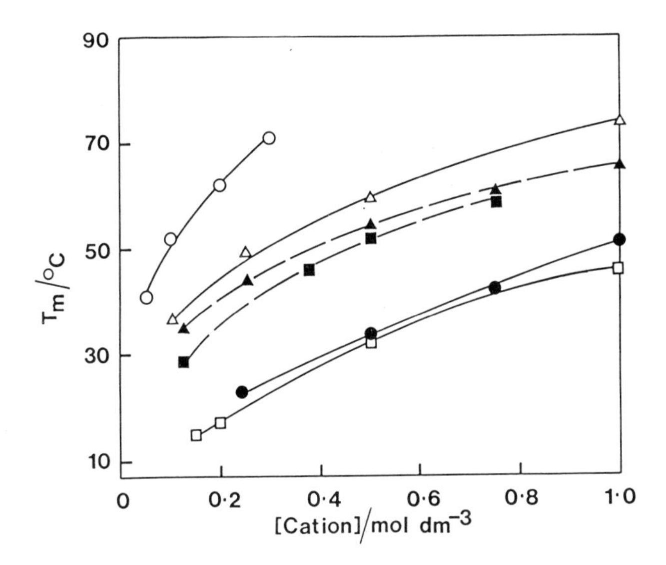

Figure 1. Transition midpoint temperature of the order-disorder transition of kappa carrageenan, as a function of the salt molarity, for various chloride salts; $N(CH_3)_4^+$ (□), Na^+ (●), NH_4 (Δ), K^+ (O), Mg^{2+} (■) and Ca^{2+} (▲).

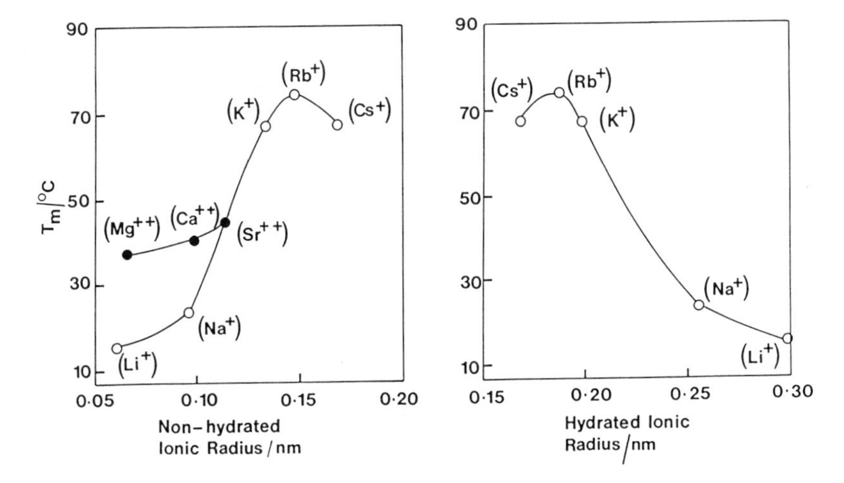

Figure 2. Comparison of the transition midpoint temperature for the order-disorder transition of kappa carrageenan for the various salts either as the hydrated or non-hydrated ionic radius (0.25M salt).

increased effectiveness of the divalent ions over monovalent ions of the same size is not surprising, The fact that a number of the monovalent ions stabilise the ordered state more than divalent ions, however, rules out any conclusion that the involvement of ions in the ordering event can simply arise through non-specific polyelectrolyte interactions or through solvent (lyotropic) effects.

Figure 2 shows that although the actual size of the ions are different if they are solvated or non-solvated the overall trend and conclusions are unaffected. The order of effectiveness of the cations is independent of the salt concentration in the range of ionic strength studied, with the increase in stability of the ordered state, and increasing aggregation and gelation following the series:

$$Rb^{+} > K^{+} \sim Cs^{+} > NH_4^{+} > Sr^{++} > Ca^{++} > Mg^{++} > Na^{+} > N(CH_3)_4^{+} > Li^{+}.$$

Similar series have been reported previously by Rochas and Rinaudo[2]

$$Rb^{+} > Cs^{+} > K^{+} > NH_4^{+} > Ca^{2+} > Sr^{2+} > Mg^{2+} >> Na^{+} \sim N(CH_3)_4^{+} > Li^{+}$$

and Morris, Rees and Robinson[1] who proposed Rb^{+}, K^{+} and Cs^{+} as gelling ions for kappa carrageenan and Li^{+} and $N(CH_3)_4^{+}$ as non-gelling ions.

Iota Carrageenan

Optical rotation results obtained on heating and cooling pure iota carrageenan at a fixed concentration of different univalent salts are shown in figure 3. From these temperature profiles it can be seen that there is a slight dependence on the cation type, and that no thermal hysteresis occurs. At all salt concentrations the effect of varying the cation has been shown to be much smaller that that observed for kappa carrageenan. For example, the increase in melting temperature on changing the salt type from LiCl to RbCl, at 0.25M salt, was ~ 4°C for iota carrageenan and ~ 60°C for kappa carrageenan. The effectiveness of the monovalent ions in stabilising the order to disorder transitions of iota carrageenan follows the lyotropic series.

In contrast to the behaviour observed for kappa carrageenan, the divalent cations are at least an order of magnitude more effective than any of the monovalent ions in stabilising the ordered state of iota carrageenan. The order disorder transition of iota carrageenan on both heating and cooling in 5mM $CaCl_2$ is shown in Figure 3. This demonstrates the importance of Ca^{2+} ions in inducing the ordered state, and suggests that control of calcium is important when using this material.

X52 'Iota Carrageenan'

The commercial samples of carrageenan used for industrial processes are often sold for gel strength properties and are therefore blends of iota and kappa carrageenan. The effect of salt-type and ionic strength for these blended materials may be expected to be quite complicated as a result of the very different responses to salt of the two pure materials. A commercial sample has been studied, and the dependence of the overall transition of salt concentration and cation-type is shown in Figure 4.

The disorder to order transition is seen to have something of an intermediate dependence on ion type, with KCl being highly effective at stabilising the ordered state, (i.e. similar to the response found for pure kappa carrageenan), but with divalent ions (Ca^{2+} and Mg^{2+}) being even more effective (as in iota

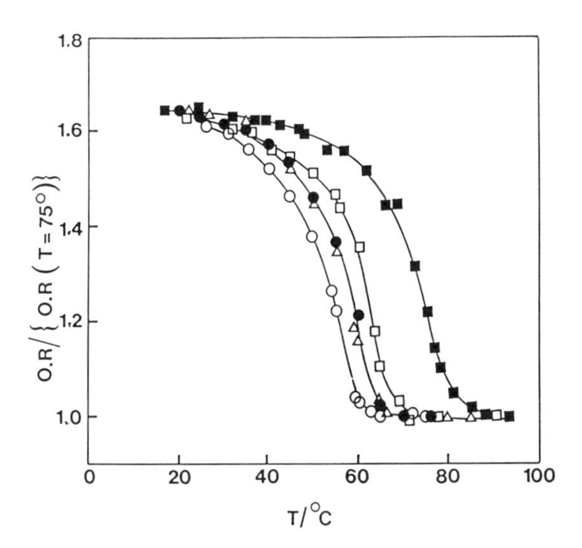

Figure 3. Temperature course of the order-disorder transition of iota carrageenan (0.3%) on heating and cooling, in the presence of 0.25M LiCl (O), KCl (Δ), RbCl (●) and CsCl (◻) and in the presence of 5mM $CaCl_2$ (■).

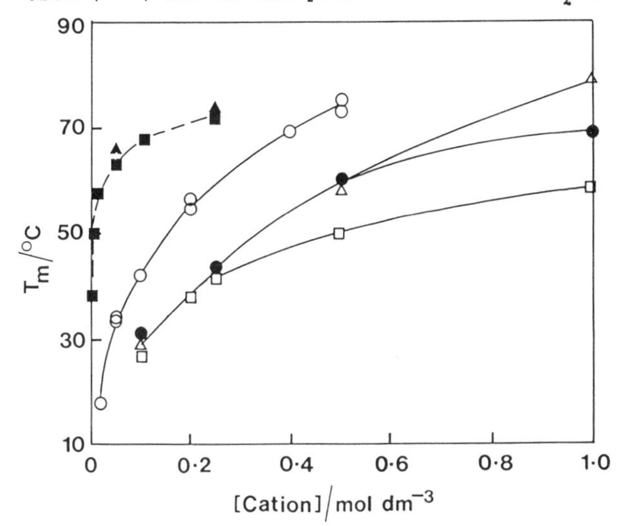

Figure 4. Transition midpoint temperature of the order-disorder transition of X52 'iota carrageenan' as a function of the salt molarity for the various chloride salts: $N(CH_3)_4^+$ (◻), Na^+ (●), NH_4^+ (Δ), K^+ (O), Mg^{2+} (■) and Ca^{2+} (▲).

carrageenan).

The disorder to order transition in the mixed system is found to have a broader temperature range than that observed for either of the individual polymers in isolation. This suggests that partial separation of the polymer occurs even if mixed double helices are also produced. In the presence of salts that stabilise the ordered state of kappa carrageenan (KCl, RbCl and CsCl) two discrete transitions are observed by both optical rotation and differential scanning calorimetry[15] indicating quite severe phase separations of the two materials to give helices involving the individual types of polymer. For the salts that have less of an effect on kappa carrageenan [NaCl, $N(CH_3)_4$ Cl] optical rotation and DSC indicate only a single transition, and high resolution ^{13}C NMR studies show concurrent loss of signal from both types of carrageenan even under conditions where kappa carrageenan in isolation would be disordered. Thus a range of effects can occur, and suggest that mixed helices containing both an iota and kappa carrageenan chain can be produced.

Figure 5 shows the ionic-size dependence of the order-disorder transition of X52 in 0.25M and 0.1M Group I chlorides. At both salt levels, the transition midpoint varies systematically with ionic radius of the monovalent cations, with the ordered state being stabilised most in the presence of RbCl. In Figure 5 the observed T_m values for X52 are compared with values calculated [Eq. (1)] from the melting data for pure iota and kappa carrageenan, on the assumption that the influence of each component on the overall properties is directly proportional to its known relative concentration.

$$T_m\ (mix) = [p(kappa)\ T_m(kappa)] + [p(iota)\ T_m(iota)] \quad (1)$$

Where p(kappa) and p(iota) are the fractions of kappa and iota carrageenan in the mixture (0.40 and 0.60 respectively for the X52 used).

There is good agreement (Fig. 5) between observed and calculated melting temperatures at both salt levels for Li^+, Na^+, K^+ and Rb^+. The greatest divergence occurs in the presence of Cs^+ (possibly indicating a tighter size-requirement for cation-binding to 'mixed' helices), but even here the standard of agreement is encouraging.

Thus using a simple linear equation it is possible to predict the melting temperature of mixtures of iota and kappa carrageenan at any salt level for the monovalent ions. In practice therefore it is feasible to blend carrageenan samples to obtain a range of melting temperatures at a fixed salt concentration and salt type. Predictions are also possible for the divalent ions although the salt concentration range is limited due to the high sensitivity of pure iota carrageenan to the more highly charged ions.

EFFECTS OF ANIONS

Kappa Carrageenan

Until recently the effects of anions on the disorder transition of these polyanion systems had been ignored. As anions will effect the solvent properties we might expect the presence of different anions to change the stability of both the ordered and disordered states of the carrageenans. The size of this change in stability and the way in which it will affect the ordering process will depend on the sensitivity of the polymer to its solvation.

Results obtained for a range of salt concentrations and anion types with kappa carrageenan is dependent upon the anionic form

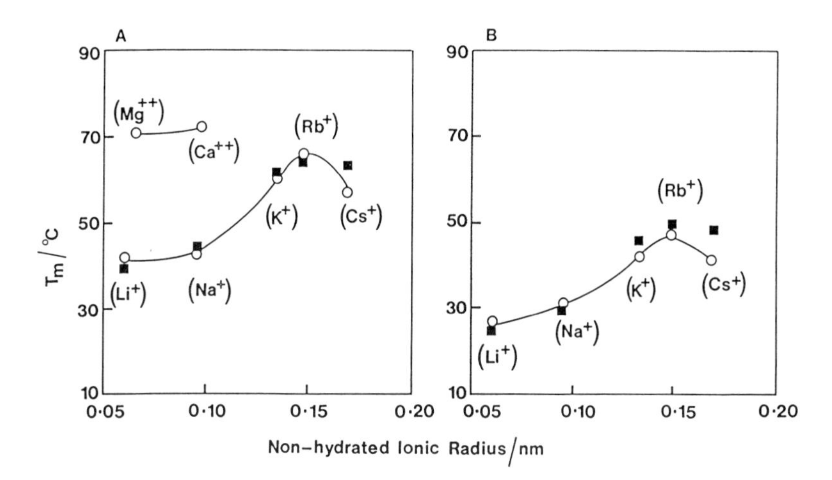

Figure 5. Comparison of the transition midpoint temperatures for the order-disorder transition of X52 'iota carrageenan' and the non-hydrated ionic radii of various cations at salt concentrations of A-0.25M and B - 0.1M. The predicted melting temperatures from combining the known melting points of the pure carrageenan material are shown (■).

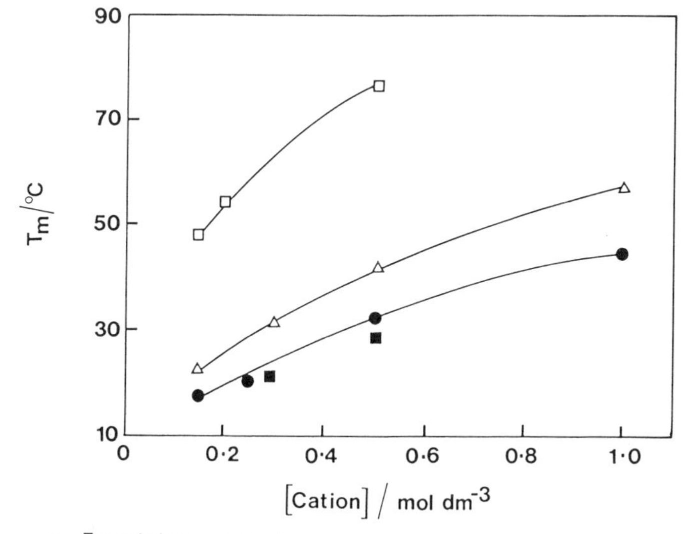

Figure 6. Transition midpoint temperature of the order-disorder transition of kappa carrageenan in the presence of tetramethylammonium sulphate (■), chloride (●), nitrate (Δ) and iodide (□).

following the lyotropic series for the anions (Figure 6). It has also been observed that in the presence of potassium iodide kappa carrageenan will produce a weak clear gel, while under the same conditions but with KCl or K_2SO_4, strong turbid gels are produced.
From these results it is apparent that the different states of kappa carrageenan are very sensitive to changes in solvent quality, suggesting that it is close to being insoluble in water. The result of this is that on adding ions that increase the ordering in water (eg. F^-) with a lowering of its ability to solvate the polymer, a larger amount of aggregation results.

Iota Carrageenan

In contrast to the observed effects of anions on kappa carrageenan, the stability and ordering of iota carrageenan are unaffected by anion type: clear elastic gels produced with the whole range of lyotropic anions, and, as illustrated in Figure 7, the transition midpoint temperature at a fixed ionic strength is independent of the co-anion. This lack of response to lyotropic changes suggests that unlike kappa carrageenan, which is hovering on the verge of insolubility and is therefore very sensitive to solvent quality, iota carrageenan (by virtue of its higher charge density) has far less tendency to insolubility and therefore shows little response to small changes in hydration.

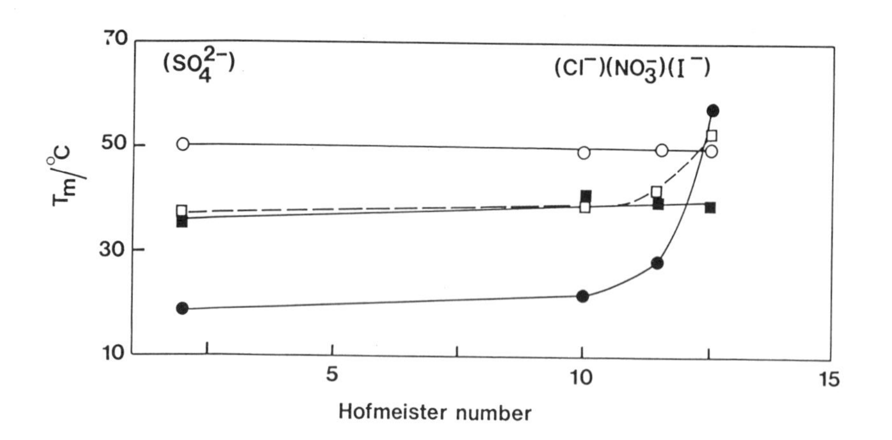

Figure 7. Dependence of the transition midpoint temperature for kappa carrageenan (●), iota carrageenan (O) and X52 'iota carrageenan' (■) on the Hofmeister number of the co-anion present. The predicted results for X52 'iota carrageenan' are also shown (□).

X52 'Iota Carrageenan'

The order-disorder transition of X52 (Figure 7) shows little dependence on anion type, with the melting temperatures obtained with Cl^-, SO_4^{2-} and NO_3^- laying between those observed for the two pure materials. The melting temperatures for the commercial sample can again be predict from Equation 1. A comparison of the observed and predicted results is made in Figure 7. For the sulphate, chloride and nitrate the predicted and observed results are in close agreement. In the case of iodide, however, the high melting temperature observed for pure kappa carrageenan is not reflected in the T_m value obtained for the commercial mixture. This suggests that the near-insolubility of kappa carrageenan under these solvent conditions can be reduced or eliminated by formation of mixed helices in which the kappa carrageenan chain is 'solubilised' by its more highly charged iota carrageenan partner.

CONCLUSIONS

1. The helical state and gel structure of kappa carrageenan can be induced by a number of different cations.
2. The stability of the ordered state of kappa carrageenan is strongly dependent upon the type of ions present at any one salt concentration. The cation dependence appears to be ion size rather than ion charge sensitive. Anions influence the stability by changing the solvent quality with the order of the effect following the lyotropic series.
3. The stability of the ordered state of iota carrageenan shows little or no response to a range of cations and anions. The small cation dependence follows the lyotropic series.
4. The melting temperature of mixed iota/kappa carrageenan blends under different conditions of salt type and ionic strength, can be predicted from the results obtained for the pure materials.

References

1. E R Morris, D A Rees and G Robinson, 'Cation-Specific Aggregation of Carrageenan Helices: Domain Model of Polymer Gel Structure'. J Mol Biol, 138 (1980), 349-362.
2. C Rochas and M Rinaudo, 'Activity Coefficients of Counterions and Conformation in Kappa Carrageenan Systems'. Biopolymers 19 (1980) 1675-1687.
3. M Rinaudo, A Kariman and M Milas, 'Polyelectrolyte Behaviour of Carrageenans in Aqueous Solutions'. Biopolymers. 18 (1979), 1673-1683.
4. P S Belton, G R Chilvers, V J Morris and S F Tanner, 'Effects of Group I Cations of the Gelation of Iota carrageenan' Int J Biol Macromol 6 (1984) 303-308.
5. I T Norton, D M Goodall, E R Morris and D A Rees, 'Role of Cations in the Conformation of Iota and Kappa carrageenan'. J Chem Soc Faraday, Trans 1, 79, (1983) 2475-2488.
6. D M Goodall and I T Norton, 'Polysaccharide Conformations and Kinetics'. Accounts of Chemical Research 20 (1987) 59-65.
7. H Grasdalen and O Smidsrod, Iodine specific Formation of kappa carrageenan Single Helices. ^{127}I NMR Spectroscopic Evidence for Selective Site Binding of Iodide Anions in the ordered Conformation'. Macromolecules 14 (1981) 1845-1847.
8. O Smidsrod, I L Anderson, H Grasdalen and T Painter, 'Evidence for a salt Promoted "Freeze-out" of Linkage Conformation in Carrageenan as a Pre-Reqisite for Gel Formation'. Carbohydr. Res 80. (1980) C11-C16
9. K R J Austen, D M Goodall and I T Norton, 'Anion Independent Conformational Ordering in Iota-carrageenan: Disorder-Order Equilibria and Dynamics' Carbohydrate Res. 140 (1985) 251-262.

10. I T Norton, E R Morris and D A Rees, 'Lyotropic Effects of Simple Anions on the Conformation and Interactions of Kappa-carrageenan'. Carbohydrates Res, 134 (1984) 89-101.
11. K J Austen, D M Goodall and I T Norton, 'Anion Effects on the Equilibria, and Kinetics of the Disorder-Order Transition of Kappa Carrageenan. Biopolymers 27 (1988) 139-155.
12. G P Lewis, W Derbyshire, S Ablett, P J Lillford and I T Norton, Investigation of the NMR Relaxation of Aqueous Gels of the Carrageenan Family and of the effects of ionic content and character.
Carbohydrate Res. 160 (1987) 397
13. J W McBain. 'Colloid Science'. Heath, Boston 1950 ch9.
14. E R Morris, D A Rees, I T Norton, D M Goodall, 'Colorimetric and Chiriptical Evidence of Aggregate Driven Delix Formation in Carrageenan systems. Carbohydrate Res. 80 (1980), 317-323.

A pulsed proton NMR study of ion effects on agarose aggregation

N.D.HEDGES, W.DERBYSHIRE[+*], P.J.LILLFORD and I.T.NORTON

Unilever Research, Colworth House, Sharnbrook, Bedfordshire MK44 1LQ, UK
[+]*Department of Physical Sciences, Sunderland Polytechnic, UK*
[*]*Now at: R.H.M. Research and Engineering, Lincoln Road, High Wycombe, Buckinghamshire, UK*

ABSTRACT

A useful tool in studying the aggregation of agarose chains is pulsed proton n.m.r. The variation of the spin-spin relaxation time of agarose gels with temperature yields useful information about the extent of aggregation. From heating and cooling spin-spin relaxation thermal profiles, the effect of ion type and concentration on agarose chain association may be investigated. Anion effects on agarose aggregation are very much larger than cation effects. They are larger in fact than anion effects seen in iota and kappa carrageenan. Indeed, high concentrations of SCN^- and I^- ions are able to almost totally inhibit agarose aggregation. The addition of high SCN^- ion concentrations to the surface of an agarose gel will solubilise it. Shifts in the mid-point temperatures of the associative and dissociative processes are linear with increasing anion concentration. The magnitude of anion effects are consistent with the position of the ion in the lyotropic series. By studying mixtures of ions we have shown that the effects are purely colligative.

INTRODUCTION

Agarose is an essentially neutral polysaccharide in which the disaccharide repeat unit is a 1,3 linked β-D-galactose residue and a 1,4 linked 3,6-anhydro-α-L-galactose residue, figure 1.

This structure is, however, idealised and the isolated polysaccharide will have a low level of sulphate, methoxyl and pyruvate groups. Arnott et al (1) and Guisley (2) have shown that these groups influence the thermal behaviour of agarose gels. Thermal hysteresis is seen on heating and cooling agarose gels (1,3) and has been interpreted by Liang et al (4) in terms of stabilisation of the ordered conformation by aggregation. X-ray fibre diffraction data from orientated agarose fibres, coupled with an optical rotation investigation, suggest that the shift in optical rotation on cooling a hot agarose solution is due to formation of double helices. A schematic representation of these events is given in figure 2.

The x-ray data for the agarose fibres is not unequivocal.

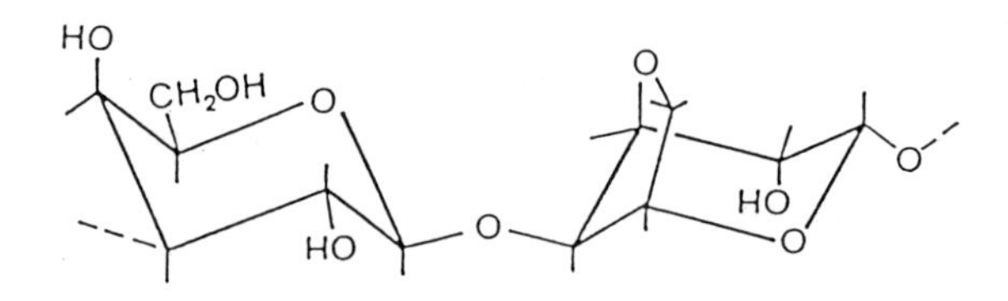

Figure 1. Idealised disaccharide repeat unit of agarose.

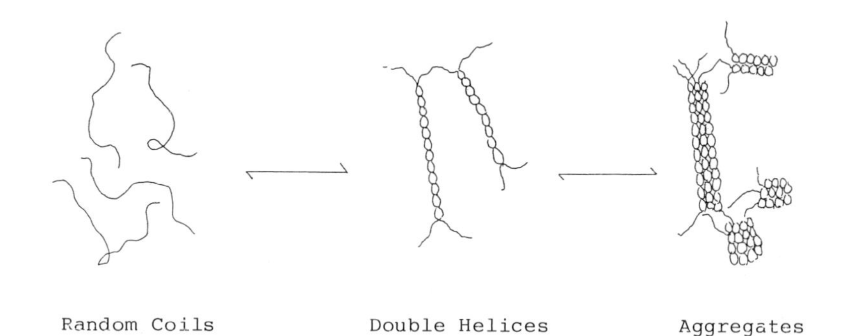

Figure 2. Schematic diagram for the ordering events occurring on cooling a hot agarose solution.

Some controversy exists concerning the exact nature of the conformational change in the random coils of agarose. An alternative mechanism (5) invokes the formation of extended single helices which aggregate into crystalline fibrils. This mechanism essentially proposes the "freezing" of the solution stereochemistry in the gelled state. There is, however, a reasonable consensus that all mechanisms yet proposed involve a conformational change in the polysaccharide chains followed by extensive aggregation into 'crystalline' fibrils.

Spin-Spin Relaxation in Agarose Gels

A model developed by Zimmerman and Brittin (6) and extended by Woessner and Zimmerman (7) to describe the variation of the observed spin-spin relaxation time with temperature, for water adsorbed onto silica, has been modified and used to explain the proton spin-spin relaxation data for water in agarose gels (8). The model comprises three phases:- a bulk water phase, a phase tightly bound to the agarose chains and a very tightly bound phase, denoted a, b and c respectively. The populations of the phases are Pa, Pb and Pc and their intrinsic relaxation times are T_2a, T_2b and T_2c. Exchange occurs between these phases and the observed spin-spin relaxation time is described by equation [1].

$$\frac{1}{T_2 obs} = \frac{Pa}{T_2 a} + \frac{Pb}{T_2 b} + \frac{Pc}{(1-Pc)(T_2 c+\tau c)} \qquad - [1]$$

τc is essentially the residence time of a proton in the c phase. Although this model has been used to quantitatively describe the variation of the spin-spin relaxation time of agarose gels with temperature (9), it requires several simplifying assumptions due to the complex nature of the bound water phases.

For this reason the bound and tightly bound phases proposed by Ablett et al (8) will be treated as one phase of intrinsic relaxation time T_2 bound. Thus to a good approximation the essential features of the variation of the observed spin-spin relaxation time of a gel, above the temperature at which the bulk water freezes and below the temperature at which the gel begins to disaggregate, may be described by;

$$\frac{1}{T_2 obs} = \frac{HC}{1-HC}\left[\frac{1}{(T_2 bound+\tau c)}\right] + \frac{1}{T_2 a} \qquad - [2]$$

C is the agarose concentration expressed as weight of agarose per weight of water. H is the amount of water molecules whose molecular motions are significantly modified by the polymer chains. Thus HC is equivalent to Pb+Pc and 1-HC is equivalent to Pa. T_2a is the spin-spin relaxation time of pure water.

T_2 bound will reflect the extent of agarose chain mobility and consequently T_2 bound, and also the observed relaxation time of the gel, will reflect the extent of agarose helix aggregation.

EXPERIMENTAL

The agarose used in this study was a Marine Colloids sample, code number BRE027 from the Gelidium species of seaweed. The methoxyl and pyruvate contents were determined from high resolution proton n.m.r. spectra of 5% solutions in D_2O acquired at 90°C. The sulphate content was determined from microanlysis data. The methoxyl content, substituted at the 6-0 position of the β-D-galactose ring was 1.7 groups per 100 disaccharide residues. The pyruvate content, as 4,6-0-(1-carboxyethylidene)-D-glucose was found to be 0.9 residues per 100 disaccharide repeat units. The sulphate content, as β-D-galactose-6-sulphate was found to be 2.3 residues per 100 disaccharide repeat units.

The gels were prepared by dilution from a 5% stock solution which was prepared by pressure cooking the agarose powder in water for 25 minutes at approximately 125°C.

The hot solutions were placed in n.m.r. tubes which were then sealed above the solution. The solutions were allowed to gel at room temperature before being stored for a minimum of 24hrs at 4°C before being measured.

Measurements were made using a Bruker CXP spectrometer operating at 60MHz. T_2 measurements were made using the Carr, Purcell, Meiboom and Gill pulse sequence.

The decays in magnetisation were analysed by a damped Newton-Raphson method and the quality of the fit assessed by a plot of the difference between actual and calculated data. Within experimental error all decays were described by a single exponential.

RESULTS AND DISCUSSION

Figure 3(a) shows the variation of T_2 with temperature for a 1% gel. Quite clearly a minimum value for the T_2 is observed at around 310K. This is believed to be due to changes in exchange rate between water protons in the phases described earlier. Evidence supporting this hypothesis comes from an ^{17}O, ^{2}H and ^{1}H spin-spin relaxation, temperature dependence study (10), and also an investigation into shifts in the T_2 minimum as a function of increasing the D_2O/H_2O ratio of the solvent (11).

At the low temperature side of the minimum the observed relaxation time is dominated by the residence time of the water protons in the tightly bound phase. On heating the gel, the exchange rate is increased so that τc is reduced and as a consequence the observed spin-spin relaxation time decreases.

As the temperature increases, further fast exchange occurs and the observed spin-spin relaxation is described by;

$$\frac{1}{T_2\,\mathrm{obs}} = \frac{HC}{1-HC}\left[\frac{1}{T_2\,\mathrm{bound}}\right] + \frac{1}{T_2\,a} \qquad - [3]$$

The observed relaxation time is dominated by the temperature dependence of T_2 bound, which increases with increasing temperature.

Duff (12) has shown that the variation of the observed T_2 with temperature may be described by an Arrhenius relationship, provided that the bulk water phase does not freeze or the gel does not begin to disaggregate.

At the low temperature side of the T_2 minimum the observed data may be fitted to;

$$\log_e \left[\frac{1}{T_2 \text{obs}} \right] = \frac{-\Delta E}{RT} + \text{constant} \qquad - [4]$$

Where ΔE is the energy describing the variation of τc with temperature, T is the thermodynamic temperature and R is the universal gas constant. The constant being sample dependent.

Similarly at the high temperature side of the T_2 minimum the observed data may be fitted to;

$$\log_e \left[\frac{1}{T_2 \text{obs}} \right] = \frac{\Delta H}{RT} + \text{constant} \qquad - [5]$$

ΔH is the enthalpy describing the variation of T_2 bound with temperature.

Equations 4 and 5 may be fitted to the observed data either side of the T_2 minimum. Figure 3(b) shows the actual and calculated values of T_2 plotted against temperature for the 1% gel data shown in figure 3(a). At the low temperature side of the T_2 minimum, the actual and calculated values are in good agreement. This suggests that the observed T_2 varies with temperature in an Arrhenius manner. Some deviation does occur on approaching the T_2 minimum temperature as the contribution of T_2 bound to the observed spin-spin relaxation time becomes significant. On the high temperature side of the T_2 minimum the actual and calculated values are coincident over only a limited temperature range. The larger deviations from the actual and calculated values above 333K are due to disaggregation of the agarose network. Thus T_2 bound is increased further by increases in agarose chain motions as the agarose helices begin to disorder.

The calculated data essentially describes the variation of the spin-spin relaxation time with temperature of a theoretical gel which does not dissociate. The cooling data prior to the onset of the associated process describes the temperature dependence of disordered form. The mid-point temperatures of the disaggregation (Tm) and aggregation (Tg) process may be defined as the intersection of a line, constructed mid-way between the calculated data and the cooling data prior to the onset of the aggregation and the heating and cooling curves.

Salt Effects

It has been recognised that simple salts may drastically alter the solubility and other solution properties of macromolecules. The relative effectiveness of different ions to flocculate agar and gelatin was investigated by Bruins (13) and were placed in a well defined order known as the lyotropic series.

These lyotropic numbers correlate with other physical chemical measurements which reflect on ions charge/size ratio and the extent of ion/water interaction (14). Ions with a large lyotropic number are believed to disrupt water structure and reduce the extent of water/water bonding whereas ions with smaller lyotropic numbers tend to increase the water structure. Table 1 shows the lyotropic numbers for some monovalent anions and cations.

Thus lithium, iodide and thiocyanate ions would be described

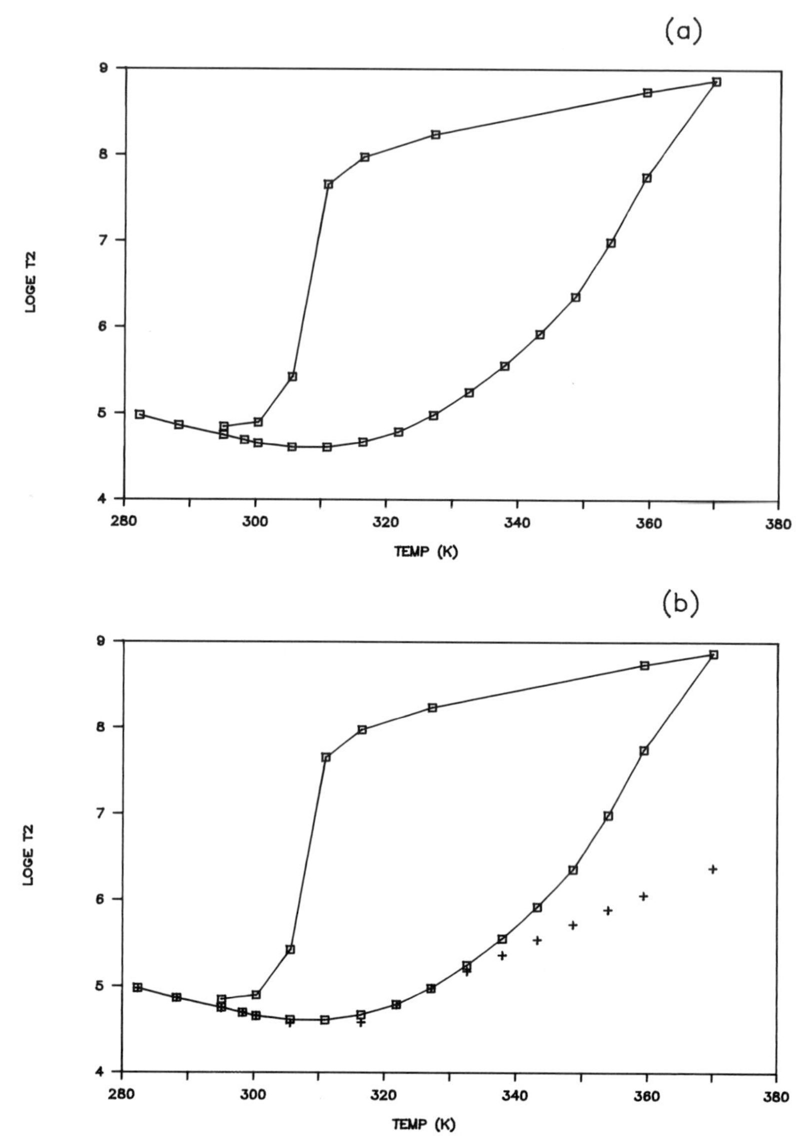

Figure 3. Variation of spin-spin relaxation time with temperature for a 1% agarose gel (a) and including the calculated spin-spin relaxation times from equations 4 and 5 (b).

as water structure disrupting ions, while F^- and K^+ would be described as water structure making ions.

Figure 4(a) shows the variation of the agarose Tm, the midpoint temperature of the dissociative process, with ionic concentration for KCl and LiCl solutions. As can be seen the effect of cations on Tm is effectively zero.

Ion	Lyotropic number
Li^+	115
Na^+	100
K^+	75
F^-	4.3
Cl^-	10.0
I^-	12.5
SCN^-	13.3

Table 1: Lyotropic numbers of some monovalent ions

Anion Effects

Figure 4(b) shows the variation of Tm with anion concentration. It is quite clear that the anion effects are much larger than cation effects and are consistent with the position of the ion in the lyotropic series. For example SCN^- ions are believed to cause large amounts of disruption to the solvent structure and cause the largest reductions in Tm. Conversely the F^- ion is a water structure making ion and increases the mid point temperature of the dissociative process.

Figure 5(a) shows the variation of T_2 with temperature for a 1% agarose gel in the presence of 2 mol dm^{-3} KSCN. Indeed 2 mol dm^{-3} KSCN is able to effectively inhibit agarose aggregation. Conversley figure 5(b) shows the variation of T_2 with temperature for a 1% agarose gel in the presence of 2.0 mol dm^{-3} KF. In this case extensive aggregation has occurred and even at a temperature of 370K the gel does not fully disaggregate.

Figure 6(b) shows the variation of Tg with anion concentration. Again the anion effects are linear over the whole anion concentration range studied, until either the gels no longer aggregate as in the case of SCN^- and I^- ions, or virtually precipitate as in the case of F^- ions. This linear response with increasing number of ions is typical of a lyotropic response. Again the cation effects are small, Figure 6(a), but consistent with the ion's position in the lyotropic series.

To assess the colligative nature of the anions on the shifts in Tm and Tg, mixtures of I^- and F^- ions were investigated. Figure 7(a) shows the variation of Tm with fluoride ion concentration in the presence and absence of 1.0 mol dm^{-3} KI. The relative change in Tm caused by a given fluoride concentration, in the range 0 to 1.0 mol dm^{-3} would seem not to be influenced by the presence of 1.0 mol dm^{-3} KI. Similarly, Figure 7(b) shows the variation of Tg with fluoride concentration in the presence and absence of 1.0 mol dm^{-3} KI. Again the shift in Tg for a given F^- concentration seems independent of the presence of the I^- ions. This result would suggest the observed value of either Tm or Tg is essentially a

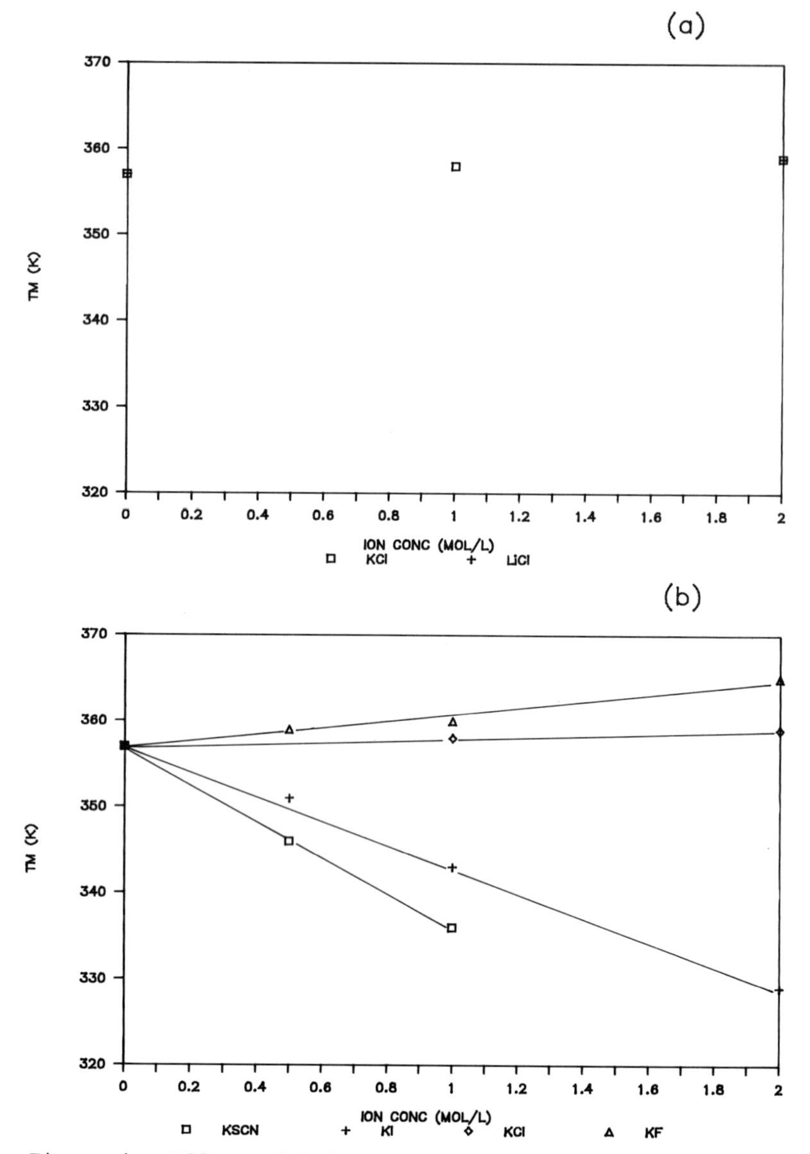

Figure 4. Effect of (a) cation type and concentration and (b) anion type and concentration on the mid-point temperature of the dissociative process Tm for a 1% agarose gel.

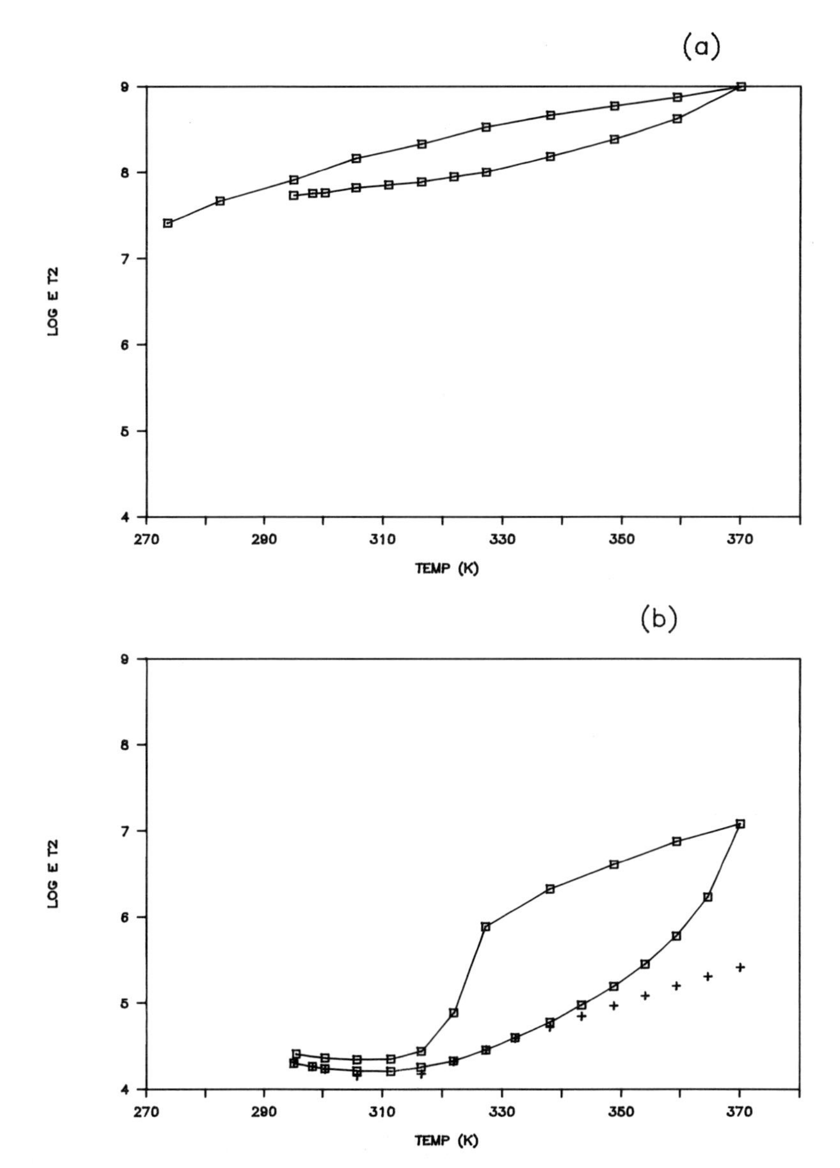

Figure 5. Variation of spin-spin relaxation time for a 1% agarose gel in the presence of; (a) 2.0 mol dm^{-3} KSCN and (b) 2.0 mol dm^{-3} KF.

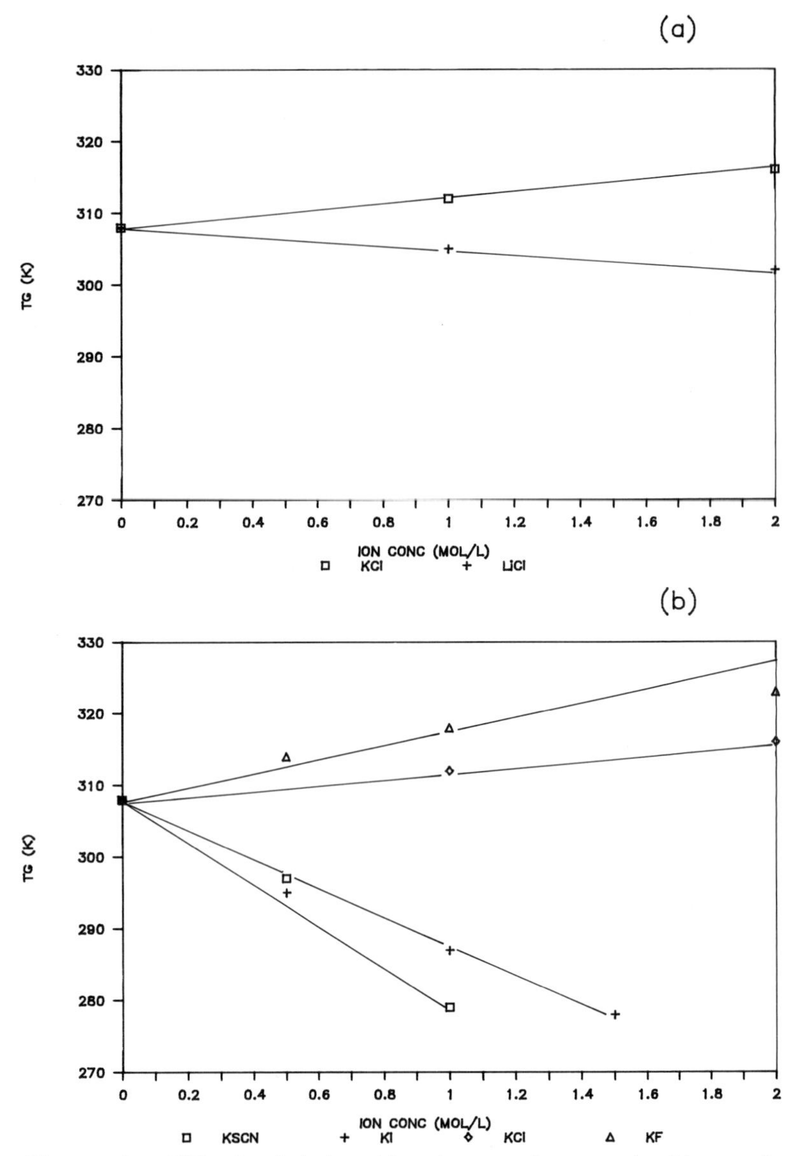

Figure 6. Effect of (a) cation type and concentration and (b) anion type and concentration on the mid-point temperature of the associative process Tg for a 1% agarose gel.

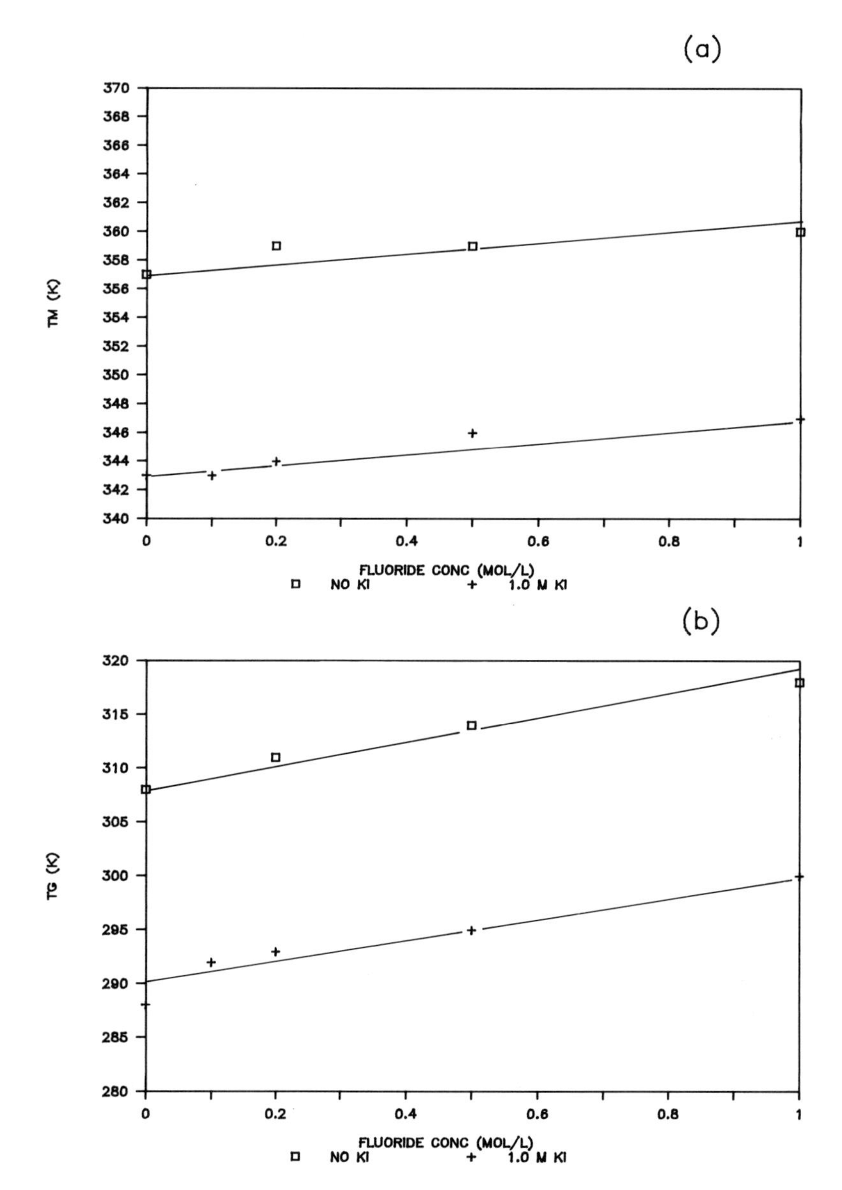

Figure 7. Variation of (a) Tm and (b) Tg with increasing fluoride ion concentration in the presence and absence of 1.0 mol dm^{-3} iodide ions.

function of the number and type of ions present.

As a further example figure 8 and shows the variation of Tm and Tg with increasing iodide level for fixed fluoride ion concentrations. The shifts in Tm and Tg appear independent of fluoride concentration. Thus for the ion mixtures investigated the values of Tm and Tg seem to be dependent on the concentration and type of ion present in solution. For the iodide/fluoride mixtures there seems to be no evidence of either ion dominating the thermal behaviour of the agarose gels.

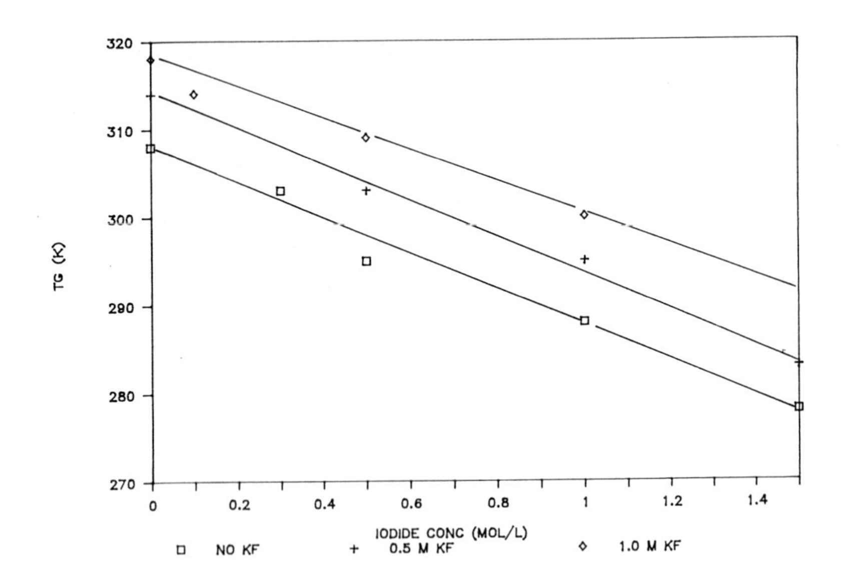

Figure 8. Variation of Tg with increasing iodide ion concentration in the presence and absence of fluoride ions.

CONCLUSIONS

(i) Cation effects are small.

(ii) Anion effects are large and consistent with the ions position in the lyotropic series.

(iii) Shifts in Tm and Tg are linear with ion concentration unless the gel fails to aggregate or begins to precipitate.

(iv) In mixed ions Tm and Tg essentially depend on the colligative effects of the ions present.

REFERENCES

1. Arnott, S., Fulmer, A., Scott, W.E., Dea, I.C.M., Moorhouse, R. and Rees, D.A. (1974) J. Mol. Biol. 90, 269.
2. Guisley, K.B. (1970) Carbohydrate Res. 13, 247.
3. Child, T.F. and Pryce, N.G. (1972) Biopolymers 11, 409.
4. Liang, J.S., Stevens, E.S., Morris, E.R. and Rees, D.A. (1979) Biopolymers 18, 327.
5. Letherby, M.R. and Young, D.A. (1981) J. Chem. Soc. Faraday Trans. 1 77, 1953.
6. Zimmerman, J.R. and Brittin, W.E. (1957) J. Phys. Chem. 61, 1328.
7. Woessner, D.E. and Zimmerman, J.R. (1963) J. Phys. Chem. 67, 1590.
8. Ablett, S., Lillford, P.J., Baghdadi, S.M.A. and Derbyshire, W. (1978) J. Colloid Interface Sci. 67, 2, 355.
9. Walker, P.M. (1988) PhD Thesis.
10. Ablett, S. and Lillford, P.J. (1977) Chem. Phys. Lett 50, 97.
11. Woessner, D.E. and Snowden, B. (1970) J. Colloid Interface Sci. 34, 283.
12. Duff, I.D. (1973) PhD Thesis.
13. Bruins, E.M. (1932) Proc. Acad. Amsterdam 35, 107.
14. McBain, J.W. (1950) Colloid Science; Heath: Boston Chapter 9.

Studies on agar from red seaweed

FASIHUDDIN B.AHMAD and HASMAH RAJI

Department of Chemistry, Universiti Kebangsaan Malaysia, Sabah Campus, Locked Bag No 62, 88996 Kota Kinabalu, Sabah, Malaysia

ABSTRACT

Physicochemical characterisation on agar samples extracted from various red seaweed were studied and their properties were compared to commercial agar. The percentage yield varied between 26 to 40% and no significant variation was observed throughout the year. The agar was found to have the following composition: galactose: 47.0 to 59.5%; 3,6 - anhydrogalactose 32.0 to 42.0%, 6-0-methyl-D-galactose 1.5 to 3.0%, sulphate 1.5 to 4.0% and glucose 1.5 to 2.6%. Alkali pretreated samples gave higher 3,6-anhydrogalactose unit and lower sulphate content. The results also show that the pretreatment with alkali also increases the gel strength and gel rigidity but decreases the viscosity. The gel strength of 1.0% agar solution extracted from *Gracilaria verrucosa* was 115 g/cm^2 and after pretreatment with alkali the gel strength increased to 200 g/cm^2.

INTRODUCTION

Agar is a complex sulphated galactan extracted from marine algae of the class Rhodophyceae (Kennedy *et al.*, 1984). It acts as a structural component in the cell walls and may also be involved in ion-exchange processes. Agar is considered to be composed of 1,3-linked- β-D-galactopyranose residues alternating with 1,4- linked 3,6-anydro-α-L-galactopyranose residues in a repeating linear sequence. The family of galactans range from the neutral polymer agarose, which may contain methoxyl groups, to a highly acidic polymer containing a substantial percentage of substituents. The structure of the D-galactose residues is variable because the existence of sulphate ester groups, methoxyl groups, 4-6-0-(l-carboxyethylidene) groups and glucuronic acid groups to certain degrees (Duckworth and Yaphe, 1971; Chapman and Chapman, 1980). The anyhdro-L-galactose residues can be replaced by L-galactose-6-sulphate or substituted by methoxyl groups (Mackie and Preston, 1974).

The major red seaweed that is being used for agar productions are from the genus *Gelidium* and *Gracilaria.* Agar is widely used in brewing, pharmaceutical, confectionery food, bacterial media, manufucture of jams and many other uses.

It has been shown that the rheological properties of agar depend upon the degree of regularity of alternation of D-galactose and 3,6-anhydro-L-galactose residues, the content and position of sulphate groups and also the molecular weight (Watase and Nishinari, 1983).

This study was performed to determine the physicochemical properties of agar extracted from various red seaweed found locally to evaluate the potential uses of the seaweed.

MATERIAL AND METHODS

Four types of red seaweeds i.e. *Gracilaria verrucosa, Gracilaria eucheumoides, Gracilaria* sp. and *Gelidium* sp. were studied. The weeds were hand picked monthly from rocks and corals during low tide from Kudat, Sabah, Malaysia from the period of January 1988 to December 1988. The seaweed collected were washed in sea water and brought to the laboratory where they were washed several times with fresh water and dried.

About 60 g of the seaweeds were extracted with 1.5 liter of water with continuous stirring for 3 hours at 130°C in an autoclave. For alkali pretreated samples about 60 g of the seaweed were placed in 0.5 - 3.0% NaOH solution made up to 1.5 liter. The resulting suspension was stirred at 75° to 80°C for one hour. After the alkali pretreatment, the algae were soaked in running water for 10 hours to wash out the excess NaOH. The algae was further extracted in 1.5 liter of water in an autoclave at 130°C for three hours.

The resulting mixture from alkali pretreated algae and without alkali pretreatment was filtered with suction through thick filter paper in a heated funnel. The filtered agar was then placed in stainless trays and allowed to gel for 24 hours. The resulting gels were first frozen at 0 to -5° for 24 hours and then at -5° to -10°C for another 24 hours. This procedure will avoid gel freezing only at the surface. The trays were allowed to thaw. This process was repeated 3 times.

During this treatment most of the soluble impurities such as sugars, pigments and salts will be removed resulting in pure agar. The thawed gel was poured into ethanol to give precipitation which was filtered and dried.

Chemical Analysis of Agar

The 3,6-anhydrogalactose residues were measured according to the method of Yaphe (1960) but Methyl 3,6-anhydro-α-D-galactose was used as a standard (Lewis *et al.*, 1963). Sulphate was determined spectrophotometrically according to Jones and Letham (1956) using 4-amino-4 -chlorodiphenyl reagent. Glucose, galactose and 6-0-methylgalactose were determined according to the method of Whyte and Englar (1980).

Physical Analysis of Agar

Intrinsic viscosity, [η] was obtained by plotting reduced viscosity η_{sp}/C versus concentration (C) and inherent viscosity lnη_{rel}/C versus C for aqueous solution in the presence of 0.01 M sodium salicylate to inhibit aggregation (Watase and Nishinari, 1983) using an Ubbelohde viscometer at 35°C.

All rheological properties of the agar were measured on 1% aqueous solutions. Gel strength, gel deformation, gel cohesion and gel rigidity were measured using an Instron Universal Testing Machine Model 1122, operating at a crosshead speed of 10 mm/min, chart speed at 50 mm/min with a 1 cm^2 stainless steel probe. The gelation and gel melting temperature were determined by the method of Whyte and Englar (1980).

RESULTS AND DISCUSSION

Table 1 gives the results on physicochemical properties of agar extracted from various algae. The percentage yield of agar depended on the type of algae used. *Gracilaria* sp. gave the highest yield about 40%. The percentage yield varies between 26 to 40%. It has been reported that yield of agar varies with species of alga, season, location and environment (Marsalolin *et al.,* 1979). The percentage yield of agar from various parts of the world varies significantly, for example Mohanty (1956) reported agar yield from 30 - 45% from *Gracilaria verrucosa* in India, in Australia, Wood (1946) reported agar yield between 27 - 33% from *Gracilaria confervoides* while Dias-Piferrer and Perez (1964) in Puerto Rico reported yield of agar from 23 - 37%. The results obtained in this study agreed well with the reported values.

Seasonal variation on yield of agar were studied and the results tabulated in Figure 1. The percentage of agar extracted from *Gracilaria verrucosa* varies from 30 to 36%, while for *Gelidium* sp. it varies from 26 to 32%. No significant seasonal variation in the yield was observed for all the samples studied. Slightly lower values were recorded from August to October probably due to lower sea temperature and short sunshine period during the rainy season.

The 3,6 anhydrogalactose content varies from 26 - 33%, while alkali pretreated algae gave higher values (Table 1). The results obtained in this study agreed well with the reported values (Duckworth,, *et al.*, 1971 and Whyte and Englar, 1981).

Results in Table 1 clearly show that pretreatment with alkali decrease the sulphate content and effect the gelation temperature. Anomalies to the basic agarose molecules (Rees, 1972) occur when a portion of the L-galactopyranose residues is substituted with 6-0-sulphate ester in place of 3,6-anhydrogalactose bridges. Pretreatment of algae with alkali will enhance the formation of the anhydro residues by the elimination of sulphate ester groups. This will result in attaining more closely the functional backbone of the agarose polymer (Rees, 1969). Pretreatment with alkali effected the 3,6-anhydrogalactose more on agar samples extracted from *Gracilaria* spp. compared to *Gelidium* sp. Therefore to obtain better quality agar from *Gracilaria* spp, it is necessary to treat the algae with alkali prior to the normal extraction.

All the algae samples pretreated with alkali gave higher values of gelation temperature and this corresponds well with the increase of 6-0-methylgalactose (Table 1). The results obtained agree with the observation reported by Akabori (1926) and Guisley (1970) that increasing methoxyl content resulted in increasing gelation temperature. Gel melting temperature is considered to be dependent on the molecular weight of the agar (Bruderer and Brossi, 1965; and Selby and Wynne, 1973). The differential in the gel melting temperature from the samples studied is consistent with the difference in the viscosity (Table 2).

Table 1 : Physicochemical properties of agar from various spesies.

Species of algae	*Gelidium* sp.	*Gracilaria verrucosa*	*Gracilaria eucheumoides*	*Gracilaria* sp.
Parameters				
Agar in algae, %	26	36	30	40
3,6-anhydrogalactose, %	30(36)	33(42)	26(32)	30(36)
sulphate, %	3.8(1.7)	3.5(1.5)	4.0(2.0)	3.9(2.5)
6-0-methylgalactose, %	1.8(2.8)	2.0(3.0)	1.5(2.0)	1.6(2.5)
galactose, %	55.0(53.0)	59.5(47.0)	58.2(54.5)	56.0(51.5)
gulucose, %	1.5	1.0	2.6	1.6
gelation temperature, 1% agar, °C.	37.2(39.9)	40.0(39.0)	35.0(37.0)	37.5(39.5)
gel melting temperature, 1% agar, °C	88.0(83.3)	82.5(79.0)	80.0(78.0)	89.0(85.0)

note : values in parentheses correspond to value for algae pretreated with alkali.

Table 2 : Physical properties of 1% agar solution

Species of algae used	*Gelidium* sp.	*Gracilaria verrucosa*	*Gracilaria eucheumoides*	*Gracilaria* sp
Properties				
Intrinsic viscosity dl/g	4.5(4.0)	4.2(3.8)	4.5(4.0)	5.0(4.3)
gel strength, g/cm^2	138(159)	115(200)	100(144)	110(171)
gel deformation, mm	7.2(3.9)	8.9(3.9)	4.6(4.2)	7.8(4.2)
gel cohesion, mm	4.9(2.7)	5.8(2.2)	3.6(2.5)	4.5(2.8)
gel rigidity, $g/cm^2/mm$	13.0(17.5)	14.1(20.5)	4.9(12.2)	12.0(16.2)

note : Values in parentheses correspond to value for algae pretreated with alkali.

Table 2 gives the intrinsic viscosity of the agar samples and rheological parameters. The intrinsic vicosity varies between 4.2 to 5.0 dl/g and the agar from alkali pretreated algae gave slightly lower viscosity due to alkaline depolymerisation.

The gel strength which is defined as the maximum load required to rupture the gel matrix (g/cm^2) and rigidity which is defined as the load per unit distance required to deform the surface of the gel was calculated from the slope of the first half of the load-deformation curve which provide a measure of the firmness or stiffness of the gel network (g/cm^2/mm) dependent to the type of algae used in the extraction of agar (Table 2). Pretreatment of the algae with alkali greatly increase the gel strength and also the regidity. The increase in the gel strength corresponds to the increase in 3,6-anhydrogalactose residues and decrease in sulphate content (Table 1). The 3,6-anhydrogalactose component present gave a measurement of the gel strength. The higher the percentage of 3,6-anhydrogalactose the higher is the gel strength.

CONCLUSION

The percentage yield of agar from the species studied is high with no significant seasonal variation on yield being observed throughout the year. The gel strength of the samples studied is comparable to the algae being used industrially. The pretreatment by alkali increased the gel strength and rigidity but the concentration of alkali should be kept below 2% to obtain high gel strength.

ACKNOWLEDGMENT

The authors wish to acknowledge MPKSN for the R&D grant No. 1-07-03-11, UKM, Mr. Mohd. Tahir Abd. Majid for revising the manuscript and Ms. Gomera Jumat for typing this manuscript.

REFERENCES

Akabori, S. (1926), Bull. Chem. Soc. Japan, 1, 125
Arnot, S., Fulmer, A. and Scott, W.E. (1974) J. Mol. Biol. *90,* 269-284.
Bruderer, H. and Brossi, A. (1965). Helv. Chim. Acta, *48,* 1945.
Chapman, V.J. and Chapman, D.J. (1980). Seaweed and their Uses. Chapman and Hall, London. pp. 148-193.
Dias-Piferrer, M. and Perez, C.C. (1964). Inst. Biol. Marina, C.A.A.M. Univ. Puerto Rico, 1-45.
Duckworth, M., Hong, K.C. and Yaphe, W. (1971). Carbohydr. Res. *18,* 1-19
Duckworth, M. and Yaphe, W. (1971) Carbohydr. Res. *16,* 189-197.
Glicksman, M. (1979). In Blanshard, J. M.V. and Mitchell, J.R. (Eds.) Polysaccharides in Food. Butterworths. pp. 185-204.
Johnes, A.S. and Letham, D.S. (1956) Analyst. *81,* 15-18.
Kennedy, J.F., Griffiths, A.J. and Atkins, D.P. (1984). In Phillips, G.O., Wedlock, D.J. and Williams, P.A. (Eds.) Gums and Stabilisers for the Food Industry 2. Pergamon Press, Oxford, pp. 420-422.
Lewis, B.A., Smith F. and Stephens, A.M. (1963). Methods in Carbohydr. Chem. 2, 172-188.
Mackie, W. and Preston, R.D. (1974). In Stewart, W.D.P (Ed.) Algal Physiology and Biochemistry. Blackwell Scientific, Oxford, p. 40.
Marsaioli, A.J., Reis F de A.M, Magalhaes, A.F., Ruveda E.A. and Kuck, A.M. (1979). Phytochemistry *18,* 165
Mohanty, G.B. (1955). J. Sci. Indust. Res. *14*(A), 36.
Rees, D.A. (1969). Adv. In. Carbohydr. Chem. & Biochem. *24,* 267-332.
Rees, D.A. (1972). Polysaccharide gels : a molecular view. Chem. and Ind. (London) pp. 636.
Watase, M. and Nishinari, K. (1983). Rheologica Acta *22,* 580-587.
Whyte, J.N.C. and Englar, J.R. (1980). Botanica Marina *23,* 277-283.

Whyte, J.N.C. and Englar, J.R. (1981a). Phytochemistry *20,* 237-240
Whyte, J.N.C. and Englar, J.R. (1981b) In. Levring, T. (Ed.) X[th] Int. Seaweed Symposium. Walter de Gruyter, Berlin, pp. 537.
Wood, E.J.F. (1943). Coun. Sci. Ind. Res. Bull. *203,* 1-43
Yaphe, W. (1960). Anal. Chem. *32,* 1327.

Influence of composition on the resistance to compression of kappa carrageenan-locust bean gum–guar gum mixed gels. Relations between instrumental and sensorial measurements

M.H.DAMASIO, S.M.FISZMAN, E.COSTELL and L.DURÁN

Instituto de Agroquímica y Tecnología de Alimentos, Jaime Roig 11, 46010 Valencia, Spain

ABSTRACT

Mechanical behaviour of kappa-carrageenan (C) - locust bean gum (LBG) gels and C-LBG-guar gum (GG) gels was studied by compression tests on an Instron Texturometer and by non-oral sensory evaluation of resistance to compression between fingers. Gum proportions varied from 10 to 70% and when both gums were used their ratio was 1:1. Two series of both C-LBG and C-LBG-GG gels were prepared with 0.5 and 0.75% total hydrocolloid concentrations in each. Effects of type of gum and of gum proportion on the different compression parameters (deformability moduli at two deformation ranges: 10-20 and 20-30 %, maximum rupture force, deformation to rupture and energy) and their interactions were significant at $p \leq 0.05$. The same effects were also significant on some non-oral sensory parameters. Regression equations were selected that satisfactorily related sensory to instrumental parameters.

INTRODUCTION

Mixtures of kappa-carrageenan and seed gums, mainly locust bean and guar gums, are widely used in industry for their convenient functional properties (1). Any of these gums produce synergistic effects in carrageenan gels. In general, they may enhance mechanical resistance (2,3,4) and decrease liquid loss through syneresis of gels (5). As to their sensorial properties, locust bean gum has been reported to render less brittle and more elastic gels when mixed with kappa-carrageenan (6). However, little information is available on the quantitative effect of small changes in composition of these mixed gels on their mechanical and sensorial characteristics.

In this paper, the effects of adding locust bean gum or a mixture of this with guar gum to kappa-carrageenan gels on their response to mechanical compression and on their sensory non-oral texture attributes are studied. Finally, relationships between instrumental and sensorial measurements are also reported.

MATERIALS AND METHODS

Samples composition and preparation

Gel samples were prepared at two total hydrocolloid concentration (HC): 0.5 and 0.75%, each of them of three different compositions: 1) only kappa-carrageenan (C), 2) mixtures of C plus locust bean gum (LBG) and 3) mixtures of C:LBG: guar gum (GG). Proportions in final formulation of either LBG or the mixture LBG+GG (1:1) were: 10, 20, 30, 40, 50, 60 and 70% of HC. Gels were prepared following a previously described procedure (3), cut in 17 x 17 mm (7) and measured at room temperature.

Instrumental measurements

4 cylindrical probes from each sample were compressed to rupture in an Instron Testing Machine model 6021 at 50 mm/min cross-head speed. Five parameters were recorded: maximum rupture force, in N, (Fmax), deformability moduli, in N/mm^2, between 10-20% (Eap_1) and 20-30 % deformation (Eap_2), deformation up to rupture, in % (Def) and energy, in J, (En).

Sensory evaluation

Cylindrical probes of gels were compressed to rupture between two fingers by 7 trained judges. 4 parameters were evaluated: deformation before rupture (DBR), firmness (FIR), resistance to rupture (RR) and type of rupture (TR) using 10 cm semistructured scales.

RESULTS AND DISCUSSION

Influence of composition on compression parameters

Two-way ANOVA analyses of data applied separately to 0.5 and 0.75% HC sets of samples showed that both effects -type of gum and percent gum substitution- and their interaction on the 5 compression parameters were all significant at $p \leq 0.05$ (table I).

In general, substitution of LBG for C was more effective than when using both gums. Considering all parameters jointly, it can be observed that percent substitution between 40 and 60 % produced maximum effects (figure 1). It is interesting to note that over 20 % substitution, LBG in 0.5 % HC samples produced more resistant to rupture gels than 0.75% HC samples substituted with the mixture LBG-GG.

Influence of composition on sensorial parameters

Three-way ANOVA analyses were applied to sensory data for samples with 0, 20, 40 and 60 % gum substitution both at 0.5 and 0.75% HC. Effects of gum type and percent gum substitution, as well as their interaction, on the 4 parameters evaluated were, in general, significant at $p \leq 0.05$ (table II). Interactions between panelists and each of the other two effects were mostly not significant. The panel was able to differentiate between types of gum by using any of the sensory parameters, but was less efficient in distinguishing between gum substitution percentages. In general, LBG substituted gels were found to be firmer and more resistant to rupture than those with the mixture LBG-GG, and higher values were given to 40 to 60% substituted samples.

Sensory-instrumental relationships

By regression analysis the main relationships found between sensory and instrumental parameters are represented by the following equations:

$$DBR = -5.803 + 0.2283\ Def \qquad (r^2 = 0.87)$$

$$FIR = -2.417 + 0.8380\ Eap_2 \qquad (r^2 = 0.76)$$

$$RR = 0.1505 + 0.3808\ Fmax \qquad (r^2 = 0.92)$$

$$TR = 15.786 - 0.2243\ Def \qquad (r^2 = 0.52)$$

It is interesting to observe that Fmax values represent 92% of the variability of the sensorially measured resistance to rupture and Eap_2 values account for 76 % of the variability of firmness.

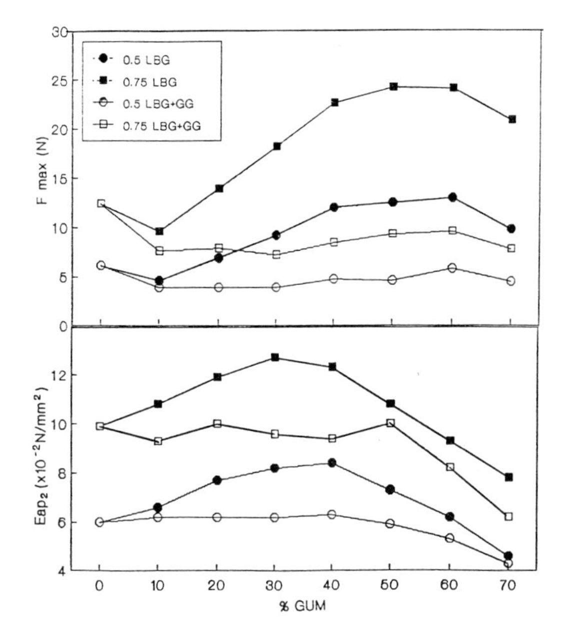

Figure 1. MAXIMUM RUPTURE FORCE (Fmax) AND DEFORMABILITY MODULUS BETWEEN 20 AND 30 % DEFORMATION (Eap_2) VALUES AS A FUNCTION OF % GUM SUBSTITUTION.

Table I. STATISTICAL F VALUES CORRESPONDING TO COMPRESSION PARAMETERS: MAXIMUM RUPTURE FORCE (Fmax), DEFORMABILITY MODULI (Eap_1 and Eap_2), DEFORMATION (DEF) AND ENERGY (EN).

EFFECT	TOTAL HYDRO-COL. CONC.	Fmax	Eap_1	Eap_2	DEF	EN
GUM TYPE		9006	678	549	4430	9866
GUM %	0.5 %	645	300	214	823	1026
TYPE X %		497	38	34	223	579
GUM TYPE		6333	1953	2688	3368	6290
GUM %	0.75 %	296	585	1028	508	464
TYPE X %		316	108	135	165	322

All F values are significant at $p \leq 0.05$.

Table II. STATISTICAL F VALUES CORRESPONDING TO SENSORIAL PARAMETERS: DEFORMATION BEFORE RUPTURE (DBR), FIRMNESS (FIR), RESISTANCE TO RUPTURE (RR) AND TYPE OF RUPTURE (TR).

	0.5 % TOTAL HYDROC. CONC.				0.75 % TOTAL HYDROC. CONC.			
EFFECTS:	DBR	FIR	RR	TR	DBR	FIR	RR	TR
GUM TYPE	29.44*	16.46*	38.95*	9.38*	27.79*	76.51*	63.34*	5.20*
GUM %	25.06*	4.69*	10.28*	38.84*	8.68*	0.54	1.14	69.99*
PANELIST	12.78*	12.70*	4.25*	5.79*	12.76*	4.51*	1.77	5.18*
TYPE X GUM %	7.69*	7.74*	7.93*	1.44	5.59*	9.80*	8.51*	0.87
TYPE X PAN.	0.37	1.07	0.97	0.62	0.61	1.44	0.32	0.98
GUM % X PAN.	3.24*	1.58	2.19	1.24*	3.15	1.82	1.29*	2.32

* Significant at $p \leq 0.05$

ACKNOWLEDGEMENTS

The authors are indebted to the CICYT for financial support (project ALI 88-0238), to the Ministerio de Educación y Ciencia of Spain for the fellowship awarded to author Fiszman and to CAPES of Brazil to author Damasio.

REFERENCES

1. Carrol V., Miles M.J. and Morris V.I.(1984) in Gums and Stabilisers for the Food Industry 2. (ed. G.O. Phillips, D.J. Wedlock and P.A. Williams) pp 501-506. Pergamon, Oxford, U.K.
2. Christensen O. and Trudsøe J. (1980). J. Texture Stud. 11, 137-147.
3. Fiszman S.M., Baidón S., Costell E. and Durán L. (1987). Rev. Agroquim. Tecnol. Aliment. 27, 519-529.
4. Cairns P., Morris V.J., Miles M.J. and Brownsey, G.J. (1986). Food Hydrocolloids, 1, 89-93.
5. Baidón S., Fiszman S.M., Costell E. and Durán L. (1987). Rev. Agroquim. Tecnol. Aliment. 27, 545-555.
6. Oakenfull D.G. (1984).CSIRO Food Research Quarterly 44(3), 49-55.
7. Durán L., Costell E. and Fiszman S.M. (1987) in Physical Properties of Food 2- (ed. Jowitt R, Escher, F, Kent, M.; McKenna, B and Roques, M.) pp 429-443. Elsevier Applied Science, Essex. U.K.

Mechanical behaviour of fruit gels. Effect of fruit content on the compression response of kappa carrageenan-locust bean gum mixed gels with sucrose

S.M.FISZMAN and L.DURÁN

Instituto de Agroquímica y Tecnología de Alimentos (CSIC), Jaime Roig 11, 46010 Valencia, Spain

ABSTRACT

Different levels (10 to 40%)of peach pulp (P) (11ºBrix) were incorporated into kappa carrageenan (C)-locust bean gum (LBG) mixed gels at several concentrations (0.5, 0.75 and 1% C; 0, 0.1, 0.2 and 0.3% LBG). 55% sucrose (SUC) was added to all gels. Fruit gels were weaker than the corresponding water gels. Increasing proportions of P produced, in general, decreases on the compression parameters measured. Syneresis, measured as the extent of exudate diffusion into filter paper, was partially inhibited by the addition of fruit pulp. No synergistic effect of LBG with C was observed in the studied composition range. The obtained results may be of practical interest in the formulation of commercial fruit gelled products of similar composition.

INTRODUCTION

Gelled sweet fruit products are mainly represented by traditional jams and jellies, though more recently products such as fruit bars, fabricated fruit pieces or pie fillings are also formulated with fruit, sugar and adequate combination of polysaccharides to get the desired texture, among other functional and sensory characteristics.

A great deal of information is available on the mechanical, rheological properties of C gels and on the synergistic effect of LBG or other seed gums (1,2,3,4).

Less information can be found on the effect of sugars on this type of gel (5,6). No data is readily available on the effect of fruit content on C-LBG gels with added SUC. The objective of this work is to study some mechanical properties of these mixed gels of similar composition to commercial products paying attention to the effects produced by addition of peach pulp.

MATERIALS AND METHODS

Sample composition and preparation

48 samples were prepared containing 3 levels of C (0.5, 0.75, 1 %), 4 levels of LBG (0, 0.1, 0.2, 0.3 %) and 4 levels of

P (10, 20, 30, 40 %). 0.5 % KCl was added to every sample. Gels were held 24 h. at 4-6ºC, cut in 17x17 mm cylinders (7) conditioned at room temperature and measured.

Instrumental measurements

3 cylinders of each sample were compressed in an Instron Universal Testing Machine model 6021 using a 50 mm diameter plunger at a cross-head speed of 50 mm/min. Tests were conducted up to rupture of gels. Two parameters were registered: maximum rupture force (Fmax) at the break point of the gels, expressed in N, and deformability or apparent Young´s modulus (Eap) measured between points corresponding to 10 and 20% deformation and expressed in N/m^2.

Syneresis index

The extent of the exudate diffusion on a filter paper after 2 hours was taken as syneresis index. Dried (105ºC, 24 h) Whatman nº 1 filter paper was used and results expressed as the difference between the diameter of the wet area and the probe diameter (ΔD) in cm. Two replicates of each sample were tested.

RESULTS AND DISCUSSION

Analyses of variance were applied to data obtained on compression parameters -Fmax and Eap- for gels with different proportions of C, LBG and P.55% SUC was added to all gels.

Effect of fruit pulp and gum contents on compression parameters. Maximum rupture force.

Addition of P produced a clear weakening effect on C gels: Fmax values for 0.75 % C gels with 10 to 40 % pulp with 55 % SUC were 2-4 N, while without pulp we previously obtained values around 20 N (for 0.75 % C with 60 % SUC) (8).

Increasing P in the studied range produced a decrease in Fmax values, this effect being clearly shown for 1 % C gels but only negligible for 0.5 % C gels, as a result of the significant P-C interaction (figure 1A, table I).

C-LBG interaction was also significant but more difficult to explain. Fmax values for gels without LBG were intermediate between those for gels with different LBG content, at all C concentrations (figure 1). As previously reported by the authors (8), presence of SUC in concentrations over 50 % might inhibit the synergistic effect of LBG added to C gels. Moreover, around 60 % SUC, this gum may impair the gelling capacity of C (table I).

Deformability modulus.

A number of 0.3 % LBG gels broke before attaining 20 % deformation. Then, Eap data for all 0.3 % LBG samples were not included in the analysis of variance.

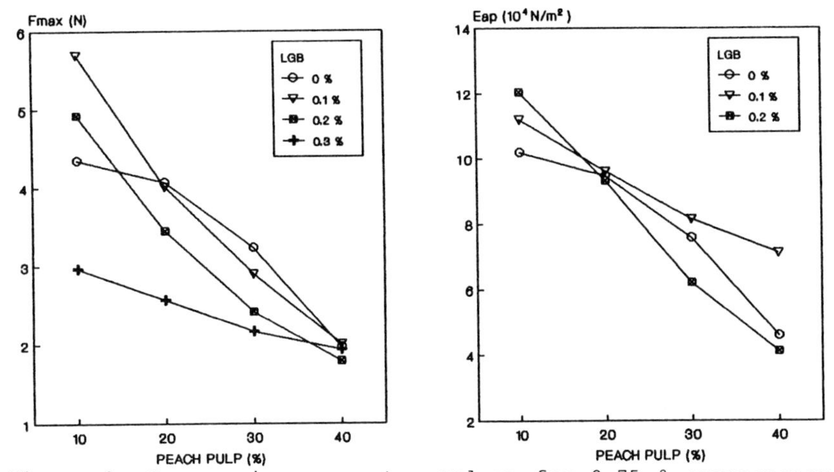

Figure 1. Compression parameters values for 0.75 % carrageenan gels as a function of peach pulp concentration at different levels of locust bean gum. A: Fmax and B: Eap.

Table I. STATISTICAL F VALUES FOR MAXIMUM RUPTURE FORCE (Fmax), DEFORMABILITY MODULUS (Eap) AND SYNERESIS (S).

Effects:	Fmax		Eap		S	
c x p x g	7.84	(2.19)	11.78	(2.47)	18.71	(2.42)
p x g	6.71	(2.64)	20.92	(3.09)	10.63	(2.83)
c x g	26.06	(3.02)	21.55	(3.62)	6.28	(3.22)
c x p	74.48	(3.02)	53.38	(3.09)	19.14	(3.22)
g	19.27	(4.02)	10.35	(4.94)	3.80	(4.24)
p	379.66	(4.02)	310.18	(4.09)	898.69	(4.24)
c	1023.46	(4.87)	1468.76	(4.94)	480.58	(5.09)
LSD values	1.38		2.22		0.36	

F tables in parentheses ($p \leq 0.01$)

All interactions and single factors studied were significant (table I). Statistical F values of single effects were highest for C concentration, followed by P content, the effect of LBG proportion being lower. Interpreting the interactions between them, it can be said that firmness of the studied gels increased with C concentration and decreased with P, this effect being negligible for 0.5 % C and very clear for 1 % C samples; LBG effect, though significant, was erratic and comparatively small, in agreement with the fact that a high level of SUC acts as a hindrance to the LBG action (figure 1B).

Effect of fruit pulp and gum content on gels syneresis

It was found that 1) all interactions were significant (table I) within the studied composition range. 2) P and C single effects were highly significant whereas LBG effect was not significant and 3) the lowest values corresponded to 1 % C gels with 40 % pulp content (with 55 % SUC) and with or without LBG.

The recognised reducing effect of LBG addition on gel syneresis was not clearly shown and either slightly increased or decreased particular values, as compared with samples without LBG.

From these results and considering the magnitude of absolute values, it should be pointed out that besides the expected favourable repressing effect of C concentration, pulp content effect is also negative and may be expected to have commercial importance.

ACKNOWLEDGEMENTS

To CICyT (Spain) for financial support of project ALI88-0238 and to the Ministerio de Educación y Ciencia (Spain) for the fellowship granted to author FISZMAN.

REFERENCES

1. Cairns, P., Morris, V.J., Miles, M.J. and Brownsey, G.J. (1986). Food Hydrocolloids 1, 89-93.
2. Fiszman, S.M., Baidón, S., Costell, E. and Durán, L. (1987). Rev. Agroquím. Tecnol. Aliment. 27, 519-529.
3. Christensen, O. and Trudsoe, J. (1980). J. Texture Stud. 11, 137-147.
4. Ainsworth, P.A. and Blanshard, J.M.V. (1980). J. Texture Stud. 11, 149-162.
5. Gerdes, D.L., Burns, E.E. and Harrow, L.S. (1987). Lebensm.-Wiss.u. Technol. 20, 282-286.
6. Rey, D.K. and Labuza, T.P. (1981). J. Food Sci. 46, 786-789, 793.
7. Durán, L., Costell, E. and Fiszman, S.M. (1987) in Physical Properties of Foods-2. (ed. Jowitt, R., Escher, F., McKenna, B. and Roques, M.) pp. 429-443. Elsevier Applied Science Publishers LTD, Essex, U.K.
8. S.M. Fiszman and L. Durán (1989). Food Hydrocolloids, in press.

Viscoelastic properties of physical gels: critical behaviour at the gel point

G.CUVELIER, C.PEIGNEY-NOURY and B.LAUNAY

E.N.S.I.A., 1, av. des Olympiades, 91305 Massy cedex, France

ABSTRACT

The viscoelastic properties of iota-carrageenan and xanthan/carob mixtures in water are studied by dynamic measurements. Both systems form thermoreversible gels and their complex modulus is determined over a 4 decades frequency range for the sol and gel states and during the sol-gel transition by stepwise increase or decrease of the temperature of measurement.
At critical conditions (temperature and/or time), a typical spectrum is observed : the storage (G') and loss (G") moduli are equal over the whole range of frequency and follow a power law, $\omega^{\sim 0.5}$. These observations are in agreement with those obtained by Winter and Chambon on a chemically gelling system and with their theoretical approach associating this rheological behaviour with the percolation threshold.
Therefore, we propose to extend the use of this rheological criterion to define incipient gelling or melting points of physical gels. this very sensitive criterion can be used to determine critical conditions of gelation. In any case, the approach of the percolation threshold and the structuration of the gel are dynamic phenomena that can be studied in a sensible way by the proposed rheological method.

INTRODUCTION

Many types of gelling systems are used in the food area. Gelling agents and mechanisms of gelation are varied : polysaccharides, proteins (globular or micellar), single or multicomponent, reversible or not, but all of them are physical gels, structured by weak bonds (hydrogen, electrostatic, hydrophobic)(1-4). The detection of the gel point and the study of the gelation conditions are of a great practical interest, but are also rather difficult for these thermodynamically unstable systems. Mechanical properties are commonly used to characterize the gels and follow their change beyond the gel point : measurements of Young's modulus (small amplitude compression test), of gel force (failure test), of the elastic component G' of the complex modulus G* (dynamic masurements) (5), measurements being made at a fixed frequency (5,6).
The aim of this work is to examine how the viscoelastic behaviour evolves on a large frequency range in the vicinity of the gel point. Two types of thermoreversible physical gels are studied : iota-carrageenan and a mixed system, xanthan/carob.

MATERIALS AND METHODS

The samples of iota-carregeenan, carob and xanthan were kindly supplied by Iranex (Rouen, France), Meyhall Chemical (Kreuzlingen, CH) and Rhône-Poulenc (France), respectively.

1% iota carrageenan (IC) solution is prepared by dispersing the gum in cold water and heating at 85°C for 15 min. 0.5 % xanthan (X) and carob (C) solutions were prepared separately by dispersing the gums in 0.1M NaCl, stirring at room temperature for 30 min, then at 80°C for 1h. The X/C mixture (50/50) was then prepared by stirring 30 min at 80°C. 500ppm of NaN_3 were added to all the solutions as antimicrobial agent.

Dynamic measurements were performed using a Rheometrics Fluids Rheometer fitted with coaxial cylinders (R1/R2=0.96). The solutions were put in the rheometer at about 70°C and then thermostated at the required temperature.

The linearity of viscoelastic behaviour was checked at different frequencies (0.1, 1, 10, 50 rad/s) at 25°C and 36°C for IC and at 25°C and 52°C for X/C.

A strain amplitude of 8% was retained for frequency sweep on IC and 3% for X/C, both in the linear domain.

RESULTS

For the two systems, the mechanical spectra obtained at 25°C are typical of structured gels, with clearly defined elastic properties, the storage modulus G' being almost independent of the frequency. As the temperature was increased stepwise, we monitored the changes in viscoelastic properties, until a behaviour typical of a macromolecular solution, corresponding to a melted state, was observed (Fig. 1).

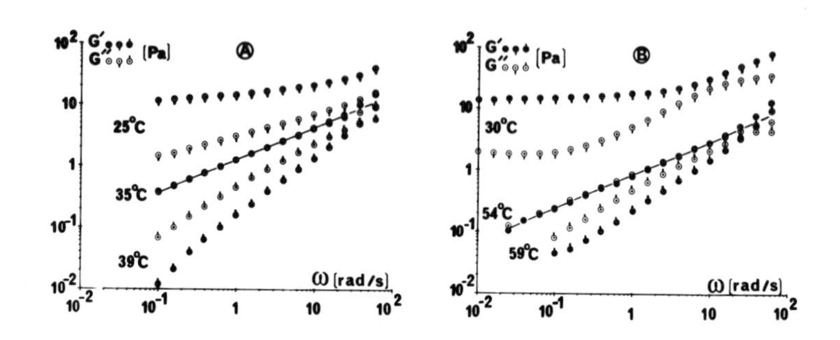

Figure 1. Changes in storage and loss moduli versus frequency curves when the temperature is increased. A : 1% IC in water (strain 8%); B : 0.5% X/C (50/50) in 0.1M NaCl (strain 3%).

At a critical temperature, about 35°C for IC (Fig. 1A) and 54°C for X/C (Fig. 1B) a characteristic spectrum is observed : the values of G'(ω) and G"(ω) become almost equal over a whole range of frequency. Moreover, at this point the dependence of the modulus on the frequency on log scale is linear with a slope of about 0.5. WINTER & CHAMBON (7) have studied in the same way the viscoelastic properties of a chemically gelling system (PDMS) for which the crosslinking reaction has been stopped at different degrees of reaction, and they observed the same behaviour at the gel point. The authors have proposed a theoretical approach to explain the power-law dependency. In more recent work (8,9) they have shown that the frequency independence of tan = G"/G' is the most generally valid gel point criterion. The experimental value of the power-law exponent is 0.5 only in stoichiometric conditions for PDMS. In the case of single component gelling systems like carrageenan, stoichiometry has no significance and even for mixed system stoichiometric conditions are very difficult to define. It is also important to consider that physical gels are never in a true equilibrium state (10-13). In given conditions, for example when the temperature is increased for IC (Fig. 2, 36°C) interactions are melted and the gel point is just reached but the system can evolve and new interactions can be formed at constant temperature, G' becoming greater than G"(Fig.2 & 3). In that case, it must be considered that the critical temperature of gelation was not reached, because the viscoelastic spectrum was modified on standing (Fig.3). For the same gel, heated at 37°C (Fig. 2), G" is only just greater than G' : the melting temperature is probably slightly lower, but the system seems to be at equilibrium.

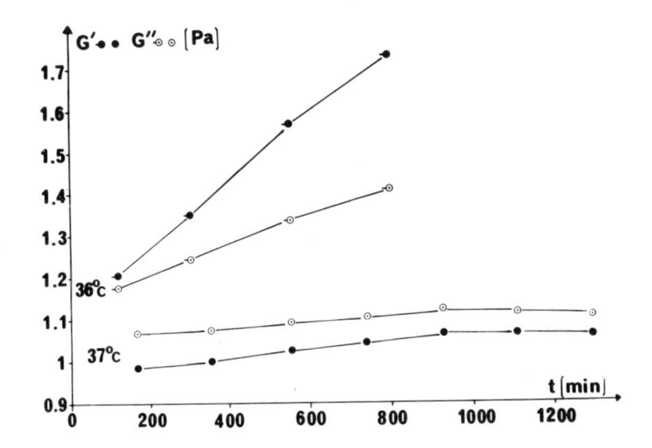

Figure 2. Effect of standing time on G' and G" (ω=1 rad/s) after heating of a gel of 1% IC at 25°C up to 36°C or 37°C.

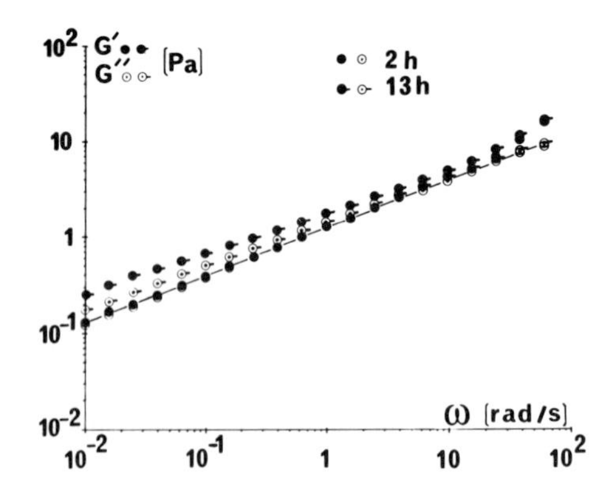

Figure 3. Storage and loss moduli versus frequency after 2h and 13h standing time at 36°C, 1% IC (see Fig.2).

In conclusion, we propose to apply to physical gels the rheological criterion G'(ω)=G"(ω) to determine the percolation threshold, whether in a transitory state or in equilibrium conditions. This criterion has been used to show that the gelation temperature of X/C (50/50, total concentration 0.5% in NaCl 0.1M) is 53°C at equilibrium (11) and to establish the critical temperature/concentration conditions for the gelation of IC in water at two melting and setting rates (13).

REFERENCES

1. Oakenfull D. (1987) Crit. Rev. Fd Sci. Nut., 26 (1), 1-25.
2. Morris V.J. (1986) *in* Functionnal Properties of Macromolecules, Mitchell J.R. & Ledward D.A. eds, Elsevier Appl. Sci. Pub., London, 121-170.
3. Ledward D.A. (1986) *in ref 2*, 171-202.
4. Clark A.H. & Lee-Tuffnell C.D. (1986) *in ref 2*, 203-272.
5. Clark A.H. & Ross-Murphy (1987). Adv. in Polym. Sci. 83, Springer Verlag, Berlin, 57-192.
6. te Nijenhuis K. (1981) Colloid Polym. Sci., 259, 522-535.
7. Winter H. H. & Chambon F. (1986) J. Rheol., 30 (2), 367-382.
8. Winter H. H. (1987) Polym. Eng. Sci., 27 (22), 1698-1702.
9. Chambon F. & Winter H. H. (1987) J. Rheol. 31 (8), 683-697.
10. Cuvelier G. & Launay B. (1986) *in* Gums and Stabilisers for the Food Industry-3, Phillips G.O., Wedlock D.J. & Williams P.A. eds, Elsevier Appl. Sci. Pub., London, 147-158.
11. Cuvelier G. (1988) Thesis, Univ. Paris XI-ENSIA
12. Cuvelier G. & Launay B. (1988) Carbohyd. Polym., 8, 271-284.
13. Peigney C. (1987) Thesis, Univ. ParisVII-XI-ENSIA.

Applications of alginates

A.IMESON
Red Carnation Gums, Sir John Lyon House, 5 High Timber Street, Upper Thames Street, London EC4V 3PA, UK

ABSTRACT

Commercial applications for alginates in foods are based on the interaction between sodium alginate and cations to generate or modify specific food rheology. To prepare thermally-stable products, calcium ions must be introduced in a controlled fashion to form the ordered junction zones needed for stable gel networks. Suitable calcium salts, sequestrants and acids are used to regulate the gelling reaction. Processes have been developed to produce a wide range of foods including encapsulated products, fruit pieces and fillings, structured vegetables and meat products. In liquid foods, interactions are regulated further by esterifying the alginate. The emulsification properties, flow characteristics and acid stability are used for dressings and sauces and in stabilising fruit drinks and beer foam.

INTRODUCTION

Alginate, or algin in the United States, is the term used to describe the various salts of the structural polymer of practically pure uronic acids which are isolated from the cell walls of brown algae (Phaeophycae). A process for extracting alginate from seaweed was patented by Stanford in 1881(1), but it was not until 1929 that commercial production commenced by Kelco in the USA. Other major alginate companies were established in the UK in 1934 and in Norway in 1939. Despite this lengthy period for investigating alginate properties and developing applications, the material continues to generate interest and find new uses. Currently, annual worldwide demand for alginates is estimated at 22-24000 tonnes and further growth is anticipated. About half of the present sales are in food, petfood and pharmaceutical applications.

The monovalent salts of alginic acid, including sodium, potassium and ammonium, are soluble in cold and hot water. Calcium, aluminium and other polyvalent ion alginates, with the exception of magnesium, are insoluble. Thus the controlled introduction of polyvalent ions into an alginate solution is able to form viscous solutions, thixotropic gels, thermally-stable gels or precipitates depending on the concentration and rate of mixing of the components. The principal commercial applications for alginates in food utilise the interaction between sodium alginate and calcium ions. Other ionic materials, such as protein, may participate to modify or generate particular food rheology.

The benefits of alginate gels to the consumer and food processor are that:

a) the crosslinks enable the gel to be formed at any temperature - for energy efficiency this is generally carried out in the cold,

b) the setting time may be adjusted from a few seconds to many minutes,

c) the gel will maintain its shape throughout thermal processing including all types of cooking and retorting.

Practical application of alginate gelation requires an understanding of the processes involved and how these may be manipulated.

ALGINATE GELATION

The mannuronic and guluronic acids in sodium alginate occur in three different segments within the linear co-polymer. These are regions of almost exclusively polymannuronic acid or polyguluronic acid and segments where the uronic acids are distributed in a near random to near alternating sequence (2). The buckled conformation of the polyguluronic acid segments permits co-ordination with calcium and this ion is able to crosslink adjacent alginate chains. As more calcium is introduced, co-operative association involving sequences of over

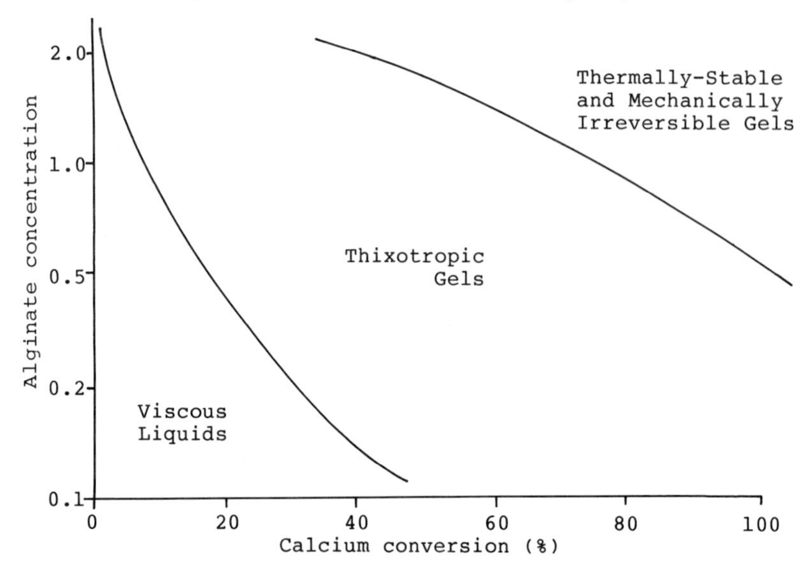

Fig. 1. Effect of calcium conversion on alginate rheology (4)

20 guluronic acid residues on adjacent polymers produces chain dimerisation which extends to form a three-dimensional network (3).

In food systems containing alginate, low levels of calcium give temporary associations between chains seen as highly viscous or thixotropic solutions. At higher calcium levels all the polyguluronate sites may be occupied producing a strong bond which is mechanically and thermally irreversible. The half stoichiometric ratio of calcium to sodium alginate is described as the degree of calcium conversion: approximately 15-30% conversion will give thixotropic gels and 60-100% conversion, stable gels as shown in Figure 1 (4).

The rate of gel formation and the time to reach equilibrium are affected by alginate type and concentration, calcium solubility, acidity and temperature.

Alginate type

Only three species of seaweed, *Macrocystis*, *Ascophyllum* and *Laminaria*, are abundant and accessible in quantities for commercial exploitation, although *Lessonia* and *Durvillea* are becoming increasingly important (5). *Ascophyllum* and *Macrocystis* are major sources of alginate rich in polymannuronic acid but relatively low in polyguluronic acid resulting in weaker, more elastic gels with good freeze-thaw stability. Alginate from *Laminaria hyperborea* stipes or stems is rich in polyguluronic acid and produces firm, brittle gels with good stability to heating. Increasing demand for this latter type of alginate has led to the greater exploitation of *Lessonia flavicans* from Chile.

Gel Formation

In solution, alginate can react instantly with free calcium ions. Consequently, for strong homogeneous gels where the majority of polyguluronate sites are occupied, calcium ions must be introduced in a controlled fashion so that all mixing can take place before the gel sets. This is achieved using a suitable calcium salt and by regulating the rate of reaction with a calcium sequestrant.

For systems using soluble calcium salts, the chloride, acetate and lactate may be employed. For neutral gels (pH 5.5-7.0), sparingly soluble calcium salts, such as calcium sulphate, produce the desired rate of release of calcium ions for gelling alginate. In acid systems with a pH below 5.5, where calcium salts have greater solubility, calcium citrate, carbonate or phosphate must be used. Table 1 indicates the solubility of different calcium salts together with their availabilty for reacting with alginate: these figures are estimated from the changes in solution viscosity caused by adding low levels of the different calcium salts to an alginate solution and expressing the change relative to the effect of adding the same level of calcium as calcium chloride.

The product pH will often determine the choice of calcium salt but some regulation of gel formation may be achieved by selecting an appropriate acid or buffer. Adipic and fumaric acids dissolve slowly whereas glucono 1,5 lactone hydrolyses in water to reduce the pH gradually. Citric acid rapidly lowers pH but sodium citrate can be used to buffer the system so that rapid calcium release and pre-setting is avoided. It is important to maintain the product pH above 3.8 to avoid forming alginic acid which prevents hydration of the alginate powder and reduces gel strength and storage stability.

Table 1. Calcium Salt Solubility and Estimated Availibilty

Calcium salt	Percent solubility at		Estimated percent of calcium reacting			
	20°C	100°C	At pH 6.0 after		At pH 4.0 after	
			5mins	2hrs	5mins	2hrs
Calcium chloride anhydrous	74.5	159	100	100	100	100
Calcium lactate tetrahydrate	3.1	7.9	100	100	100	100
Calcium sulphate dihydrate	0.24	0.22	80	90	90	100
Calcium citrate tetrahydrate	0.085	0.10	15	30	70	80
Dicalcium phosphate dihydrate	0.03	0.07	10	60	90	100
Dicalcium phosphate anhydrous	nil	nil	1	1	45	90

Further regulation of gel formation is achieved using a calcium sequestrant or, occasionally, temperature control. In this latter case, thermal motion of the alginate prevents chain alignment and gelation. However, this is only suitable for relatively low levels of free calcium, such as in milk products. With higher levels of soluble calcium, and for general use at room temperature, citrates and phosphates are used to sequester or bind calcium during initial stages of reaction preventing premature gelation of the alginate. The sequestrant also binds the free calcium found in hard water. If this is not controlled it may prevent or hinder solution preparation. Furthermore, calcium from hard water will interact with alginate, contributing to the final gel strength and modifying the texture of the end-product.

Generally, phosphate sequestrants are more efficient than citrates: their sequestering power increases with chain length. Sequestrant efficiency also varies with acidity. The comparison shown in Table 2 is based on the change in viscosity when the sequestrant is added to an alginate solution containing low levels of calcium chloride.

For a given level of alginate and calcium salt a progressive increase in the amount of sequestrant reduces the strength of the gel and eventually prevents gelation.

Table 2. Sequestrant Efficiency at Different pH Values

Sequestrant type	Grams of sequestrant required to sequester 1g calcium at: pH 6.0	pH 4.0
Citrate	13	40
Orthophosphate	123	80
Pyrophosphate	30	28
Hexametaphosphate	7	5

The components of the alginate gelling system depend on the product being manufactured and the process being employed.

GELATION PROCESSES

There are three main methods used to form gels: diffusion, internal or bulk setting and gelation by cooling. These are described in turn and illustrated with examples of current applications and developments.

Diffusion Setting

By this method, gels are formed simply by diffusing calcium ions into an alginate solution. The gel provides the main textural features in the finished product and alginate concentrations of 1% and over are required to give the necessary gel strength. The high alginate levels also give suitable viscosities to assist shape retention during extrusion and setting although cheaper thickeners, such as guar gum, are also added to control solution viscosity and syneresis during storage.

A critical stage in the gelling process is correct hydration of the alginate. Water soluble polymers tend to lump and hydrate poorly if dispersion is poor. Physical separation of the gum particles, by blending the alginate with other powdered ingredients, such as sugar, starch or low levels of salt, or by slurrying the alginate in oil or alcohol, gives an effective dispersion and facilitates hydration. High speed or high shear mixers and eductors use energy coupled with controlled dosing to separate particles so that hydration can proceed effectively.

Readily-soluble calcium salts are used to prepare the setting baths. The most common material is calcium chloride used at 5-10% as this is cheap and dissolves easily. However, it can impart an unpleasant after-taste if excess salt is carried over into the end-product. Calcium lactate has a more acceptable flavour but it has a lower solubility of about 5% although this is sufficient for most processes.

The simplest commercial application involves dropping or extruding a sodium alginate solution into a setting bath to form beads or fibres. Currently, the dried material is being developed in medical applications where the haemostatic properties of alginate are utilised for wound dressings. The dressing is removed simply by washing with a sequestrant solution and this avoids disturbing the healing wound.

Alginate gels are one of the most widely used carrier systems for immobilised cells (6). The cold-setting gel is non-toxic and low molecular weight substrates readily diffuse through pores in the beads. This process is being developed commercially (7) to encapsulate yeast cells for use in the secondary fermentation of champagne and other sparkling drinks.

For other food products, any ingredient may be incorporated into an alginate solution provided it does not react causing premature setting or completely inhibit gelation. A major process in use for many years involves adding concentrated pimiento puree to the alginate and forming a gelled sheet (8). This can be cut into small, resilient strips suitable for automatic filling into olives.

Onion rings are produced from the entire onion by mixing a blend of sodium alginate and starch into onion puree and extruding an anulus of the viscous mix into a calcium chloride setting bath (9). The short residence time of 2-5 seconds gives a film of calcium alginate around the ring. This holds the product shape until flash-frying gelatinises the starch which provides the main textural properties of the final product.

Sophisticated processing may be used to give more complex products. Liquid-centred fruits, such as blackcurrants, blueberries and redcurrants, are made by pulsing a thickened fruit puree containing a low level of calcium lactate through a continuous film of sodium alginate (10). The bead is dropped into a calcium lactate setting bath where diffusion of calcium ions into the alginate gives a strong skin accurately reproducing the texture of the natural fruit. Such elaborate processing enables suitable flavours, freeze-thaw stability and a soft texture for frozen foods to be created by incorporating sugars into the fruit during manufacture.

A further variation for this gelling technique is to diffuse acid into a mixture of alginate and an insoluble calcium salt. The drop in pH dissolves the calcium salt which, in turn, gels the alginate. This sequence is used to form a confectionery product where droplets of glucose syrup, sugars, sodium alginate and calcium phosphate are gelled by the diffusion of lactic acid.

Although diffusion systems are suitable for small products, ion transfer rates limit the dimensions of products which can be made on a commercial scale. Internal or bulk setting is used to form larger gelled pieces within a shorter time period.

Internal setting

In this method the calcium is released under controlled conditions simultaneously throughout the system. The most frequently used calcium salts are calcium sulphate dihydrate in neutral gels and dicalcium phosphate in acid products. In most instances a sequestrant is included to eliminate the effects of water hardness on alginate hydration and to control calcium release during the early stages of processing. As with diffusion set systems, vegetable and meat purees which do not affect the gelling mechanism can be used in place of some of the water used for the alginate solution. The practical limit for these purees is set by the viscosity of the alginate: for a high viscosity alginate the gum concentration, based on the free water available for hydration, should be 3.0-3.5% maximum and up to about 5.0% for a medium viscosity alginate. If higher levels of puree are used, the alginate only partially hydrates in the limited water available and the viscous mix is difficult to handle.

Structured fruit analogues which have an homogeneous texture, such as apple and peach, are made by rapidly mixing a blend of fruit puree and acid with a solution of a high gel strength alginate and dicalcium phosphate (11). The calcium salt does not interact at neutral pH but dissolves when mixed with the acid. A low level of disodium hydrogen orthophosphate sequesters the initial release of free calcium and gelation occurs within a few minutes to form the fruit slab or shape.

Macerated alginate gels are used for fruit fillings. Each gel fragment retains the characteristics of an alginate gel giving stability during baking and retaining moisture during shelf-life. In addition, the puree has a high viscosity suitable for suspending fruit and dosing into bakery products.

New developments for structured meat products use sequential addition of fine mesh sodium alginate, calcium carbonate and glucono 1,5 lactone to bind coarse-minced meat pieces together (12). The alginate gel binds chilled and frozen products and contributes to the cooked meat texture where a strong protein gel provides most of the final texture. Buffering from the proteins prevents acidic flavours in the cooked meat.

A major application, using many hundreds of tonnes of alginate annually, is the preparation of meat analogues for petfood. Meat slurries are mixed with a high gel strength alginate solution followed by a calcium sulphate slurry (13). Portions of this blend are dropped into setting baths or mixed with small quantities of calcium chloride to form an alginate skin. This keeps the gelling mixture in separate pieces whilst internal setting proceeds. Once gelled, the pieces are filled into cans with a meat slurry. During retorting, the alginate chunks remain intact. A reversible gelling system, usually based on carrageenan and locust bean gum, allows rapid heat penetration throughout the can during retorting and forms a gelled meat matrix around the chunks on cooling.

Gel Formation by Cooling

Most alginate gels are prepared and used at room temperature. Occasionally, it is advantageous to use a hot solution which contains all the components to form a gel: alginate, calcium salt, sequestrant and acid, which sets only when the mixture is cooled. This is achieved by formulating the mix to ensure gelation does not occur at elevated temperatures: thermal energy of the alginate prevents chain alignment and only after cooling can the calcium-induced associations take place. Relatively weak gels are produced in these formulations as the calcium ion levels are restricted to low levels. However, once set, the gel will not melt. It should be noted that hard water can contribute significant levels of calcium which affect the gelation temperature and strength.

By carefully adjusting the levels of calcium used in these formulations it is possible to produce a gel which sets at about 50-60°C. This is suitable for a jelly dessert prepared at ambient temperatures. Reducing the calcium further can give a mixture which gels when frozen but which does not remelt when brought to room temperature, ideal as a jelly centre for a frozen dessert.

Syneresis

In all systems set by diffusion and internal setting an excess of the gelling ingredient is used. This produces an imbalance in the gel as the crosslinks continue to form, squeezing water out of the gel. This phenomenon of syneresis will be seen as separation of free liquid and shrinkage of the product. To some extent this may be controlled by including other water binding agents, such as guar, or by using high levels of soluble solids. Adjusting the level of calcium or ratio of calcium to sodium ions in the setting bath may also reduce syneresis and shrinkage and modify the texture of the final gel. In some instances, controlled syneresis is useful to assist in demoulding the gel and for giving an attractive surface gloss.

Alginate gels formed by cooling are unusual as they exhibit minimal syneresis. This is because the calcium ions are equally available to all the alginate chains during gelation giving a very stable network.

LIQUID FOODS

In liquid foods, it is inappropriate to alter product texture significantly as a result of alginate gelation. Nevertheless, stabilisation using limited association of alginate chains is used in a range of products including flavour emulsions, dressings, sauces, fruit drinks and beer.

Propylene glycol alginate (PGA) is used to achieve the required stability in these foods. This material is made by reacting moist alginic acid with propylene oxide under pressure to yield a range of products varying in molecular weight or viscosity and in degree of substitution.

Commercial applications utilise the inherent properties of the substituent sidechains and exploit the controlled interaction with calcium. Esterifying the alginate provides some primary emulsification to prevent separation of oil in water emulsions. The sidechains also confer improved stability to acid degradation during prolonged storage. Weak polymer associations generate high viscosities at low shear to give long shelf-life and desirable sensory properties.

Between 0.5 and 2.0% of a highly esterified PGA is used alone or in combination with other thickeners and stabilisers for fish and citrus oil emulsions, depending on shelf life, rheology and cost limitations. Dressings, sauces and syrups require lower levels of 0.2-0.5% PGA in combination with starches and other stabilisers, such as guar gum and xanthan gum, to give the required product characteristics. In sauces containing significant levels of calcium salts within vegetable, fruit or dairy components, ions slowly diffuse and interact with the PGA in solution. This gives enhanced viscosity, improved mouthfeel properties and extended shelf-life.

Other cationic material is able to associate with the PGA to provide a stabilising network. Only 0.05-0.15% of a medium esterified, high viscosity PGA is used in fruit drinks to prevent pulp sedimentation and oil separation. Pulp aggregation involves protein interacting with pectin carboxyls formed by enzymic changes during storage. Residual carboxyl groups in the PGA interact with protein avoiding subsequent interaction with pectin. Crosslinking by calcium and protein provides a very weak stabilising alginate network.

A major worldwide application for PGA is for stabilising the foam on beers and lagers. This foam is important for making beer look attractive and freshly poured. Flavour components are concentrated in the foam and, during consumption, foam cling to the glass walls (lacing) is desirable. Polymeric materials, such as polysaccharides, which increase viscosity in the bubble wall extend foam stability and enhance lacing (14). Non-dispersed lipid introduced into the beer during serving or consumption destroys the foam by entering the bubble wall and introducing a break into the stabilising polypeptide film. Highly esterified, low viscosity PGA is prepared as a 1% stock solution, filtered and dosed into beer at 40-80 ppm. The PGA associates with the amino groups on the peptides in the bubble wall protecting the foam against the destabilising effects of lipid whilst avoiding peptide precipitation and haze formation produced with other charged polysaccharides and proteins.

In conclusion, the interaction between alginates and calcium ions may be used to thicken, stabilise and gel a wide range of food products. Utilisation of the reaction requires consideration of the contributions of acid, calcium and other cations to alginate rheology. These factors may be manipulated for use in different food products and processes to provide a versatile mechanism for controlling food rheology with benefits for processors and consumers.

ACKNOWLEDGMENTS

I should like to thank Kelco International Ltd and Protan (UK) Ltd. for their helpful contributions in the preparation for this manuscript and for their assistance with visual aids for presenting this paper at the Wrexham Conference.

REFERENCES

1 Stanford, E.C.C. (1881) British Patent 142.

2 Morris, E.R. (1989) in Gums and Stabilisers for the Food Industry 5, (Eds Phillips, G.O., D.J. Wedlock and P.J.Williams) IRL Press, Oxford, UK.

3 Morris, E.J., D.A. Rees and E.J. Welsh (1977) J. Supramol. Struct.,6, 259.

4 Sime, W.J. (1981) in Gums and Stabilisers for the Food Industry 2, (Eds Phillips, G.O., D.J. Wedlock and P.J.Williams) pp 177-188 Pergamon Press, Oxford, UK.

5 Wynne, M.J. (1981) in The Biology of Seaweeds (Eds Lobban, C.S. and M.J. Wynne) pp 52 Blackwell Scientific Press, Oxford, UK.

6 Bucke, C. and A. Wiseman (1981) Chemy Ind., 234-241.

7 Moet and Chandon (1980) French Patent 1086332.

8 Riuz-Granados, M.R. (1974) Spanish Patent 426978.

9 Smadar, Y., H. Roth, J.P. McCarthy and J.H. Noyer (1969) U.S. Patent 4126704.

10 Young, R., M.F. Woods and F.W. Wood (1973) U.K. Patent 1302275.

11 Sneath, M.E. (1973) U.K. Patent 1369198.

12 Schmidt, G.R. and I.K. Means (1986) U.S. Patent 4603054.

13 Enright, C.E. (1977) U.K. Patent 1474629.

14 Jackson, G., R.S. Roberts and T. Wainwright (1980) J. Inst. Brew. (1980), 86, 34-37.

Commercialisation of new synergistic applications of carrageenan

G.J.SHELSO

FMC Corporation Marine Colloids Division, Box 308, Rockland, Maine 04841, USA

ABSTRACT

Synergistic associations have been utilized for years, a common one being that between kappa-carrageenan and locust bean gum. An even stronger synergism is to be found between carrageenan and konjac flour. Konjac is a remarkable material with unique functionalities such as a very strong synergism with carrageenan and xanthan, unique film-forming abilities, and stability of its alkaline gels -- even under autoclaving conditions. Although used for centuries as a food in the Orient, few applications for this hydrocolloid have been seen in the West. This paper discusses the commercialization of this traditional Oriental food: understanding and characterizing its properties, defining its limits, and finding how to use it with carrageenan and other hydrocolloids.

INTRODUCTION

Kappa-carrageenan and the galactomannan locust bean gum (LBG) are a well-known synergistic pair whose properties make them useful to the food industry. A less well-known synergistic pair is kappa-carrageenan and the glucomannan known as konjac. As the Instron stress-strain curves show in the following Fig. 1, the LBG / carrageenan combination gives about 4 times the break force of an equal weight of carrageenan, and the konjac / carrageenan synergism is twice as strong as that. The gel is not only stronger, but more elastic with less syneresis than carrageenan alone. Whereas the LBG / carrageenan combination gives a peak gel strength at a 50 / 50 ratio, Fig. 2 shows that the optimum ratio for konjac / carrageenan is closer to 40 / 60.

As with a kappa-carrageenan gel, the K+ cation is necessary for the optimum gel strength. The gels used in Figures 1 and 2 used a total K+ concentration of about 0.04M, and were broken with a 1.075 cm diameter plunger travelling at 16 cm/min. into a 4.5 cm high gel at 25° C.

Konjac is not synergistic with iota- or lambda-carrageenans.

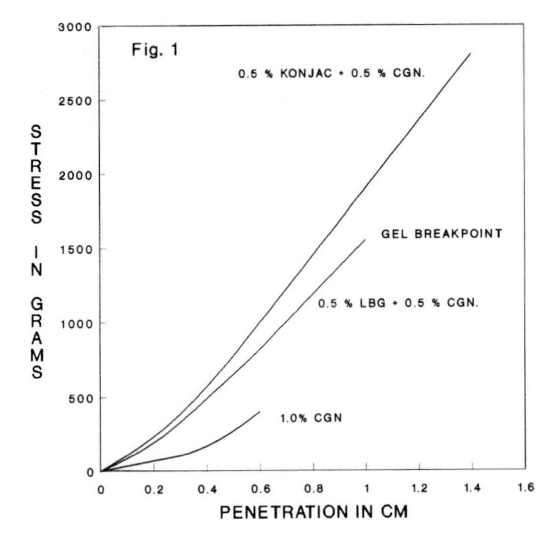

Figure 1. Synergism of kappa-cgn./konjac: Instron stress-strain curves terminated at the break point.

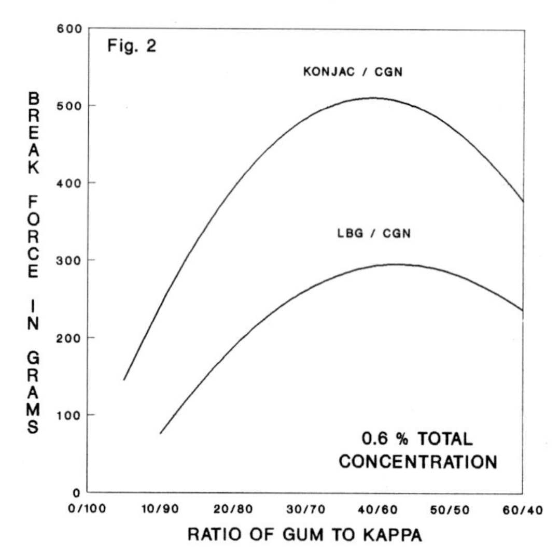

Figure 2. Synergism of konjac and LBG with kappa-carrageenan The effect of varying the ratio of hydrocolloids

However, before one can commercialize on the konjac / kappa-carrageenan synergism, a supplier must consider many important questions concerning the less well-known partner in that union. Some considerations could be --

-- What is konjac?
-- What are its properties in water such as dispersibility, hydration rate, viscosity, the effect of soluble solids and salts?
-- Does it gel on its own?
-- Is it synergistic with anything else?
-- What is its availability?
-- What is the regulatory status?

SOME CONSIDERATIONS

Konjac description

Konjac is a perennial herb of the amorphophallus species (i.e. - A. riveri, A. konjac) grown in semitropical to temperate climates. They produce a tuber which can be dried and subsequently milled to yield a glucomannan flour consisting of rather uniformly-sized individual particles that are roughly 100 to 500 microns in size with few fines. In terms of the U.S. Standard Sieve Series, that is a granulation range of about 35 to 140 mesh.

Konjac is a glucomannan made up of random, generally repeating sequences of glucose and mannose in the ratio of 1:1.6 connected with beta-(1-4) linkages. Every 9 sugar units, on the average, has an acetyl group attached in the -6 position, the importance of which will be discussed later in the section on alkaline gelation. Weight average molecular weight ranges from 100,000 to 1 million, but generally speaking is about half a million.

Properties in water

Dispersion. The relatively large particle size of konjac flour is responsible for its easy dispersion in water. The particles immediately begin to imbibe water, and swell until they disgorge the glucomannan, building a very high viscosity.
Hydration rate. With minimal agitation, it may take up to 4 hours to develop maximum viscosity at room temperature. As you would expect, increasing the mixing shear and increasing the temperature will increase the rate of hydration. High shear or cooking near the boil will give maximum viscosity within 15 minutes. This viscosity is very nearly the same as that attained by the room temperature hydration (assuming all viscosities are measured at the same temperature).
Viscosity. Konjac flour does give a very high viscosity yield -- possibly the highest of the food-grade hydrocolloids. Rheologically speaking, it exhibits pseudoplastic flow without thixotropy or yield point. Once fully-hydrated, increasing the sol temperature will cause a lowering of viscosity, but lowering the temperature to the same starting point will give back the original viscosity. As with many hydrocolloids, the viscosity / concentration relationship is an exponential function (see the following Fig. 3).

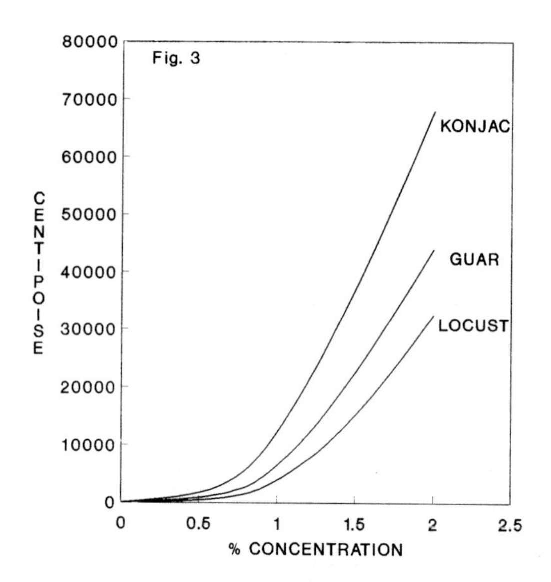

Figure 3. Effect of concentration on the viscosity of konjac flour, guar gum, and locust bean gum. Brookfield 20 rpm centipoise (mPa.s) at 25 C., fully hydrated sols.

Solubility in sugar. Because konjac flour is useful in desserts, you need to know how soluble it is in concentrated sugar solutions. We have found that konjac will effectively hydrate and thicken sugar solutions up to 25 % solids if the sugar is dissolved first, and up to 30 % if the konjac is hydrated first or the solution is heated. Evidently at the higher solids levels, there is too much competition for the available water for the konjac to reach maximum viscosity. This is not uncommon for high viscosity hydrocolloids.
Solubility in salts. Since konjac is a non-ionic hydrocolloid, salts as used in foods do not affect the hydration and viscosity development. We have tested it in up to 1 % cation with no adverse effects. Alkaline salts are a special case which will be covered in the following section on gels.

Konjac gels
Thermally-stable gels using mild alkali. Konjac has been used as a food in the Orient for centuries because of its unusual ability to form thermally-stable gels which can be boiled in water and even subjected to autoclaving temperatures. Konnyaku noodles are an example. If you mix a konjac sol with limewater or other mild alkali and then heat it, a strong gel forms. These conditions deacetylate the glucomannan, and the lack of steric hindrance allows these basically linear molecules to come more closely together forming more crystalline areas. However, if the concentration is high enough -- approximately

1.5 % or greater -- these very long molecules are so intertwined that they cannot come out of solution, but form a gel. Not only is it thermally-stable, but it stands up to virtually any alkalinity or acidity found in foods. The gel is even stronger hot than when it is cold.

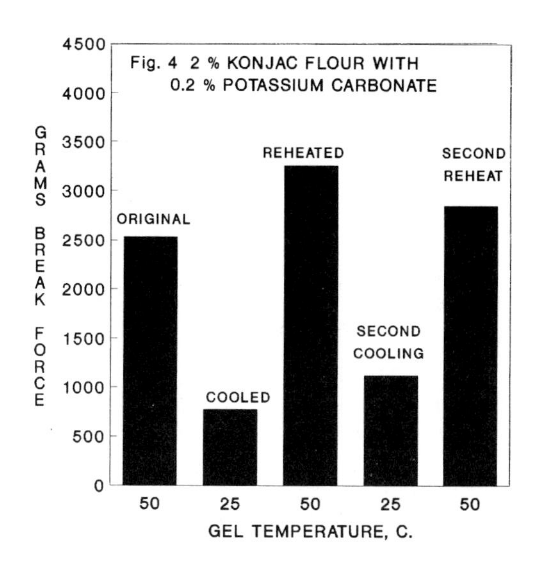

Figure 4. Effect of gel temperature on gel strength of thermally-stable konjac gels made with potassium carbonate. Gels are cycled between heating and cooling. Same Instron procedure as used in Figure 1.

Thermally-stable gels by freezing. Heat stable gels can be made by freezing and thawing a konjac sol without alkaline treatment. However, even though they are softer gels, they can be firmed up by steeping in warm alkali. Curiously, if a thermally-stable gel prepared with alkali is frozen, it forms a sponge-like matrix from which the water can be squeezed and reabsorbed.

Thermally-stable, neutral or acidic gels by autoclaving. Autoclaving temperatures allow gel formation with konjac on the acid side of neutral as long as there is a source of hydroxyl ions available from a buffer system. An important finding is that a synergistic gel of konjac / kappa-carrageenan can also be made thermally stable by autoclaving, giving some very interesting properties.

Film formation. Unique, thermally-stable films can be made with konjac flour. Merely casting a film of 1 % konjac in water containing 1 % glycerine and allowing it to dry will form a flexible film which will become soft and gelatinous if immersed in boiling water. The glycerine acts as a plasticizer / humectant to keep the films supple even in low humidity conditions and elevated temperatures.

To make films that are strong, supple, and elastic even in boiling water, you add a mild alkali before casting the film. Calcium carbonate, for example, gives an opaque film and potassium carbonate gives a transparent film. Inclusion of glycerine will also keep these films flexible in low humidity conditions.

Other synergisms.
Konjac / xanthan synergism. Compared to LBG / xanthan, the konjac / xanthan synergism is very strong and very elastic.

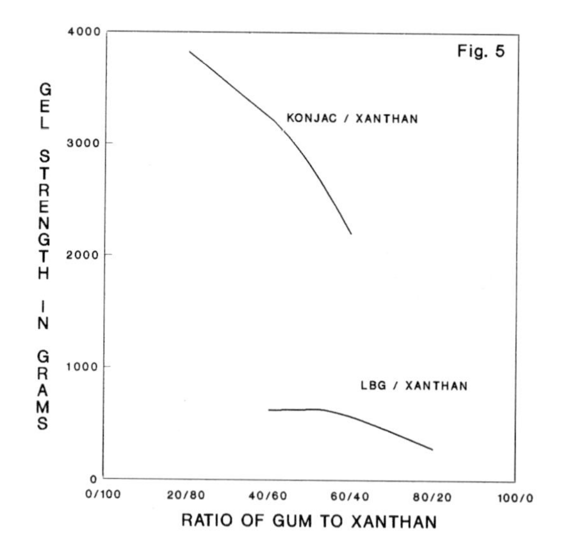

Figure 5. Synergism of konjac / xanthan compared to LBG / xanthan at 1 % total solids. Instron at 50 mm./min., 1.075 cm. D. plunger, distilled water, 25 C.

Even though Figure 5 shows the break force getting higher as the konjac level goes down, it ends at a 20 / 80 ratio of konjac / xanthan because the gel was getting so flexible that the plunger reached the bottom of the gel dish (about 45 mm.) before the gel broke.

As discussed at the 4th International Conference in Wrexham (1), the xanthan component cannot be in the helix-coil configuration for it to interact with the glucomannan. We find the easiest way to make a synergistic, thermally-reversible gel is to disperse the 2 gums together, then cook to about 85 C. for 15 min. before cooling.

Konjac / starch interaction. The addition of a small amount of konjac flour (10 % of the starch) greatly enhances the viscosity of selected starches. Particularly interactive are native corn, modified high amylose corn, modified waxy maize, and modified tapioca starches. It enhances cling and reduces flow (as measured with a Bostwick Consistometer), yet doesn't affect the gel formation (i.e. - retrogradation).

Availability

There are several commercial-sized sources available. It grows wild in Indonesia, Thailand, and many other Southeast Asian countries, but is cultivated in Thailand, China (PRC), and Japan. Although it is a perennial plant, the first-year tubers do not have a high enough glucomannan content to be harvestable. The plant produces a useable tuber in 2 years, but 3 years is preferred. Because konjac has been used as a food for centuries, patterns of cultivation and separation are firmly established. In other words, it is not herbal curiosity which must be developed into useable quantities.

Regulatory

In China and Japan (and likely most other Asian countries), konjac flour is considered a natural food ingredient, and therefore does not require special regulatory consideration.

In the United States, konjac flour is Generally Recognized As Safe (GRAS), as that term is defined in the U.S. Federal Food, Drug, and Cosmetic Act (Section 201(S)). It has a history of use as food in the USA since at least 1900 and, as mentioned here previously, has been in general use in the Far East for centuries. Furthermore, konjac flour has been the subject of many tests in various animal model systems and has also been tested in humans. The tests have included acute oral toxicity, eye and skin irritation, dermal toxicity, and tests for mutagenicity. No adverse reactions or toxic effects have been observed.

REFERENCE

1. Brownsey, G.J., Cairns, P., Miles, M.J., and Morris, V.J., "Mechanism of gelation for mixed gels", (1988) "Gums and Stabilizers for the Food Industry 4", ed.Phillips, G.O., Wedlock,D.J. and Williams, P.A. IRL Press publ. p.163.

Studies on kappa carrageenan – konjac mannan mixed gels

D.H.DAY, M.J.LANGDON, G.O.PHILLIPS and P.A.WILLIAMS
The North East Wales Institute, Deeside, Clwyd CH5 4BR, UK

Abstract

Solutions of kappa carrageenan and konjac mannan were mixed in varying amounts under solvent conditions which would promote gelation of a) either the kappa carrageenan or konjac mannan only or b) both kappa carrageenan and konjac mannan together. Homogeneous rigid gels were obtained in the presence of KCl where conditions promoted gelation of the carrageenan. However, in the presence of NaOH where the conditions favoured gelation of the konjac mannan alone, an inhomogeneous system was obtained with an aqueous phase rich in carrageenan and a precipitated gel phase rich in konjac mannan. In the presence of KOH and K_2CO_3 where both kappa carrageenan and konjac mannan gel, homogeneous gels were obtained which differed in turbidity depending on the gelling agent and concentration of konjac mannan in the gel. Studies on the breakforce using a Stevens LFRA texture analyser revealed that the optimum ratio of kappa carrageenan to konjac mannan in gels formed in KCl, KOH and K_2CO_3 was 50:50 respectively and the gels formed in KCl had twice the breakforce of the gels obtained in KOH and K_2CO_3. Studies on the elastic properties of the gels revealed an optimum kappa carrageenan to konjac mannan ratio of 40:60 respectively with the gels prepared in the presence of KCl again having twice the elasticity of gels obtained in KOH and K_2CO_3.

INTRODUCTION

Kappa carrageenan is used extensively in the food industry because of its ability to form thermoreversible gels[1]. It is often used in combination with galacto mannans such as locust bean, carob and tara gums[2-4] with which it exhibits considerable synergy. Dea et al[5] suggested that there was a specific interaction occurring between the carrageenan double helix and the mannan chain of galactomannans and also with glucomannans[6]. The gluco mannan or galactomannan are thought to promote helix formation and thus enhance gelation. However, Brownsey et al[7,8] using X-ray diffraction have shown that there is no specific interaction occurring between kappa carrageenan and glucomannan or galactomannan. They instead proposed that gelation occurs as a consequence of the mutual incompatibility of the two polymers. In this study mixed gels of kappa carrageenan and konjac mannan have been prepared under various ionic environments, and their properties monitored.

Experimental

Preparation of mixed solutions of Kappa carrageenan and konjac mannan

a) Without additional electrolyte

3g of kappa carrageenan (FMC Marine Colloids Div) was dissolved in 500 cm^3 of distilled water at 80°C using a mechanical stirrer. 3g of konjac mannan (FMC Marine Colloids Div) was dispersed in 500 cm^3 of distilled water at room temperature for five minutes and then dissolved by mixing at 80°C for two hours using a mechanical stirrer, replacing any water lost due to evaporation. The solutions were allowed to cool to 25°C before mixing aliquots in varying proportions in 60 cm^3 glass jars. The jars were then placed in a water bath at 80°C and left to equilibrate for ten minutes before cooling to 25°C.

b) In the presence of NaOH

Aqueous solutions of kappa carrageenan and konjac mannan were prepared by dissolving 3g in 500 cm^3 of water as above. A suitable volume of NaOH was added to the individual solutions at 25°C to give a final concentration of 15m mol dm^{-3}. The solutions were then mixed in varying proportions and equilibrated at 80°C for ten minutes before cooling to 25°C.

c) In the presence of KCl

Aqueous solutions of kappa carrageenan and konjac mannan were prepared as above but were maintained at 80°C. A suitable volume of KCl (2.5 mol dm^{-3}) was added to each of the solutions to give a final KCl concentration of 50m mol dm^{-3}. The solutions were then mixed whilst hot in varying proportions in order to avoid premature gelation of the kappa carrageenan and equilibrated in a water bath at 80°C for ten minutes before cooling to 25°C.

d) In the presence of KOH and K_2CO_3

Aqueous solutions of kappa carrageenan and konjac mannan were prepared as above. Aliquots of either KOH or K_2CO_3 were pipetted into 60 cm^3 glass jars to give a final concentration of 50 mol dm^{-3} and 25 mol dm^{-3} respectively. The kappa carrageenan and konjac mannan at 80°C and 20°C respectively, were added in varying proportions to the jars with thorough mixing to minimise errors due to premature gelation. The samples were then equilibrated in a water bath at 80°C for ten minutes and then cooled to 25°C.

Gel properties

Gels formed were quantified using a Stevens LFRA texture analyser. The elasticity was qantified as the force applied during the loading period[9] for a 3mm deformation at a probe speed of 0.2 mm/sec using a 1 cm diameter probe. The gel strength was determined at a probe speed of 0.5 mm/sec using a 1 cm diameter probe.

Results and Discussion

Gelation did not occur on simply mixing solutions of the two polysaccharides in the absence of added electrolyte and on mixing in the presence of NaOH the konjac mannan was found to form an insoluble precipitate. However, on mixing in the presence of KCl, KOH or K_2CO_3 homogeneous gels were obtained which were thermoreversible.

The elasticity of gels obtained in the presence of KCl, KOH and K_2CO_3 are given in figure 1. Elasticity is a maximum in all three systems when the kappa carrageenan/konjac mannan ratio is 40:60. The maximum value of E in the presence of KCl is almost double that in KOH or K_2CO_3. Similarly the breakforce of the gels in KCl as given in figure 2 is almost double that in the presence of KOH and K_2CO_3 with the maximum value being obtained when the kappa carrageenan:konjac mannan ratio is 50:50. On examining the penetration distance at the break point for gels in the presence of KCl (fig. 3) it can be seen that when kappa carrageenan is in excess the gels break at lower penetration distances and hence are more brittle than the gels when konjac mannan is in excess. The gels obtained were homogeneous but varied in opacity due to precipitation of the konjac and/or precipitation of calcium (present in the carrageenan) as the hydroxide or carbonate.

The differences in the properties of the mixed gels prepared in the presence of KCl to those prepared in the presence of KOH or K_2CO_3 is not completely understood. In the presence of KCl only the carrageenan is expected to form a gel network and the konjac mannan will contribute to the gel properties either through specific interaction with the carrageenan double helices or through the mutual incompatibility of the polymer chains[5-8]. In the presence of KOH or K_2CO_3 carrageenan gelation will occur but in addition the konjac mannan may gel or more probably precipitate as was the case for carrageenan/konjac mannan mixtures in the presence of NaOH. Such precipitation would lead to a reduction in the interactions between the carrageenan/konjac mannan chains and would be reflected in the overall gel properties. An additional factor which may also contribute is the fact that the carrageenan itself in the absence of konjac mannan forms slightly stronger gels in the presence of KCl as compared to KOH or K_2CO_3 (see fig. 2).

FIG 1. Elasticity measurements on mixed gels containing Konjac Mannan and Kappa Carrageenan in the presence of:

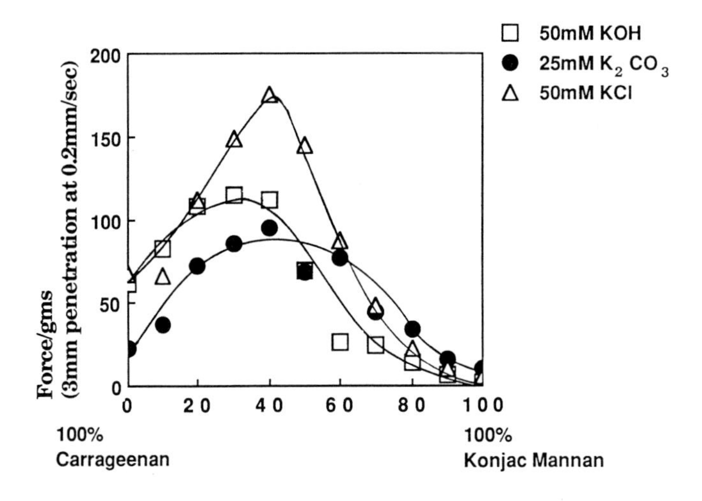

FIG 2. Breakforce measurements on mixed gels containing Konjac Mannan and Kappa Carrageenan in the presence of:

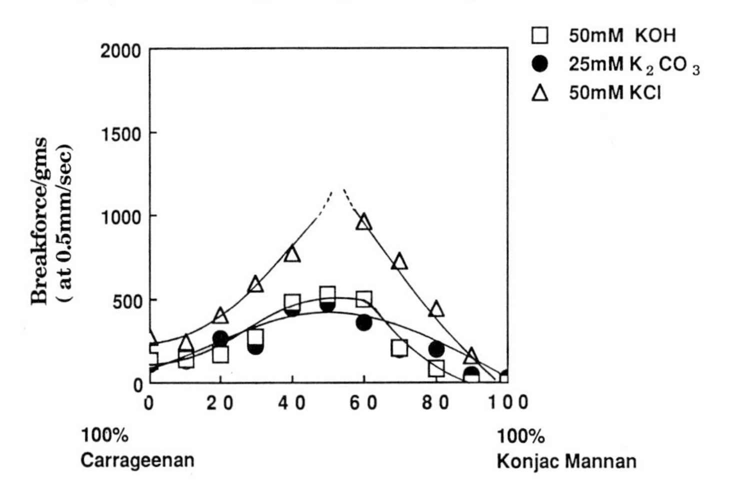

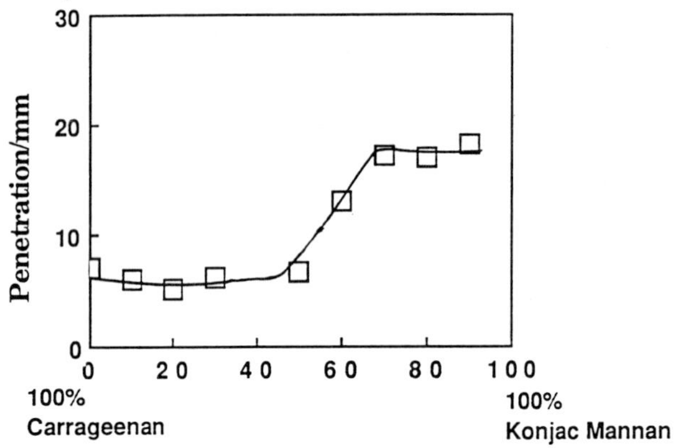

FIG 3. Penetration distance at break for mixed gels of Konjac mannan/ Kappa Carrageenan(0.6% total polymer concentration) in the presence of 50mM dm KCl at 25 C

Conclusions

Mixed gels of kappa carrageenan and konjac mannan can be prepared in the presence of potassium ions alone or in the presence of potassium ions together with hydroxide or carbonate ions. The elasticity and breakforce of such mixed gels are much greater than might be anticipated from the properties of gels of the individual polysaccharides.

Acknowledgement

The authors are indebted to FMC Marine Colloids Division for their support and guidance in undertaking this work.

References

1. Guisely K.B., Stanley N.F. and Whitehouse P.A., 1980 "Handbook of Water Soluble Gums and Resins" (ed. R.L. Davidson) McGraw Hill.

2. Copenhagen Pectin Factory Bulletin B1.

3. Cairns P., Morris V. J., Miles M. J., and Brownsey G. J. 1986 "Gums and Stabilisers for the Food Industry 3" (ed. Phillips, G. O. Wedlock, D. J. and Williams, P. A.) Elsevier Appl Science.

4. Ainsworth, P. A., and Blanshard J. M. V. 1980 J. Text Studies 11 149-162.

5. Dea, I. C. M., McKinnon A. A. and Rees, D. A. 1972 J. Mol. Biol. 68 153-172.

6. Dea, I. C. M., 1981 "Solution Properties of Polysaccharides" 1st ed. (ed D. A. Brandt) ACS Symp 150

7. Carrol, N., Miles, M. J. and Morris, V. J. 1986 "Gums and Stabilisers for the food industry 3" (ed Phillips, G. O., Wedlock, D. J. and Williams, P. A.) Elsevier Appl Science.

8. Cairns P., Miles M. J., and Morris, V. J., 1988 Carbohydrate Polymers 8 99-104.

9. Comby, S., Doublier, J. L. and Lefebure, J. "Gums and Stabilisers for the Food Industry 3" (ed. Phillips G. O., Wedlock D. J. and Williams P. A.) Elsevier appl science.

Simultaneous filling layered dessert gel with food hydrocolloids

S.OHASHI and H.IIDA

SAN—EI Chemical Industries Ltd, 1−1 Sanawa-Cho 1-Chome, Toyonaka, Osaka 561, Japan

ABSTRACT

The method for preparing layered dessert gels is summarised. Xanthan gum/locust bean gum blend and carrageenan are used for each layer. Total gum content is 1.0% (w/w). Each hydrocolloid solution is poured into the mould at 70°C and cooled. Two layers are clearly separated. Other galactomannans such as guar gum or tara gum and xanthan gum blend also make two layers with carrageenan but sols or gels are not firm and not practical.

INTRODUCTION

Hydrocolloids, such as carrageenan, gelatin and agar are commonly used as gelling agents for dessert gels in the world food industry. But recently, Japanese consumers have been demanding a variety of unusual and distinctive finished products. Dessert gels containing sarcocarp, fruit juice, dietary fiber or coffee extract, for example. Sparkling carbonated dessert gels, drinks with small globules of gel and so on. The character, texture and taste of these products are determined primarily by the composition of the hydrocolloids.

Another of these products is layered dessert gel, made up of different kinds of color, taste, flavor and texture. For example, the top layer might be an elastic yogurt-flavored gel, with the bottom layer a brittle, orange flavoured gel and so on. Layered gels are generally made by one of the following methods:

1. Pre-gelatinization filling method:

   The mould is filled half full with the solution for the lower layer which is then cooled until it gels. The solution for the upper layer is poured in on top and cooled until it gels, producing a layered dessert gel.

2. Filling method using the difference in the soluble solid contents of the upper and lower layers.

This method works by creating a difference in the soluble solid content of the upper and lower layers of the gel. A layered dessert gel is produced by making the soluble solid contents of the lower layer greater than that of the upper.

With the first method, the two layers are clearly separated, but, because the lower layer is gelled first, the two layers come apart when the finished product is eaten. In addition, the double cooling process results in a large energy loss which increases manufacturing costs. The second method is more popular, although it has the disadvantage that the flavor, color and texture of the two layers often get mixed up together.

The simultaneous filling method is completely different to these two methods. With no need for a double cooling process, the dessert gel where a clear layer is formed is prepared by pouring a dissolved hydrocolloid heated to 70°C into the mould.

Figure 1 shows, in stylized form, the method for preparing these dessert gels.

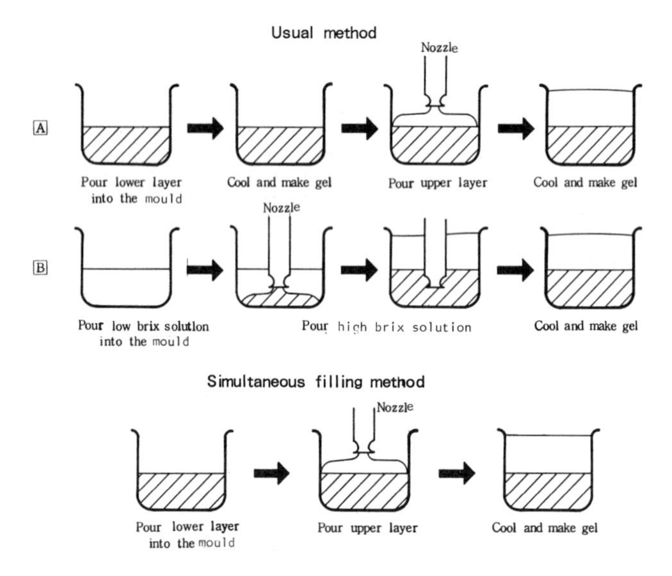

The outstanding feature of the simultaneous filling method is that it uses the hydrocolloids in general use in the food industry - xanthan gum and locust bean gum for the top layer, carrageenan, agar or pectin for the lower layer. By layering a gel which uses the synergistic interaction of xanthan gum and locust bean gum to give an "elastic" sensation and a "brittle" gel like carrageenan or agar, you can enjoy the sensation of eating two different types of gel at once.

MATERIALS AND METHODS

Hydrocolloid samples were commercial and used without purification, xanthan gum and galactomannan dry blends were added to water and stirred at 80°C for 10 minutes until dissolved. Carrageenan was dissolved by the same method. After dissolving, the pH value was adjusted to pH 4 with citric acid.

The separation of the two layers was checked by adding colors to each layer and then observing the layering condition. Each solution was maintained at a temperature of 75°C until ready to be poured into the mould and then poured into the mould at a temperature of 70°C. After filling, the mould was put into a refrigerator at 4°C to prepare the gels.

RESULTS AND DISCUSSION

Various blend ratios of xanthan gum and galactomannan were examined for layer separation. Total gum content was 1.0% (w/w) for upper and lower layers. Xanthan gum/galactomannan ratios which were suitable for layer separation was 25:75 ~ 75:25. Xanthan gum/locust bean gum blends were made and gave an elastic gel but xanthan gum/guar gum or tara gum were not practical.

When the pH of the xanthan gum/galactomannan was below 3, layer separation was not clear. This is because xanthan gum/galactomannan synergistic interaction is dependent on pH. Gel strength or viscosity of xanthan gum/galactomannan is weakened with decrease of pH.

Locust bean gum and carrageenan interaction is well known and locust bean gum can regulate brittle textures of carrageenan. Carrageenan and locust bean gum blend (blend ratio for example 1:1) total gum content 1% (w/w) did not make layers with carrageenan (1% w/w). This simple phenomenon suggests the difference of the interaction mechanism between xanthan gum/locust bean gum and carrageenan/locust bean gum.

Characteristics of whipping cream caused by tempering

MASAYUKI NODA and YUKIO SOGO
Technical Research Institute, Snow Brand Milk Products Co. Ltd,
1−1−2 Minamidai, Kawagoe, Saitama 350, Japan

ABSTRACT

The effects of tempering conditionson the increase of viscosity and whipping properties of a whipping cream were examined. When a whipping cream was kept under the tempering condition in the temperature range of ±4(℃) from the melting point of cream fats and then the cream was cooled slowly, the viscosity of the cream increased, and whipping time and overrun decreased. This behavior was caused by coagulation of the fat globules and the coagulation seemed to be caused by the growth of fat crystals. Also, the effects of emulsifiers and stabilisers on the viscosity after tempering were examined. It was found that as the concentration of the lecithin , carrageenan and xanthan increased the viscosity of the cream also increased while it decreased with increase of the concentration of sucrose mono ester and gum arabic.

INTRODUCTION

It is well known that the viscosity of a whipping cream increases and occasionally the cream solidifies after warming and cooling(called tempering). Mulder et al.(1) reported that the viscosity of the cream increased by rebodying in the presence of phospholipid. Also, Prentice et al.(2) pointed out that the viscosity of the cream during aging depended on homogenization, temperature of pasteurization and cooling temperature. Many workers have examined the change of physical properties of the cream, however, the mechanism of the increase of viscosity has not been explained.

In the present paper, the following examinations were carried out in order to clarify the mechanism of the viscosity increase of the cream after tempering: the effect of the tempering conditions on the characteristics of the whipping cream and on the physical properties of the cream fats; and the effects of emulsifiers and stabilisers on the viscosity of the cream.

MATERIALS AND METHODS

Reconstituted whipping cream with a composition of 40.0(%) milk fat, 4.6(%) solid fat, 0.6(%) emulsifiers and stabiliser and 54.8(%) water was used. The melting point of the milk fat was 35.5 (℃). The cream was sterilized with UHT(Ultra High Temperature) and homogenized in two stages,at 6.0(MPa) and 2.0(MPa).

The appearance of the whipping cream was observed after various tempering conditions, various warming temperatures and cooling rates.

After tempering the cream was whipped at room temperature with a domestic electric mixer (3).

Thermal properties of the cream fats after tempering were examined by DSC(Differencial Scanning Calorimetry). Polymorphism of the cream fats after tempering was examined by X-ray diffractometry.The crystal behavior of cream fats was observed with the polarized microscope.

The effect of tempering on the increase of viscosity was examined using emulsifiers and stabilisers.

RESULTS AND DISCUSSION

The appearances of the whipping cream after tempering are shown in Table 1. The slow cooling after tempering promoted the increase of viscosity, while rapid cooling(over 2.2 deg/min) had no influence on it. Temperature which caused the viscosity increase, ranges from 30(℃) to 39(℃). Table 2 shows whipping time and overrun decreased after 30(℃) tempering and slow cooling. The changes of physical properties of the cream after tempering were caused by coagulation of the fat globules and the coagulation seemed to be caused by the changes of physical properties of cream fats. Therefore, DSC melting curves, X-ray diffraction pattens and crystal behaviors of the cream fats were examined.

With respect to the thermal properties of the cream fats determined by DSC, no difference was found between slow and rapid cooling rate after 35(℃) tempering. In polymorphism the cream fats was determined by X-ray diffractometry after 35(℃) tempering, the fat crystal form was β' at both the slow and rapid cooling rate. The growth of fat crystals of the cream fats is shown in Fig.1. Fat crystals were minute at the all tempering temperature under rapid cooling, while with slow cooling, when cream was held at tempering temperature not lower than 37(℃) macro crystals grew, and minute crystals appeared below 37 (℃). Diameters of macro crystals were about 0.3 (mm).

The effect of emulsifiers and stabilisers on the increase of viscosity of the whipping cream were examined using a model cream. Results obtained are shown in Table 3. Lecithin , carrageenan and xanthan accelerated the viscosity increase of the cream while sucrose mono ester and gum arabic decreased it.

Table 1 Degree of gelling of the whipping cream

Cooling rate	Tempering Temperature(℃)												
(deg/min)	05	10	15	20	25	30	35	39	40	45	50	55	60
0.43	n	n	n	n	n	s	g	g	n	n	n	n	n
1.16					n	s	g		n				
2.20		n		n	n	n	n		n		n		n
12.00					n	n	n		n				

UHT whipping creams were cooled down to 5 (℃) with given cooling rates after warming at various temperaturesfor a hour. In this table, n, s and g indicate not gelling, slightly gelling and gelling respectively.

Table 2 Whipping properties of the whipping cream after tempering

Whipping properties	Tempering Temperature(℃)											
	10		20		30		40		50		60	
	R	S	R	S	R	S	R	S	R	S	R	S
Whipping time (min)	7.8	6.3	4.3	4.5	6.7	3.0	7.8	9.2	5.9	6.4	5.0	5.7
Overrun (%)	176	174	173	171	91	82	156	154	152	151	152	156

In this table, R indicates rapid cooling(2.2 deg/min) and S indicates slow cooling(0.43 deg/min).

CONCLUSION

The viscosity of the whipping cream depends on the tempering temperature and the cooling rate. It increases in the temperature range of ±4(℃) from the melting point of cream fats and at the slow cooling rate. The mechanism of the viscosity increase seems to be that existence of the unmelted cream fats at a tempering temperature initiates the growth of macro crystals which break down the fat globules, as a result, the coagulation of the fat globules occurs and the viscosity increase is induced.

Sucrose mono ester and gum arabic were effective at preventing increase in the viscosity of whipping cream after tempering.

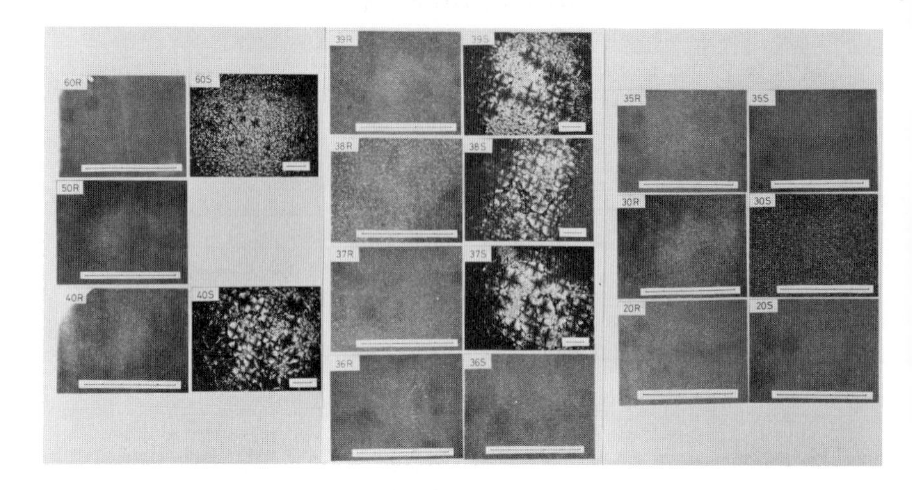

Fig.1 The growth of the fat crystals for the preparations of cream fats after tempering (Polalized observation)All line segments in the photographs represent 0.5(mm). Figures express the tempering temperature and S indicates slow cooling and R, rapid cooling.

Table 3 Change of the viscosity for model creams after tempering when various emulsifiers and stabilisers were used

Emulsifiers and	Concentration (%)					
Stabilisers	0	0.1	0.2	0.3	0.4	0.6
Sucrose mono ester		688	301	275		
Lecithine			119		233	908
Carrageenan	600	804	1731			
Xanthan	442	985	1723			
Gum Arabic	11.1*)	4.0*)	3.3*)			

Figures indicate the viscosity after tempering,starred figures indicate viscosity ratio(after tempering/before tempering).

REFERENCES

(1) Mulder H. and Walstra P.(1975) The Milk Fat Globule, 202

(2) Prentice J.H. and Chapman H.R.(1975) J. Dairy Res., 36, 269

(3) Noda M. and Shiinoki Y.(1986) J. Texture Studies, 17, 189-204

LIST OF PARTICIPANTS

AARS S.	Nycomed A/S, Nycovn 1-2, PO Box 4220 Torshov, N-0401, Oslo 4, NORWAY
AHMAD F.B.	Universiti Kebangsaan Malaysia, Kampus Sabah, Beg Berkunci No.62, 88996 Kota Kinabalu, Sabah, MALAYSIA
AHRENS B.	G.Lipman & Geffken, Hans Duncker Str.13, D-2050, Hamburg 80, WEST GERMANY
ALLEN J.C.	The North East Wales Institute, Connah's Quay, Deeside, CLWYD, CH5 4BR
ALLONCLE M.	Sanofi-Bio-Industries, INRA, LPCM, BP527, 44026 Nantes Cedex 03, FRANCE
ANNEFORS S.	AB Draco, Box 34, S-22100 Lund, SWEDEN
ARMISEN R.	Hispanger SA, Poligono Industrial de Villa, Lonquejar, Apartado 392, Burgos, SPAIN
AUTIO K.	Technical Research Centre of Finland, Biologinknja 1, 02150 Espoo, FINLAND
BAAL H.	Suiker Unie Research, Oostelyke Havendyk I5, 4704 RA Roosendaal, THE NETHERLANDS
BALDUCK P.	Sanofi Bio Industries, Chaussee de Charleroi 123A, Boite 3, 1060 Bruxelles, BELGIUM
BALL A.	Dept.of Applied Biochemistry and Food Science, University of Nottingham, School of Agriculture, Sutton Bonington, Loughborough, LEICS LE12 5RD
BARBER G.A.	17 Harvey Avenue, Nantwich, CHESHIRE CW5 6LE
BARKER S.A.	Chemistry Department, University of Birmingham, PO BOX 363, BIRMINGHAM B15 2TT

BAXENDALE P	Kelco International, Westminster Tower, 3, Albert Embankment, LONDON
BECKETT J.	Cesalpinia (UK) Ltd, 92 Hastings St., Luton, BEDS. LU1 5BH
BEST E.T.	Rowntree plc, YORK YO1 1XY
BISGAARD, P.	Litex a/s, Risingevej 1, DK-2665 Vallensbaek, Strand, DENMARK
BLAKEMORE W.	FMC Corporation, 200 East Randolph Drive, Chicago, Illinois 60601, USA
BOLLINGER H.	Herbafood Nahrungsmittel GmbH, Industriestrasse 1, D-7556 Oetigheim, WEST GERMANY
BOTTGER L.	Grindsted Products A/S, Edwin Rahrs Vej 38, DK-8220, Braband, DENMARK
BOWLER P.	RMH Research and Engineering Ltd, Lincoln Road, High Wycombe, BUCKS, HP12 3QR
BOYAR M.M.	Quaker Oats Ltd, PO Box 24, Bridge Road, Southall, MIDDLESEX UB2 4AG
BRADSHAW T.	British Sugar Technical Centre, Colney Lane, Colney, Norwich, NORFOLK NR4 7UB
BROOKS C.	PFW (UK) Ltd, PO Box 18, 9 Wandsworth Road, Greenford, MIDDLESEX UB6 7JH
BROOKS W.J.	Winthrop Laboratories, Edgefield Avenue, Fairdon, NEWCASTLE-UPON-TYNE NE3 3TT
BUCKLEY K.	KB Food Research Services, 2, Baldocks Lane, Melton Mowbray, LEICS. LE13 1EN
BUHL S.	The Copenhagen Pectin Factory, DK 4623, Little Skensved, DENMARK
BULIGA G.	Kraft Inc. 801 Waukegan Road, Glenview, IL 60025, USA
BURGER J.J.	Quest International, PO Box 2, 1400 CA Bassum, THE NETHERLANDS

CHALLEN I.	Kelco International, Westminster Tower, 3, Albert Embankment, LONDON SE1 7R2
CHRISTIANSON D.D.	USDA-ARS-NRRC, 1815 N.University, Peoria, Illinois 61604, USA
CLARE K.	Kelco Division of Merck and Co.Inc. 8225 Aero Drive, San Diego, CA 92123, USA
COLLINS P.	Overseal Foods Ltd, Swan's Park, Park Road, Overseal, Burton-on-Trent, STAFFS. E12 6JX
CONGDON J.	Grindsted Products Ltd, Northern Way, Bury St.Edmunds, SUFFOLK, IP32 6NP
COOKE D.	Unilever Research, Colworth House, Sharnbrook, BEDFORDSHIRE MK44 1LQ
COWBURN P.	National Starch Ltd, Ashburton Road East, Trafford Park, MANCHESTER
CURRIE A.	Kelco International Ltd, Waterfield, Tadworth, SURREY KT20 5HQ
CUVELIER G.	Food Science Dept. ENSIA, 1 Avenue des Olympiades, 91305 Massy, FRANCE
DALBE B.	Rhone Poulenc, 25 Quai Paul Doumer, 92408 Courbevoie Cedex, FRANCE
DAS K.V.C.	Quest International Food Ingredients Division, Lindtsedijk 8, 3336 LE, Zwijndrecht, THE NETHERLANDS
De BRUYCKER M.	Amylum, Burchstraat 10, 9300 Aalst, BELGIUM
De CONINCK V.	Cerestar, Research and Development Centre, Havenstraat 84, B-1800, Vilvoorde, BELGIUM
De VRIES J.	Avebe, Klaas Nieboerweb 12, 9607 PN Foxhol, THE NETHERLANDS
DEA I.C.M.	Leatherhead Food Research Assoc. Randalls Road, Leatherhead, SURREY KT22 7RY

DEEKS J.B.	Sanofi-Bio-Industries, 9-14 Cheap Street, Newbury, BERKS
DELEST P.	Sanofi-Bio-Industries, 66 Avenue Marcedau, 75008 Paris, FRANCE
DESSAUX F.	Iranex Colloides Naturels, 129 Chemin de Croisset, BP 4151, 76723 Rouen, FRANCE
DICKINSON E.	Procter Dept.of Food Science, University of Leeds, LEEDS LS2 9JT
DIKEMAN R.	The NutraSweet Co. 601 E, Kensington Road, Mount Prospect, Illinois, USA
DONGOWSKI G.	Academy of Sciences of the GDR, Central Institute of Nutrition, 1505 Potsdam-Rehbrucke, Arthur Schnert Alle 114/116, GERMAN DEMOCRATIC REPUBLIC
DUNFORD P.	Lucas Ingredients Ltd, Moravian Road, Kingswood, BRISTOL BS15 2NG
END L.	BASF, Polymer Research Division, 2 KM-G 201, D-6700 Ludwigshafen/Rhein, WEST GERMANY
ENDERS G.	Miles Inc. PO Box 932, CD001, 1127 Myrtle Street, Elkhart, Indiana 46515, USA
ENDRESS H.U.	Herbstreith KG Pectinfabrik, Postfach PO Box 1261, D-7540 Neuerburg, WEST GERMANY
FAULDS C.	AFRC Food Research Inst. Colney Lane, NORWICH NR4 7UA
FENYO J.C.	Lab.de Echanges Cellulaires, UA CNRS No.203, Faculte des Sciences et Techniques, Universite de Rouen, 76130 Mont Saint Aignan, FRANCE
FIELDING A.	Tunnel Avebe Starches, Avebe House, Otterham Quay, Rainham, Gillingham, KENT ME8 7UU

FISZMAN S.	Instituto de Agroguimica y tecnologia de Alimentos, Consejo Superior de Investigacions Cientificas, Calle Jaime Roig 11, 46010 Valencia, SPAIN
FOX J.E.	G.C.Hahn and Co. Stabilisierungstechnik GmbH, Aegidienstrasse 22, D-2400 Lubeck 1, WEST GERMANY
FOYLE R.	Food Technology Dept, Cannington College, Nr.Bridgewater, SOMERSET
FREI G.	NutraSweet AG Zug, Innere Guterstr.2-4, 6304 Zug, SWITZERLAND
GARTI N.	Casali Institute, Hebrew University, Jerusalem 91904, ISRAEL
GIBSON W.	Kelco International Ltd, Waterfield, Tadworth, SURREY, KT20 5HQ
GIDLEY M.	Unilever Research, Colworth House, Sharnbrook, BEDFORD MK44 1LQ
GOODALL D.	University of York, YORK, Y01 5DD
GORDON C.	H.J.Heinz Ltd, Hayes Park, Hayes, MIDDLESEX
GRADY N.	Biocon (UK) Ltd, Eardiston, Nr.Tenbury Wells, WORCS WR15 8JJ
GREGORY D.	Grindsted Products, Northern Way, Bury St.Edmonds, SUFFOLK, IP32 6NP
GRISTWOOD C.	Colman's of Norwich, Carrow, NORWICH NR1 2DD
HAGERMAN C-G.	Extraco AB, Stidsvig, S-26400 Klippan, SWEDEN
HARDIMAN B.	Pedigree Petfoods, Mill Street, Melton Mowbray, LEICESTERSHIRE
HARDING S.E.	Dept.of Applied Biochemistry and Food Science, University of Nottingham, Sutton Bonington, LEICS.LE12 5RD

HARRIS P.
Unilever Research, Colworth House, Sharnbrook, BEDFORD MK44 1LQ

HARROP R.
The North East Wales Institute, Connah's Quay,Deeside, CLWYD CH5 4BR

HEDGES R.
Unilever Research, Colworth House, Sharnbrook, BEDFORD MK44 1LQ.

HEWEDY M.M.
Department of Dairy Technology, Cairo University, Cairo, EGYPT

HILL M.A.
Dept.of Food & Nutritional Sciences, Kings College London, Kensington Campus, Campden Hill Road, LONDON W8 7AH

HILL S.
Dept.of Applied Biochemistry and Food Science, School of Agriculture, Sutton Bonington, Loughborough, LEICS LE12 5RD

HODGSON I.
Kelco International Ltd, Westminster Tower, 3 Albert Embankment, LONDON, SE1 7RZ

HOLE M.
Humberside College of HE, Nuns Corner, Grimsby, S.HUMBERSIDE DN34 5BQ

HOMLER B.
The NutraSweet Company, 601 East Kensington Road, Mount Prospect, Illinois 60056, USA

HOPKINS R.M.W.
Meyhall Chemicals Ltd, 6E Church Road, Bebington, Wirral, MERSEYSIDE, L63 7PG

HOWLING D.
Cerestar (UK) Ltd, Trafford Park, MANCHESTER M17 1PA

HUGHES H.
The North East Wales Institute, Connah's Quay, DEESIDE CH5 4BR

HYDE-SMITH J.J.
H.P.Bulmer Pectin Ltd, Plough Lane, HEREFORD HR4 0LE

IIDA H.
San-Ei Chemical Industries Ltd, 1-1 Sanwa-Cho 1-Chome, Toyonaka, Osaka 561, JAPAN

IMESON A.
Red Carnation Gums, Sir John Lyon House, 5 High Timber Street, Upper Thames Street, LONDON EC4V 3PA

JACKSON B.	Cerestar (UK) Ltd, Trafford Park, MANCHESTER M17 1PA
JACKSON P.	Kelco International Ltd, 54-68 Ferndell Street, South Granville, NSW 2142, AUSTRALIA
JOHNSTON-BANKS F.	Gelatine Products Ltd, Sutton Weaver, Runcorn, CHESHIRE
JORDAN G.W.	Gelatine Products Ltd, Sutton Weaver, Runcorn, CHESHIRE
JUD B.	Unipektin AG, CH,8264 Eschenz, SWITZERLAND
KATIR A.	Nexxel Inc. One Bridge Plaza, Suite No.615, Fort Lee, New Jersey, 07024, USA
KHALID A.S.	Food Research Centre, PO Box 213, Khartoum, SUDAN
KHALID S.A.	Pharmacognosy Dept,Faculty of Pharmacy, University of Khartoum, PO Box 1996, Khartoum, SUDAN
KITAMURA S.	Kyoto Prefectural University, Shimogamo, Kyoto 606, JAPAN
KLEPP R.	Jungbunzlauer AG, Schwarzenbergplatz 18, 1010 Wien, AUSTRIA
KRAMP E.	Ilford AG, Industriestr. 15, CH-1701, Fribourg, SWITZERLAND
KRAVTCHENCKO T.	Dept.of Food Chemistry, Agricultural University, Bomenweg 2, 6703, HD Wageningen, THE NETHERLANDS
KUGE T.	Kyoto Prefectural University, Shimogamo, Kyoto 606, JAPAN
LAUGHTON D.	Biocon Ltd, Kilnagleary, Carrigaline, Co.Cork, EIRE
LAST S.	Grinsted Products, Northern Way, Bury St.Edmonds, SUFFOLK IP32 6NP
LAWSON P.	Cerestar UK Ltd, Trafford Park, MANCHESTER M17 1PA

LEDWARD D.A.

Dept.of Applied Biochemistry and Food Science, School of Agriculture, University of Nottingham, Sutton Bonington, Loughborough, LEICS 1Z 5RD

LEHRIAN D.W.

Hershey Foods Corporation, Technical Centre, 1025 Reese Avenue, Hershey, PA 17033-0805, USA

LENSEN H.

European Patent Office, Patentlaan 2, PO Box 5818, 2280 HV Rijswijk, THE NETHERLANDS

LESLIE I.

Poldy's Fresh Foods Ltd, Naas Industrial Estate, Naas, Co.Kildare, EIRE

LIPS A.

Unilever Research, Colworth House, Sharnbrook, BEDFORD MK44 1LQ

LUYTEN J.M.J.G.

Department of Food Science, Wageningen Agricultural University, Bomenweg 2, 6703 HD Wageningen, THE NETHERLANDS

MALFAIT T.

Katholieke Universiteit Leuven, B-8500 Kortrijk, BELGIUM

MANNION R.

Faculty of Agricultural Sciences, University of Nottingham, Sutton Bonington, Nr.Loughborough, LEICS LE12 5RD

MAURICE T.

Ault Foods Ltd, 75,Bathurst Street, London, Ontario, CANADA N6B 1N8

MAY C.D.

H.P.Bulmer Pectin Ltd, Plough Lane, HEREFORD HR4 0LE

McBURNEY S.

Chemistry Dept, University of York, Heslington, YORK YO1 5DD

McCAFFREY M.

Roche Products Ltd, PO Box 8, Welwyn Garden City, HERTS AL7 3AY

McCLEARY B.V.

Biological and Chemical Research Inst.NSW, Department of Agriculture, PMB 10, Rydalmere 2116 NSW, AUSTRALIA

McGINLEY E.J.

FMC Corporation, Box 8, Princeton, New Jersey 08540, USA

MERRICK W.	Quest International, Lindtsedijk 8, 3336, Lezwijndrecht, THE NETHERLANDS
MEYER P.D.	Suiker Unie Research, PO Box 1308, 4700 BH Roosendaal, THE NETHERLANDS
MILLANE R.P.	The Whistler Centre for Carbohydrate Research, South Hall, Purdue University, West Lafayette, Indiana 47907, USA
MILLER E.	Herbstreith KG, Pectinfabrik, PO Box 1261, D-7540 Neuenbuerg, WEST GERMANY
MITCHELL J.R.	Faculty of Agricultural Sciences, University of Nottingham, Sutton Bonington, Nr.Loughborough, LEICS LE12 5RD
MLOTKIEWICZ J.	Spillers Foods, Station Road, CAMBRIDGE CB1 2JN
MODLISZEWSKI J.	FMC Corporation, Box 8, Princeton, New Jersey 08540, USA
MOOREHOUSE R.	Kelco Division of Merck & Co.Inc. PO Box 23576, 8225 Aero Drive, San Diego, CA 92123-1718, USA
MOEBUS K.	Arthur Branwell & Co.Ltd, Bronte House, 58-62 High Street, Epping, ESSEX CM16 4AE
MORLEY R.	Delphi Consultant Services, 948 Cabot Court, Stone Mountain, Georgia 30083, USA
MORRIS E.R.	Cranfield Institute of Technology, Silsoe College, Silsoe, BEDFORD
MORRIS V.J.	AFRC Food Research Institute, Colney Lane, NORWICH NR4 7UA
MURRAY J.C.F.	PFW (UK) Ltd, PO Box 18, 9, Wadsworth Road, Perivale, MIDDLESEX UB6 7JH
MURRAY K.	Pedigree Petfoods, Melton Mowbray, LEICESTERSHIRE
K. NISHINARI	National Food Research Institute, 2-1-2 Kannondai, Yatabe, Macho, Tsukuba-gun, Ibaraki-Ken 305, JAPAN

NODA M.	Snow Brand Milk Products Ltd, 1-1-2 Minamidai, Kawagoe, Saitama, JAPAN
NORTON I.T.	Unilever Research, Colworth House, Sharnbrook, BEDFORD MK44 1LQ
O'LEARY M.	Interfood Ltd, Westmead, Swindon, WILTSHIRE SN5 7YT
O'MAHONY M.	Kelco International Ltd, Westminster Tower, 3, Albert Embankment, LONDON SE1 7RZ
OAKENFULL D.	CSIRO Division of Food Processing, PO Box 52, North Ryde, NSW 2113, AUSTRALIA
OATES G.	Dept.of Biochemistry, National University of Singapore, 10 Kent Ridge Crescent, SINGAPORE 0511
OHASHI S.	San-El Chemical Industries Ltd, 1-1 Sanwa-Cho 1-Chome, Toyonaka, Osaka, 561, JAPAN
ONIONS A.	Honeywill & Stein Ltd, Times House, Throwley Way, Sutton, SURREY SM1 4AF
ONSOYEN E.	Protan A/S, PO Box 420, N-3002, Drammen, NORWAY
OWEN G.	General Foods Ltd, Banbury, OXON
PAINTING B	Hoechst UK Ltd, Hoechst House, Salisbury Road, Hounslow, MIDDLESEX
PEDERSON J.K.	The Copenhagen Pectin Factory, Lille Skensved, DK 4623, DENMARK
PERSSON P.	Sveriges Starkelse Product Center, S-291 91 Kristianstad, SWEDEN
PHILLIPS G.O.	The North East Wales Institute, Connah's Quay, Deeside, CLWYD, CH5 4BR
PILNIK W.	Wageningen Agricultural University, Postbus 8129, 6700 EV Wageningen, THE NETHERLANDS
PROCTER A.A.	Cerestar UK Ltd, Trafford Park, MANCHESTER M17 1PA

QUEMENER B.	INRA, Lab.Biochimie et Technologie des Glucides, BP 527, 44026 Nantes Cedex 03, FRANCE
RAATTAMAA S.	F.Ahlgrens Tek.Fab.AB, Box 622, S-801 26 Gavle, SWEDEN
REARDON P.	Premier Brands UK Ltd, The Orchard, Chivers Way, Histon, CAMBRIDGE CB4 4NR
REICHMAN D.	Adumin Chemicals Ltd, Mishor Adumin, ISRAEL 90610
RILEY P.	Tunnel Avebe Starches, Avebe House, Otterham Quay, Rainham, Gillingham, KENT ME8 7UU
RIZOTTI R.	Sanofi Bio Industries, Usine de Baupte, 50500 Carentan, FRANCE
SABINE D.	Biocon (UK) Ltd, Eardiston, Nr.Tenbury Wells, WORCS, WR15 8JJ
SALARI C.	Cesalpinia (UK) Ltd, 92 Hasting Street, Luton, BEDFORDSHIRE LU1 5BH
SANCHEZ-RAEL F.	Hispanager SA, Poligone Industrial de Villa Lonquejar, Apartado 392, Burgos, SPAIN
SANDERSON G.	Kelco Division of Merck & Co.Inc. 8355 Aero Drive, San Diego, CA 92123, USA
SCHOENTJES M.	Sanofi Bio Industries, 66 Avenue Marceau, Paris, FRANCE
SCHOETLER W.	Kelco International GmbH, Rodingsmarkt 52, D-2000 Hamburg 11, WEST GERMANY
SCHULTZE K.H.	Kelco International GmbH, Rodingsmarkt 52, D-2000 Hamburg, 11, WEST GERMANY
SCHWITZGUEBEL T.	Dow Europe SA, Bachtobelstrasse 3, 8810 Horgen, SWITZERLAND
SHARPE V.	Fibrisol Service Ltd, Colville Road, LONDON W3 8TE
SHAW C.J.	Smith & Nephew Medical Ltd, 101 Hessle Road, HULL, HU3 2BN

SHELSO G.J.	FMC Corporation, Marine Colloids Division, PO Box 308, Rockland, Maine 04841, USA
SHERIDAN M.	Pedigree Petfoods, Mill Street, Melton Mowbray, LEICS.LE13 1BB
SKLAVOS M.	Biocon GmbH, Rosenheimstrasse 9, Kolbermoor 8208, WEST GERMANY
SMITH J.	Kitchens of Sara Lee UK Ltd, Lancaster Road, Carnaby Industrial Estate, Carnaby, Bridlington, NORTH HUMBERSIDE YO15 3QY
SOMMERVILLE F.	European Patent Office, Patentlaan 2, PO Box 5818, 2280 HV Rijswijk (2H), HOLLAND
SPIERS C.	Pedigree Petfoods, Mill Street, Melton Mowbray, LEICS. LE13 1BB
STAINSBY G.	Procter Department of Food Science, University of Leeds, LEEDS LS2 9JT
STELFOX H S.	Kelco International Ltd, Waterfield, Tadworth, SURREY KT20 5HQ
STEPHEN A.M.	CSIR Carbohydrate Research Unit, Dept.of Organic Chemistry, University of Cape Town, Rondebosch 7700, SOUTH AFRICA
TAKERKART M.G.	Sanofi Bio Industries, Moulins Premiers, Boite Postale 23, 84800 L'Isle sur la Sorgue, FRANCE
TANNER N.	Red Carnation Gums Ltd, Sir John Lyon House, 5 High Timber St.Upper Thames Street, LONDON EC4V 3PA
TEMPEST J.M.	Spillers Foods Ltd, Research and Technology Centre, Station Road, CAMBRIDGE CB1 2JN
TOLSTOGUZOV V.B.	Institute of Organo-Element Compounds of the USSR, Academy of Sciences, Vavilov Street 28, MOSCOW, USSR
VAISHNAV V.	Bunge Foods Group, 3582 McCall Place, Atlanta, Georgia 30340, USA
VAN COILLIE R.	Aqualon BV, BP 12, 27460 Alizay, FRANCE

VAN DER GEIT T. Sanofi Bio Industries, Kanzlerstrasse 6, Postfach 330480, D-4000 Dusseldorf, WEST GERMANY

VANNESTE K. Katholieke Universiteit Leuven, Celestijnenlaan 200F, 3030 Heverlee, BELGIUM

VILLAUDY B. Sanofi Bio Industries, Usine de Baupte, 50500 Carentan, FRANCE

VORAGEN A.G.J. Agricultural University, Dept.of Food Science, PO Box 8129, 6700 EV Wageningen, THE NETHERLANDS

WALLNER M. Junbunzlauer AG, Schwarzenbergplatz 18, 1010 Wien, AUSTRIA

WAREING M. Arthur Branwell & Co. Ltd, 58-62 High Street, Epping, ESSEX

WEDLOCK D.J. Shell Research Ltd, Sittingbourne Research Centre, Sittingbourne, KENT ME9 8AG

WEISS H.O. Biocon GmbH, Rosenheimerstrasse 9, Kolbermoor 8208, WEST GERMANY

WESTON N.J. Kelco International Ltd, Waterfield, Tadworth, SURREY KT20 5HQ

WIELINGA W.C. Meyhall Chemicals AG, CH8280 Kreuzlingen, SWITZERLAND

WILLIAMS C. PB Gelatins UK Ltd, Treforest, MID GLAMORGAN CF37 5SU

WILLIAMS G.R. Masterdry Ltd, Bason Bridge, Highbridge, SOMERSET TA9 4RP

WILLIAMS P.A. The North East Wales Institute, Connah's Quay, Deeside, CLWYD, CH5 4BR

WILLIAMSON G. AFRC Institute of Food Research, Colney Lane, NORWICH NR4 7UA

WINWOOD R.J. Tunnel Avebe Starches, Avebe House, Otterham Quay, Rainham, Gillingham, KENT ME8 7UU

WOLF P. Pepsico Inc., 100 Stevens Avenue, Valhalla, NY 10595, USA

WOLTERS M.G.E. Civo-Analysis-TNO, Box 360, 3700 AJ Zeist, THE NETHERLANDS

WULF S. The NutraSweet Co. Box 730, 1419 Lake Cook Road, Deerfield, Illinois 60015, USA

WYNANS G. Suiker Unie Research, Oostelijke Havendijk 15, 4704 RA Roosendaal, THE NETHERLANDS

SUBJECT INDEX